# Handbook of
# Thin Plate
# Buckling
## and
# Postbuckling

# Handbook of
# Thin Plate Buckling
# and Postbuckling

## Frederick Bloom

*Department of Mathematics*
*Northern Illinois University*
*DeKalb, Illinois*

## Douglas Coffin

*Institute of Paper Science and Technology*
*Atlanta, Georgia*

**CRC Press**
Taylor & Francis Group
Boca Raton London New York

CRC Press is an imprint of the
Taylor & Francis Group, an **informa** business

A CHAPMAN & HALL BOOK

## Library of Congress Cataloging-in-Publication Data

Bloom, Frederick, 1944-
    Handbook of thin plate buckling and postbuckling / Frederick Bloom, Douglas Coffin.
       p.  cm.
    Includes bibliographical references and index.
    ISBN 1-58488-222-0
    1. Plates (Engineering). 2. Buckling (Mechanics). I. Coffin, Douglas. II. Title.
TA660.P6 B55 2000
624.1'7765—dc21                                           00-050441
                                                              CIP

© 2001 by Chapman & Hall/CRC

No claim to original U.S. Government works
International Standard Book Number 1-58488-222-0
Library of Congress Card Number 00-050441
Printed in the United States of America 1 2 3 4 5 6 7 8 9 0
Printed on acid-free paper

# Contents

# Preface

In the Spring of 1996, the authors initiated a research program at the Institute of Paper Science and Technology (Atlanta, Ga.) whose goal was to model cockle, i.e., the hygroscopic buckling of paper. In the course of researching the aforementioned problem, an extensive examination was made of the expansive literature on the buckling and postbuckling behavior of thin plates in an effort to understand how such phenomena were controlled by the wide variety of factors listed below. A comparative study of plate buckling was then produced and published as a pair of IPST reports; this book presents the essential content of these reports.

The essential factors which influence critical buckling loads, initial mode shapes, and postbuckling behavior for thin plates are studied, in detail, with specific examples, throughout this treatise; among the factors discussed which affect the occurrence of initial buckling and the initial mode shapes into which plates buckle are the following:

(i) **Aspect Ratios:** Included in our discussions will be examples of buckling for rectangular plates, circular plates, and annular plates; we will look at how, e.g., critical buckling loads for the various types of plates are affected by changes in the aspect ratio, associated with the plate, given that all other factors (support conditions, load conditions, etc.) are held fixed. For a rectangular plate, the aspect ratio is just the ratio of its sides; for a circular plate it is just the plate radius while, for an annular region, it is the ratio of the inner to the outer radius.

(ii) **Support Conditions:** The edge(s) of a plate may be supported in many different ways; mathematically, an edge support condition is reflected in a constraint on the out-of-plane deflection of that plate along its edge(s). The basic types of support that will be considered are those which correspond to having the edges(s) of the plate clamped, free, or simply supported. For a rectangular plate or an annular plate, one edge (or pair of edges) may be supported in one manner while the other edge (or pair) of edges can be supported in a different manner.

(iii) **Load Conditions:** The edge(s) of a thin plate may be subjected to compressive, tensile, or shear loadings or to some combinations of such loadings along different edges or pairs of edges. The effect of using different load conditions on critical buckling loads and mode shapes for rectangular, circular, and annular plates are illustrated throughout the book.

(iv) **Material Symmetry:** For thin plates exhibiting a particular constitutive response, the material symmetry exhibited by the plate may have a marked effect on, e.g., the magnitude of the lowest buckling load, when all other factors, such as aspect ratio or edge support conditions are held constant. The basic types of material symmetry that will be examined in this work are isotropic symmetry, rectilinearly orthotropic symmetry, and cylindrically orthotropic symmetry.

(v) **Constitutive Behavior:** Our examination of the influence of constitutive response on initial buckling of thin plates was guided by the behavior observed with respect to the deformation of paper, i.e., that under various combinations of loading conditions, paper may exhibit linear elastic, nonlinear elastic, viscoelastic, or elastic-plastic. Examples of all four kinds of such constitutive response are presented throughout this treatise.

(vi) **Hygroscopic or Thermal Stress Distributions:** Thin plates are often exposed to variations in temperature or moisture or hygroexpansive strains which result in local compressive stress distributions that can cause buckling even in the absence of applied external loading.

Beyond the factors delineated above, which influence both critical buckling loads and mode shapes for thin plates and also influence the subsequent postbuckling behavior, two other important factors come into play with respect to postbuckling behavior.

(vii) **Initial Imperfections:** Thin plates often possess some small initial transverse deflection (e.g., a paper sheet with an initial cockle which is wetted, dried, and experiences further local buckling) or may be subjected to a small normal loading (transverse) to the initially flat unbuckled state of the plate; the presence of either or both of these conditions amounts to an "initial imperfection" which may have a profound effect on the subsequent postbuckling behavior of the plate.

(viii) **Secondary Buckling:** After a thin plate has buckled into an initial mode shape, it is possible that, with increasing load, instead of obtaining a buckled shape which deforms continuously during postbuckling, as the magnitude of the loading increases, a critical load larger than the initial buckling load is reached at which the buckle pattern changes suddenly and a new mode shape appears; such phenomena have been exhibited with respect to plate buckling and will be examined with particular emphasis on the unsymmetric wrinkling that has been observed in circular plates.

Throughout this treatise, several results appear which cannot be found in earlier books on plate buckling, e.g., the discussion of the Johnson-Urbanik generalization of the von Karman Equations for plates exhibiting nonlinear elastic behavior; other results appear which have not previously been published elsewhere, most notably the analysis of initial buckling for annular rectilinearly orthotropic thin plates, which was presented in the Ph.D. Thesis of the second author.

The authors wish to express their gratitude to the Institute of Paper Science and Technology and its member companies for the support which made possible the production of the IPST reports that form the basis for the present work.

Frederick Bloom          Douglas Coffin
DeKalb, Illinois         Atlanta, Georgia
April 2000               April 2000

# CHAPTER 1

# Introduction: Plate Buckling and the von Karman Equations

## 1.1 Buckling Phenomena

Problems of initial and postbuckling represent a particular class of bifurcation phenomena; the long history of buckling theory for structures begins with the studies by Euler [1] in 1744 of the stability of flexible compressed beams, an example which we present in some detail below, to illustrate the main ideas underlying the study of initial and postbuckling behavior. Although von Karman formulated the equations for buckling of thin, linearly elastic plates which bear his name in 1910 [2], a general theory for the postbuckling of elastic structures was not put forth until Koiter wrote his thesis [3] in 1945 (see, also, Koiter [4], [5]); it is in Koiter's thesis that the fact that the presence of imperfections could give rise to significant reductions in the critical load required to buckle a particular structure first appears. General theories of bifurcation and stability originated in the mathematical studies of Poincaré [6], Lyapunov [7], and Schmidt [8] and employed, as basic mathematical tools, the inverse and implicit function theorems, which can be used to provide a rigorous justification of the asymptotic and perturbation type expansions which dominate studies of buckling and postbuckling of structures. Accounts of the modern mathematical approach to bifurcation theory, including buckling and postbuckling theory, may be found in many recent texts, most notably those of Keller and Antman, [9], Sattinger [10], Iooss and Joseph [11], Chow and Hale [12], and Golubitsky and Schaeffer [13], [14]. Among the noteworthy survey articles which deal specifically with buckling and postbuckling theory are those of Potier-Ferry [15], Budiansky [16], and (in the domain of elastic-plastic response) Hutchinson [17]. Some of the more recent work in the general area of bifurcation theory is quite sophisticated and deep from a mathematical standpoint, e.g., the work of Golubitsky and Schaeffer, cited above, as well as [18] and [19], which employ singularity theory for maps, an outgrowth of the catastrophe theory of Thom [20]). Besides problems in buckling and postbuckling of structures and, in particular, the specific problems associated with the buckling and postbuckling behavior of thin plates, general ideas underlying bifurcation and nonlinear stability theory have been used to study problems in fluid dynamics related to the instabilities of viscous flows as well as branching problems in nonlinear heat transfer, superconductivity, chemical reaction theory, and many other areas of mathematical physics. Beyond the references already listed, a study of various fundamental issues in branching theory, with applications to a wide variety of problems in physics and engineering, may be found in references [21]-[58], to many of which we will have occasion to refer throughout this work.

To illustrate the phenomena of bifurcation within the specific context of buckling and postbuckling of structures, we will use the example of the buckling of a thin rod under compression, which is due to Euler, op. cit, 1744, and which is probably the simplest and oldest physical example which illustrates this phenomena. In Fig. 1.1, we show a homogeneous thin rod, both of whose ends are pinned, the left end being fixed while the right end is free to move along the $x$-axis. In its unloaded state, the rod coincides with that portion of the $x$-axis between $x = 0$ and $x = 1$.

Under a compressive load $P$, a possible state of equilibrium for the rod is that of pure compression; however, experience shows that, for sufficiently large values of $P$, transverse deflections can occur. Assuming that the buckling takes place in the $x, y$ plane, we now investigate the equilibrium of forces on a portion of the rod which includes its left end; the forces and moments are taken to be positive, as indicated in Fig. 1.2.

Let $X$ be the original $x$-coordinate of a material point located in a portion of the rod depicted in Fig. 1.2. This point moves, after buckling, to the point with coordinates $(X + u, v)$. We let $\varphi$ be the angle between the tangent to the buckled rod and the $x$-axis and $s$ the arc length along a portion of the rod (measured from the left end). Although more general constitutive laws may be considered, we restrict ourselves here to the case of an inextensible rod in which the Euler-Bernoulli law relates the moment $M$ acting on a cross-section with the curvature $d\varphi/ds$. Thus $s = X$ while

$$M = -EI\frac{d\varphi}{ds} \tag{1.1}$$

with $EI$ the (positive) bending stiffness. The constitutive relation (1.1) is supplemented by the geometric relation

$$\frac{dv}{ds} = \sin\varphi \tag{1.2}$$

and the equilibrium condition

$$M = Pv \tag{1.3}$$

Combining the above relations, we obtain the pair of first order non-linear differential equations

$$\begin{cases} \lambda v = -\dfrac{d\varphi}{ds}, & \lambda = P/EI \\ \dfrac{dv}{ds} = \sin\varphi \end{cases} \tag{1.4}$$

with associated boundary conditions

$$v(0) = v(\ell) = 0 \tag{1.5}$$

A solution of (1.4), (1.5) is a triple $(\lambda, v, \varphi)$ and any solution with $v(s) \neq 0$ is called a buckled state; we note that $\lambda = 0$ implies that $v \equiv 0$ and cannot, therefore, generate a buckled state. When $\lambda \neq 0$, (1.4), (1.5) is equivalent to the boundary value problem

$$\frac{d^2\varphi}{ds^2} + \lambda\sin\varphi = 0, \quad 0 < s < \ell \tag{1.6a}$$

$$\varphi'(0) = \varphi'(\ell) = 0 \tag{1.6b}$$

The actual lateral deflection $v(s)$ can then be calculated from $\varphi$ by using the first equation in (1.4). We note that the rod has an associated potential energy of the form

$$V = \frac{1}{2}EI\int_0^\ell \left(\frac{d\varphi}{ds}\right)^2 ds - P\left[\ell - \int_0^\ell \cos\varphi\, ds\right] \tag{1.7}$$

and that setting the first variation $\delta V = V'(\varphi)\delta\varphi = 0$ yields the differential equation (1.6a) with the natural boundary conditions (1.6b).

The linearized version of (1.6a,b) for small deflections $v$, and small angles $\varphi$, is obtained by substituting $\varphi$ for $\sin\varphi$ (a precise mathematical justification for considering the linearized problems so generated will be considered below); the linearized problem for $v$ then becomes

$$\begin{cases} \dfrac{d^2v}{ds^2} + \lambda v = 0 \\ v(0) = v(\ell) = 0 \end{cases} \tag{1.8}$$

which has eigenvalues $\lambda_n = \dfrac{n^2\pi^2}{\ell^2}$, $n = 1, 2, \cdots$ with corresponding eigenfunctions $v_n = c\sin(\dfrac{n\pi x}{\ell})$. It is desirable to be able to plot $v$ versus $\lambda$ but, unfortunately, $v$ is itself a function (an infinite-dimensional vector) so we content ourselves instead, in Fig. 1.3, with a graph of the maximum deflection $v_{\max}$ versus $\lambda = P/EI$. For loads below the first critical load $P_1 = \dfrac{\pi^2 EI}{\ell^2}$ no buckling is possible. At the load $P_1$, buckling can take place in the mode $c\sin\left(\dfrac{\pi x}{\ell}\right)$ but the size of the deflection is undetermined due to the presence of the arbitrary constant $c$. Now, if an analysis of the buckled behavior of the rod is based entirely on the linearized problem, then we see that, as the load slightly exceeds $P_1$, the rod must return to its unbuckled state until the second critical load $P_2 = \dfrac{4\pi^2 EI}{\ell}$ is reached, when buckling can again occur, but now in the new mode $c\sin\left(\dfrac{2\pi x}{\ell}\right)$, which is still of undetermined size. Except at the critical loads $P_n = \dfrac{n^2\pi^2 EI}{\ell^2}$ there is no buckling. *On physical grounds the picture provided only by the linearized problem is clearly unsatisfactory.*

The nonlinear problem (1.4), (1.5), gives a more reasonable prediction of the buckling phenomena. Again, no buckling is possible until the compressive load reaches the first critical load of the linearized theory (the reason for this being discussed below) and as this value is exceeded the possible buckling deformation is completely determined except for sign. For values of the load between the first two critical loads there are therefore three possible solutions as shown in Fig. 1.4. When $P$ exceeds the second critical load a new, determinate, pair of nontrivial solutions appears, and so forth. The values of $\lambda$ corresponding to these critical loads are said to be branch points (or bifurcation points) of the trivial solution because new solutions, initially of small size, appear at these points.

A diagram like Fig. 1.4 is called a branching (or bifurcation diagram) and we may justify the statements, above, about the buckling behavior for the problem of the compressed thin rod by exhibiting the closed form solutions of (1.6a,b). The deflection $v$ can then be obtained from the first relation in (1.4). We first note certain simple properties of the boundary value problem (1.4), (1.5), namely, if any one of the triplets $(\lambda, v, \varphi), (\lambda, v, \varphi + 2n\pi), (\lambda, -v, -\varphi)$, or $(-\lambda, -v, \varphi + \pi + 2n\pi)$, is a solution so are the other three; all four solutions yield congruent deflections. In fact, the first two are identical, while the third is a reflection of the first about the $x$-axis and the fourth is a reflection of the first about the origin; in this last case, a reversal in the sign of the load is accounted for by interchanging the roles of the left-hand and right-hand ends of the rod. Similar remarks apply to the system (1.6a,b).

All nontrivial lateral deflections $v(s)$ of the rod are therefore generated by those solutions of the initial value problem

$$\begin{cases} \dfrac{d^2\varphi}{ds^2} + \lambda \sin\varphi = 0, & \lambda > 0 \\ \varphi(0) = \alpha, & \varphi'(0) = 0, \quad 0 < \alpha < \pi \end{cases}$$

(1.9)

which have a vanishing derivative at $s = \ell$.

The initial value problem (1.9) has one and only one solution and this solution may be interpreted as representing the motion of a simple pendulum, with $x$ being the time and $\varphi$ the angle between the pendulum and the (downward) vertical position. In view of the initial conditions, $\varphi(s)$ will be periodic. If we multiply the equation in (1.9) by $d\varphi/ds$, then this equation becomes a first order equation which can be solved (explicitly) by elliptic integrals, i.e.

$$\varphi(s) = 2 \arcsin\left[k \; sn(\sqrt{\lambda}s + K)\right]$$

(1.10)

with

$$\begin{cases} k = \sin\left(\dfrac{\alpha}{2}\right) \\ K = \displaystyle\int_0^{\frac{\pi}{2}} \dfrac{d\zeta}{\sqrt{1 - k^2 \sin^2\zeta}} \end{cases}$$

(1.11)

and

$$sn\; u = u - (1 + k^2)\dfrac{u^3}{3!} + (1 + 14k^2 + k^4)\dfrac{u^5}{5!}$$

$$-(1 + 135k^2 + 135k^4 + k^6)\dfrac{u^7}{7!} + \cdots$$

(1.12)

The parameter $K$ takes on values between $\dfrac{\pi}{2}$ and $+\infty$ and there is a one-to-one correspondence between these values and those of $\alpha$ in $0 < \alpha < \pi$. The period of oscillation for $\varphi$ is $4K$ and the condition $\varphi'(\ell) = 0$ is satisfied if and only if

$$K = \left(\frac{\ell}{2n}\right)\sqrt{\lambda}, \ n = 1, 2, \cdots \tag{1.13}$$

As $K > \dfrac{\pi}{2}$, in order for nontrivial solutions to exist, we must have $\lambda > \pi^2/\ell^2$. Also, given a value of $\lambda > \pi^2/\ell^2$ there are as many distinct nontrivial solutions $\varphi(s)$, with $0 < \varphi(0) < \pi$, as there are integers $n$ such that $(\ell/2n)\sqrt{\lambda} > \pi/2$. Thus, we have for

$$\begin{cases} 0 \leq \lambda \leq \pi^2/\ell^2, & \text{only the trivial solution} \\ \pi^2/\ell^2 < \lambda \leq 4\pi^2/\ell^2, & \text{one nontrivial solution} \\ n^2\pi^2/\ell^2 < \lambda \leq (n+1)^2\pi^2/\ell^2, & n \text{ nontrivial solutions} \end{cases}$$

If we then also take into account other values of $\varphi(0)$, and the possibility of negative $\lambda$, we obtain Fig. 1.5, which depicts the maximum value of $\varphi(s)$ versus $\lambda$; the only (physically) significant portion of this figure is, of course, the part drawn with solid (unbroken) curves, i.e., for $\lambda > 0$ and $-\pi < \varphi_{max} < \pi$.

The broken curves for $\lambda > 0$ are those attributable to angles $\varphi$ that are determined only by modulo $2\pi$, while the dotted curves for $\lambda < 0$ correspond to having the free end of the rod at $x = -\ell$. The deflection $v(s)$ is then obtained from the first equation in (1.4) and yields the bifurcation diagram in Fig. 1.4, where it is known that the various branches do not extend to infinity because $v_{max}$ does not grow monotonically (at large loads the buckled rod may form a knot so that the maximum deflection can decrease, although the maximum slope must increase.)

We have presented a complete solution of the buckling problem for a compressed thin rod which is based on classical analysis of an associated initial value problem for the boundary value problem (1.6a,b) first treated by Euler in 1744. The equivalent problem for the deflection $v(s)$ is (1.4), (1.5) and the results of the analysis, based on a closed form solution using elliptic integrals, seem to indicate a prominent position for the linearized (eigenvalue) problem (1.8), or, equivalently, for the linearization of (1.6a,b) namely,

$$\begin{cases} \dfrac{d^2\varphi}{ds^2} + \lambda\varphi = 0, \quad 0 < s < \ell \\ \varphi'(0) = \varphi'(\ell) = 0 \end{cases} \tag{1.14}$$

Indeed, this system governs the initial buckling of the rod for reasons that are explained below; our explanation will be phrased in just enough generality so as to make our remarks directly applicable to the study of buckling and postbuckling behavior for thin plates. For the case of linear elastic response, the von Karman system of nonlinear partial differential equations, with associated boundary conditions, which govern plate buckling, are derived in § 1.2.

As we will specify below, the nonlinear boundary value problem (1.6a, b), which governs the behavior of the compressed rod, can be put in the form

$$G(\lambda, u) = 0 \tag{1.15}$$

where $\lambda$ is a real number, $u$ is an element of a real Banach space $\mathcal{B}$ with norm $\|\cdot\|$, and $G$ is a nonlinear mapping from $\mathcal{R} \times \mathcal{B}$ into $\mathcal{B}, \mathcal{R}$ being the real numbers. The restriction to real $\lambda$ and a real Banach space $\mathcal{B}$ is based on the needs in applications where only real branching is of interest. Strictly speaking, a solution of (1.15) is an ordered pair $(\lambda, u)$ but we often refer to $u$ itself as the solution (either for fixed $\lambda$, or depending parametrically on $\lambda$); to study branching (or bifurcation) we must have a simple, explicitly known solution $u(\lambda)$ of (1.15). We may make the assumption that

$$G(\lambda, 0) = 0, \text{ for all } \lambda, \tag{1.16}$$

so that $u(\lambda) = 0$ is a solution of (1.15) for all $\lambda$ and this solution is then known as the basic solution (for the problem considered above, this solution corresponds to the compressed, unbuckled rod). The main problem is to study branching from this basic solution (i.e., the unbuckled state) although within the context of plate buckling we will also discuss, in §5, branching from nontrivial solutions of (1.15), i.e., secondary buckling of thin plates. Thus, the goal is to find solutions of (1.15) which are of small norm (small "size" in the relevant Banach space $\mathcal{B}$); this motivates the following:

**Definition.** We say that $\lambda = \overset{\circ}{\lambda}$ is a branch point of (1.15) (equivalently a bifurcation point or, for the case in which (1.15) represents the equilibrium problem for a structure, such as the compressed rod considered above, a "buckling load") if every neighborhood of $(\overset{\circ}{\lambda},0)$, in $\mathcal{R} \times \mathcal{B}$, contains a solution $(\lambda, u)$ of (1.15) with $\|u\| \neq 0$.

We note that the above definition is restricted only to small neighborhoods of $(\overset{\circ}{\lambda},0)$; also, the definition is equivalent to the existence of a sequence of solutions $(\lambda^n, u^n)$ of (1.15), with $\|u^n\| \neq 0$ for each $n$, such that $(\lambda^n, u^n) \rightarrow (\overset{\circ}{\lambda},0)$, as $n \rightarrow \infty$. Before placing the compressed rod problem within the context of the general formulation (1.15), and then explaining the reason for the apparent principal position of the associated linearized problem, it is worthwhile to list some of the principal problems of bifurcation theory (equivalently, buckling and postbuckling theory for structures); these are covered by the following questions:

1. Where are the branch-points? What is the relation of the branch points to the eigenvalues of some appropriately defined linearized problem?

2. How many distinct branches emanate from a branch-point? Is the branching to the left or right?

3. Can we describe the dependence of the branches on $\lambda$, at least in a neighborhood of the branch point?

4. In a physical problem, which branch does the physical system follow?

5. How far can branches be extended? If we are dealing with a Banach space $\mathcal{B}$, consisting of functions belonging to a certain class, can we guarantee that a particular branch represents a positive function. Does secondary branching occur?

Answers to questions 1-3, above, are the only ones which clearly fall within the domain of bifurcation theory; problem 4 is related to stability theory, while problem 5 requires, for an answer, techniques of global analysis which fall outside the strict domain of branching theory.

**Remarks**: Quite often (such as will be the case for the buckling of thin plates governed by the von Karman equations, or some modification thereof) the functional equation $G(\lambda, \mu) = 0$, in a Banach space $\mathcal{B}$, subject to the condition (1.16), will assume the particular form

$$Au - \lambda u = 0, \text{ with } A0 = 0 \qquad (1.17)$$

with $A$ a nonlinear operator from $\mathcal{B}$ into $\mathcal{B}$.

We now want to indicate how the problem governing buckling of a compressed rod, which we treated above, can be cast in the form (1.15). The situation is actually quite simple; we set

$$G(\lambda, \varphi) = \frac{d^2\varphi}{ds^2} + \lambda \sin \varphi \qquad (1.18)$$

and for the Banach space $\mathcal{B}$ take

$$\mathcal{B} = \{\varphi \in C^2[0, \ell], \text{ with } \varphi'(0) = \varphi'(\ell) = 0\} \qquad (1.19)$$

where $C^2[0, \ell]$ is the set of all functions defined and continuous on $[0, \ell]$ which have continuous derivatives up to order 2 on the open interval $(0, \ell)$. The mapping $G$, as given by (1.18), is a nonlinear differential operator defined on $\mathcal{B} \times \mathcal{R}^+$ where $\mathcal{R}^+$ is the set of all nonnegative real numbers.

Now, suppose for the sake of simplicity, that $G(\lambda, u) = 0$ were actually a scalar equation (like (1.6a)) and suppose, also, that there exists an equilibrium solution of the form $(\lambda_0, u_0)$; by hypothesis (1.16), $(\lambda, 0)$ is such an equilibrium solution for any $\lambda$. Thus $G(\lambda_0, u_0) = 0$. One may ask the question: does there exist, in a small neighborhood of $(\lambda_0, u_0)$, a unique solution of $G(\lambda, u) = 0$ in the form $u = u(\lambda)$? An answer to this question is provided by the well-known implicit-function theorem of mathematical analysis for whose statement we require the concept of Frechét derivative.

The concept of Frechét derivative for mappings between Banach or Hilbert spaces is defined as in numerous texts; an explanation of this concept, which belongs to the engineering literature on buckling and postbuckling behavior of elastic structures, may be found in the survey paper of Budiansky [16]. To begin with, we have a Banach or Hilbert space ($\mathcal{B}$ or $\mathcal{H}$, respectively) with a norm $\| \cdot \|$ that is used to measure the size of elements in this space; for example, for the function space $\mathcal{B}$, given by (1.19), in the problem associated with buckling of a compressed rod, for $\varphi$ in $\mathcal{B}$ we may take

$$\|\varphi\| = \sqrt{\int_0^\ell \varphi^2(s)ds} \qquad (1.20)$$

Then, if $f$ is a nonlinear mapping on the (function) space in question, we say that $f$ has a Frechét derivative $f_u$ (or $f'(u)$), at an element (function) $u$ in the space, if $f'(u)$ is a linear map such that

$$\lim_{\|v\| \to 0} \frac{\|f(u + v) - f(u) - f'(u)v\|}{\|v\|} = 0 \qquad (1.21)$$

In most cases of importance in applications, the concept of Frechét derivative $f'(u)$ is actually equivalent to the definition of the so-called Gateaux derivative of $f$ at $u$, namely, for $v$ any other element (i.e., function) in the space in question, $f'(u)$ is that linear mapping which is defined by

$$f'(u)v \equiv \lim_{\varepsilon \to 0} \frac{f(u + \varepsilon v) - f(u)}{\varepsilon} \tag{1.22}$$

or, equivalently,

$$f'(u)v = \frac{\partial}{\partial \epsilon}[f(u + \varepsilon v)]|_{\varepsilon=0} \tag{1.23}$$

For our purposes in this treatise, we may think of the Frechét derivative of a nonlinear mapping $G$, such as the one in (1.18), (which depends on the parameter $\lambda$ ), as being given by (1.22) or (1.23), e.g., at $\lambda = \lambda_0$, $\varphi = \varphi_0$, the Frechét derivative of $G(\lambda, u)$, as defined by (1.18), is given by

$$G_\varphi(\lambda_0, \varphi_0)\Psi = \frac{\partial}{\partial \varepsilon}[G(\lambda_0, \varphi_0 + \epsilon\Psi)]|_{\varepsilon=0} \tag{1.24}$$

with $\Psi \in \mathcal{B}$, as defined in (1.19). With the concept of Frechét derivative in hand, we may state the following version of the implicit function theorem which, in essence, underlies all of bifurcation (or buckling) theory:

**Theorem**: Let $G$ (parametrized by $\lambda$, as in (1.18)) be a Frechét differentiable, nonlinear mapping, on some Banach (or Hilbert) space and suppose that $(\lambda_0, u_0)$ is an (equilibrium) solution, i.e, $G(\lambda_0, u_0) = 0$. If the *linear map* $G_u(\lambda_0, u_0)$ has an *inverse* which is a *bounded* (i.e. continuous) mapping then *locally*, for $|\lambda - \lambda_0|$ *sufficiently small*, there exists a differentiable mapping (i.e. function) $u(\lambda)$ such that $G(\lambda, u(\lambda)) = 0$. Furthermore, in a sufficiently small neighborhood of $(\lambda_0, u_0)$, $(\lambda, u(\lambda))$ is the only solution of $G(\lambda, u) = 0$

**Remarks**: Under the hypothesis (1.16), $(\lambda, 0)$ is an equilibrium solution for all $\lambda$ (note that the mapping $G$ as given by (1.18) satisfies this condition). Thus, if $G_u(\lambda, 0)$ is a (linear) map with an inverse, which is bounded, then $u(\lambda) = 0$, for all $\lambda$, is the only solution of $G(\lambda, u(\lambda)) = 0$ and one may conclude that no branching (or bifurcation) of solutions can occur from the trivial solution $u(\lambda) = 0$.

As a consequence of the implicit function theorem, one may conclude that branching of solutions can only occur if the linear mapping $G_u$, evaluated at a specific (equilibrium) solution $(\lambda_c, u_0)$ is singular, so that $G_u(\lambda_c, u_0)$ does not have a bounded inverse; such a point $(\lambda_c, u_0)$ is then a candidate for a branch (or bifurcation) point.

**Example** (Consider the compressed thin rod, as governed by $G(\lambda, \varphi) = 0$, $G$ defined by (1.18), with $\varphi$ in the space $\mathcal{B}$ given by (1.19). Using the definition (1.23) of Frechét derivative, we easily find that for any $\varphi_0$, in the space given by (1.19), the derivative $G_\varphi(\lambda, \varphi_0)$ is the linear mapping $L$, which acts on a function $\Psi$ in $\mathcal{B}$ as follows:

$$L\Psi = G_\varphi(\lambda, \varphi_0)\Psi \equiv \frac{d^2\Psi}{ds^2} + \lambda\Psi\cos\varphi_0 \qquad (1.25)$$

so that, in particular, for the equilibrium solution $(\lambda, \varphi_0 = 0)$

$$L\Psi = G_\varphi(\lambda, 0)\Psi \equiv \frac{d^2\Psi}{ds^2} + \lambda\Psi \qquad (1.26)$$

where, as $\Psi \in \mathcal{B}$, $\Psi$ must satisfy $\Psi'(0) = \Psi'(\ell) = 0$. But $L$ is invertible if and only if the only solution of $L\Psi = 0$, for $\Psi$ in $\mathcal{B}$, is given by $\Psi \equiv 0$. Thus, $L$ is not invertible if there are values $\lambda = \lambda_c$ such that $L\Psi = 0$, for $\Psi \in \mathcal{B}$, has nontrivial solutions, i.e., if there exists $\lambda = \lambda_c$ such that

$$\begin{cases} \dfrac{d^2\Psi}{ds^2} + \lambda_c\Psi = 0 \\ \Psi'(0) = \Psi'(\ell) = 0 \end{cases} \qquad (1.27)$$

has at least one nontrivial solution $\Psi$; such solutions (of 1.27) are, of course, eigenfunctions, corresponding to the eigenvalue $\lambda = \lambda_c$, which then becomes a candidate for being a branch (or bifurcation) point for the boundary value problem (1.6a,b) associated with the compressed (thin) rod. The eigenvalue problem (1.27) is the *linearized* problem associated with (1.6a,b); in the parlance of buckling theory for elastic structures, $\lambda = \lambda_c$, such that there exists a nontrivial solution $\Psi = \Psi_c(s)$ of (1.27), is a possible buckling "load" – actually the buckling load divided by EI in this case – and the eigenfunction $\Psi_c$ is the associated buckling mode. In all cases of interest in classical buckling theory, the linearization of an equilibrium equation (or set of equations), such as (1.15), about an equilibrium solution $(\lambda_0, u_0)$, leads, in the manner described above, to an eigenvalue-eigenfunction problem

$$G_u(\lambda_c, u_0)v_c = 0 \qquad (1.28)$$

for the buckling "loads" $\lambda_c$ and the associated buckling modes $v_c$; this will, of course, be the case for the von Karman equations which we derive in §1.2.

We remark that it is possible to construct equations (and systems of equations) of the form $G(\lambda, u) = 0$, with $G(\lambda, 0) = 0$, for all $\lambda$, and $G_u(\lambda, 0)$ not invertible, for which branching does not take place from each eigenvalue of the linear operator $L = G_u(\lambda, 0)$. However, there are general theorems of bifurcation theory, which apply, e.g., to branching of solutions of the von Karman equations, and which guarantee that eigenvalues of the linear operator $L = G_u(\lambda, 0)$ are not only, by virtue of the implicit function theorem, candidates for branch points, but are, indeed, values of $\lambda$ where bifurcation occurs; we state only one such theorem: Let $G(\lambda, u) = Au - \lambda u$, with $A$ a nonlinear operator such that $A(0) = 0$, and suppose that $A$ is completely continuous (or compact). Then if $\overset{\circ}{\lambda} \neq 0$ is an eigenvalue of odd multiplicity for the linearized operator $L = A'(0)$, $\overset{\circ}{\lambda}$ is a branch-point of the basic solution $u(\lambda) \equiv 0$ of the nonlinear problem.

The theorem stated above is a classical result which is due to Leray and Schauder (see, e.g., [26]): the proof is not constructive (being based on topological degree theory) and thus little information is obtained about the structure of a bifurcating branch. The definition of completely continuous for a nonlinear operator $A$ on a Banach or Hilbert space is technical and may be found in any text (e.g. [26] ) on functional analysis; suffice it to say that the operators which are generated in the study of buckling and postbuckling of elastic (and, often, viscoelastic) structures do conform, in the proper mathematical setting, to the definition of a completely continuous operator.

---

## 1.2 The von Karman Equations for Linear Elastic Isotropic and Orthotropic Thin Plates

In this section we will derive the classical von Karman equations which govern the out-of-plane deflections of thin isotropic and orthotropic linear elastic plates as well as the linearized equations which mediate the onset of buckling; the equations will be presented in both rectilinear coordinates and in polar coordinates. We begin with a derivation, in rectilinear coordinates, of the von Karman equations for linear elastic, isotropic, (and then orthotropic), behavior.

## 1.2.1 Rectilinear Coordinates

We begin with the case of isotropic symmetry. Our derivation follows that in the text by Troger and Steindl [66] in which the authors begin with the derivation of the equilibrium equations for shallow shells, undergoing moderately large derivations, and then specialize to the case of the plate equations (which follow as a limiting case corresponding to zero initial curvature); such an approach is of particular interest to us in as much as we will want, later on, to look at imperfection buckling of thin plates.

In Fig. 1.6, we depict the undeformed middle surface of a shallow shell; this middle surface is represented in Cartesian coordinates by the function $w = W(x, y)$ and the displacements of the middle surface corresponding to the $x, y, z$ directions are denoted by $u, v$, and $w$ respectively.

To derive the nonlinear shell equations we

(i) First obtain a relationship between the displacements and the strain tensor; this is done by extending the usual assumptions made in the linear theory of plates and shells, and retaining nonlinear terms in the membrane strains, while still assuming a linear relationship for the bending strains.

(ii) Relate the strain tensor to the stress tensor by using the constitutive law representing linear, isotropic elastic behavior.

(iii) Derive the equilibrium equations which relate the stress tensor to the external loading.

We begin (see Fig. 1.7) by considering an infinitesimal volume element $dV = hdxdy$ of the shell. Let $\vec{r}'$ be the position vector to a point $P'$, in the interior of the shell, which is located at a distance $\zeta$ from the middle surface; then with $\vec{e}_i, i = 1, 2, 3$, the unit vectors along the coordinate axes

$$\vec{r}' = x\, \vec{e}_1 + y\, \vec{e}_2 + (W + \zeta)\, \vec{e}_3 \tag{1.29}$$

If we employ a form of the Kirchhoff hypothesis, i.e., that sections $x = const., y = const.$, of the undeformed shell, remain plane after deformation, and also maintain their angle with the deformed middle surface, then in terms of the displacement components $u, v, w$ of the middle surface of the shell, with $w_{,x} = \dfrac{\partial w}{\partial x}$, etc.,

$$u(\zeta) = u - \zeta w_{,x}, \quad v(\zeta) = v - \zeta w_{,y}, \quad w(\zeta) = w \tag{1.30}$$

are the displacement components of $P'$ when it moves, after the displacement, (which arises, e.g., because of the application of middle surface forces at the boundary, or variations in temperature or moisture content) to a point $\hat{P}$. The position vector $\vec{\hat{r}}$ to the point $\hat{P}$ is then

$$\vec{\hat{r}} = (x + u - \zeta w_{,x}) \, \vec{e_1} + (y + v - \zeta w_{,y}) \, \vec{e_2}$$
$$+ (W + \zeta + w) \, \vec{e_3} \tag{1.31}$$

The components of the strain tensor for this deformation of the shell are then obtained by computing the difference of the squares of the lengths of the differential line elements $\vec{dr}$ and $\vec{d\hat{r}}$; a length computation based on (1.31) and (see Fig. 1.6) $\vec{r} = x \, \vec{e_1} + y \, \vec{e_2} + W \, \vec{e_3}$ yields

$$\frac{1}{2} \left[ |\vec{d\hat{r}}\,|^2 - |\vec{dr}\,|^2 \right] = \varepsilon_{xx} dx^2 + 2\varepsilon_{xy} dx dy + \varepsilon_{yy} dy^2 \tag{1.32}$$

where

$$\begin{cases} \varepsilon_{xx} = u_{,x} + \dfrac{1}{2} \left( u_{,x}^2 + v_{,x}^2 + w_{,x}^2 \right) + W_{,x} \, w_{,x} \\ \qquad\quad - \zeta w_{,xx} + \cdots \\[2mm] \varepsilon_{yy} = v_{,y} + \dfrac{1}{2} \left( u_{,y}^2 + v_{,y}^2 + w_{,y}^2 \right) + W_{,y} \, w_{,y} \\ \qquad\quad - \zeta w_{,yy} + \cdots \\[2mm] \gamma_{xy} \equiv 2\varepsilon_{xy} = u_{,y} + v_{,x} + u_{,x} \, v_{,y} + w_{,x} \, w_{,y} \\ \qquad\quad + W_{,x} \, w_{,y} + W_{,y} \, w_{,x} - 2\zeta w_{,xy} + \cdots \end{cases} \tag{1.33}$$

and terms of at least third order have been neglected. If we now take into consideration the fact that, for stability problems connected with thin-walled structures, the displacement $w$, which is orthogonal to the middle surface, is much larger than the displacements $u, v$ in the middle surface, then the terms quadratic in $u$ and $v$, in (1.33), may be neglected in comparison with those in $w$; we thus obtain the (approximate) kinematical relations

$$\begin{cases} \varepsilon_{xx} = u_{,x} + \dfrac{1}{2} w_{,x}^2 + W_{,x} \, w_{,x} - \zeta w_{,xx} \\[2mm] \varepsilon_{yy} = v_{,y} + \dfrac{1}{2} w_{,y}^2 + W_{,y} \, w_{,y} - \zeta w_{,yy} \\[2mm] \gamma_{xy} = 2\varepsilon_{xy} = u_{,y} + v_{,x} + w_{,x} \, w_{,y} + W_{,x} \, w_{,y} \\[2mm] \qquad\quad + W_{,y} \, w_{,x} - 2\zeta w_{,xy} \end{cases} \tag{1.34}$$

For the particular case in which the shallow shell is a plate whose middle surface, prior to buckling, occupies a region in the $x, y$ plane (so that $W \equiv 0$) (1.34) reduces further to

$$
\begin{cases}
\epsilon_{xx} = u_{,x} + \dfrac{1}{2} w_{,x}^2 - \zeta w_{,xx} \\[2mm]
\epsilon_{yy} = v_{,y} + \dfrac{1}{2} w_{,y}^2 - \zeta w_{,yy} \\[2mm]
\gamma_{xy} \equiv 2\epsilon_{xy} = u_{,y} + v_{,x} + w_{,x}\, w_{,y} - 2\zeta w_{,xy}
\end{cases}
\tag{1.35}
$$

We now use the generalized Hooke's law to describe the material behavior of the isotropic, linear elastic, shallow shell and we make the usual assumption for thin-walled structures that $\sigma_{zz} \simeq 0$. The stress components, $\sigma_{xz}$ and $\sigma_{yz}$, which will appear in the equilibrium equations, do not contribute to the constitutive relationship as the corresponding strains are zero (due to the Kirchhoff hypothesis). Thus, a plane strain problem over the thickness $h$ of the shell is obtained for which the relevant constitutive equations may be written in the form

$$
\begin{cases}
hE\left(u_{,x} + \dfrac{1}{2} w_{,x}^2 + W_{,x}\, w_{,x}\right) = N_x - \nu N_y \\[2mm]
hE\left(v_{,y} + \dfrac{1}{2} w_{,y}^2 + W_{,y}\, w_{,y}\right) = N_y - \nu N_x \\[2mm]
Gh(u_{,y} + v_{,x} + w_{,x}\, w_{,y} + \\[2mm]
\quad W_{,x}\, w_{,y} + W_{,y}\, w_{,x}) = N_{xy}
\end{cases}
\tag{1.36}
$$

where $E$ is Young's modulus, $\nu$ is the Poisson number, $G = E/2(1+\nu)$ is the shear modulus, and

$$
\begin{cases}
N_x = \displaystyle\int_{-h/2}^{h/2} \sigma_{xx} dz \\[2mm]
N_y = \displaystyle\int_{-h/2}^{h/2} \sigma_{yy} dz \\[2mm]
N_{xy} = \displaystyle\int_{-h/2}^{h/2} \sigma_{xy} dz
\end{cases}
\tag{1.37}
$$

are the *averaged stresses* over the shell thickness $h$, which is assumed to be small. The *bending moment* $M_x$ is defined to be

$$
M_x = \int_{-h/2}^{h/2} \sigma_{xx} \xi d\xi
\tag{1.38}
$$

and by using (1.34) and (1.36) this may be computed to be

$$M_x = \int_{-h/2}^{h/2} \frac{E}{1-\nu^2} (\epsilon_{xx} + \nu\epsilon_{yy}) \xi d\xi$$

(1.39)

$$= -K(w_{,xx} + \nu w_{,yy})$$

with the *plate stiffness* $K$ given by

$$K = \frac{Eh^3}{12(1-\nu^2)}$$

(1.40)

With analogous definitions (and computations) for the bending moments $M_y$ and $M_{xy}$ we find that

$$\begin{cases} M_y = -K(w_{,yy} + \nu w_{,xx}) \\ M_{xy} = -(1-\nu)Kw_{,xy} \end{cases}$$

(1.41)

Although, in a nonlinear theory, the equations of equilibrium for the stresses and loads must be calculated in the deformed geometry, in the nonlinear theory of shallow shells one makes the approximation that the undeformed geometry can still be used for this purpose; this assumption then restricts the validity of the equations we obtain to moderately large deformations. We will, therefore, use the equilibrium equations of linear shell and plate theory, namely, the three force equilibrium equations

$$\begin{cases} N_{x,x} + N_{xy,y} = 0 \\ N_{xy,x} + N_{y,y} = 0 \end{cases}$$

(1.42)

and

$$Q_{xz,x} + Q_{yz,y} + N_x(W+w)_{,xx}$$
$$+ N_y(W+w)_{,yy} + 2N_{xy}(W+w)_{,xy} + t = 0$$

(1.43)

where $t(x,y)$ is a distributed normal loading, and the two moment equilibrium equations

$$\begin{cases} M_{xy,x} + M_{y,y} - Q_{yz} = 0 \\ \\ M_{yx,y} + M_{x,x} - Q_{xz} = 0 \end{cases} \qquad (M_{xy} = M_{yx})$$

(1.44)

where the indicated forces and moments are depicted in figures 1.8 and 1.9 and the moment equilibrium equation about the $z$-axis is satisfied identically.

We now proceed by differentiating the first equation in (1.44) with respect to $y$, and the second equation with respect to $x$, and then inserting the resulting expressions for $Q_{xz,x}$ and $Q_{yz,y}$ in (1.43) so as to obtain the following equation from which the shear forces have been eliminated:

$$M_{x,xx} + 2M_{xy,xy} + M_{y,yy}$$
$$+N_x(W+w)_{,xx} + 2N_{xy}(W+w)_{,xy} \qquad (1.45)$$
$$+N_y(W+w)_{,yy} + t = 0$$

The next step consists of introducing the Airy stress function $\Phi(x,y)$, which is defined so as to satisfy

$$N_x = \Phi_{,yy}, \quad N_y = \Phi_{,xx}, \quad N_{xy} = -\Phi_{,xy} \qquad (1.46)$$

With the above definitions, both equilibrium equations in (1.42) are satisfied identically. Finally, if we substitute (1.39), (1.41) and (1.46) into the remaining equilibrium equation, (1.45), we obtain the nonlinear partial differential equation

$$K\Delta^2 w = \Phi_{,yy}(W+w)_{,xx} + \Phi_{,xx}(W+w)_{,yy} \qquad (1.47)$$
$$-2\Phi_{,xy}(W+w)_{,wy} + t(x,y)$$

for the two unknowns, the (extra) deflection $w(x,y)$ and the Airy stress function $\Phi(x,y)$, where $\Delta^2$ denotes the biharmonic operator in rectilinear Cartesian coordinates, i.e.,

$$\Delta^2 w = \frac{\partial^4 w}{\partial x^4} + 2\frac{\partial^4 w}{\partial x^2 \partial y^2} + \frac{\partial^4 w}{\partial y^4} \qquad (1.48)$$

To obtain a second partial differential equation for $w(x,y)$ and $\Phi(x,y)$, we make use of the identity

$$(u_{,x})_{,yy} + (v_{,y})_{,xx} - (u_{,y} + v_{,x})_{,xy} = 0 \qquad (1.49)$$

into which we substitute, from the constitutive relations (1.36), for the displacement derivatives $u_{,x}$, $v_{,y}$, and $u_{,y} + v_{,x}$. One then makes use of the definition of the Airy function $\Phi$ to replace the stress resultants $N_x, N_y$, and $N_{xy}$; there results the following equation

$$\Delta^2\Phi = Eh\Big[(w^2_{,xy} - w_{,xx}\,w_{,yy})$$
$$\qquad (1.50)$$
$$+2W_{,xy}\,w_{,xy} - W_{,xx}\,w_{,yy} - W_{,yy}\,w_{,xx}\Big]$$

If we introduce the nonlinear (bracket) differential operator by

$$[f,g] \equiv f_{,yy}\, g_{,xx} - 2f_{,xy}\, g_{,xy} + f_{,xx}\, g_{,yy} \tag{1.51a}$$

so that

$$[f,f] = 2(f_{,xx}\, f_{,yy} - f_{,xy}^2) \tag{1.51b}$$

then the system consisting of (1.47) and (1.50) can be written in the more compact form

$$K\Delta^2 w = [\Phi, W + w] + t \tag{1.52a}$$

$$\frac{1}{Eh}\Delta^2 \Phi = -\frac{1}{2}[w,w] - [W,w] \tag{1.52b}$$

In particular, for the deflection of a thin plate, which in its undeflected configuration occupies a domain in the $x, y$ plane, so that $W \equiv 0$, (1.52a), (1.52b) reduce to the *von Karman plate equations* for an isotropic linear elastic material, namely,

$$K\Delta^2 w = [\Phi, w] + t \tag{1.53a}$$

$$\frac{1}{Eh}\Delta^2 \Phi = -\frac{1}{2}[w,w] \tag{1.53b}$$

An important alternative form for the bracket $[\Phi, w]$, which reflects the effect of middle surface forces on the deflection, is

$$[\Phi, w] = N_x \frac{\partial^2 w}{\partial x^2} + 2N_{xy}\frac{\partial^2 w}{\partial x \partial y} + N_y \frac{\partial^2 w}{\partial y^2} \tag{1.54}$$

**Remarks**: The curvatures of the plate in planes parallel to the $(x, z)$ and $(y, z)$ planes, are usually denoted by $\kappa_x$ and $\kappa_y$, respectively, while the twisting curvature is denoted by $\kappa_{xy}$. Strictly speaking, the curvature $\kappa_x$, e.g., is given by

$$\kappa_x = \frac{-\dfrac{\partial^2 w}{\partial x^2}}{\left\{1 + \left(\dfrac{\partial w}{\partial x}\right)^2\right\}^{\frac{3}{2}}} \tag{1.55}$$

where the minus sign is introduced so that an increase in the bending moment $M_x$ results in an increase in $\kappa_x$. As $w_{,x}^2$ is assumed to be small, even within the context of the nonlinear theory of shallow shells, the curvatures are usually approximated by, e.g. $\kappa_x = -w,_{xx}$, in which case for linear isotropic response

$$
\begin{cases}
\kappa_x = (M_x - \nu M_y)/\left\{(1-\nu^2)K\right\} \\
\kappa_y = (M_y - \nu M_x)/\left\{(1-\nu^2)K\right\} \\
\kappa_{xy} = (1+\nu)M_{xy}/\left\{(1-\nu)K\right\}
\end{cases}
\tag{1.56}
$$

**Remarks:** For a plate which is initially stress-free, but possesses an initial *imperfection* in the form of a built-in deflection, whose mid-surface is given by the equation $z = w_0(x,y)$, the appropriate modification of the von Karman equations (1.53a,b) for a plate exhibiting linear elastic isotropic response may be obtained from the shallow shell equations (1.52a,b) by replacing $W(x,y)$ by $w_0(x,y)$. Alternatively, if we set

$$
\tilde{w}(x,y) = w(x,y) + w_0(w,y)
\tag{1.57}
$$

then we obtain from (1.53a,b) the following imperfection modification of the usual von Karman equations:

$$
\begin{cases}
K\Delta^2(\tilde{w} - w_0) = [\Phi, \tilde{w}] + t \\
\dfrac{1}{Eh}\Delta^2\Phi = -\dfrac{1}{2}[\tilde{w} - w_0, \tilde{w} - w_0] \\
\qquad\qquad\quad -[w_0, \tilde{w} - w_0]
\end{cases}
\tag{1.58}
$$

In (1.58), $\tilde{w}(x,y)$ represents the net deflection, and (1.58) reduces to (1.53a,b) if $w_0 \equiv 0$. Other modifications of (1.53a,b) are needed if the plate is subject to thermal or hygroexpansive strains, or if the stiffness $K$, or the thickness $h$ of the plate are not constant.

**Remarks:** For a normally loaded (linear elastic, isotropic) plate with a rigid boundary, the in-plane boundary conditions are specified by the vanishing of the displacements $u, v$, rather than by specification of the middle surface forces; in such cases, there may be advantages to expressing the large deflexion equations for an initially flat plate, of constant thickness, in terms of the displacements $u, v, w$. The displacement equations, with some straightforward work, can be shown to have the following form

$$
\begin{aligned}
\dfrac{h^2}{12}\left(\Delta^4 w - \dfrac{t}{K}\right) = {}& w,_{xx}\left\{u,_x + \nu v,_y + \dfrac{1}{2}w,_x^2 + \dfrac{1}{2}\nu w,_y^2\right\} \\
& + w,_{yy}\left\{v,_y + \nu u,_x + \dfrac{1}{2}w,_y^2 + \dfrac{1}{2}\nu w,_x^2\right\} \\
& + (1-\nu)w,_{xy}\left(u,_y + v,_x + w,_x\, w,_y\right)
\end{aligned}
\tag{1.59a}
$$

$$\frac{\partial}{\partial x}\left[u_{,x} + v_{,y} + \frac{1}{2}\left\{w_{,x}^2 + w_{,y}^2\right\}\right]$$
$$+ \left(\frac{1-\nu}{1+\nu}\right)(\nabla^2 u + w_{,y}\,\nabla^2 w) = 0 \qquad (1.59b)$$

and

$$\frac{\partial}{\partial y}\left[u_{,x} + v_{,y} + \frac{1}{2}\left\{w_{,x}^2 + w_{,y}^2\right\}\right] \qquad (1.59c)$$
$$+ \left(\frac{1-\nu}{1+\nu}\right)(\nabla^2 v + w_{,y}\,\nabla^2 w) = 0$$

where $\nabla^2$ is the usual Laplacian operator.

We now proceed (still in rectilinear coordinates) to derive the appropriate form of the von Karman plate equations for the case of a thin plate which exhibits linear elastic behavior and has (rectilinear) orthotropic symmetry. Thus, consider an orthotropic thin plate for which the $x$ and $y$ axes coincide with the principal directions of elasticity; then the constitutive equations have the form

$$\left\{\begin{array}{c} \sigma_{xx} \\ \sigma_{yy} \\ \sigma_{xy} \end{array}\right\} = \left[\begin{array}{ccc} c_{11} & c_{12} & 0 \\ c_{21} & c_{22} & 0 \\ 0 & 0 & c_{66} \end{array}\right]\left\{\begin{array}{c} \epsilon_{xx} - \beta_1 \Delta H \\ \epsilon_{yy} - \beta_2 \Delta H \\ \gamma_{xy} \end{array}\right\} \qquad (1.60)$$

where we have included in the strain components possible hygroexpansive strains $\beta_i \Delta H, i = 1, 2$ where the $\beta_i$ are the hygroexpansive coefficients and $\Delta H$ represents a humidity change; alternatively, we could replace the $\beta_i \Delta H$ by thermal strains $\alpha_i \Delta T$ with the $\alpha_i$ thermal expansion coefficients and $\Delta T$ a change in temperature. In (1.60)

$$\begin{cases} c_{11} = E_1/(1 - \nu_{12}\nu_{21}) \\ c_{12} = E_2\nu_{21}/(1 - \nu_{12}\nu_{21}) \\ c_{21} = E_1\nu_{12}/(1 - \nu_{12}\nu_{21}) \\ c_{22} = E_2/(1 - \nu_{12}\nu_{21}) \\ c_{66} = G_{12} \end{cases} \qquad (1.61)$$

with $E_1\nu_{12} = E_2\nu_{21}$ so that $c_{12} = c_{21}$. In (1.61), $E_1, E_2, \nu_{12}, \nu_{21}$, and $G_{12}$ are, respectively, the Young's moduli, Poisson's ratios, and shear modulus associated with the principal directions. The constants

$$D_{ij} = c_{ij}\frac{h^3}{12} \qquad (1.62)$$

are the associated rigidities (or stiffness ratios) of the orthotropic plate, specifically,

$$D_{11} = \frac{E_1 h^3}{12(1 - \nu_{12}\nu_{21})}, \quad D_{22} = \frac{E_2 h^3}{12(1 - \nu_{12}\nu_{21})} \qquad (1.63)$$

are the bending rigidities about the $x$ and $y$ axes, respectively, while

$$D_{66} = \frac{G_{12} h^3}{12} \qquad (1.64)$$

is the twisting rigidity. The ratios $D_{12}/D_{22}$, $D_{12}/D_{11}$ are often called reduced Poisson's ratios. For the thin orthotropic plate under consideration, the strains $\epsilon_{xx}, \epsilon_{yy}$, and $\gamma_{xy}$, the averaged stresses (or stress resultants) $N_x, N_y$, and $N_{xy}$, and the bending moments $M_x, M_y$, and $M_{xy}$ are still given by (1.35), (1.37), (1.38), and the relevant expressions for $M_y$ and $M_{xy}$, which are analogous to (1.38). Thus, with

$$\begin{cases} \sigma_{xx} = c_{11}(\epsilon_{xx} - \beta_1 \Delta H) + c_{12}(\epsilon_{yy} - \beta_2 \Delta H) \\ \sigma_{yy} = c_{21}(\epsilon_{xx} - \beta_1 \Delta H) + c_{22}(\epsilon_{yy} - \beta_2 \Delta H) \\ \sigma_{xy} = c_{66}\gamma_{xy} \end{cases}$$

we have

$$\sigma_{xx} = c_{11}(u_{,x} + \tfrac{1}{2}w_{,x}^2 - \zeta w_{,xx}) \qquad (1.65a)$$

$$+ c_{12}(v_{,y} + \tfrac{1}{2}w_{,y}^2 - \zeta w_{,yy})$$

$$- (c_{11}\beta_1 \Delta H + c_{12}\beta_2 \Delta H)$$

$$\sigma_{yy} = c_{21}(u_{,x} + \tfrac{1}{2}w_{,x}^2 - \zeta w_{,xx}) \qquad (1.65b)$$

$$+ c_{22}(v_{,y} + \tfrac{1}{2}w_{,y}^2 - \zeta w_{,yy})$$

$$- (c_{21}\beta_1 \Delta H + c_{22}\beta_2 \Delta H)$$

and

$$\tau_{xy} = c_{66}\left(\tfrac{1}{2}(u_{,y} + v_{,x}) + w_{,x}w_{,y} - 2\zeta w_{,xy}\right) \qquad (1.65c)$$

Using the expressions in (1.65 a,b,c), we now compute the bending moments $M_x$, $M_y$, and $M_{xy}$ to be

$$
\left\{
\begin{aligned}
M_x = &\; c_{11} \int_{-h/2}^{h/2} \left[ \xi \left( u_{,x} + \frac{1}{2} w_{,x}^2 \right) - \xi^2 w_{,xx} \right] d\xi \\
&+ c_{12} \int_{-h/2}^{h/2} \left[ \xi \left( v_{,y} + \frac{1}{2} w_{,y}^2 \right) - \xi^2 w_{,yy} \right] d\xi \qquad \text{(1.66a)} \\
&- \int_{-h/2}^{h/2} \xi \left( c_{12} \beta_1 \Delta H + c_{22} \beta_2 \Delta H \right) d\xi
\end{aligned}
\right.
$$

$$
\left\{
\begin{aligned}
M_y = &\; c_{21} \int_{-h/2}^{h/2} \left[ \xi \left( u_{,x} + \frac{1}{2} w_{,x}^2 \right) - \xi^2 w_{,xx} \right] d\xi \\
&+ c_{22} \int_{-h/2}^{h/2} \left[ \xi \left( v_{,y} + \frac{1}{2} w_{,y}^2 \right) - \xi^2 w_{,yy} \right] d\xi \qquad \text{(1.66b)} \\
&- \int_{-h/2}^{h/2} \xi \left( c_{11} \beta_1 \Delta H + c_{22} \beta_2 \Delta H \right) d\xi
\end{aligned}
\right.
$$

and

$$
\left\{
\begin{aligned}
M_{xy} = \\
c_{66} \int_{-h/2}^{h/2} \left[ \xi \left( \frac{1}{2} \left[ u_{,y} + v_{,x} \right] + w_{,x}\, w_{,y} \right) - 2\xi^2 w_{,xy} \right] d\xi \quad \text{(1.66c)}
\end{aligned}
\right.
$$

At this stage of the calculation, sufficient flexibility exists to handle any dependence of the (possible) hygroexpansive strains $\beta_i \Delta H$, $i = 1, 2$, on the variable $z$; one could, e.g., assume that the $\beta_i \Delta H$, $i = 1, 2$ are independent of $z$, depend either linearly or quadratically on $z$, or are even represented by convergent power series of the form

$$
\left\{
\begin{aligned}
\beta_1 \Delta H = \alpha_0 + \sum_{m=1}^{\infty} \alpha_m z^m \\
\beta_2 \Delta H = \beta_0 + \sum_{m=1}^{\infty} \beta_m z^m
\end{aligned}
\right.
$$

If, however, $\beta_i \Delta H = f_i(z)$, $i = 1, 2$, with the $f_i(z)$ even functions of $z$, i.e., $f_i(z) = f_i(-z)$, $i = 1, 2$, then $\xi \beta_i \Delta H$ will be an odd function of $\xi$, $-h/2 < \xi < h/2$, and the integrals in both (1.66a) and (1.66b), which involve the hygroexpansive strains, will vanish. For the sake of simplicity, we will proceed here by assuming that the hygroexpansive strains are constant through the thickness of the plate; this assumption will be relaxed in the discussion of hygroexpansive/thermal buckling in Chapter 6. Then

$$M_x = -c_{11}w_{,xx} \left. \frac{\xi^3}{3} \right]_{-h/2}^{h/2} - c_{12}w_{,yy} \left. \frac{\xi^3}{3} \right]_{h/2}^{\frac{h}{2}}$$
$$= -(c_{11}w_{,xx} + c_{12}w_{,yy}) \cdot \frac{h^3}{12}$$

or

$$M_x = -(D_{11}w_{,xx} + D_{12}w_{,yy}) \tag{1.67a}$$

while, in an analogous fashion,

$$M_y = -(D_{21}w_{,xx} + D_{22}w_{,yy}) \tag{1.67b}$$

and

$$M_{xy} = -2c_{66} \cdot w_{,xy} \left. \frac{\xi^3}{3} \right]_{\frac{-h}{2}}^{\frac{h}{2}} \tag{1.67c}$$
$$= -2D_{66}w_{,xy}$$

We now set $W \equiv 0$, $t \equiv 0$ in (1.45), to reflect the fact that we are dealing with an initially flat plate which is not acted on by a distributed normal load, introduce the Airy stress function through (1.46), and employ the results in (1.67a,b,c), so as to deduce that

$$-(D_{11}w_{,xx} + D_{12}w_{,yy})_{,xx} - 4D_{66}w_{,xxyy}$$
$$-(D_{21}w_{,xx} + D_{22}w_{,yy})_{,yy} + \Phi_{,yy} \, w_{,xx}$$
$$-2\Phi_{,xy} \, w_{,xy} + \Phi_{,xx} \, w_{,yy} = 0$$

or

$$D_{11}w_{,xxxx} + [D_{12} + 4D_{66} + D_{21}]w_{,xxyy} \tag{1.68}$$
$$+ D_{22}w_{,yyyy} - [\Phi, w] = 0$$

The corresponding modification of the first von Karman equation for the deflection of a thin, linearly elastic, orthotropic plate, when there exists an initial deflection $z = w_0(x, y)$, is easily obtained from (1.45) and (1.67a,b,c) by setting $W = w_0$ and defining, as in (1.57), $\tilde{w} = w + w_0$. To obtain the appropriate modification of the second von Karman equation (1.53b) for the case of a linearly elastic, thin, orthotropic plate, we once again begin with the identity (1.49). We then use the constitutive relations (1.65 a,b,c) to compute the average stresses (or stress resultants) $N_x, N_y$, and $N_{xy}$, introduce the Airy function through (1.46), and solve the resulting equations for $u_{,x}$, $v_{,y}$, and $u_{,y} + v_{,x}$; these, in turn, are substituted into (1.49) and there results the partial differential equation

$$\frac{1}{E_1 h}\Phi_{,yyyy} + \frac{1}{h}\left(\frac{1}{G_{12}} - \frac{2\nu_{12}}{E_2}\right)\Phi_{,xxyy} \qquad (1.69)$$

$$+ \frac{1}{E_2 h}\Phi_{,xxxx} = -\frac{1}{2}[w, w]$$

if one makes use of the fact that $E_2\nu_{21} = E_1\nu_{12}$.

The complete set of von Karman equations in rectilinear Cartesian coordinates thus consists of (1.68) and (1.69). In an isotropic plate $E_1 = E_2 = E$, $G_{12} = G = E/2(1 + \nu)$, with $\nu_{12} = \nu_{21} = \nu$, and the system of equations (1.68), (1.69) reduces to the system (1.53a,b), with $t \equiv 0$, where $K$ is the common value of the principal rigidities $D_1 = D_{11}, D_2 = D_{22}$, and $D_3 = D_{12}\nu_{12} + 2D_{66}$.

**Remarks:** The system of equations (1.68), (1.69), as well as their specializations to (1.53 a,b), for the case of linear, isotropic, elastic response, may be obtained from a variational (minimum energy) principle based on the potential energy

$$U = \frac{1}{2}\int_A \int_{-h/2}^{h/2} \{c_{11}(\epsilon_{xx} - \beta_1\Delta H)^2 \qquad (1.70)$$

$$+ 2c_{12}(\epsilon_{xx} - \beta_1\Delta H)(\epsilon_{yy} - \beta_2\Delta H)$$

$$+ c_{22}(\epsilon_{yy} - \beta_2\Delta H)^2 + c_{66}\gamma_{xy}^2 \} dz dA$$

where the outer integral is computed over the area A occupied by the plate. In (1.70), or its equivalent for the case where the plate exhibits isotropic response, we must first substitute from (1.35) in order to express the integrand as a polynomial expression in the displacement derivatives.

**Remarks:** In the case of very thin plates, which may have deflections many times their thickness, the resistance of the plate to bending can, often, be neglected; this amounts, in the case of a plate exhibiting isotropic response, to taking the stiffness $K = 0$, in which case the problem reduces to one of finding the deflection of a flexible membrane. The equations which apply in this case were obtained by A. Föppl [69] and turn out, of course, to be just the von Karman equations (1.53a,b) with $K$ set equal to zero.

## 1.2.2  Polar Coordinates

In this section we will present the appropriate versions of the von Karman equations in polar coordinates (actually, cylindrical coordinates must be used for the equilibrium equations) for a thin linearly elastic plate. We will present these equations for the following cases: a plate exhibiting isotropic symmetry, both for the general situation as well as for the special situation in which the deformations are assumed to be radially symmetric, a plate which exhibits cylindrically orthotropic behavior (to be defined below), and a plate which exhibits the rectilinear orthotropic behavior that was specified in the last subsection. We begin with the simplest case, that of a linearly elastic isotropic plate.

It is well known that the operators present in the von Karman equations (1.53a,b) are invariant with respect to changes in the coordinate system; in particular, the von Karman equations in polar coordinates for a thin, linearly elastic, isotropic plate may be obtained from (1.53a,b) by transforming the bracket and biharmonic operators into their equivalent expressions in the new coordinate system. If $u_r$, $u_\theta$ denote, respectively, the displacement components in the middle surface of the plate, while $w = w(r, \theta)$ denotes the out of plane displacement, then the strain components $e_{rr}, e_{r\theta}$, and $e_{\theta\theta}$ are given by ($\gamma_{r\theta} = 2e_{r\theta}$) :

$$\begin{cases} e_{rr} = \dfrac{\partial u_r}{\partial r} + \dfrac{1}{2}\left(\dfrac{\partial w}{\partial r}\right)^2 - \zeta\dfrac{\partial^2 w}{\partial r^2} \\[2mm] e_{\theta\theta} = \dfrac{u_r}{r} + \dfrac{1}{r}\dfrac{\partial u_\theta}{\partial \theta} + \dfrac{1}{2r^2}\left(\dfrac{\partial w}{\partial \theta}\right)^2 - \zeta\left(\dfrac{1}{r}\dfrac{\partial w}{\partial r} + \dfrac{1}{r^2}\dfrac{\partial^2 w}{\partial \theta^2}\right) \\[2mm] e_{r\theta} = \dfrac{\partial u_\theta}{\partial r} - \dfrac{u_\theta}{r} + \dfrac{1}{r}\left(\dfrac{\partial u_r}{\partial \theta}\right) + \dfrac{1}{r}\left(\dfrac{\partial w}{\partial r}\right)\left(\dfrac{\partial w}{\partial \theta}\right) \\[2mm] \qquad -2\zeta\left(\dfrac{1}{r}\dfrac{\partial^2 w}{\partial r\partial \theta} - \dfrac{1}{r^2}\dfrac{\partial w}{\partial \theta}\right) \end{cases} \qquad (1.71)$$

With the components of the stress tensor $\sigma_{rr}, \sigma_{r\theta}, \sigma_{\theta\theta}, \sigma_{zz}, \sigma_{rz}$ and $\sigma_{\theta z}$ as shown in Fig. 1.10, ($\sigma_{r\theta} = \sigma_{\theta r}$), ($\sigma_{rz} = \sigma_{zr}$), and $F_r, F_\theta$ the components of the applied body force in the radial and tangential directions, the equilibrium equations are

$$\begin{cases} \dfrac{\partial \sigma_{rr}}{\partial r} + \dfrac{1}{r}\dfrac{\partial \sigma_{r\theta}}{\partial \theta} + \dfrac{\sigma_{rr} - \sigma_{\theta\theta}}{r} + \dfrac{\partial \sigma_{rz}}{\partial z} + F_r = 0 \\[2mm] \dfrac{\partial \sigma_{r\theta}}{\partial r} + \dfrac{1}{r}\dfrac{\partial \sigma_{\theta\theta}}{\partial \theta} + \dfrac{2}{r}\sigma_{r\theta} + \dfrac{\partial \sigma_{\theta z}}{\partial z} + F_\theta = 0 \end{cases} \tag{1.72a}$$

and

$$\sigma_{rr}\left(\frac{1}{r}w_{,r} + w_{,rr}\right) + \sigma_{\theta\theta}\left(\frac{1}{r^2}w_{,\theta\theta}\right)$$

$$+\sigma_{r\theta}\left(\frac{2}{r}w_{,r\theta}\right) + \frac{1}{r}\sigma_{rz} + \frac{\partial \sigma_{rr}}{\partial r}w_{,r}$$

$$+\frac{\partial \sigma_{\theta\theta}}{\partial \theta}\left(\frac{1}{r^2}w_{,\theta}\right) + \frac{\partial \sigma_{r\theta}}{\partial r}\left(\frac{1}{r}w_{,\theta}\right) \tag{1.72b}$$

$$\frac{\partial \sigma_{r\theta}}{\partial \theta}\left(\frac{1}{r}w_{,r}\right) + \frac{\partial \sigma_{rz}}{\partial r} + \frac{\partial \sigma_{rz}}{\partial z}w_{,r}$$

$$+\frac{1}{r}\frac{\partial \sigma_{\theta z}}{\partial \theta} + \frac{\partial \sigma_{\theta z}}{\partial z}\left(\frac{1}{r}w_{,\theta}\right) + F_z = 0$$

The transformation of the stress components in Cartesian coordinates to those in polar coordinates is governed by the formulas:

$$\begin{cases} \sigma_{rr} = \sigma_{xx}\cos^2\theta + \sigma_{yy}\sin^2\theta + 2\sigma_{xy}\sin\theta\cos\theta \\ \sigma_{\theta\theta} = \sigma_{xx}\sin^2\theta + \sigma_{yy}\cos^2\theta - 2\sigma_{xy}\sin\theta\cos\theta \\ \sigma_{r\theta} = (\sigma_{yy} - \sigma_{xx})\sin\theta\cos\theta + \sigma_{xy}\left(\cos^2\theta - \sin^2\theta\right) \end{cases} \tag{1.73}$$

with analogous transformation formulae for $\sigma_{rz}$ and $\sigma_{\theta z}$. If we set, for the deflection $w$ and the Airy stress function $\Phi$, $\bar{w}(r,\theta) = w(r\cos\theta,$ $r\sin\theta)$, $\bar{\Phi}(r,\theta) = \Phi(r\cos\theta, r\sin\theta)$, and then drop the superimposed bars in the polar coordinate system, it can be shown directly that the stress resultants (or averaged stresses) $N_r$, $N_\theta$, and $N_{r\theta}$ are given in terms of $\Phi$ by

$$\begin{cases} N_r = \dfrac{1}{r}\Phi_{,r} + \dfrac{1}{r^2}\Phi_{,\theta\theta} \\[2mm] N_\theta = \Phi_{,rr} \\[2mm] N_{r\theta} = \dfrac{1}{r^2}\Phi_{,\theta} - \dfrac{1}{r}\Phi_{,r\theta} \end{cases} \tag{1.74}$$

while the operators $\Delta^2$ and $[\ ,\ ]$ are given by

$$
\begin{cases}
\Delta^2 w = w_{,rrrr} + \dfrac{2}{r} w_{,rrr} - \dfrac{1}{r^2} w_{,rr} \\[2mm]
\qquad + \dfrac{2}{r^2} w_{,rr\theta\theta} + \dfrac{1}{r^3} w_{,r} - \dfrac{2}{r^3} w_{,r\theta\theta} \\[2mm]
\qquad + \dfrac{1}{r^4} w_{,\theta\theta\theta\theta} + \dfrac{4}{r^4} w_{,\theta\theta}
\end{cases}
\qquad (1.75a)
$$

and

$$
\begin{aligned}
[w, \Phi] = \ &w_{,rr} \left( \frac{1}{r} \Phi_{,r} + \frac{1}{r^2} \Phi_{,\theta\theta} \right) \\
&+ \left( \frac{1}{r} \cdot w_{,r} + \frac{1}{r^2} w_{,\theta\theta} \right) \Phi_{,rr} \\
&- 2 \left( \frac{1}{r} w_{,r\theta} - \frac{1}{r^2} w_{,\theta} \right) \left( \frac{1}{r} \Phi_{,r\theta} - \frac{1}{r^2} \Phi_{,\theta} \right)
\end{aligned}
$$

or, in view of (1.74),

$$
\begin{aligned}
[w, \Phi] = \ &N_r w_{,rr} - 2 N_{r\theta} \left( \frac{1}{r^2} w_{,\theta} - \frac{1}{r} w_{,r\theta} \right) \\
&+ N_\theta \left( \frac{1}{r} \cdot w_{,r} + \frac{1}{r^2} w_{,\theta\theta} \right)
\end{aligned}
\qquad (1.75b)
$$

For the special case of a radially symmetric deformation, in which $u_r = u_r(r)$, $u_\theta = 0$, and $w = w(r)$, the expressions in (1.75a) and (1.75b) for the biharmonic and bracket operators reduce to

$$
\Delta^2 w = w_{,rrrr} + \frac{2}{r} w_{,rrr} - \frac{1}{r^2} w_{,rr} + \frac{1}{r^3} w_{,r}
\qquad (1.76)
$$

and

$$
[w, \Phi] = N_r w_{,rr} + N_\theta \frac{1}{r} w_{,r}
\qquad (1.77)
$$

Thus, for the von Karman equations for a thin, linearly elastic, isotropic plate, in polar coordinates, we have (with $t \equiv 0$):

$$K\left[w_{,rrrr} + \frac{2}{r}w_{,rrr} - \frac{1}{r^2}w_{,rr} + \frac{2}{r^2}w_{,rr\theta\theta}\right.$$

$$\left. + \frac{1}{r^3}w_{,r} - \frac{2}{r^3}w_{,r\theta\theta} + \frac{1}{r^4}w_{,\theta\theta\theta\theta} + \frac{4}{r^4}w_{,\theta\theta}\right]$$

$$= w_{,rr}\left(\frac{1}{r}\Phi_{,r} + \frac{1}{r^2}\Phi_{,\theta\theta}\right) \qquad (1.78)$$

$$+ \left(\frac{1}{r}w_{,r} + \frac{1}{r^2}w_{,\theta\theta}\right)\Phi_{,rr}$$

$$-2\left(\frac{1}{r}w_{,r\theta} - \frac{1}{r^2}w_{,\theta}\right)\left(\frac{1}{r}\Phi_{,r\theta} - \frac{1}{r^2}\Phi_{,\theta}\right)$$

and

$$\frac{1}{Eh}\left[\Phi_{,rrrr} + \frac{2}{r}\Phi_{,rrr} - \frac{1}{r^2}\Phi_{,rr} + \frac{2}{r^2}\Phi_{,rr\theta\theta}\right.$$

$$\left. + \frac{1}{r^3}\Phi_{,r} - \frac{2}{r^3}\Phi_{,r\theta\theta} + \frac{1}{r^4}\Phi_{,\theta\theta\theta\theta} + \frac{4}{r^4}\Phi_{,\theta\theta}\right] \qquad (1.79)$$

$$= -\left\{w_{,rr}\left(\frac{1}{r}w_{,r} + \frac{1}{r^2}w_{,\theta\theta}\right) - \left(\frac{1}{r}w_{,r\theta} - \frac{1}{r^2}w_{,\theta}\right)^2\right\}$$

while, for the special case of radial symmetry, these reduce to

$$\left\{ \begin{array}{c} K[w_{,rrrr} + \dfrac{2}{r}w_{,rrr} - \dfrac{1}{r^2}w_{,rr} + \dfrac{1}{r^3}w_{,r}] \\[2mm] = \dfrac{1}{r}w_{,rr}\,\Phi_{,r} + \dfrac{1}{r}w_{,r}\,\Phi_{,rr} \end{array} \right. \qquad (1.80)$$

and

$$\frac{1}{Eh}\left[\Phi_{,rrrr} + \frac{2}{r}\Phi_{,rrr} - \frac{1}{r^2}\Phi_{,rr} + \frac{1}{r^3}\Phi_{,r}\right] \qquad (1.81)$$

$$= -\frac{1}{r}w_{,r}\,w_{,rr}$$

**Remarks:** The second product on the right-hand side of equation (1.75b) may be written in the more compact form

$$-2\frac{\partial}{\partial r}\left(\frac{1}{r}\frac{\partial w}{\partial \theta}\right) \cdot \frac{\partial}{\partial r}\left(\frac{1}{r}\frac{\partial \Phi}{\partial \theta}\right)$$

**Remarks:** For the case of a radially symmetric deformation, it may be easily shown that isotropic symmetry for the linearly elastic material yields the relations

$$N_r = \frac{Eh}{1-\nu^2}\left(e^0_{rr} + \nu e^0_{\theta\theta}\right)$$
$$= \frac{Eh}{1-\nu^2}\left[\frac{du_r}{dr} + \frac{1}{2}\left(\frac{dw}{dr}\right)^2 + \nu\frac{u_r}{r}\right] \tag{1.82a}$$

and

$$N_\theta = \frac{Eh}{1-\nu^2}\left(e^0_{\theta\theta} + \nu e^0_{rr}\right)$$
$$= \frac{Eh}{1-\nu^2}\left[\frac{u_r}{r} + \nu\frac{du_r}{dr} + \frac{\nu}{2}\left(\frac{dw}{dr}\right)^2\right] \tag{1.82b}$$

$$N_{r\theta} = 2Ghe^0_{r\theta} \tag{1.82c}$$

**Remarks:** In lieu of (1.78), a useful (equivalent) form for the first von Karman equation (especially for our later discussion of the buckling of annular plates) is

$$K\Delta^2 w = N_r w_{,rr} - 2N_{r\theta}\left(\frac{1}{r^2}w_{,\theta} - \frac{1}{r}w_{,r\theta}\right) \tag{1.83}$$
$$+ N_\theta\left(\frac{1}{r}w_{,r} + \frac{1}{r^2}w_{,\theta\theta}\right)$$

If $\sigma_{rr}$, $\sigma_{r\theta}$, and $\sigma_{\theta\theta}$ are independent of the variable $z$, in the plate, then $N_r$, $N_{r\theta}$, and $N_\theta$ in (1.83), as well as in all the other expressions prior to (1.83), where these stress resultants appear, may be replaced, respectively, by $h\sigma_{rr}$, $h\sigma_{r\theta}$, and $h\sigma_{\theta\theta}$.

**Remarks:** It is easily seen that, for the case of an axially symmetric deformation of the plate, the relevant equations, i.e., (1.80), (1.81) may be rewritten in the form

$$K\left\{\frac{1}{r}\frac{d}{dr}r\frac{d}{dr}\frac{1}{r}\frac{d}{dr}r\frac{dw}{dr}\right\} =$$
$$\frac{1}{r}\frac{d\Phi}{dr}\frac{d^2w}{dr^2} + \frac{1}{r}\frac{d^2\Phi}{dr^2}\frac{dw}{dr} \tag{1.84}$$

and

$$\frac{1}{Eh}\left\{\frac{1}{r}\frac{d}{dr}r\frac{d}{dr}\frac{1}{r}\frac{d}{dr}r\frac{dw}{dr}\right\} = -\frac{1}{r}\frac{dw}{dr}\frac{d^2w}{dr^2} \qquad (1.85)$$

**Remarks**: From the structure of the operators $\Delta^2$ and $[\,,\,]$, in the polar coordinate system, i.e., (1.75a), (1.75b), it is clear that a troublesome singularity arises at $r = 0$; the boundary conditions which must be imposed to deal with this difficulty at $r = 0$ will be discussed in the next subsection.

The quantities

$$\kappa_r = \frac{d^2w}{dr^2}, \quad \kappa_\theta = \frac{1}{r}\frac{dw}{dr}, \qquad (1.86)$$

for the case of radially symmetric deformations of a plate, are the middle-surface curvatures. If the plate is circular, with radius $R$, then the strain-energy of bending for the isotropic, linearly elastic plate is

$$\mathcal{V}_B = \frac{1}{2}K \int\!\!\int_A (\kappa_r^2 + 2\nu\kappa_r\kappa_\theta + \kappa_\theta^2)dA$$
$$= \pi K \int_0^R \left[ w_{,rr}^2 + 2\nu w_{,rr} \cdot \frac{w_{,r}}{r} + \left(\frac{1}{r}w_{,r}\right)^2 \right] r\,dr \qquad (1.87a)$$

while the strain-energy of stretching is

$$\mathcal{V}_S = \frac{Eh}{2(1-\nu^2)} \int\!\!\int_A (e_{rr}^2 + 2\nu e_{rr}e_{\theta\theta} + e_{\theta\theta}^2)\,dA$$
$$= \frac{\pi Eh}{1-\nu^2} \int_0^R \left[ (u_{r,r} + \frac{1}{2}w_{,r}^2)^2 \right.$$
$$\left. +2\nu\left(u_{r,r} + \frac{1}{2}w_{,r}^2\right)\frac{u_r}{r} + \left(\frac{u_r}{r}\right)^2 \right] r\,dr \qquad (1.87b)$$

If, e.g., we are considering symmetric deformations of a circular plate of radius $R$, which is compressed (symmetrically) by a uniformly distributed force $P$ per unit length, around its circumference, so that the net potential energy of loading is

$$\mathcal{V}_L = -\pi P \int_0^R w_{,r}^2 r\,dr \qquad (1.88)$$

then the total potential energy of the plate is

$$W(u, w, P) \equiv \mathcal{V}_B + \mathcal{V}_S + \mathcal{V}_L \qquad (1.89)$$

We now turn to the equations, in polar coordinates, for a linearly elastic, orthotropic, body with cylindrical anisotropy; in this case, there are three planes of elastic symmetry, one of which is normal to the axis of anisotropy, the second of which passes through that axis, and the third of which is orthogonal to the first two. For a plate, we choose the first plane of elastic symmetry to be parallel to the middle plane of the plate; in this case the constitutive relations assume the form

$$
\begin{cases}
e_{rr} = \dfrac{1}{E_r}\sigma_{rr} - \dfrac{\nu_\theta}{E_\theta}\sigma_{\theta\theta} \\[2mm]
e_{\theta\theta} = -\dfrac{\nu_r}{E_r}\sigma_{rr} + \dfrac{1}{E_\theta}\sigma_{\theta\theta} \\[2mm]
\gamma_{r\theta} = \dfrac{1}{G_{r\theta}}\sigma_{r\theta}
\end{cases}
\tag{1.90}
$$

with $E_r$, $E_\theta$ being the Young's moduli for tension (or compression) in the radial and tangential directions, respectively, $\nu_r$ and $\nu_\theta$ the corresponding Poisson's (principal) ratios, and $G_{r\theta}$ the shear modulus which characterizes the change of angle between the directions $r$ and $\theta$. As $E_r\nu_\theta = E_\theta\nu_r$, the constitutive equations (1.90) can be recast in the form

$$
\begin{cases}
e_{rr} = \dfrac{1}{E_r}(\sigma_{rr} - \nu_r\sigma_{\theta\theta}) \\[2mm]
e_{\theta\theta} = \dfrac{1}{E_\theta}(\sigma_{\theta\theta} - \nu_\theta\sigma_{rr}) \\[2mm]
\gamma_{r\theta} = \dfrac{1}{G_{r\theta}}\sigma_{r\theta}
\end{cases}
\tag{1.91}
$$

so that

$$
\begin{cases}
\sigma_{rr} = \dfrac{E_r}{1 - \nu_r\nu_\theta}e_{rr} + \dfrac{\nu_r E_\theta}{1 - \nu_r\nu_\theta}e_{\theta\theta} \\[2mm]
\sigma_{\theta\theta} = \dfrac{E_r\nu_\theta}{1 - \nu_r\nu_\theta}e_{rr} + \dfrac{E_\theta}{1 - \nu_r\nu_\theta}e_{\theta\theta} \\[2mm]
\sigma_{r\theta} = G_{r\theta}\gamma_{r\theta}
\end{cases}
\tag{1.92}
$$

One may compute the strains $e_{rr}$, $e_{\theta\theta}$, and $\gamma_{r\theta}$ by using (1.71) and, then, by employing (1.92), the stresses $\sigma_{rr}$, $\sigma_{\theta\theta}$, and $\sigma_{r\theta}$. The equations of equilibrium which apply in this situation are (1.72 a,b) and these then produce stress components $\sigma_{rz}$ and $\sigma_{\theta z}$. The stresses in the cylindrically orthotropic plate then lead to stress resultants $N_r$, $N_\theta$, and $N_{r\theta}$ and bending and twisting moments $M_r$, $M_\theta$, and $M_{r\theta}$. By employing straightforward calculations, we are led to the following results for a cylindrically orthotropic plate:

$$
\begin{cases}
M_r = -D_r\left[w_{,rr} + \nu_\theta\left(\dfrac{1}{r}w_{,r} + \dfrac{1}{r^2}w_{,\theta\theta}\right)\right] \\[2mm]
M_\theta = -D_\theta\left[\nu_r w_{,rr} + \left(\dfrac{1}{r}w_{,r} + \dfrac{1}{r^2}w_{,\theta\theta}\right)\right] \\[2mm]
M_{r\theta} = M_{\theta r} = -2\tilde{D}_{r\theta}\left(\dfrac{w}{r}\right)_{,r\theta}
\end{cases}
\tag{1.93}
$$

and

$$
\begin{cases}
N_r = \dfrac{E_r h}{1 - \nu_r \nu_\theta}\left(\dfrac{\partial u_r}{\partial r} + \dfrac{1}{2}w_{,r}^2\right) \\[3mm]
\quad + \dfrac{\nu_r E_\theta h}{1 - \nu_r \nu_\theta}\left(\dfrac{u_r}{r} + \dfrac{1}{r}\dfrac{\partial u_\theta}{\partial \theta} + \dfrac{1}{2r^2}w_{,\theta}^2\right) \\[3mm]
N_\theta = \dfrac{\nu_\theta E_r h}{1 - \nu_r \nu_\theta}\left(\dfrac{\partial u_r}{\partial r} + \dfrac{1}{2}w_{,r}^2\right) \\[3mm]
\quad + \dfrac{E_\theta}{1 - \nu_r \nu_\theta}\left(\dfrac{u_r}{r} + \dfrac{1}{r}\dfrac{\partial u_\theta}{\partial \theta} + \dfrac{1}{2r^2}w_{,\theta}^2\right) \\[3mm]
N_{r\theta} = G_{r\theta} h\left(\dfrac{\partial u_r}{\partial r} - \dfrac{u_r}{r} + \dfrac{1}{r}\dfrac{\partial u_r}{\partial \theta} + \dfrac{1}{r}w_{,r}w_{,\theta}\right)
\end{cases}
\tag{1.94}
$$

with $D_r$ and $D_\theta$, respectively, the bending stiffnesses around axes in the $r$ and $\theta$ directions, passing through a given point in the plate, and $\tilde{D}_{r\theta}$ the twisting rigidity; these are given by

$$
D_r = \frac{E_r h^3}{12(1 - \nu_r \nu_\theta)}, \quad D_\theta = \frac{E_\theta h^3}{12(1 - \nu_r \nu_\theta)}
\tag{1.95}
$$

and

$$
\tilde{D}_{r\theta} = \frac{G_{r\theta} \cdot h^3}{12}
\tag{1.96}
$$

while

$$
D_{r\theta} = D_r \nu_\theta + 2\tilde{D}_{r\theta}
\tag{1.97}
$$

Using the expressions for $M_r, M_\theta, M_{r\theta}$ in (1.94), those for $N_r, N_\theta$, and $N_{r\theta}$ in (1.74), and a compatibility equation for the displacements in polar coordinates, we find the following form of the von Karman equations for a linearly elastic, thin plate exhibiting cylindrically orthotropic symmetry, (where we have once again introduced the Airy stress function through the relations (1.74)):

$$D_r w_{,rrrr} + 2D_{r\theta}\frac{1}{r^2}w_{,rr\theta\theta} + D_\theta\frac{1}{r^4}w_{,\theta\theta\theta\theta} \qquad (1.98)$$

$$+2D_r\frac{1}{r}w_{,rrr} - 2D_{r\theta}\frac{1}{r^3}w_{,r\theta\theta} - D_\theta\frac{1}{r^2}w_{,rr}$$

$$+2(D_\theta + D_{r\theta})\frac{1}{r^4}w_{,\theta\theta} + D_\theta\frac{1}{r^3}w_{,r}$$

$$= \left(\frac{1}{r}\Phi_{,r} + \frac{1}{r^2}\Phi_{,\theta\theta}\right)w_{,rr} + \Phi_{,rr}\left(\frac{1}{r}w_{,r} + \frac{1}{r^2}w_{,\theta\theta}\right)$$

$$+2\left(\frac{1}{r^2}\Phi_{,\theta} - \frac{1}{r}\Phi_{,r\theta}\right)\left(\frac{1}{r}w_{,r\theta} - \frac{1}{r^2}w_{,\theta}\right)$$

and

$$\frac{1}{E_\theta}\Phi_{,rrrr} + \left(\frac{1}{G_{r\theta}} - \frac{2\nu_r}{E_r}\right)\frac{1}{r^2}\Phi_{,rr\theta\theta} \qquad (1.99)$$

$$+\frac{1}{E_r}\frac{1}{r^4}\Phi_{,\theta\theta\theta\theta} + \frac{2}{E_\theta}\frac{1}{r}\Phi_{,rrr}$$

$$-\left(\frac{1}{G_{r\theta}} - \frac{2\nu_r}{E_r}\right)\frac{1}{r^3}\Phi_{,r\theta\theta} - \frac{1}{E_r}\frac{1}{r^2}\Phi_{,rr}$$

$$+\left(2\frac{1-\nu_r}{E_r} + \frac{1}{G_{r\theta}}\right)\frac{1}{r^4}\Phi_{,\theta\theta} + \frac{1}{E_r}\frac{1}{r^3}\Phi_{,r}$$

$$= -h\left[w_{,rr}\left(\frac{1}{r}w_{,r} + \frac{1}{r^2}w_{,\theta\theta}\right)\right.$$

$$\left. - \left(\frac{1}{r}w_{,r\theta} - \frac{1}{r^2}w_{,\theta}\right)^2\right]$$

Equations (1.98), (1.99) may also be obtained directly from many sources in the literature, e.g., the paper by Uthgenannt and Brand [70].

Equations (1.98), (1.99), which govern the general deflections of a cylindrically orthotropic, linearly elastic, thin plate reduce to those which govern the deflections, in polar coordinates, of an isotropic plate, i.e. (1.78), (1.79) when

$$D_r = D_\theta = D_{r\theta} = K \equiv \frac{Eh^3}{12(1-\nu^2)} \qquad (1.100a)$$

and

$$E_r = E_\theta = E, \ \nu_r = \nu_\theta = \nu \qquad (1.100b)$$

Also, for the case of a cylindrically orthotropic plate, undergoing axisymmetric deformations, with the assumption of radial symmetry then implying that all derivatives with respect to $\theta$ in (1.98) and (1.99) vanish, these equations reduce to

$$\Delta_r^2 w = \frac{1}{D_r} F(w, \Phi) \qquad (1.101)$$

and

$$\Delta_r^2 \Phi = -\frac{1}{2} E_\theta h \cdot F(w, w) \qquad (1.102)$$

where

$$\begin{cases} \Delta_r^2 = \dfrac{d^4}{dr^4} + \dfrac{2}{r} \dfrac{d^3}{dr^3} - \dfrac{\beta}{r^2} \dfrac{d^2}{dr^2} + \dfrac{\beta}{r^3} \dfrac{d}{dr} \\[2mm] \beta = E_\theta / E_r \ (\text{ the 'orthotropy ratio'}) \\[2mm] F(w, \Phi) = \dfrac{1}{r} \left[ \dfrac{d}{dr} \left( \dfrac{d\Phi}{dr} \dfrac{dw}{dr} \right) \right] \end{cases} \qquad (1.103)$$

For an isotropic plate undergoing axisymmetric deformations, $D_r = D, E_\theta = E, \beta = 1$, and (1.101), (1.102) specialize to the form given in (1.84), (1.85)

**Remarks**: It is often useful to have available the inverted form of the constitutive relations (1.91) for a cylindrically orthotropic, linearly elastic, thin plate, namely,

$$\begin{pmatrix} \sigma_{rr} \\ \sigma_{\theta\theta} \end{pmatrix} = \frac{E_r}{1 - \nu_r \nu_\theta} \begin{pmatrix} 1 & \nu_\theta \\ \nu_\theta & \beta \end{pmatrix} \begin{pmatrix} e_{rr} \\ e_{\theta\theta} \end{pmatrix} \qquad (1.104a)$$

and

$$\sigma_{r\theta} = G_{r\theta} \gamma_{r\theta} \qquad (1.104b)$$

The last case to be considered in this subsection concerns the situation in which the linearly elastic, thin plate exhibits rectilinear orthotropic behavior; thus, the constitutive relations (1.60) apply, (as they would, e.g., in the case of a linearly elastic paper sheet) but, because we are interested in studying deflections of circular or annular regions, it is more appropriate to formulate the corresponding von Karman equations in polar coordinates instead of rectilinear coordinates. In this situation, we have a mismatch between the elastic symmetry which is built into the form of the constitutive relations, and the geometry of the region undergoing buckling; this greatly complicates the structure of the von Karman equations. It is worth noting that if we make use of the transformation

$$
\begin{cases}
e_{rr} = e_{xx} \cos^2 \theta + e_{yy} \sin^2 \theta + \gamma_{xy} \cos \theta \sin \theta \\[2mm]
e_{\theta\theta} = e_{xx} \sin^2 \theta + e_{yy} \cos^2 \theta - \gamma_{xy} \cos \theta \sin \theta \\[2mm]
\gamma_{r\theta} = 2(e_{yy} - e_{xx}) \cos \theta \sin \theta + \gamma_{xy}(\cos^2 \theta - \sin^2 \theta)
\end{cases}
\tag{1.105}
$$

of the principal strains to the polar coordinate system, in conjunction with the analogous result (1.73) for the stress components, and the rectilinear orthotropic constitutive relations (1.60), we may transform these constitutive equations directly into polar coordinate form; the polar coordinate form of the constitutive relations will, indeed, be indicated below. However, it is worth noting that the first of the von Karman equations for this situation has, essentially, been derived in Coffin [71]and involves, of course, using the polar coordinate equivalent for (1.45) with $W \equiv 0, t \equiv 0$, namely,

$$
\frac{1}{r}(rM_r)_{,rr} + \frac{1}{r^2}M_{\theta,\theta\theta} - \frac{1}{r}M_{\theta,r} + \frac{1}{r}M_{r\theta,r\theta} + N_r w_{,rr}
\tag{1.106}
$$

$$
+ N_\theta \left(\frac{1}{r}w_{,r} + \frac{1}{r^2}w_{,\theta\theta}\right) + 2N_{r\theta}\left(\frac{w}{r}\right)_{,r\theta} = 0
$$

where the stress resultants $N_r, N_\theta$, and $N_{r\theta}$ are, once again, given by (1.74) in terms of the Airy function $\Phi(r, \theta)$; we note that the sum of the last three terms in (1.106) is (again) identical with the right-hand side of (1.75b), i.e., with $[w, \Phi]$. From the work in [71], we deduce the following expressions for the bending moments (which may, of course, be obtained by directly transforming the expressions in (1.67a), (1.67b) and (1.67c) into polar coordinates):

$$M_r = -\tilde{D}_1 w_{,rr} - \tilde{D}_{12}\left[\frac{1}{r^2}w_{,\theta\theta} + \frac{1}{r}w_{,r}\right]$$
$$-2\tilde{D}_{16}\left(\frac{1}{r}w\right)_{,r\theta}$$

(1.107a)

$$M_\theta = -\tilde{D}_{12}w_{,rr} - \tilde{D}_2\left[\frac{1}{r^2}w_{,\theta\theta} + \frac{1}{r}w_{,r}\right]$$
$$-2\tilde{D}_{26}\left(\frac{1}{r}w\right)_{,r\theta}$$

(1.107b)

$$M_{r\theta} = -\tilde{D}_{16}w_{,rr} - \tilde{D}_{26}\left[\frac{1}{r^2}w_{,\theta\theta} + \frac{1}{r}w_{,r}\right]$$
$$-2\tilde{D}_6\left(\frac{1}{r}w\right)_{,r\theta}$$

(1.107c)

where,

$$\begin{cases} \tilde{D}_1 = D_1\cos^4\theta + D_3\cos^2\theta\sin^2\theta + D_2\sin^4\theta \\ \tilde{D}_2 = D_1\sin^4\theta + D_3\cos^2\theta\sin^2\theta + D_2\cos^4\theta \\ \tilde{D}_{12} = \nu_1 D_2 + (D_1 + D_2 - 2D_3)\cos^2\theta\sin^2\theta \\ \tilde{D}_6 = D_{66} + (D_1 + D_2 - 2D_3)\cos^2\theta\sin^2\theta \\ \tilde{D}_{16} = \left[(D_2 - D_3)\sin^2\theta - (D_1 - D_3)\cos^2\theta\right]\cos\theta\sin\theta \\ \tilde{D}_{26} = \left[(D_2 - D_3)\cos^2\theta - (D_1 - D_3)\sin^2\theta\right]\cos\theta\sin\theta \end{cases}$$

(1.108)

with $D_1, D_2$ and $D_3$ the principal rigidities, $D_1 = D_{11}, D_2 = D_{22}, D_3 = D_2\nu_{12} + 2D_{66}$, as defined by (1.63), (1.64). If we now substitute from (1.107a,b,c) into the equilibrium equation (1.106), we obtain the first of the von Karman partial differential equations governing the out-of-plane deflection of a rectilinear orthotropic, thin, elastic plate in polar coordinates:

$$\tilde{D}_1 w_{,rrrr} + \tilde{D}_3 \left[ \frac{2}{r} w_{,rrr} + \frac{2}{r} \left( \frac{1}{r} w_{,r} \right)_{,r\theta\theta} - \frac{1}{r} \left( \frac{1}{r} w_{,r} \right)_{,r} + \frac{4}{r^4} w_{,\theta\theta} \right]$$

$$+ \tilde{D}_2 \frac{1}{r^4} w_{,\theta\theta\theta\theta} + 4 \tilde{D}_{16} \cdot \frac{1}{r} w_{,rrr\theta} + 4 \tilde{D}_{26} \frac{1}{r^4} w_{,r\theta\theta\theta}$$

$$+ \frac{12}{r} \left( \tilde{D}_{16} - \tilde{D}_{26} \right) \left( \frac{1}{r} w \right)_{,rr\theta}$$

$$+ \left\{ \tilde{D}_2 - \tilde{D}_1 + \left( \tilde{D}_{26} - \tilde{D}_{16} \right) \cot 4\theta \right\}$$

$$\left\{ \frac{3}{2r} \left( \frac{1}{r} w_{,r} \right)_{,r} + \frac{3}{r^3} w_{,r\theta\theta} - \frac{4}{r^4} w_{,\theta\theta} \right\}$$

$$+ \left\{ (\tilde{D}_2 - \tilde{D}_1) \tan 2\theta + 2(\tilde{D}_{26} - \tilde{D}_{16}) \right\}$$

$$\left\{ \frac{3}{2r^4} w_{,\theta\theta\theta} + \frac{1}{r^3} w_{,r\theta} \right\} = [w, \Phi] \tag{1.109}$$

where $[w, \Phi]$ is given by (1.74) and (1.75b), and

$$\tilde{D}_3 = \tilde{D}_{12} + 2\tilde{D}_6 \tag{1.110}$$

**Remarks:** Inasmuch as the expressions for the moments $M_r, M_\theta$, and $M_{r\theta}$ in (1.107 a,b,c) may be obtained from the expressions for $M_x, M_y$, and $M_{xy}$ in (1.67 a,b,c), and these latter expressions have been derived by assuming that any existing hygroexpansive strains $\beta_i \Delta H, i = 1, 2$, (equivalently, thermal strains $\alpha_i \Delta T$) are constant throughout the thickness $h$ of the plate, if the $\beta_i \Delta H$ vary with $z$ in any manner except as an odd function of $z$ (with respect to the middle plane of the plate) then the expressions for the moments in (1.67a,b,c), and their polar coordinate counterparts in (1.107a,b,c) would have to be rederived; the new expressions obtained for $M_r, M_\theta$, and $M_{r\theta}$ must then be substituted back into (1.106) so as to obtain the appropriate modification of (1.109), which applies in the presence of hygroexpansive strains.

**Remarks:** The second of the von Karman equations which apply to the problem of studying the out-of-plane deflections of a linear elastic, rectilinearly orthotropic, thin, plate in polar coordinates does not appear in [71] because the primary focus in that work was on studying the initial buckling problem; nor does the relevant form of this second of the von Karman equations in polar coordinates for the case of a linearly elastic, rectilinearly orthotropic plate appear to have been derived anywhere else in the literature. The calculation, however, which is required to obtain the equation which complements (1.109) may be carried out in one of two ways: first of all, by transforming the three fourth order partial derivatives $\Phi_{,xxxx}$, $\Phi_{,xxyy}$, and $\Phi_{,yyyy}$ which appear on the left-hand side of (1.69); the right-hand side of (1.69), in polar coordinates, will be identical with the right-hand side of (1.79).

Alternatively, to obtain the form of the second of the von Karman equations, we may rewrite the constitutive relations for a linearly elastic, rectilinearly orthotropic material, i.e.,

$$
\begin{cases}
e_{xx} = \dfrac{1}{E_1}\sigma_{xx} - \dfrac{\nu_{21}}{E_2}\sigma_{yy} \\[2mm]
e_{yy} = -\dfrac{\nu_{12}}{E_1}\sigma_{xx} + \dfrac{1}{E_2}\sigma_{yy} \\[2mm]
\gamma_{xy} = \dfrac{1}{G_{12}}\sigma_{xy}
\end{cases}
\tag{1.111}
$$

(where we have, for now, not considered the presence of possible hygroexpansive or hygrothermal strains) in the polar coordinate form

$$
\begin{cases}
e_{rr} = a_{11}\sigma_{rr} + a_{12}\sigma_{\theta\theta} + a_{13}\tau_{r\theta} \\[2mm]
e_{\theta\theta} = a_{21}\sigma_{rr} + a_{22}\sigma_{\theta\theta} + a_{23}\tau_{r\theta} \\[2mm]
\gamma_{r\theta} = a_{31}\sigma_{rr} + a_{32}\sigma_{\theta\theta} + a_{33}\tau_{r\theta}
\end{cases}
\tag{1.112}
$$

where the $a_{ij} = a_{ij}(\theta)$, in contrast to the case of a cylindrically orthotropic material, i.e. (1.91), in which the constitutive "coefficients" are $\theta$-independent. The strain components in (1.112) are given by the relations (1.71) in terms of the displacements $u_r, u_\theta$, and $w$, where $-\dfrac{h}{2} < \zeta < \dfrac{h}{2}$; if we think in terms of averaging the constitutive relations (1.112) over the thickness of the plate we may, in essence, ignore the expressions involving $\zeta$ in (1.71). For the in-plane stress distribution (prebuckling), $w = 0$, in which case

$$
e_{rr} = \frac{\partial u_r}{\partial r}, \quad e_{\theta\theta} = \frac{1}{r}\frac{\partial u_\theta}{\partial \theta} + \frac{1}{r}u_r, \quad \gamma_{r\theta} = \frac{1}{r}\frac{\partial u_r}{\partial \theta} + \frac{\partial u_\theta}{\partial r} - \frac{1}{r}u_\theta \tag{1.113}
$$

In polar coordinates, the compatibility equation for the strains assumes the form

$$\frac{\partial^2 e_{rr}}{\partial\theta^2} + r\frac{\partial^2}{\partial r^2}(re_{\theta\theta}) - \frac{\partial^2}{\partial r\partial\theta}(r\gamma_{r\theta}) - r\frac{\partial e_{rr}}{\partial r} \equiv \mathcal{L}(u_r, u_\theta) = 0 \quad (1.114)$$

when $w = 0$. If $w \neq 0$, then the strain compatibility relation yields

$$\mathcal{L}(u_r, u_\theta) = -\frac{r^2}{2}[w, w] \qquad (1.115)$$

where

$$[w, w] = 2\left\{ w_{,rr}\left(\frac{1}{r}w_{,r} + \frac{1}{r^2}w_{,\theta\theta}\right) - \left(\frac{1}{r}w_{,r\theta} - \frac{1}{r^2}w_{,\theta}\right)^2 \right\} \qquad (1.116)$$

The essential idea behind the derivation of the second of the von Karman equations is to compute the polar coordinate form (1.112) of the constitutive relations (1.111), substitute (1.112) into (1.115), and then set

$$\begin{cases} \sigma_{rr} = \dfrac{1}{r}\dfrac{\partial\Phi}{\partial r} + \dfrac{1}{r^2}\dfrac{\partial^2\Phi}{\partial\theta^2} \\[2mm] \sigma_{\theta\theta} = \dfrac{\partial^2\Phi}{\partial r^2} \\[2mm] \tau_{r\theta} = -\dfrac{\partial^2}{\partial r\partial\theta}\left(\dfrac{1}{r}\Phi\right) \end{cases} \qquad (1.117)$$

in the resulting equation. We begin by recalling the transformation (1.105) of the principal strains $e_{xx}, e_{yy}$, and $\gamma_{xy}$ to the polar coordinate system and the analogous transformation of the principal stresses, i.e.,

$$\begin{cases} \sigma_{rr} = \sigma_{xx}\cos^2\theta + \sigma_{yy}\sin^2\theta + 2\sigma_{xy}\sin\theta\cos\theta \\[1mm] \sigma_{\theta\theta} = \sigma_{xx}\sin^2\theta + \sigma_{yy}\cos^2\theta - 2\sigma_{xy}\sin\theta\cos\theta \\[1mm] \sigma_{r\theta} = (\sigma_{yy} - \sigma_{xx})\sin\theta\cos\theta + \sigma_{xy}(\cos^2\theta - \sin^2\theta) \end{cases} \qquad (1.118)$$

Using the transformation (1.105) of the principal strains, in conjunction with the constitutive relations (1.111), we find that

$$\begin{aligned} e_{rr} = &\left(\frac{1}{E_1}\cos^2\theta - \frac{\nu_{12}}{E_1}\sin^2\theta\right)\sigma_{xx} \\ &+ \left(\frac{1}{E_2}\sin^2\theta - \frac{\nu_{21}}{E_2}\cos^2\theta\right)\sigma_{yy} \\ &+ \frac{\sin 2\theta}{G_{12}}\sigma_{xy} \end{aligned} \qquad (1.119)$$

$$e_{\theta\theta} = \left( \frac{1}{E_1} \sin^2 \theta - \frac{\nu_{12}}{E_1} \cos^2 \theta \right) \sigma_{xx}$$

$$+ \left( \frac{1}{E_2} \cos^2 \theta - \frac{\nu_{21}}{E_2} \sin^2 \theta \right) \sigma_{yy} \qquad (1.120)$$

$$- \frac{\sin 2\theta}{G_{12}} \sigma_{xy}$$

and

$$e_{r\theta} = -\frac{(1 + \nu_{12})}{2} \sin 2\theta \cdot \sigma_{xx} + \frac{(1 + \nu_{21})}{2} \sin 2\theta \cdot \sigma_{yy}$$

$$+ \frac{\cos 2\theta}{G_{12}} \cdot \sigma_{xy} \qquad (1.121)$$

By solving the relations in (1.118) for $\sigma_{xx}, \sigma_{yy}$, and $\sigma_{xy}$ in terms of $\sigma_{rr}, \sigma_{\theta\theta}$, and $\sigma_{r\theta}$ and then substituting these results for $\sigma_{xx}, \sigma_{yy}$, and $\sigma_{xy}$ into (1.119), (1.120), and (1.121), and simplifying, we obtain (1.112) with

$$a_{11} = \frac{\cos^2 \bar{\theta}}{E_1} (\cos^2 \bar{\theta} - \nu_{12} \sin^2 \bar{\theta}) + \frac{\sin^2 \bar{\theta}}{E_2} (\sin^2 \bar{\theta} - \nu_{21} \cos^2 \bar{\theta})$$

$$\frac{1}{4G_{12}} \sin^2 2\bar{\theta} \qquad (1.122a)$$

$$a_{12} = \frac{\sin^2 \bar{\theta}}{E_1} (\cos^2 \bar{\theta} - \nu_{12} \sin^2 \bar{\theta}) + \frac{\cos^2 \bar{\theta}}{E_2} (\sin^2 \bar{\theta} - \nu_{21} \cos^2 \bar{\theta})$$

$$- \frac{1}{4G_{12}} \sin^2 2\bar{\theta} \qquad (1.122b)$$

$$a_{13} = \sin 2\bar{\theta} \left\{ \frac{1}{E_1} (\nu_{12} \sin^2 \bar{\theta} - \cos^2 \bar{\theta}) + \frac{1}{E_2} (\sin^2 \bar{\theta} - \nu_{21} \cos^2 \bar{\theta}) \right.$$

$$\left. + \frac{\cos 2\bar{\theta}}{2G_{12}} \right\} \qquad (1.122c)$$

$$a_{21} = \frac{\cos^2 \bar{\theta}}{E_1} (\sin^2 \bar{\theta} - \nu_{12} \cos^2 \bar{\theta}) + \frac{\sin^2 \bar{\theta}}{E_2} (\cos^2 \bar{\theta} - \nu_{21} \sin^2 \bar{\theta})$$

$$- \frac{\sin^2 2\bar{\theta}}{4G_{12}} \qquad (1.122d)$$

$$a_{22} = \frac{\sin^2 \bar{\theta}}{E_1}(\sin^2 \bar{\theta} - \nu_{12}\cos^2 \bar{\theta}) + \frac{\cos^2 \bar{\theta}}{E_2}(\cos^2 \bar{\theta} - \nu_{21}\sin^2 \bar{\theta})$$
$$+ \frac{\sin^2 2\bar{\theta}}{4G_{12}} \tag{1.122e}$$

$$a_{23} = \left\{ \frac{1}{E_1}(\nu_{12}\cos^2 \bar{\theta} - \sin^2 \bar{\theta}) + \frac{1}{E_2}(\cos^2 \bar{\theta} - \nu_{21}\sin^2 \bar{\theta}) \right.$$
$$\left. - \frac{\cos 2\bar{\theta}}{2G_{12}} \right\} \tag{1.122f}$$

$$a_{31} = \sin \bar{\theta} \cos \bar{\theta} \left[ \frac{(1 + \nu_{21})}{E_2}\sin^2 \bar{\theta} - \frac{(1 + \nu_{12})}{E_1}\cos^2 \bar{\theta} \right.$$
$$\left. + \frac{\cos 2\bar{\theta}}{2G_{12}} \right] \tag{1.122g}$$

$$a_{32} = \sin \bar{\theta} \cos \bar{\theta} \left[ \frac{(1 + \nu_{21})}{E_2}\cos^2 \bar{\theta} - \frac{(1 + \nu_{12})}{E_1}\sin^2 \bar{\theta} \right.$$
$$\left. - \frac{\cos 2\bar{\theta}}{2G_{12}} \right] \tag{1.122h}$$

$$a_{33} = \sin 2\bar{\theta} \left\{ \sin \bar{\theta} \cos \bar{\theta} \left[ \frac{(1 + \nu_{12})}{E_1} + \frac{(1 + \nu_{21})}{E_2} \right] \right\}$$
$$+ \frac{\cos^2 2\bar{\theta}}{2G_{12}} \tag{1.122i}$$

To summarize, the second of the von Karman equations in polar co-ordinates, for a linearly elastic, thin plate, possessing rectilinear orthotropic symmetry, is obtained by first inserting the $a_{ij}(\theta)$, as given by (1.122a)-(1.122i) into the constitutive relations

$$e_{rr} = a_{11} \left( \frac{1}{r}\Phi_{,r} + \frac{1}{r^2}\Phi_{,\theta\theta} \right) + a_{12}\Phi_{,rr}$$
$$- a_{13} \left( \frac{1}{r}\Phi \right)_{,r\theta} \tag{1.123a}$$

$$e_{\theta\theta} = a_{21} \left( \frac{1}{r}\Phi_{,r} + \frac{1}{r^2}\Phi_{,rr} \right) + a_{22}\Phi_{,rr}$$
$$- a_{23} \left( \frac{1}{r}\Phi \right)_{,r\theta} \tag{1.123b}$$

$$\gamma_{r\theta} = a_{31}\left(\frac{1}{r}\Phi_{,r} + \frac{1}{r^2}\Phi_{,rr}\right) + a_{32}\Phi_{,rr} - a_{33}\left(\frac{1}{r}\Phi\right)_{,r\theta} \qquad (1.123c)$$

and then substituting (1.123a,b,c) into the compatibility relation

$$\frac{\partial^2 e_{rr}}{\partial \theta^2} + r\frac{\partial^2}{\partial r^2}(re_{\theta\theta}) - \frac{\partial^2}{\partial r\partial\theta}(r\gamma_{r\theta})$$

$$-r\frac{\partial e_{rr}}{\partial r} = -\left\{w_{,rr}\left(\frac{1}{r}w_{,r} + \frac{1}{r^2}w_{,\theta\theta}\right) - \left(\frac{1}{r}w_{,r\theta} - \frac{1}{r^2}w_{,\theta}\right)^2\right\}$$

$$(1.124)$$

A comprehensive study of postbuckling for rectilinearly orthotropic plates with circular geometries will not be attempted in the present work; therefore, we will forgo carrying out the remainder of the derivation of the second of the von Karman equations for this situation leaving, instead, the straightforward calculations as an exercise for the reader. The initial buckling of rectilinearly orthotropic (circular) annular plates will be treated in Chapter 4 and the in-plane displacement differential equations associated with the buckling of rectilinearly orthotropic circular plates will be obtained in Chapter 3.

Along the edge of the plate, at $r = a$, we prescribe, in general, the radial and tangential components $p_r(\theta)$ and $p_\theta(\theta)$, respectively, of the applied traction where

$$\begin{cases} p_r(\theta) = h\left[\sigma_{xx}\cos^2\theta + \sigma_{yy}\sin\theta\cos\theta + \sigma_{xy}\sin 2\theta\right]\Big|_{r=a} \\ p_\theta(\theta) = h\left[(\sigma_{yy} - \sigma_{xx})\sin\theta\cos\theta - \sigma_{xy}\cos 2\theta\right]\Big|_{r=a} \end{cases} \qquad (1.125)$$

i.e.

$$p_r(\theta) = h\sigma_{rr}|_{r=a}, \quad p_\theta(\theta) = h\sigma_{r\theta}|r = a \qquad (1.126)$$

and regularity conditions must, in addition, be prescribed at the center of the plate, i.e., at $r = 0$, with respect to both the Airy function $\Phi$ and the transverse deflection $w$. Boundary conditions for all the classes of buckling problems we have introduced to this point are discussed, in detail, in the next section of this Chapter.

## 1.3 Boundary Conditions

For a linearly elastic, isotropic, (or orthotropic) thin plate, we have seen that the von Karman equations, which govern the out-of-plane deflections of the plate, form a coupled system of nonlinear partial differential equations for the deflection $w$ and the Airy stress function $\Phi$, both in polar as well as in rectilinear coordinates; in this subsection we will formulate the specific boundary conditions which must be considered in conjunction with these equations for the cases of both isotropic and orthotropic symmetry in rectilinear coordinates (for rectangular plates) in polar coordinates (for plates with circular geometries, i.e., annular plates). We will consider first the boundary conditions which apply to the defection $w$ and then somewhat later on, those which apply to the Airy function $\Phi$; those conditions which apply to the Airy function are, in fact, best considered in conjunction with the discussion that will follow in section 1.4.

### 1.3.1 Boundary Conditions on the Deflection

Consider, as in Fig. 1.11, a thin, linearly elastic plate which occupies a region $\Omega$ in the $x, y$ plane with a smooth (or piecewise smooth) boundary $\partial\Omega$. By piecewise smooth we mean that $\partial\Omega$ is the union of a finite number of smooth arcs or pieces of curves, e.g., a rectangle. We denote by $\vec{n}$ the unit normal to the boundary, at any arbitrary point on the boundary, and by $\vec{t}$ the unit tangent vector to the boundary at that point. Derivatives of functions $f$, defined on $\partial\Omega$, in the direction of the normal to the boundary will be denoted by $\dfrac{\partial f}{\partial n}$, while those in the direction of the tangent to the boundary are denoted by $\dfrac{\partial f}{\partial s}$, with $s$ a measure of arc length along the boundary. For example, if $\Omega$ is a circle of radius $R$ in the $x, y$ plane, centered at $(0,0)$, i.e.

$$\Omega = \left\{ (x,y) \,|\, x^2 + y^2 \leq R^2 \right\} \tag{1.127a}$$

so that

$$\partial\Omega = \left\{ (x,y) | x^2 + y^2 = R^2 \right\} \tag{1.127b}$$

and $f = f(r, \theta)$ is defined on $\Omega$, with first partial derivatives continuous up to the boundary $\partial\Omega$, then

$$\frac{\partial f}{\partial n} = f_{,r} \quad \text{and} \quad \frac{\partial f}{\partial s} = \frac{1}{r}\frac{\partial f}{\partial \theta} \tag{1.128}$$

the latter result being a consequence of the fact that $s = r\theta$, so that $\frac{\partial}{\partial \theta} = \frac{\partial}{\partial s}\frac{\partial s}{\partial \theta}$. Although there are many different types of (physical) boundary conditions which can be considered for the deflection $w$ of a thin plate, three types are most prevalent in the literature on plate buckling: clamped edges, simply supported edges, and free edges. With respect to the general geometry shown in Fig. 1.11, and irrespective of whether we are dealing with isotropic or orthotropic symmetry, these three sets of boundary conditions lead to the following requirements on $\partial\Omega$:

($i$)      $\partial\Omega$ is clamped: $w = 0$ and $\dfrac{\partial w}{\partial n} = 0$, on $\partial\Omega$      (1.129a)

($ii$)      $\partial\Omega$ is simply supported: $w = 0$

                        and $M_n = 0$, on $\partial\Omega$   (1.129b)

($iii$)      $\partial\Omega$ is Free: $M_n = 0$

                and $Q_n + \dfrac{\partial M_{tn}}{\partial s} = 0$, on $\partial\Omega$      (1.129c)

where $M_n$ is the bending moment on $\partial\Omega$ in the direction normal to $\partial\Omega$, $M_{tn}$ is the twisting moment on $\partial\Omega$, with respect to the tangential and normal directions on $\partial\Omega$, and $Q_n$ is the shearing force associated with the direction normal to $\partial\Omega$. We now specify the conditions in (1.129 a,b,c) for the cases of isotropic and orthotropic symmetry in both rectilinear and circular geometries.

### 1.3.1.1   Isotropic Response: Rectilinear Geometry

We take for $\Omega$ the rectangle occupying the domain $0 \le x \le a, \ 0 \le y \le b$.

**(i) $\partial\Omega$ is Clamped.**
In this case, as a consequence of (1.129a), we have

$$\begin{cases} w(0,y) = 0, & 0 \leq y \leq b \\ w(a,y) = 0, & 0 \leq y \leq b \\ w(x,0) = 0, & 0 \leq x \leq a \\ w(x,b) = 0, & 0 \leq x \leq a \end{cases} \tag{1.130a}$$

and

$$\begin{cases} w_{,x}\,(0,y) = 0, & 0 \leq y \leq b \\ w_{,x}\,(a,y) = 0, & 0 \leq y \leq b \\ w_{,y}\,(x,0) = 0, & 0 \leq x \leq a \\ w_{,y}\,(x,b) = 0, & 0 \leq x \leq a \end{cases} \tag{1.130b}$$

where $w_{,x}\,(0,y) \equiv \dfrac{\partial w(x,y)}{\partial x}|_{x=0}$, etc.

**(ii)   $\partial\Omega$ is Simply Supported**

In this case, as a consequence of (1.129b), we have, first of all, the conditions (1.130a), because $w = 0$ on $\partial\Omega$. The condition $M_n = 0$ translates, in this case, into $M_x = 0$, for $x = 0$, $x = a$, $0 \leq y \leq b$, and $M_y = 0$ for $y = 0$, $y = b$, $0 \leq x \leq a$, or, in view of the expressions for the bending moments $M_x, M_y$ in (1.39) and (1.41), respectively,

$$\begin{cases} w_{,xx} + \nu w_{,yy}\,|_{x=0} = 0, & 0 \leq y \leq b \\ w_{,xx} + \nu w_{,yy}\,|_{x=a} = 0, & 0 \leq y \leq b \\ w_{,yy} + \nu w_{,xx}\,|_{y=0} = 0, & 0 \leq x \leq a \\ w_{,yy} + \nu w_{,xx}\,|_{y=b} = 0, & 0 \leq x \leq a \end{cases} \tag{1.131}$$

The conditions in (1.131) may be simplified somewhat: as $w(0,y) = 0$, $0 \leq y \leq b$, we have $w_{,y}\,(0,y) = 0$ and $w_{,yy}\,(0,y) = 0$, for $0 \leq y \leq b$. Thus, by the first equation in (1.131), $\dfrac{\partial^2 w}{\partial x^2}(x,y)|_{x=0} = 0$, $0 \leq y \leq b$, so, in fact

$$\Delta w|_{x=0} = \left[\frac{\partial^2 w}{\partial x^2}(x,y) + \frac{\partial^2 w}{\partial y^2}(w,y)\right]_{x=0} = 0 \tag{1.132}$$

for $0 \leq y \leq b$. In other words, if the edge $x = 0$, $0 \leq y \leq b$, is simply supported, then along this edge we have

$$w|_{x=0} = \Delta w|_{x=0} = 0, \ 0 \leq y \leq b \tag{1.133}$$

**(iii) $\partial\Omega$ is Free**

If all four edges of the rectangular plate are free then, first of all, as a consequence of the condition $M_n = 0$, on $\partial\Omega$, in (1.129c), the four relations in (1.131) must hold. The second relation in (1.129c) becomes

$$\begin{cases} Q_{yz} + \dfrac{\partial M_{xy}}{\partial x}\Big|_{y=0} = 0,\ 0 \le x \le a \\[2mm] Q_{yz} + \dfrac{\partial M_{xy}}{\partial x}\Big|_{y=b} = 0,\ 0 \le x \le a \\[2mm] Q_{xz} + \dfrac{\partial M_{yx}}{\partial y}\Big|_{x=0} = 0,\ 0 \le y \le b \\[2mm] Q_{xz} + \dfrac{\partial M_{yx}}{\partial y}\Big|_{x=a} = 0,\ 0 \le y \le b \end{cases} \qquad (1.134)$$

or, if we employ the moment equilibrium equations (1.44):

$$\begin{cases} M_{y,y} + 2M_{xy,x}\big|_{y=0} = 0, 0 \le x \le a \\[2mm] M_{y,y} + 2M_{xy,x}\big|_{y=b} = 0, 0 \le x \le a \\[2mm] M_{x,x} + 2M_{yx,y}\big|_{x=0} = 0, 0 \le y \le b \\[2mm] M_{x,x} + 2M_{yx,y}\big|_{x=a} = 0, 0 \le y \le b \end{cases} \qquad (1.135)$$

However, by virtue of (1.41),

$$M_{y,y} + 2M_{xy,x} = -K\left[w_{,yyy} + (2-\nu)w_{,xxy}\right] \qquad (1.136\text{a})$$

while by (1.39), (1.41), and the fact that $M_{xy} = M_{yx}$

$$M_{x,x} + 2M_{yx,y} = -K\left[w_{,xxx} + (2-\nu)w_{,xyy}\right] \qquad (1.136\text{b})$$

Thus, if all four edges of the plate were free, we would have

$$\begin{cases} w_{,yyy} + (2-\nu)w_{,xxy}\,\big|_{y=0} = 0,\ 0 \le x \le a \\[2mm] w_{,yyy} + (2-\nu)w_{,xxy}\,\big|_{y=b} = 0,\ 0 \le x \le a \\[2mm] w_{,xxx} + (2-\nu)w_{,xyy}\,\big|_{x=0} = 0,\ 0 \le y \le b \\[2mm] w_{,xxx} + (2-\nu)w_{,xyy}\,\big|_{x=0} = 0,\ 0 \le y \le b \end{cases} \qquad (1.137)$$

**Remarks:** In any actual problem that one would want to consider with respect to buckling or postbuckling of a (rectangular) thin, linearly elastic, isotropic plate, there would usually be a mixing of the various boundary conditions delineated above along parallel pairs of edges. For example, if the edges along $x = 0, x = a, 0 \le y \le b$, were clamped while those along $y = 0, y = b, 0 \le x \le a$, were simply supported then, by (1.130a,b) and (1.131), the full set of boundary conditions would read as follows:

$$w(0, y) = w(a, y) = w_{,x}(0, y) = w_{,x}(a, y) = 0, \qquad (1.138a)$$

for $0 \le y \le b$, and

$$\begin{cases} w(x, 0) = w(x, b) = 0, & 0 \le x \le a \\ w_{,yy} + \nu w_{,xx}|_{y=0} = 0, & 0 \le x \le a \\ w_{,yy} + \nu w_{,xx}|_{y=b} = 0, & 0 \le x \le a \end{cases} \qquad (1.138b)$$

Of course, as $w(x, 0) = 0$, $0 \le x \le a$, $w_{,xx}(x, 0) = 0$, $0 \le x \le a$, so that, by virtue of the second equation in (1.138b), we have $\frac{\partial^2 w(x, y)}{\partial y^2}\Big|_{y=0}$ $= 0$, $0 \le x \le a$, in which case $\Delta w|_{y=0} = 0$, $0 \le x \le a$. Thus, (1.138b) may be replaced by

$$\begin{cases} w|_{y=0} = \Delta w|_{y=0} = 0, & 0 \le x \le a \\ w|_{y=b} = \Delta w|_{y=b} = 0, & 0 \le x \le a \end{cases} \qquad (1.138c)$$

The number of possible combinations of different boundary conditions along parallel pairs of edges on the rectangle is, of course, quite large and we do not delineate them all here; we will reference, however, both initial and postbuckling results for several such different combinations in Chapter 2. We now turn our attention to results in the polar coordinate geometry which conforms naturally to regions with circular symmetry.

### 1.3.1.2  Isotropic Response: Circular Geometry

We take for $\Omega$ the annular region occupying the domain $a \le r \le b$, where $a \ge 0, b > a$, and $r = \sqrt{x^2 + y^2}$; if $a = 0$, the annulus degenerates into a circle of radius $b$ and, because of singularities which can develop in solutions of the von Karman equations (1.78), (1.79), which apply in this case, regularity conditions with respect to the deflection (as well as the Airy function) must be satisfied at $r = 0$

### (i) $\partial\Omega_i$ is Clamped
We take for $\partial\Omega_i$, $\{(x, y) | x^2 + y^2 = R_i\}$, $i = 1, 2$ where $R_1 \equiv a$, $R_2 \equiv b$; then $\partial\Omega$ is the union of $\partial\Omega_1$ with $\partial\Omega_2$ and, if $R_1 \equiv a = 0$, then $\partial\Omega$ is just

the circle of radius $R_2 = b$. If $\partial\Omega_1$ is clamped, then as a consequence of (1.129a)

$$w(R_i, \theta) = 0, \quad w_{,r}(R_i, \theta) = 0, \quad 0 < \theta \le 2\pi \qquad (1.139)$$

## (ii) $\partial\Omega_i$ is Simply Supported

In this case, the first condition $w(R_i, \theta) = 0$, $0 < \theta \le 2\pi$, in (1.139) still holds, but the second condition, according to (1.129b), is replaced by $M_r = 0$ on $\partial\Omega_i$; although we obtained (1.78), (1.79) without calculating $M_r$ directly for the isotropic case, in polar coordinates, we may easily obtain $M_r$ for the present situation by specializing the first result in (1.94), for a cylindrically orthotropic thin plate, to the case of isotropic symmetry. Thus, the second of the simply supported boundary conditions for $w$ reads

$$\left[ w_{,rr} + \nu \left( \frac{1}{r} w_{,r} + \frac{1}{r^2} w_{,\theta\theta} \right) \right]_{r=R_i} = 0, \qquad (1.140)$$

for $0 < \theta \le 2\pi$.

## (iii) $\partial\Omega_i$ is Free

If $\partial\Omega_i$ is free then (1.140) applies, for $0 < \theta \le 2\pi$, because, as in the simply supported case, we still have $M_r = 0$ at $r = R_i$. By (1.129c), the other condition at $r = R_i$ is

$$\left[ Q_r + \frac{1}{r} M_{r\theta,\theta} \right]_{r=R_i}$$
$$= \left[ (\Delta w)_{,r} + \frac{1-\nu}{r} \left( \frac{1}{r} w_{,\theta} \right)_{,r\theta} \right]_{r=R_i} = 0 \qquad (1.141)$$

where

$$\Delta w = w_{,rr} + \frac{1}{r} w_{,r} + \frac{1}{r^2} w_{,\theta\theta} \qquad (1.142)$$

**Remarks:** As was the case for a rectangular plate, for a thin, linearly elastic, isotropic, annular plate, one may mix and match the various sets of boundary conditions delineated above, e.g., if the outer radius at $r = b$ is clamped, while the inner radius at $r = a$ is free, the boundary conditions would read as follows:

$$w(b, \theta) = w_{,r}(b, \theta) = 0, \quad 0 < \theta \le 2\pi \qquad (1.143a)$$

$$\left[ w_{,rr} + \nu \left( \frac{1}{r} w_{,r} + \frac{1}{r^2} w_{,\theta\theta} \right) \right]_{r=a} = 0, \quad 0 < \theta \le 2\pi \quad (1.143b)$$

$$\left[ (\Delta w)_{,r} + \frac{1-\nu}{r} \left( \frac{1}{r} w_{,\theta} \right)_{,r\theta} \right]_{r=a} = 0, \quad 0 < \theta \le 2\pi \quad (1.143c)$$

**Remarks:** Suppose that $a = 0$, so that the annular plate degenerates to a circular plate of radius $b$; If the boundary at $r = b$ is clamped, then (1.139) holds with $i = 2$ and $R_2 = b$. If the boundary at $r = b$ is simply supported, then $w(b, \theta) = 0$, $0 \le \theta \le 2\pi$, and, in addition, (1.140) applies with $i = 2$ and $R_2 = b$. Finally, if the boundary at $r = b$ is free, then (1.140), with $i = 2$, $R_2 = b$ holds, as well as (1.141), with $i = 2, R_2 = b$. For any of these three situations, for the circular plate of radius $b$, we have a fourth order equation for $w$ (either (1.78), or its specialization, (1.84), to the case of axially symmetric deformations) and only two boundary conditions (at $r = b$). The missing boundary conditions which must be imposed arise because of the singularity which is inherent in the von Karman system (1.78), (1.79)—or its axially symmetric form (1.84), (1.85)—at $r = 0$; the usual assumptions are either that

$$w|_{r=0} < \infty \quad \text{and} \quad \frac{\partial w}{\partial r}\Big|_{r=0} = 0 \qquad (1.144)$$

or that

$$w|_{r=0} < \infty \quad \text{and} \quad \frac{\partial}{\partial r} \left( \frac{1}{r} \frac{\partial w}{\partial r} \right)\Big|_{r=0} = 0 \qquad (1.145)$$

The first condition, in either (1.144) or (1.145), that of a finite deflection at the center of the plate, is obvious, while justification of the second conditions in each set is based on requirements of regularity, i.e., continuity of a certain number of derivatives of $w$ with respect to the radial coordinate (see, e.g., the paper of Friedrichs and Stoker [72]). Conditions with respect to the behavior of the Airy stress function $\Phi$ must also be prescribed at $r = 0$, in the case of a circular plate, but a discussion of these will be left for section 1.3.

### 1.3.1.3 Rectilinear Orthotropic Response: Rectilinear Geometry

We again take for $\Omega$ the domain $\{(x, y) | 0 \le x \le a, 0 \le y \le b\}$ but now the constitutive relations (1.60) hold, with the hygroexpansive strains $\beta_i \Delta H$, (equivalently, the thermal strains $\alpha_i \Delta T$) $1, 2$, assumed to be constant through the thickness $h$ of the plate.

**(i) $\partial\Omega$ is Clamped**

In this case, the change from isotropic to orthotropic symmetry is inconsequential; the general conditions in (1.129a) once again translate into (1.130a,b).

**(ii) $\partial\Omega$ is Simply Supported**

Because we still have $w = 0$ on $\partial\Omega$, the conditions delineated in (1.130a) still apply in this case. As a consequence of the second condition in (1.129b), however, we have, in lieu of (1.131), the following statements, which are, by virtue of (1.67a,b), equivalent to $M_x = 0$, for $x = 0$, $x = a$, $0 \leq y \leq b$, and $M_y = 0$, for $y = 0$, $y = b$, $0 \leq x \leq a$:

$$\begin{cases} D_{11}w_{,xx} + D_{12}w_{,yy} |_{x=0} = 0, & 0 \leq y \leq b \\ D_{11}w_{,xx} + D_{12}w_{,yy} |_{x=a} = 0, & 0 \leq y \leq b \\ D_{21}w_{,xx} + D_{22}w_{,yy} |_{y=0} = 0, & 0 \leq x \leq a \\ D_{21}w_{,xx} + D_{22}w_{,yy} |_{y=b} = 0, & 0 \leq x \leq a \end{cases} \qquad (1.146)$$

However, by (1.61) - (1.63), and the fact that $E_1\nu_{12} = E_2\nu_{21}$:

$$\frac{D_{12}}{D_{11}} = \frac{C_{12}}{C_{11}} = \frac{E_2\nu_{21}/(1 - \nu_{12}\nu_{21})}{E_1/(1 - \nu_{12}\nu_{21})}$$

and

$$\frac{D_{21}}{D_{22}} = \frac{C_{21}}{C_{22}} = \frac{E_1\nu_{12}/(1 - \nu_{12}\nu_{21})}{E_2/(1 - \nu_{12}\nu_{21})}$$

or

$$\frac{D_{12}}{D_{11}} = \nu_{12} \quad \text{and} \quad \frac{D_{21}}{D_{22}} = \nu_{21} \qquad (1.147)$$

in which case, (1.146) may be rewritten as

$$\begin{cases} w_{,xx} + \nu_{12}w_{,yy} |_{x=0} = 0, & 0 \leq y \leq b \\ w_{,xx} + \nu_{12}w_{,yy} |_{x=a} = 0, & 0 \leq y \leq b \\ \nu_{21}w_{,xx} + w_{,yy} |_{y=0} = 0, & 0 \leq x \leq a \\ \nu_{21}w_{,xx} + w_{,yy} |_y = b = 0, & 0 \leq x \leq a \end{cases} \qquad (1.148)$$

which, clearly, reduce to (1.131) for the isotropic case when $\nu_{12} = \nu_{21} = \nu$.

**(iii) $\partial\Omega$ is Free**

In this case, because we still have $M_x = 0$ at $x = 0$, $x = a$, for $0 \leq y \leq b$, and $M_y = 0$, at $y = 0$, $y = b$, for $0 \leq x \leq a$, the conditions in (1.148) hold along each of the respective edges of the rectangle. The second (general) condition in (1.129c) again becomes (1.134), which reduces to (1.135); for the case of orthotropic symmetry, however, we must now use, in (1.135), the expressions (1.67a,b,c) for $M_x$, $M_y$, and $M_{xy}$, respectively. Thus

$$M_{y,y} + 2M_{xy,x} = -D_{21}w_{,xxy} - D_{22}w_{,yyy} - 4D_{66}w_{,xxy}$$

so that we have, for $0 \leq x \leq a$,

$$(D_{21} + 4D_{66})\, w_{,xxy} + D_{22}w_{,yyy}\, |_{y=0} = 0 \qquad (1.149)$$

and

$$(D_{21} + 4D_{66})\, w_{,xxy} + D_{22}w_{,yyy}\, |_{y=b} = 0 \qquad (1.150)$$

for $0 \leq x \leq a$. Also,

$$M_{x,x} + 2M_{yx,y} = -D_{11}w_{,xxx} - D_{12}w_{,xyy} - 4D_{66}w_{,xyy}$$

so that, for $0 \leq y \leq b$,

$$D_{11}w_{,xxx} + (D_{12} + 4D_{66})w_{,xyy}\, |_{x=0} = 0 \qquad (1.151)$$

and

$$D_{11}w_{,xxx} + (D_{12} + 4D_{66})w_{,xyy}\, |_{x=a} = 0 \qquad (1.152)$$

**Remarks:** To check that the boundary conditions (1.149)–(1.152), for free edges on a rectangular orthotropic plate, reduce to those in (1.137), for an isotropic plate, we may note, e.g., that

$$\frac{D_{21} + 4D_{66}}{D_{22}} = \frac{E_1\nu_{12}/(1 - \nu_{12}\nu_{21}) + 4G_{12}}{E_2/(1 - \nu_{12}\nu_{21})}$$

$$= \nu_{21} + \frac{4G_{12}(1 - \nu_{12}\nu_{21})}{E_2}$$

so that with isotropic symmetry

$$\frac{D_{21} + 4D_{66}}{D_{22}} = \nu + \frac{4G(1 - \nu^2)}{E} \equiv 2 - \nu$$

if we use the fact that $G = \dfrac{E}{2(1 + \nu)}$. Thus, with the assumption of isotropic symmetry, (1.149), (1.150) reduce to the first two conditions in (1.137) and a similar reduction applies to (1.151), (1.152).

### 1.3.1.4 Cylindrical Orthotropic Response: Circular Geometry

In this situation, $\Omega$ is again the annulus defined by $a \leq r \leq b$, $a \geq 0$, $b > a$, $r = \sqrt{x^2 + y^2}$, with the circular domain of radius $b$ corresponding to $a = 0$. The appropriate constitutive response is given by (1.90) or, equivalently, (1.91), with bending moments and stress resultants given as in (1.94) and (1.95). The bending stiffnesses and twisting rigidity appear in (1.96), while $D_{r\theta}$ is defined by (1.97). Finally, the von Karman equations for a thin linearly elastic plate possessing cylindrically orthotropic symmetry are exhibited in (1.98) and (1.99). As in the case of isotropic response, we set $\partial\Omega_i = \{(x, y)|x^2 + y^2 = R_i\}$, $i = 1, 2$ with $R_1 = a$, $R_2 = b$ so that $\partial\Omega = \partial\Omega_1 \cup \partial\Omega_2$. The relevant boundary data is as follows:

**(i) $\partial\Omega_i$ is Clamped**
As in the case of isotropic response and a circular geometry, the boundary conditions with respect to $w(r, \theta)$ reduce to (1.139). If $R_1 \equiv a = 0$, we may impose the regularity conditions (1.144) or (1.145) at $r = 0$, while (1.139) holds for $i = 2$, i.e., at $r = b$.

**(ii) $\partial\Omega_i$ is Simply Supported**
In this case, the condition $w(R_i, \theta) = 0$, $i = 1, 2$, $0 < \theta \leq 2\pi$, still applies but the second condition in (1.139) must be replaced by $M_r = 0$, at $r = a$, $r = b$, which, according to (1.94), means that

$$\begin{cases} \left[ w_{,rr} + \nu_\theta \left( \frac{1}{r} w_{,r} + \frac{1}{r^2} w_{,\theta\theta} \right) \right]_{r=a} = 0 \\ \left[ w_{,rr} + \nu_\theta \left( \frac{1}{r} w_{,r} + \frac{1}{r^2} w_{,\theta\theta} \right) \right]_{r=b} = 0 \end{cases} \tag{1.153}$$

**Remarks:** Once again, combined sets of boundary data are possible, e.g., for a cylindrically orthotropic, thin, annular plate, which is linearly elastic, and has its edge at $r = a$ simply supported, while the edge at $r = b$ is clamped, we would have

$$w(a, \theta) = 0, \quad \left[ w_{,rr} + \nu_\theta \left( \frac{1}{r} w_{,r} + \frac{1}{r^2} w_{,\theta\theta} \right) \right]_{r=a} = 0 \tag{1.154a}$$

$$w(b, \theta) = 0, \quad \frac{\partial w(r, \theta)}{\partial r}\Big|_{r=b} = 0, \tag{1.154b}$$

for $0 < \theta \leq 2\pi$.

**(iii)** $\partial\Omega_i$ **is Free**

With $\partial\Omega_i$ free, $i = 1, 2$, (1.153) will still apply but, in lieu of the vanishing of either $w$ or $w_{,r}$ at $r = a, b$, we must impose the condition

$$\left[ Q_r + \frac{1}{r} M_{r\theta,\theta} \right]_{r=R_i} = 0, \quad i = 1, 2 \tag{1.155}$$

A study of the literature would seem to indicate that the general form of the free edge boundary condition for a cylindrically orthotropic plate has not been written down; rather, because of the complicated form that the von Karman equations (1.98), (1.99) take, in the most general situation, where the deflection can depend on $\theta$, most (if not all) authors, to date, have been content to deal with the axisymmetric form of these equations (and, thus, with the corresponding form of the free edge boundary condition). The axisymmetric form of the free edge boundary condition (i.e., the condition that $Q_r|_{r=R_i} = 0$, with $w_{,\theta} = 0$) is

$$w_{,rrr} + \frac{1}{r} w_{,rr} - \left( \frac{E_\theta}{E_r} \right) \frac{w_{,r}}{r^2} \Big|_{r=R_i} = 0 \tag{1.156}$$

for $i = 1, 2$. Clearly, for the isotropic, axisymmetric situation, (1.155) reduces to (1.141) because of (1.142), the fact that $E_\theta = E_r$, and the assumption that $w$ is independent of $\theta$.

### 1.3.1.5 Rectilinear Orthotropic Response: Circular Geometry

Once again $\Omega$ is the annulus $a \leq r \leq b$, $a \geq 0$, $b > a$, $r = \sqrt{x^2 + y^2}$, with $a = 0$ yielding a circular plate of radius $b$. The relevant constitutive response is defined by (1.60), (with the $c_{ij}$ as in (1.61) and the $\beta_i \Delta H$ (or, equivalently, the $\alpha_i \Delta T$) taken to be constant through the thickness of the plate); these relations must be reformulated in polar coordinates, because of the assumed circular geometry of the plate, through the use of (1.73) and the analogous transformation for the components of the strain tensor (1.105), e.g., the constitutive relations will be the obvious modifications of (1.123a,b,c). The first von Karman equation is given by (1.106), with bending moments as defined by (1.107a,b,c), (1.108); this yields the partial differential equation (1.109). The second of the von Karman equations for this case comes about, either by directly transforming (1.69) into polar coordinates or by substituting the constitutive relations, in polar coordinate form, into (1.124). The relevant boundary conditions in this case are now as follows:

**(i)  $\partial\Omega_i$ is Clamped**

These conditions are, once again, identical with (1.139), for $i = 1, 2$.

**(ii)  $\partial\Omega_i$ is Simply Supported**

We again have $w(R_i, \theta) = 0$, $0 < \theta \le 2\pi$, $i = 1, 2$ and, in addition, must require that $M_r = 0$ for $r = a$ and $r = b$, $0 < \theta \le 2\pi$; by virtue of (1.107a), this is equivalent to the following statements for any $\theta$, $0 < \theta \le 2\pi$,

$$\left[ \tilde{D}_1 w_{,rr} + \tilde{D}_{12} \left( \frac{1}{r^2} w_{,\theta\theta} + \frac{1}{r} w_{,r} \right) \right.$$
$$\left. -2\tilde{D}_{16} \left( \frac{1}{r} w \right)_{,r\theta} \right]_{r=a} = 0 \qquad (1.157a)$$

$$\left[ \tilde{D}_1 w_{,rr} + \tilde{D}_{12} \left( \frac{1}{r^2} w_{,\theta\theta} + \frac{1}{r} w_{,r} \right) \right.$$
$$\left. -2\tilde{D}_{16} \left( \frac{1}{r} w \right)_{,r\theta} \right]_{r=b} = 0 \qquad (1.157b)$$

where $\tilde{D}_1$, $\tilde{D}_{12}$, and $\tilde{D}_{16}$ are defined as in (1.108)

**(iii)  $\partial\Omega_i$ is Free**

If the boundaries $\partial\Omega_i$ are free, then both (1.157a) and (1.157b) must hold and, in addition, we have (1.155) where $M_{r\theta}$ is given by (1.107c) and (1.108); to the best of the author's knowledge $Q_r$ has not been computed for this situation, to date, and will need to be calculated in the course of future work on such problems.

**Remarks:** In general, for the rectilinearly orthotropic, annular plate one will mix different types of boundary conditions along the edges at $r = a$ and $r = b$, e.g., if the inner boundary of the region is simply supported, while the outer boundary is clamped, then we would have $w(a, \theta) = 0$, $0 < \theta \le 2\pi$, together with (1.157a) and $w(b, \theta) = 0$, $\dfrac{\partial w(r, \theta)}{\partial r}\big|_{r=b} = 0$, for $0 < \theta \le 2\pi$.

## 1.4 The Linear Equations for Initial Buckling

In section 1.1 we considered a general system $G(\lambda, u) = 0$ of equilibrium equations, parametrized by the real number $\lambda$, and defined for $u$ in some Banach or Hilbert space. We indicated the connection which exists between the possibility of branching from an equilibrium solution $(\lambda_0, u_0)$ and the existence of a bounded inverse for the linear map $G_u(\lambda_0, u_0)$. In this section, we will indicate how one forms the linearized equations which govern the onset of buckling in a thin, linearly elastic, plate, i.e., we will show how to obtain the linearized equations which control initial buckling of a plate from the various sets of nonlinear von Karman equations we have presented, in both rectilinear and polar coordinates, for isotropic and orthotropic response; in the course of our discussion we will present several sets of boundary conditions for the Airy function (equivalently, for the forces specified by the various derivatives of the Airy function on the edge, or edges, of the plate.)

Although we may proceed with a direct discussion of the application of Frechét differentiation to the von Karman equations, as a means of generating the linearized equations of buckling, we note that these equations may also be generated by observing that in all the cases considered thus far, in both rectilinear and polar coordinates, the von Karman equations enjoy a special structure. Therefore, suppose that, with reference once again to Fig. 1.11, which depicts a thin elastic plate that occupies a region $\Omega$ in the $x, y$ plane, $\bar{\Phi}_0(x, y)$ is the stress function produced in the plate, under the action of applied forces on $\partial\Omega$ and/or specified boundary conditions with respect to $w$, when the plate is not allowed to deflect (i.e., $\bar{\Phi}_0(x, y)$ is associated with a state of generalized plate stress in $\Omega$). Suppose further that it is possible to characterize the class of possible loadings of the plate that we are interested in by a single parameter $\lambda$, which one may think of as being a measure of the strength of the applied edge forces; in this case we may set

$$\bar{\Phi}_0(x, y) = \lambda \Phi_0(x, y) \qquad (1.158)$$

where, in writing down (1.158), we are, clearly, thinking of the generalized state of plane stress (that is represented by the Airy function) as depending linearly on $\lambda$. We comment later on the fact that it is not always possible to express $\bar{\Phi}_0(x, y)$ in the form (1.158).

**Example**: Consider the rectangular plate of length $a$ and width $b$ which is depicted in Fig. 1.12. Here a compressive thrust of magnitude $-\lambda hb$

is applied normal to the edges at $x = 0$, $x = a$, for $0 \leq y \leq b$. If, e.g. all four edges are simply supported, and the thin plate is isotropic and linearly elastic, then, along all four edges of the plate, $w = \Delta w = 0$. Referring to the von Karman equations, (1.53a,b), which apply in this case, with $t \equiv 0$, for a state of generalized plane stress (1.53a) is satisfied identically while (1.53b) reduces to

$$\Delta^2 \bar{\Phi}_0 = 0; \quad 0 < x < a, \quad 0 < y < b \tag{1.159}$$

subject to the boundary conditions specified above. The solution of this plane stress boundary value problem may be taken to be

$$\bar{\Phi}_0(x, y) = -\lambda h \frac{y^2}{2} \tag{1.160}$$

inasmuch as we do not care about linear and constant terms in $\bar{\Phi}_0$ (because the expressions involving $\Phi$ in (1.53a,b) always appear as second derivatives in the Airy function.) From (1.160) it is clear that we may take, in accordance with (1.158), $\Phi_0(x, y) = -\frac{h}{2} y^2$. From (1.160) and (1.46) we see that, $\bar{N}_x^0 = -\lambda h$, $\bar{N}_y^0 = 0$, $\bar{N}_{xy}^0 = 0$.

Returning to the general situation we make the following observations:

(i) In every case considered in section 1.2, with respect to the buckling of a thin, linearly elastic plate, either for isotropic or orthotropic response, and whether it be for the case of rectilinear or circular geometry, the structure of the von Karman equations is as follows:

$$L_1 w = [\Phi, w] \tag{1.161a}$$

$$L_2 \Phi = -\frac{1}{2} [w, w] \tag{1.161b}$$

where $L_1$ and $L_2$ are (usually), variable coefficient, linear differential operators whose precise structure is determined by the type of coordinate system we are working in and the nature of the symmetry associated with the constitutive relations, i.e., isotropic or orthotropic, while the brackets $[\Phi, w]$ and $[w, w]$ are nonlinear in structure, but depend only on whether we are working in rectilinear Cartesian coordinates or in polar coordinates. Specifically, in Cartesian coordinates

$$\begin{cases} [\Phi, w] = \Phi_{,yy}\, w_{,xx} - 2w_{,xy}\, \Phi_{,xy} + \Phi_{,xx}\, w_{,yy} \\[4pt] \qquad\quad \equiv w_{,xx}\, N_x + 2w_{,xy} \cdot N_{xy} + w_{,yy}\, N_y \\[4pt] [w, w] = 2(w_{,xx}\, w_{,yy} - w^2_{,xy}) \end{cases} \qquad (1.162)$$

while in polar coordinates

$$\begin{cases} [\Phi, w] = \Phi_{,rr}\left(\dfrac{1}{r}w_{,r} + \dfrac{1}{r^2}w_{,\theta\theta}\right) \\[8pt] \qquad\quad + \left(\dfrac{1}{r}\Phi_{,r} + \dfrac{1}{r^2}\Phi_{,\theta\theta}\right) w_{,rr} \\[8pt] \qquad\quad - 2\left(\dfrac{1}{r}w_{,r\theta} - \dfrac{1}{r^2}w_{,\theta}\right)\left(\dfrac{1}{r}\Phi_{,r\theta} - \dfrac{1}{r^2}\Phi_{,\theta}\right) \\[8pt] \qquad\quad = w_{,rr}\, N_r - 2\left(\dfrac{1}{r^2}w_{,\theta} - \dfrac{1}{r^2}w_{,r\theta}\right) N_{r\theta} \\[8pt] \qquad\quad + \left(\dfrac{1}{r}w_{,r} + \dfrac{1}{r^2}w_{,\theta\theta}\right) N_\theta \\[8pt] [w, w] = -2\left\{ w_{,rr}\left(\dfrac{1}{r}w_{,r} + \dfrac{1}{r^2}w_{,\theta\theta}\right)\right. \\[8pt] \qquad\qquad\left. - \left(\dfrac{1}{r}w_{,r\theta} - \dfrac{1}{r^2}w_{,\theta}\right)^2 \right\} \end{cases} \qquad (1.163)$$

(ii)      We set

$$\varphi(x, y) = \Phi(x, y) - \lambda\Phi_0(x, y)$$

and note that as $\Phi_0(x, y)$ represents the state of plane stress in the plate corresponding to the deflection $w \equiv 0$, it must, therefore, (see 1.161a,b) satisfy

$$L_2\Phi_0 = 0, \quad \text{in } \Omega, \qquad (1.164)$$

with some appropriate boundary conditions on $\partial\Omega$. Then, (1.161a,b) may be put in the form

$$\begin{cases} L_1 w - \lambda\,[\Phi_0, w] = [\varphi, w] \\[4pt] L_2\varphi = [w, w] \end{cases} \qquad (1.165)$$

with $\varphi(x, y)$ the "extra" Airy stress (function) generated by the out-of-plane deflection $w$.

**Remarks**: The term $[\Phi_0, w]$ may also be written in terms of the stress resultants (or averaged stresses) $N_x^0$, $N_y^0$, $N_{xy}^0$ (rectilinear Cartesian coordinates) or $N_r^0$, $N_\theta^0$, $N_{r\theta}^0$ (polar coordinates) corresponding to the state of generalized plane stress in the plate when $w \equiv 0$.

From our general discussion in section 1.1, we know that branching from an equilibrium solution $(\lambda_0, u_0)$ of a general system of equations $G(\lambda, u) = 0$ can occur only if $G_u(\lambda_c, u_0)$ is not invertible (as a linear map), i.e., if the eigenvalue-eigenfunction problem

$$G_u(\lambda, u_0)v = 0 \qquad (1.166)$$

has a nontrivial solution $v_c$ (a "buckling mode") for some value of $\lambda$, say, $\lambda_c$ (a "buckling load"). For the general von Karman system given by (1.165), which includes all the cases we have considered in section 1.2, the problem equivalent to (1.166) may be formulated as follows:

We take, for the "vector" $u$ the pair $(w, \varphi)$ and we write

$$\begin{aligned} G^1(\lambda, u) &\equiv G^1\left(\lambda, (w, \varphi)\right) \\ &= L_1 w - \lambda\left[\Phi_0, w\right] - [\varphi, w] \end{aligned} \qquad (1.167)$$

and

$$\begin{aligned} G^2(\lambda, u) &\equiv G^2\left(\lambda, (w, \varphi)\right) \\ &= L_2 \varphi - [w, w] \end{aligned} \qquad (1.168)$$

so that

$$G(\lambda, u) = \begin{pmatrix} G^1\left(\lambda, (w, \varphi)\right) \\ G^2\left(\lambda, (w, \varphi)\right) \end{pmatrix} \qquad (1.169)$$

The von Karman equations, (1.165), are then equivalent to $G(\lambda, u) = \begin{pmatrix} 0 \\ 0 \end{pmatrix}$. The equilibrium solution $(\lambda, u_0)$, which holds for all $\lambda$, is given by

$$u_0 = (0, 0) \qquad (1.170)$$

i.e., by $w = 0$ and $\Phi = \overline{\Phi}_0 = \lambda\Phi_0$. Computing the Frechét derivative of the mapping in (1.169), at $(\lambda, u_0) \equiv (\lambda, (0,0))$, and setting $v = (\hat{w}, \hat{\varphi})$, we find that

$$G_u(\lambda, u_0)v = \begin{pmatrix} L_1\hat{w} - \lambda\left[\Phi_0, \hat{w}\right] \\ L_2\hat{\varphi} \end{pmatrix} \qquad (1.171)$$

so that $G_u(\lambda, u_0)v = 0$ is equivalent to the system of equations

$$\begin{cases} L_1\hat{w} - \lambda \left[\Phi_0, \hat{w}\right] = 0 \\[2mm] L_2\hat{\varphi} = 0 \end{cases} \qquad (1.172)$$

The linear equations in (1.172) are subject to the boundary conditions on $\partial\Omega$ for $w$ and $\varphi$. However, the boundary conditions with respect to $\Phi(x,y)$ are *always* chosen identical to the ones that are satisfied by the state of generalized plane stress $\lambda\Phi_0(x,y)$ and, therefore, $\varphi$ satisfies homogeneous boundary conditions on $\Omega$, e.g. for the example considered in this section, of compression of the rectangular plate, $\varphi = \Delta\varphi = 0$ on $\partial\Omega$. As the second equation in (1.172) is linear and homogeneous, $\hat{\varphi} \equiv 0$ in $\Omega$ and we are left, therefore, with the eigenvalue–eigenfunction problem

$$\begin{cases} L_1 w_c - \lambda_c \left[\Phi_0, w_c\right] = 0 \text{ ( in } \Omega) \\[2mm] + \begin{cases} \text{appropriate boundary conditions on} \\ \partial\Omega \text{ corresponding to clamped,} \\ \text{simply supported, or free edges,} \\ \text{or some combinations, thereof.} \end{cases} \end{cases} \qquad (1.173)$$

If we compare the general structure of the linearized problem (1.173), which governs initial buckling of the plate, for all the cases we have considered so far in this report, with the full set (1.161a,b) of von Karman equations we may note the following simple algorithm:

For the cases considered in sections 1.2, 1.3, the eigenvalue–eigenfunction problem (1.173) for the buckling loads $\lambda_c$, and the corresponding buckling modes $w_c$, may be obtained from the von Karman equations (1.161a, b) by

(i)    Suppressing the second equation, i.e., (1.161b)

and

(ii)    Setting $\Phi(x,y)$, in (1.161a), equal to $\lambda\Phi_0(x,y) \equiv \bar{\Phi}_0(x,y)$, where $\bar{\Phi}_0(x,y)$ is the Airy stress function corresponding to the state of generalized plane stress in the plate (i.e. $w \equiv 0$) which is generated by whatever loading conditions are in effect along the edge (or edges) or the plate (i.e., along $\partial\Omega$), the parameter $\lambda$ having been chosen to gauge the magnitude of that loading.

**Remarks**: For certain types of loadings we will encounter, in Chapter 2, for example, we may not be able to express $\bar{\Phi}_0$ directly in the form $\lambda\Phi_0$, with $\lambda$ representing the magnitude of a loading along a particular edge, e.g., compressive loading of the rectangular plate along the edges parallel to both the $x$ and $y$ axes with different loading magnitudes along the different pairs of parallel sides. However, the algorithm we have delineated above is still valid: we simply replace $\Phi(x,y)$ in (1.161a) by $\bar{\Phi}_0(x,y)$.

**Example**: We return to the linearly elastic, isotropic plate depicted in Fig. 1.12; the plate is simply supported along all four edges and is subjected to a compressive loading of magnitude $\lambda$ along the sides $x = 0$, $x = a$, for $0 \le y \le b$. We have already seen, in this case, that $\bar{\Phi}_0$ is given by (1.160), so that $\Phi_0 = -\dfrac{h}{2}y^2$. The relevant von Karman equations are (1.53a,b). Suppressing (1.53b), and replacing $\Phi$ in (1.53a) by $\bar{\Phi}_0 = \lambda\Phi_0$, we have

$$
\begin{aligned}
K\Delta^2 w &= \lambda\,[\Phi_0, w] \\
&= \lambda\Big\{ \Phi_{0,yy}\,w_{,xx} + \Phi_{0,xx}\,w_{,yy} - 2\Phi_{0,xy}\,w_{,xy} \Big\} \\
&= -\lambda h w_{,xx}
\end{aligned}
$$

so that the eigenvalue–eigenfunction problem we are interested in is

$$
\begin{cases}
K\Delta^2 w + \lambda h w_{,xx} = 0, & \text{in } \Omega \\
w = \Delta w = 0, & \text{on } \partial\Omega
\end{cases}
\tag{1.174}
$$

with $\Omega = \{(x,y)\,|\,0 \le x \le a,\ \ 0 \le y \le b\}$. Other examples of linearized problems governing the initial buckling of thin plates will be considered as they arise in Chapters 2–5; in certain of these later examples, we will want to modify the above discussion to account for the presence of initial imperfections or for constitutive behavior which is other than linear elastic; however, the same basic logic which took us, in this section, from the full set of von Karman equations to the (linear) eigenvalue–eigenfunction problem governing initial plate buckling, will still apply.

## 1.5 Figures: Plate Buckling and the von Karman Equations

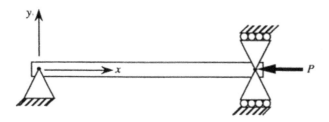

**FIGURE 1.1**
Thin rod in compression.

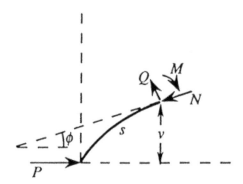

**FIGURE 1.2**
Forces and moments acting on a portion of a buckled thin rod.

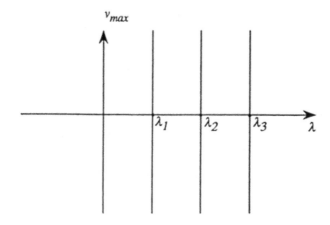

**FIGURE 1.3**
Buckling of the thin compressed rod based only on the linearized equations.

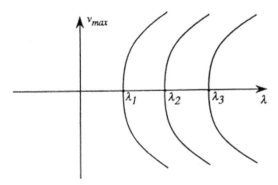

**FIGURE 1.4**
Buckling of the thin compressed rod based on the nonlinear equations.

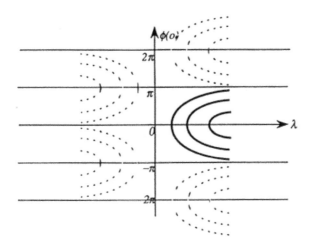

**FIGURE 1.5**
Graph of $\varphi_{max}$ vs. $\lambda$ for various ranges of the initial angle $\varphi(0)$.

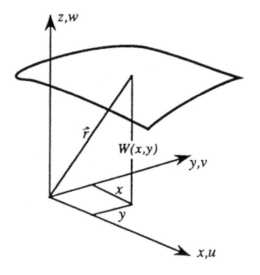

**FIGURE 1.6**
The middle surface of the shallow shell in rectilinear cartesian coordinates.

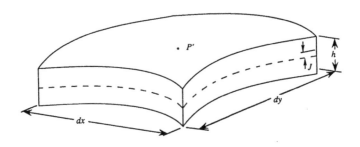

**FIGURE 1.7**
An infinitesimal volume element $dV$ of the shell; the coordinate $\zeta$ measures the distance from the middle surface.

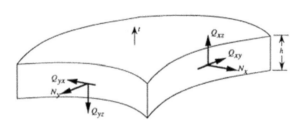

**FIGURE 1.8**
Stress resultants and distributed loading $t(x, y)$ acting on a
differential shell element.

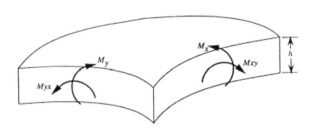

**FIGURE 1.9**
Moments acting on a differential shell element ($M_{xy} = M_{yx}$).

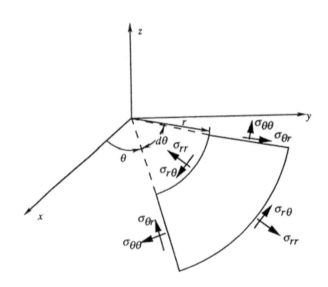

**FIGURE 1.10**
Components of the stress tensor in polar coordinates.

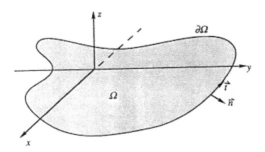

**FIGURE** 1.11
A thin elastic plate which occupies the region $\Omega$ in the $x, y$ plane.

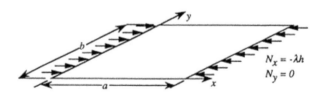

**FIGURE 1.12**
A rectangular plate under an in-plane loading.

CHAPTER 2

# Initial and Postbuckling Behavior of (Perfect) Thin Rectangular Plates

In this chapter, we consider some of the basic results which exist in the literature relative to initial buckling and postbuckling behavior of perfect, thin, rectangular plates (in the analysis of elastic-plastic buckling of rectangular plates, in section 2.3, some results on imperfection-sensitivity will be presented, although a full discussion will be relegated to section 3.1. A completely exhaustive treatment would require that we consider every conceivable combination of loading and support conditions relative to each of the four edges of a rectangular plate; in lieu of such a treatment we will present, instead, the results for the most common combinations of loading and support conditions and, in particular, will focus on those combinations which should prove to be most useful in terms of experiments that are planned in support of work on the hygroexpansive buckling of paper. We begin our delineation of the results for rectangular plates by considering the most frequently considered case of linear elastic response, both in the situation where isotropic material symmetry applies, and for the situation which is more directly related to the constitutive behavior of paper, namely, rectilinear orthotropic symmetry. We then consider the generalization [73], [74] of the von Karman equations by Johnson and Urbanik which was proposed to cover the case of nonlinear elastic response; while certain of the hypotheses employed in [73] may be considered to be suspect, the theory devised in these papers was formulated directly with the nonlinear elastic, rectilinearly orthotropic response of paper in mind and appears to give reasonable results for various combinations of loading and support conditions, at least for the problem of initial buckling. The Johnson–Urbanik theory [73], [74], which we discuss in section 2.2, does not appear to have been extended yet so as to enable one to consider postbuckling behavior, nor do results for initial buckling appear to be available, within the context of this theory, for nonlinearly elastic circular or annular plates, even within the context of isotropic response. In section 2.3, we present some results on initial buckling and postbuckling behavior of rectangular plates exhibiting either elastic-plastic or viscoelastic response. Throughout our discussions in sections 2.1 through 2.3 of this chapter, we present the most pertinent graphs that we have found in the literature to describe the specific initial or postbuckling behavior under consideration; these graphs then form the basis for our discussion in section 2.4 where we compare the initial and postbuckling behavior of rectangular plates with respect to variations in constitutive response, material symmetry, and boundary conditions.

## 2.1  Plates with Linear Elastic Behavior

Rectangular plates exhibiting linear elastic behavior may be classified according to whether the plate (which is assumed to be thin and without imperfections, i.e., there is no initial deflection and no distributed force normal to the midplane of the plate) possesses isotropic or orthotropic material symmetry.

### 2.1.1  Isotropic Symmetry: Initial Buckling

We begin by presenting, for several of the most common cases, the results for initial buckling of a rectangular plate which occupies the region in the $x, y$ plane defined by
$$0 \leq x \leq a, \quad 0 \leq y \leq b.$$

**(i)  Compression in the $x$-Direction with the Loaded Edges Simply Supported**

In this case, the situation depicted in Fig. 1.12 applies with $\sigma_{xx} = \lambda, \lambda > 0$, along the edges at $x = 0$ and $x = a$; also, the first pair of boundary conditions in (1.131) apply, along these edges, together with $w = 0$ along $x = 0$ and $x = a$ for $0 \leq y \leq b$. The relevant eigenvalue problem is governed by the equation in (1.174), i.e. by

$$\frac{\partial^4 w}{\partial x^4} + 2\frac{\partial^4 w}{\partial x^2 \partial y^2} + \frac{\partial^4 w}{\partial y^4} = -\frac{\lambda h}{K}\frac{\partial^2 w}{\partial x^2} \qquad (2.1)$$

For future reference, we recall that $K \equiv \dfrac{Eh^3}{12(1 - \nu^2)}$ and we set

$$\tilde{K} = \lambda_{cr} \frac{b^2 h}{\pi^2 K} \tag{2.2}$$

where $\lambda_{cr}$ will represent an eigenvalue (i.e. buckling load) for (2.1), subject to the conditions of simple support along the loaded edges, and various combinations of boundary conditions along the longitudinal edges at $y = 0$ and $y = b$.

**(a) The Edges $y = 0$, $y = b$, $0 \le x \le a$ are Simply Supported.**
In this subcase $w = 0$ along $y = 0$, $y = b$, $0 \le x \le a$, and the second set of boundary conditions in (1.131) also applies along these edges. We assume, because of the nature of the loading, and the conditions of simple support along the loaded edges, that the plate buckles into $m$ sinusoidal half-waves in the direction of loading; this leads, naturally, to a search for solutions of the boundary-value problem associated with (2.1) of the form

$$w(x, y) = f(y) \sin \frac{m \pi x}{a} \tag{2.3}$$

Substitution of (2.3) in (2.1) shows that $f$ must have the form

$$f(y) = A_1 \cosh \alpha y + A_2 \sinh \alpha y + A_3 \cos \beta y + A_4 \sin \beta y$$

and that the critical values $\lambda_{cr}$ of the compressive stress must satisfy the characteristic equation

$$(r^2 + s^2)^2 \sinh p \sin q = 0 \tag{2.4}$$

where

$$p/q = \sqrt{\frac{m \pi^2}{\phi}} \quad \left( \sqrt{\tilde{K}} \ne \frac{m}{\phi} \right) \tag{2.5}$$

with

$$r^2 = p^2 - \nu \frac{m^2 \pi^2}{\phi^2}, \quad s^2 = q^2 + \nu \frac{m^2 \pi^2}{\phi^2} \tag{2.6}$$

$\phi = \dfrac{a}{b}$, and $\tilde{K}$ given by (2.2). For $m = 1$, the solution of (2.4) is given by

$$\tilde{K} = \phi^{-2} + \phi^2 + 2 \tag{2.7}$$

which is graphed in Fig. 2.1. With $m = 1$, the value of $\tilde{K}$, and, thus, of $\lambda_{cr}$, increases monotonically as $\phi = \dfrac{a}{b}$ increases from $\phi = 1$ (square plate) and for large $\phi$ (a long, narrow plate), $\tilde{K} \sim \phi^2$. If we consider the case of two half-waves in the $x$-direction, then with the same definitions as in (2.5), (2.6), it can be shown that the value of $\lambda_{cr}$ is unchanged if $m/\phi$ is replaced by $\dfrac{Nm}{N\phi}$ for any integer $N$. As a consequence, it may, for example, be shown that $\lambda_{cr}$ for a plate with aspect ratio $2\phi$, buckling into the two half-wave mode $(m = 2)$, is the same as that for a plate with aspect ratio $\phi$ buckling into the one half-wave mode $(m = 1)$. The $\tilde{K}$ versus $\phi$ curves, in general, are shown in Fig. 2.2 for various values of $m$, where

$$\tilde{K} \equiv \tilde{K}_m = m^2\phi^{-2} + m^{-2}\phi^2 + 2 \qquad (2.8)$$

Note, e.g., that for aspect ratio $\phi = 2$, the minimum value of $\lambda_{cr}$ is associated with buckling into the two-half-wave mode $(m = 2)$ and, in fact, in this case $(\lambda_{cr})_{\min}$ corresponds to the value $\tilde{K} = 4$. Also, a transition occurs from buckling into $m$ half-waves to buckling into $m+1$ half-wave whenever successive curves intersect. From (2.8) it follows that $\tilde{K}_m = \tilde{K}_{m+1}$ provided

$$\phi^2 = m(m + 1) \qquad (2.9)$$

For example, the curves corresponding to $m = 1$ and $m = 2$ in Fig. 2.2 intersect when the plate aspect ratio $\phi = \sqrt{2}$. In this case, the buckling mode corresponding to $\lambda_{cr}$ would consist of a linear combination of functions of the form (2.3), with $m = 1$ and $m = 2$.

**(b) The Edges $y = 0$, $y = b$, $0 \le x \le a$, are Clamped.**
In this subcase (see (1.130a), (1.130b)), we have $w(x, 0) = w(x, b) = 0$, $0 \le x \le a$, and $w_{,y}(x, 0) = w_{,y}(x, b) = 0$, $0 \le x \le a$, as well as the conditions of simple support along the loaded edges at $x = 0$, $x = a$, $0 \le y \le b$, that we had in subcase (a). Solutions of the boundary-value problem still have the form (2.3), with $f(y)$ the same as in subcase (a), and imposition of the boundary conditions along the four edges leads to the system

$$\begin{cases} (A_1 + A_3) \sin \dfrac{m\pi x}{a} = 0 \\[2mm] (pA_2 + qA_4) \sin \dfrac{m\pi x}{a} = 0 \\[2mm] (A_1 \cosh p + A_2 \sinh p + A_3 \cos q + A_4 \sin q) \sin \dfrac{m\pi x}{a} = 0 \\[2mm] (A_1 p \sinh p + A_2 p \cosh p - A_3 q \sin q + A_4 q \cos q) \sin \dfrac{m\pi x}{a} = 0 \end{cases}$$

$$(2.10)$$

and, thus, to the characteristic equation

$$2pq(1 - \cosh p \cos q) + (p^2 - q^2) \sinh p \sin q = 0 \qquad (2.11)$$

A numerical solution of (2.11) for the case of $m = 1$, i.e., buckling into one half-wave in the $x$-direction, is depicted in Fig. 2.3. By using the energy method, one may produce the dashed curve in Fig. 2.2, for which

$$\tilde{K} = \phi^{-2} + \frac{16}{3}\phi^2 + 8/3 \qquad (2.12)$$

**(c) The Edge $y = 0$ is Clamped and the Edge $y = b$ is Simply Supported.**
In this case, along the edge at $y = 0$ we have $w(x,0) = w_{,y}(x,0) = 0$, $0 \le x \le a$, while along the edge at $y = b$, $w(x,b) = 0$, $0 \le x \le a$, and $w_{,yy} + \nu w_{,xx}|_{y=b} = 0, 0 \le x \le a$ (i.e., see (1.131)). The general form of the deflection which occurs upon initial buckling is the same as in subcases (a) and (b) and this leads to the characteristic equation

$$(r^2 + s^2)(q \sinh p \cos q - p \cosh p \sin q) = 0 \qquad (2.13)$$

whose solution, i.e., $\tilde{K}$ versus $\phi$ graph, for the case $m = 1$, is shown in Fig. 2.4; we may read off from this graph that the minimum value of $\lambda_{cr}$, in this case, is approximately the average of the values of $(\lambda_{cr})_{min}$ for the problems treated in subcases (a) and (b).

**(d) The Edge $y = 0$ is Clamped and the Edge $y = b$ is Free**
As in subcase (c), along the edge at $y = 0$ we again have $w(x,0) = w_{,y}(x,0) = 0$, $0 \le x \le a$, but now, along the edge at $b$, the last of the relations in (1.131) applies, as well as the second of the relations in (1.137). Using the same expression for the form of the eigenfunction $w$ as we employed in subcases (a)–(c), we find that the $A_i$, $i = 1, \dots, 4$, must satisfy

$$\begin{cases} A_1 ps^2 \sinh p + A_2 ps^2 \cosh p + A_3 qr^2 \sin q - A_4 q^2 r^2 \cos q = 0 \\ A_1 r^2 \cosh p + A_2 r^2 \sinh p - A_3 s^2 \cos q - A_4 s^2 \sin q = 0 \\ A_1 + A_3 = 0 \\ pA_2 + qA_4 = 0 \end{cases}$$

which leads to the characteristic equation

$$2pqr^2 s^2 + pq(r^4 + s^4)\cosh p \cos q$$
$$+(q^2 r^4 - p^2 s^4)\sinh p \sin q = 0 \qquad (2.14)$$

The solution to (2.14), for $m = 1$, is depicted in Fig. 2.5 and an energy analysis yields the result

$$\tilde{K} = \phi^{-2} + \frac{\pi\phi^2}{16(3\pi - 8)} + \frac{\pi - 4\nu}{2(3\pi - 8)} \qquad (2.15)$$

**(e) The Edge $y = 0$ is Simply Supported and the Edge $y = b$ is Free**

As in subcase (d), along the edge at $y = b$, the last one of the relations in (1.131) and the second of the relations in (1.137) hold, while along the edge at $y = 0$ we have $w(x, 0) = 0$, $0 \le x \le a$, as well as the first of the relations in (1.137). With the same general form for $w$ as in subcases (a)–(d), we find that the $A_i$, $i = 1, \ldots, 4$, satisfy

$$\begin{cases} A_1 + A_3 = 0 \\ r^2 A_1 - S^2 A_3 = 0 \\ A_1 p s^2 \sinh p + A_2 p s^2 \cosh p + A_3 q r^2 \sin q - A_4 q r^2 \cos q = 0 \\ A_1 r^2 \cosh p + A_2 r^2 \sinh p - A_3 s^2 \cos q - A_4 s^2 \sin q = 0 \end{cases}$$

which leads to the characteristic equation

$$(r^2 + s^2)(q r^4 \sinh p \cos q - p s^4 \cosh p \sin q) = 0 \qquad (2.16)$$

The solution of (2.16), for $m = 1$, is shown in Fig. 2.6 . We note that, in this case, $\tilde{K}$ appears to asymptote to a constant value as $\phi$ increases; thus, the minimum value of $\lambda_{cr}$ is always associated with $m = 1$ so that the compressed rectangular plate, which is simply supported at $x = 0$ and $x = a$, simply supported along $y = 0$, and free along the edge $y = b$, always buckles initially in the one-half-wave mode irrespective of the aspect ratio.

**(ii) Compression in the $x$-Direction with the Loaded Edges Clamped**

Changing the constraint from simply supported to clamped along the loaded edges, does not alter the form (2.1) of the equation governing initial buckling of the plate. Along $x = 0$ and $x = a$, for $0 \le y \le b$, the first pairs of conditions, in both (1.130a) and (1.130b), apply.

**(a) The Edges $y = 0$, $y = b$, $0 \le x \le a$, are Simply Supported.**

In this case $w = 0$ along $y = 0$, $y = b$, for $0 \le x \le a$, and the second set of relations in (1.131) must also be satisfied. It is now assumed that the

plate buckles into one sinusoidal half-wave in the transverse direction so
that the eigenfunctions have the form

$$w(x, y) = f(x) \sin \frac{\pi y}{b} \qquad (2.17)$$

Substitution of (2.17) into (2.1) yields

$$f(x) = A_1 \cos \alpha x + A_2 \cos px + A_3 \sin \alpha x + A_4 \sin px \qquad (2.18a)$$

with

$$\begin{cases} \alpha + \dfrac{\pi \phi}{2a} \left[ \sqrt{\tilde{K}} + \sqrt{\tilde{K} - 4} \right] \\[3mm] \beta = \dfrac{\pi \phi}{2a} \left[ \sqrt{\tilde{K}} - \sqrt{\tilde{K} - 4} \right] \end{cases} \qquad (2.18b)$$

and imposition of the boundary conditions along the four edges produces
a characteristic equation relating $\tilde{K}$ and $\phi$ (and hence, $\lambda_{cr}$ and $\phi$) whose
solution is depicted in Fig. 2.7; it may be noted that modes with even
numbers of half-waves in the $x$-direction are antisymmetric. The solution
to this problem was first obtained by Schleicher [75]. As $\phi$ increases, the
value of $\tilde{K}$ approaches that of a plate simply supported on all four sides,
because the half-wave length of the buckles in the $y$-direction approaches
a value equal to the width of the plate. Finally, the transition from $m$
to $m+1$ half waves, in the $x$-direction, takes place when $\phi^2 = m(m+2)$.

**(b) The Edges $y = 0$, $y = b$, $0 \leq x \leq a$, are Clamped.**
The solution for this subcase is attributed to Levy [76] who took the
plate to be simply supported and then made the edge slopes zero by
introducing a suitable distribution of edge-bending moments. As all
four edges are clamped, the full sets of relations in (1.130a) and (I.130b)
apply. Levy [76] developed a solution for the eigenvalue- eigenfunction
problem for this case (i.e., (2.1) subject to the boundary conditions
(1.130a), (I.130b)) in the form

$$w = \sum_{\substack{m = 1, 3, 5 \ldots \\ \text{or } 2, 4, 6 \ldots}}^{\infty} \sum_{n = 1, 3, 5 \ldots}^{\infty} \frac{(nk_m + mt_n) \sin \dfrac{m\pi x}{a} \sin \dfrac{n\pi y}{b}}{\pi^4 K \left[ \left( \dfrac{m^2}{a^2} + \dfrac{n^2}{b^2} \right)^2 \right]^2 - \dfrac{m^2 \pi^2 \lambda h}{a^2}}$$

with $m$ having odd values for buckling into an odd number of waves
in the $x$-direction, and even values for buckling into an even number of

waves; the conditions in (I.130b) are now imposed upon this form of the solution which generates the system of equations

$$
\begin{cases}
\displaystyle\sum_{\substack{m = 1,3,5,\ldots \\ \text{or } 2,4,6\ldots}}^{\infty} \dfrac{mnk_m + m^2 t_n}{\left[\dfrac{m^2}{\phi^2} + n^2\right]^2 - \dfrac{m^2 \tilde{K}}{\phi^2}} = 0 \ (n = 1,3,5,\ldots) \\[4ex]
\displaystyle\sum_{n = 1,3,5\ldots}^{\infty} \dfrac{n^2 k_m + nmt_n}{\left[\dfrac{m^2}{\phi^2} + n^2\right]^2 - \dfrac{m^2 \tilde{K}}{\phi^2}} = 0 \quad \left(\begin{array}{l} m = 1,3,5\ldots \\ \text{or } 2,4,6 \end{array}\right)
\end{cases}
\tag{2.19}
$$

The system of homogeneous, linear, algebraic equations for $k_m$ and $t_n$, (2.19) produces a characteristic equation whose solution is shown in Fig. 2.8, which is taken from Levy [76]; on this graph we also depict the results for the earlier case, where the loaded edges at $x = 0$, $x = a$, $0 \leq y \leq b$, were simply supported, and the edges along $y = 0$, $y = b$, $0 \leq x \leq a$, are clamped. From the graph in Fig. 2.8 it follows that as $\phi$ increases, $\tilde{K}$ approaches the value that it has for a long plate, which is simply supported along its loaded edges, and clamped along the edges $y = 0$ and $y = b$. In the present case (loaded edges clamped), the change from $m = 1$ to $m = 2$ half-waves in the $x$-direction takes place at the intersection of the $m = 1$ and $m = 3(\tilde{K}$ versus $\phi)$ curves which apply in the earlier case where the loaded edges were simply supported and, it is clear from Fig. 2.8, that this pattern continues as we transition from $m = 2$ to $m = 3$ half-waves in the $x$-direction, etc.

**(iii) Compression in both the $x$ and $y$ Directions.**
In [32] the author credits Bryan [77] with the earliest discussions of this problem; the situation is depicted in Fig. 2.9, where we have set $\lambda = \sigma_{xx}|_{x=0,a}$ and $|\xi = \sigma_{yy}|_{y=0,b}$.
In this case, we must first solve (1.161b), with $L_2 = \nabla^4$, and $w = 0$, to obtain the in-plane equilibrium solution $\bar{\Phi}_0(x, y)$, which is not of the (one-parameter) form $\lambda\Phi_0(x, y)$ here, and then substitute the result for $\bar{\Phi}(x, y)$ into (1.161a) so as to obtain the partial differential equation which governs the initial buckling problem; the result is that the initial deflection must satisfy

$$
K\left(\frac{\partial^4 w}{\partial x^2} + 2\frac{\partial^4 w}{\partial x^2 \partial y^2} + \frac{\partial^4 w}{\partial y^4}\right) + h\left(\lambda\frac{\partial^2 w}{\partial x^2} + \xi\frac{\partial^2 w}{\partial y^2}\right) = 0 \tag{2.20}
$$

**(a) The Edges at $x = 0$, $x = a$, $0 \leq y \leq b$, and $y = 0$, $y = b$, $0 \leq x \leq a$, are Simply Supported.**

In this subcase, the boundary conditions are given by (1.130a) and (1.131); a solution of the eigenvalue–eigenfunction problem is assumed to have the form

$$w(x,y) = A_{mn} \sin \frac{m\pi x}{a} \sin \frac{n\pi y}{b} \qquad (2.21)$$

in which case, substitution into (2.20) produces

$$K \left[ \frac{m^2}{a^2} + \frac{n^2}{b^2} \right]^2 = \frac{h}{\pi^2} \left[ \lambda \frac{m^2}{a^2} + \xi \frac{n^2}{b^2} \right] \qquad (2.22)$$

If we define the buckling coefficient $\tilde{K}_x$ (which is related to the compressive stress in the $x$-direction) by

$$\tilde{K}_x = (\lambda b^2 h)/\pi^2 K \qquad (2.23)$$

then (2.21) may be put in the form

$$\tilde{K}_x = \left[ \frac{m^2}{\phi^2} + n^2 \right]^2 \bigg/ \left( \frac{m^2}{\phi^2} + \frac{\xi n^2}{\lambda} \right) \qquad (2.24)$$

It may be shown that $(\tilde{K}_x)_{min}$ occurs when $m = n = 1$ and that

$$(\tilde{K}_x)_{min} = 4 \left( 1 - \frac{\xi}{\lambda} \right), \quad 0 < \frac{\xi}{\lambda} < \frac{1}{2} \qquad (2.25)$$

when

$$\phi \equiv \phi_{min} = \left( 1 - \frac{2\xi}{\lambda} \right)^{-1/2}, \quad 0 < \frac{\xi}{\lambda} < \frac{1}{2} \qquad (2.26)$$

For values of $\lambda$ and $\xi$ such that $\left( \frac{\xi}{\lambda} \right) > \frac{1}{2}$, the corresponding values of $\phi_{min}$ are imaginary. The graph of $\tilde{K}_x$ versus the aspect ratio $\phi$ for the plate is depicted in Fig 2.10; from this figure we may deduce that the graph of $\tilde{K}_x$, as a function of $\phi$, is asymptotic to the horizontal line $\tilde{K}_x = \lambda/\xi$. Also, for the case in which the plate is square (i.e., $\phi = 1$) and is subjected to uniform compression on all sides ($\lambda = \xi$) we have $\tilde{K}_x = 2$, which is one-half of the value of $\tilde{K}$ for a simply supported square plate which is compressed in only one direction. From (2.25) and (2.26), it follows that for $\xi < 0$ (i.e. if a tensile force acts on the edges $y = 0$, $y = b$, $0 \le x \le a$) $\phi_{min}$ is not imaginary and $(\tilde{K}_x)_{min}$ may be larger than 4; this implies that the presence of a tensile stress along the edges $y = 0$, $y = b$ can increase the initial buckling load for a simply supported plate which is acted upon by compressive stresses applied to the edges at $x = 0$ and $x = a$. An alternative rendering of the results for this problem may be found in the work of Przemieniecki [78]; these results are summarized in Fig. 2.11.

**(b) The Edges at $x = 0$, $x = a$, $0 \leq y \leq b$, and $y = 0$, $y = b$, $0 \leq x \leq a$, are Clamped.**
In this subcase, we are still dealing with (2.20), but this equation is now subjected to the boundary conditions in (1.130a), (I.130b). An approximate solution for this situation was derived by Timoshenko [64] who assumed that $\phi \approx 1$(i.e., the plate does not differ much from a square) and that $\lambda \approx \xi$; the deflection of the buckled plate is sought in the form

$$w(x, y) = \frac{1}{4}A \left(1 - \cos \frac{2\pi x}{a}\right)\left(1 - \cos \frac{2\pi y}{b}\right) \qquad (2.27)$$

Using the hypothesized form of the eigenfunction given by (2.27), and equating the work done by the applied compressive loads to the strain energy of bending, Timoshenko [64] shows that

$$\lambda + \phi^2 \xi = \frac{\pi^2 K}{b^2 h}\left[4\phi^{-2} + 4\phi^2 + \frac{8}{3}\right] \qquad (2.28)$$

so that, by virtue of the definition (2.23),

$$\tilde{K}_x = \frac{4\phi^{-2} + 4\phi^2 + 8/3}{1 + \left(\dfrac{\xi}{\lambda}\right)\phi^2} \qquad (2.29)$$

For the case $\phi = 1$, $\xi/\lambda = 1$ (i.e. a square plate under uniform compression) $\tilde{K}_x = 5.33$.

**(c) The Edges at $x = 0$, $x = a$, $0 \leq y \leq b$, are Simply Supported and the Edges at $y = 0$, $y = b$, $0 \leq x \leq a$ are Clamped.**
This subcase has also been treated by Przemieniecki in [78] ; the boundary conditions are the first pair of relations in (1.130a), and the first pair of relations in (1.131), along $x = 0$, $x = a$, for $0 \leq y \leq b$, as well as the second set of relations in (1.130a), and the second set of relations in (I.130b), along $y = 0$, $y = b$, for $0 \leq x \leq a$. The results in [78] are displayed graphically in Fig. 2.12 in the form of curves which relate $\lambda/\lambda_{cr}$

with $\phi$ for different values of $\xi/\xi_{cr}$, where $\lambda_{cr}$ is the (critical) buckling load for a simply supported plate under unidirectional compression, i.e.

$$\lambda_{cr} = \frac{4\pi^2 E}{12(1-\nu^2)}\left(\frac{h}{b}\right)^2 \tag{2.30}$$

**(iv) Compressed in the $x$-Direction by Linearly Varying Edge Forces**

As noted in [32], "there is no exact analytical solution to any plate stability problem when the edge loading is non-uniform"; in these cases, one uses an assumed form for the deflection (eigen- or initial buckling mode) and then employs an energy method. As depicted in Fig. 2.13, the stress $\sigma_{xx}$ at a distance $y$ from the origin is related to the maximum applied stress at $y = 0$, i.e. $\lambda$, by $\sigma_{xx} = \lambda\left(1 - \frac{c}{b}y\right)$ with $c \geq 0$.

**(a)  All Four Edges are Simply Supported**
Taking

$$w = \sin\frac{m\pi x}{a}\sum_{n=1}^{\infty} A_n \sin\frac{n\pi y}{b}, \tag{2.31}$$

under the hypothesis that the buckled form of the plate consists of $m$ half-waves in the $x$-direction, we compute the strain energy of bending to be

$$U = \frac{Kab\pi^2}{8}\sum_{n=1}^{\infty} A_n^2\left(\frac{1}{a^2}+\frac{n^2}{b^2}\right)^2 \tag{2.32}$$

while, for the work done by the compressive edge forces, we have

$$T = \frac{1}{2}\lambda\left(1 - \frac{c}{b}y\right)h\int_o^a\int_0^b\left(\frac{\partial w}{\partial x}\right)^2 dx dy \tag{2.33}$$

In view of (2.31), this becomes

$$T = \frac{\pi^2\lambda bh}{8a}\sum_{n=1}^{\infty} A_n$$
$$-\frac{\pi^2\lambda ch}{4ab}\cdot\left[\frac{b^2}{4}\sum_{n=1}^{\infty} A_n^2 - \frac{8b^2}{\pi^2}\sum_{n=1}^{\infty}\sum_{q=1}^{\infty}\frac{A_n A_q \cdot nq}{(n^2-q^2)^2}\right] \tag{2.34}$$

where only values for $q$ are summed over such that $n + q$ is odd. Equating $U$ and $T$ then produces an expression for $\lambda_{cr}$, which contains the coefficients $A_n$, and these must now be chosen so as to minimize the value of $\lambda_{cr}$; standard max–min calculations yield the following system of equations

$$A_n \left[ \left( 1 + n^2 \phi^2 \right)^2 - \tilde{K} \phi^2 \left( 1 - \frac{c}{2} \right) \right]$$

$$- \frac{8c}{2} \phi^2 \tilde{K} \sum_{q=1}^{\infty} \frac{A_q \cdot nq}{(n^2 - q^2)^2} = 0 \tag{2.35}$$

For $n = 2$, we have the associated characteristic equation

$$\left\| \begin{array}{cc} \left[ \left( 1 + \phi^2 \right)^2 - \tilde{K} \phi^2 \left( 1 - \frac{c}{2} \right) \right] & -\frac{16c}{9\pi^2} \phi^2 \tilde{K} \\ -\frac{16c}{9\pi^2} \phi^2 \tilde{K} & \left[ \left( 1 + 4\phi^2 \right)^2 - \tilde{K} \phi^2 \left( 1 - \frac{c}{2} \right) \right] \end{array} \right\| = 0 \tag{2.36}$$

whose solutions for $m = 1$, and various values of $c$, are depicted in Fig. 2.14; the buckling coefficients $\tilde{K}_m$, for $m > 1$, can be obtained by multiplying the abscissae of the $\tilde{K}$ versus $\phi$ curves in Fig. 2.14 by $m$. The solution described above was first elaborated by Timoshenko [64].

**(b) The Edges at $x = 0$, $x = a$, $0 \leq y \leq b$, are Simply Supported and the Edges at $y = 0$, $y = b$, $0 \leq x \leq a$, are Clamped.**
The analytical form for the boundary conditions in this case have already been described; according to [32] this problem was first solved by Nölke [79] who took the transverse deflection in the form

$$f(y) = \sum_{n=1}^{\infty} \left\{ A_n \left( \cos \frac{p_n y}{b} + \cosh \frac{p_n y}{b} \right) \right.$$

$$+ B_n \left( \cos \frac{p_n y}{b} - \cosh \frac{p_n y}{b} \right) + C_n \left( \sin \frac{p_n y}{b} + \sinh \frac{p_n y}{b} \right) \tag{2.37}$$

$$\left. + D_n \left( \sin \frac{p_n y}{b} - \sinh \frac{p_n y}{b} \right) \right\}$$

with $p_n$ a root of $\cos p_n \cosh p_n = 1$; the overall form for the initial buckling mode is still that of (2.3). Nölke's results [79] are depicted in Fig. 2.15 for those plane equilibrium stress distributions corresponding to $c$ values of $0.5, 1.0, 1.5$, and $2.0$.

**(v) Rectangular Isotropic Plates Subjected to Shearing Loads**

The situation is depicted in Fig. 2.16, and the corresponding partial differential equation which governs the initial-buckling problem is

$$\frac{\partial^4 w}{\partial x^4} + 2\frac{\partial^4 w}{\partial x^2 \partial y^2} + \frac{\partial^4 w}{\partial y^4} = -\frac{2\zeta h}{K}\frac{\partial^2 w}{\partial x \partial y} \qquad (2.38)$$

where $\zeta = \sigma_{xy}|_{y=0,b}$.

(a)  All Four Edges are Simply Supported.
The analysis, which is due to Timoshenko [64], is similar to that employed in the case of compression (in the $x$-direction) with linearly varying edge forces. Since one cannot make, in this case, assumptions about the number of half-waves in the $x$-direction, the initial buckling mode is sought in the form

$$w = \sum_{m=1}^{\infty}\sum_{n=1}^{\infty} A_{mn} \sin\frac{m\pi x}{a} \sin\frac{n\pi y}{b} \qquad (2.39)$$

The strain energy of bending is

$$U = \frac{Kab\pi^4}{8} \sum_{m=1}^{\infty}\sum_{n=1}^{\infty} A_{mn}^2 \left(\frac{m^2}{a^2} + \frac{n^2}{b^2}\right) \qquad (2.40a)$$

while the work done by the edge forces is

$$T = \zeta \cdot h \int_0^a \int_0^b \frac{\partial w}{\partial x}\frac{\partial w}{\partial y} dx dy \qquad (2.40b)$$

In view of (2.39),

$$T = -4\zeta \cdot h \sum_{m=1}^{\infty}\sum_{n=1}^{\infty}\sum_{p=1}^{\infty}\sum_{q=1}^{\infty} A_{mn} A_{pq} \frac{mnpq}{(m^2 - p^2)(q^2 - n^2)} \qquad (2.41)$$

where $m + p$, $q + n$ are odd numbers. Equating $U$ and $T$, we obtain a value of $\zeta_{cr}$ involving the coefficients $A_{mn}$ and $A_{pq}$, and choosing these coefficients so as to minimize $\zeta_{cr}$ leads to a system of equations of the form

$$A_{mn}(m^2 + n^2\phi^2)^2 + \frac{32\phi^3 \tilde{K}}{\pi^2} \sum_{p=1}^{\infty}\sum_{q=1}^{\infty} A_{pq} \frac{mnpq}{(m^2 - p^2)(n^2 - q^2)} = 0 \quad (2.42)$$

It is noted in [32], that the system (2.42) may be split into two subsystems of equations each of which can be solved separately; one subsystem contains constants $A_{mn}$ for which $(m+n)$ is odd, and represents antisymmetric buckling, while the other contains constants for which $(m+n)$ is even and represents symmetric buckling. Some of the solutions obtained by Stein and Neff [80] for this case are depicted in Fig. 2.17; according to this figure, the lowest value of $\tilde{K}$ for $.5 < \dfrac{1}{\phi} < 1.0$ is associated with one buckle (symmetric), and for $.3 < \dfrac{1}{\phi} < .5$ with two buckles (antisymmetric), while for $\dfrac{1}{\phi} \to 0$, $\tilde{K}$ approaches its value for an infinitely long plate. In analyzing the plate subjected to shearing forces along its edges, with the edges simply supported, we may interchange the roles of the sides of dimensions $a$ and $b$, respectively; thus, it makes no sense to consider values of $\dfrac{1}{\phi}$ in Fig. 2.17 which exceed one. A reasonable parabolic approximation to the graph in Fig. 2.17 is given by

$$\tilde{K} = 5.34 + 4\phi^{-2} \ (\phi > 1) \tag{2.43}$$

**(b)   All Four Edges are Clamped**
This case was analyzed by Budiansky and Connor [81] and their results are displayed in Fig. 2.18; the graph may be curve-fitted by the parabolic relation

$$\tilde{K} = 8.98 + 5.6\phi^{-2} \ (\phi > 1) \tag{2.44}$$

**(c)    The Edges at $x = 0$, $x = a$, $0 \leq y \leq b$ are Simply Supported and the Edges at $y = 0$, $y = b$, $o \leq x \leq a$, are Clamped.**
Assuming that the edges along $y = 0$, $y = b$ are the longer edges (i.e., $\phi > 1$), Cook and Rockey [82] produced the graph depicted in Fig. 2.19; on this graph we also include, for comparative reference, the $\tilde{K}$ versus $\phi$ graphs for the case where the edges $x = 0$, $x = a, 0 \leq y \leq b$, are clamped, and the edges $y = 0$, $y = b$, $0 \leq x \leq a$, are simply supported as well as the results for the subcases (a) and (b).

**(vi)    Rectangular Isotropic Plates Subject to a Combined Loading: Shear and Compression in the $x$-Direction**
We provide here one example of initial buckling under a combined loading, for which a comparative example is readily available in the orthotropic case, namely, shearing forces on all four edges of the plate, and compression in the $x$-direction, with all four edges simply supported; the partial differential equation governing initial buckling is

$(\zeta = \sigma_{xy}|_{y=0,b}, \ \lambda = \sigma_{xx}|_{x=0,a})$ :

$$K \left\{ \frac{\partial^4 w}{\partial x^4} + 2\frac{\partial^4 w}{\partial x^2 \partial y^2} + \frac{\partial^4 w}{\partial y^4} \right\}$$

$$+2\zeta h \frac{\partial^2 w}{\partial x^2 \partial y^2} + \lambda h \frac{\partial^2 w}{\partial x^2} = 0 \qquad (2.45)$$

and it is assumed that $\phi < 1$. The strain energy of bending is computed as in the case of pure shear, while the work done by the external forces is

$$T = \zeta \cdot h \int_0^a \int_0^b \frac{\partial w}{\partial x} \frac{\partial w}{\partial y} dx dy + \frac{\lambda \cdot h}{2} \int_0^a \int_0^b \left(\frac{\partial w}{\partial x}\right)^2 dx dy \qquad (2.46)$$

Upon using (2.39) we find that

$$T = -4\zeta \cdot h \sum_{m=1}^{\infty} \sum_{n=1}^{\infty} \sum_{p=1}^{\infty} \sum_{q=1}^{\infty} \frac{A_{mn} A_{pq} \cdot mnpq}{(m^2 - p^2)(q^2 - n^2)}$$

$$+\frac{\pi \lambda h b}{8a} \sum_{m=1}^{\infty} \sum_{n=1}^{\infty} m^2 A_{mn}^2 \qquad (2.47)$$

Equating $U$ and $T$ provides an expression involving $\tilde{K}_x$ and $\tilde{K}_{xy}$, the buckling coefficients in pure end compression and pure shear; choosing (as in the previously discussed cases) the coefficients $A_{mn}$ so as to minimize the buckling coefficients, leads to a system of homogeneous linear equations of the form

$$A_{mn} \left[ (m^2 + n^2 \phi^2)^2 - \tilde{K}_x m^2 \phi^2 \right]$$

$$+\frac{32\phi^3 \tilde{K}_{xy}}{\pi^2} \sum_{p=1}^{\infty} \sum_{q=1}^{\infty} A_{pq} \frac{mnpq}{(m^2 - p^2)(n^2 - q^2)} = 0 \qquad (2.48)$$

Again, two subgroups of equations representing, respectively, symmetric and antisymmetric buckling can be extracted from (2.48) and the associated characteristic equations delineated. For subgroups of ten equations each, for the symmetric and antisymmetric cases, Batdorf and Stein [83] produced the graph depicted in Fig. 2.20 in which $\lambda/\lambda_{cr}$ is plotted against $\zeta/\zeta_{cr}$. for various values of $\phi$ in the range $0 \leq \phi \leq 1$.

## 2.1.2 Rectilinear Orthotropic Symmetry: Initial Buckling

For several cases analogous to the ones considered in the isotropic case, we now consider the results which are available for initial buckling of a rectilinearly orthotropic rectangular plate which, again, occupies the region in the $x, y$ plane defined by $0 \leq x \leq a$, $0 \leq y \leq b$. From (1.68), the first of the von Karman equations in this case reads

$$D_{11}\frac{\partial^4 w}{\partial x^4} + 2(D_{12} + 2D_{66})\frac{\partial^4 w}{\partial x^2 \partial y^2} + D_{22}\frac{\partial^4 w}{\partial y^4} = [\Phi, w] \qquad (2.49)$$

where, by (1.63)

$$\begin{cases} D_{11} = \dfrac{E_1}{1 - \nu_{12}\nu_{21}} \cdot \dfrac{h^3}{12} \\[3mm] D_{22} = \dfrac{E_2}{1 - \nu_{12}\nu_{21}} \cdot \dfrac{h^3}{12} \end{cases}$$

and

$$\begin{cases} D_{12} = \dfrac{E_2\nu_{21}}{1 - \nu_{12}\nu_{21}} \cdot \dfrac{h^3}{12} = \dfrac{E_1\nu_{12}}{1 - \nu_{12}\nu_{21}} \cdot \dfrac{h^3}{12} = D_{21} \\[3mm] D_{66} = G_{12} \cdot \dfrac{h^3}{12} \end{cases}$$

The partial differential equation governing the eigenvalue–eigenfunction problem, for initial buckling of the plate, is then, by virtue of our discussion in Chapter 1, given by (2.49) with $\Phi$, the total Airy function, replaced by $\bar{\Phi}$; we recall that $\bar{\Phi}$ is the solution of (1.69), for $w \equiv 0$, subject to the appropriate edge loading conditions, i.e., $\bar{\Phi}$ satisfies

$$\frac{1}{E_1 h}\frac{\partial^4 \bar{\Phi}}{\partial y^4} + \frac{1}{h}\left(\frac{1}{G_{12}} - \frac{\partial \nu_{12}}{E_1}\right)\frac{\partial^4 \bar{\Phi}}{\partial x^2 \partial y^2} + \frac{1}{E_2 h}\frac{\partial^4 \bar{\Phi}}{\partial x^4} = 0 \qquad (2.50)$$

Equations (2.49), with $\Phi = \bar{\Phi}$, and (2.50), apply in the domain $0 < x < a$, $0 < y < b$.

### (i) Compression in the $x$-Direction with the Loaded Edges Simply Supported

Just as in subcase (i), for the isotropic situation, Fig. 1.12 applies with $\sigma_{xx} = -\lambda$, $\lambda > 0$, along the edges at $x = 0$ and $x = a$; also, along these edges, for $0 \leq y \leq b$, the first pair of boundary conditions in (1.146) hold together with $w(0, y) = w(a, y) = 0$, $0 \le y \le b$. The relevant eigenvalue problem is governed by

$$D_{11}\frac{\partial^4 w}{\partial x^4} + 2(D_{12} + 2D_{66})\frac{\partial^4 w}{\partial x^2 \partial y^2} + D_{22}\frac{\partial^4 w}{\partial y^4} = -\lambda h \frac{\partial^2 w}{\partial x^2} \quad (2.51)$$

We note that the first two boundary conditions in (1.146) are equivalent to the first pair of conditions in (1.148).

**(a)  The Edges $y = 0$, $y = b$, $0 \le x \le a$ are Simply Supported.**
In this subcase $w = 0$ along $y = 0$, $y = b$, $0 \le x \le a$, and the second pair of boundary conditions in (1.148) apply as well along these edges. Following the analysis in Lekhnitskii, [51] or [52], we note that the boundary conditions are satisfied by trial eigenfunctions of the form

$$w(x, y) = A_{mn} \sin\frac{m\pi x}{a} \sin\frac{n\pi y}{b} \quad (2.52)$$

for arbitrary constants $A_{mn}$, with the $m$ and $n$ integers. Substituting from (2.52) into (2.51) yields

$$A_{mn}\left\{\left[D_{11}\left(\frac{m}{a}\right)^4 + 2(D_{12} + 2D_{66})\left(\frac{mn}{ab}\right)^2 \right.\right.$$
$$\left.\left. +D_{22}\left(\frac{n}{b}\right)^4\right] - \lambda h\pi^2\left(\frac{m}{a}\right)^2\right\} = 0 \quad (2.53)$$

in which case, we have eigenvalues

$$\lambda \equiv \lambda_{mn} = \frac{\pi^2\sqrt{D_{11}D_{22}}}{b^2 h}\left[\sqrt{\frac{D_{11}}{D_{22}}}\left(\frac{m}{\phi}\right)^2\right.$$
$$\left. +\frac{2(D_{12} + 2D_{66})}{\sqrt{D_{11}D_{22}}}n^2 + \sqrt{\frac{D_{22}}{D_{11}}}\left(\frac{\phi}{m}\right)^2 n^4\right] \quad (2.54)$$

The critical value of $\lambda$ in (2.54) is the smallest of the $\lambda_{mn}$ and it is obtained for $n = 1$, which corresponds to one-half sine wave in the $y$ direction; thus, we need to look at the values of

$$\lambda \equiv \lambda_{m1} = \frac{\pi^2\sqrt{D_{11}D_{22}}}{b^2 h}\left[\sqrt{\frac{D_{22}}{D_{11}}}\left(\frac{m}{\phi}\right)^2\right.$$
$$\left. +\frac{2(D_{12} + 2D_{66})}{\sqrt{D_{11}D_{22}}} + \sqrt{\frac{D_{22}}{D_{11}}}\left(\frac{\phi}{m}\right)^2\right] \quad (2.55)$$

in order to determine $\lambda_{cr}$. The various cases which present themselves are as follows:

1) If the aspect ratio satisfies $\phi = m' \sqrt[4]{\dfrac{D_{11}}{D_{22}}}$, for $m'$ an integer, then $m = m'$ provides the minimal $\lambda_{m1}$ in (2.55), in which case $\lambda_{cr}$ is given by

$$\lambda_{cr} = \frac{2\pi^2 \sqrt{D_{11}D_{22}}}{b^2 h}\left(1 + \frac{[D_{12} + 2D_{66}]}{\sqrt{D_{11}D_{22}}}\right) \qquad (2.56)$$

Following the point of view taken in [51], another way to look at this is as follows: for $m = 1$ in (2.55) we have

$$\lambda_{11} = \frac{\pi^2}{b^2 h}\left[\frac{D_{11}}{\phi^2} + 2[D_{12} + 2D_{66}] + D_{22}\phi^2\right] \qquad (2.57)$$

in which case, the smallest value of $\lambda_{11}$ occurs for the aspect ratio $\phi = \sqrt{\dfrac{D_{11}}{D_{22}}}$ and this smallest value is just given by (2.56). The relation (2.56) can be expressed in the form

$$\lambda_{cr} = \frac{\tilde{K}_{min}\pi^2 \sqrt{D_{11}D_{22}}}{b^2 h} \qquad (2.58)$$

where

$$\tilde{K}_{min} = 2\left(1 + \frac{[D_{12} + 2D_{66}]}{\sqrt{D_{11}D_{22}}}\right) \qquad (2.59)$$

2) If

$$\phi = \sqrt{m(m+1)}\sqrt[4]{\frac{D_{11}}{D_{22}}} \equiv \phi_m \qquad (2.60)$$

for $m$ any integer, then at the same critical load there are two possible buckling modes, namely,

$$w = A_{m1}\sin\frac{m\pi x}{a}\sin\frac{\pi y}{b} \qquad (2.61)$$

and

$$w = A_{m+1,1}\sin\frac{(m+1)\pi x}{a}\sin\frac{\pi y}{b} \qquad (2.62)$$

and the $\phi_m$ represent, therefore, the aspect ratios where a transition from buckling in the mode (2.61) to buckling in the mode (2.62) occurs.

From (2.60) we deduce directly that, at $\lambda = \lambda_{cr}$, if

$$
\begin{cases}
0 < \phi < 1.41 \sqrt[4]{\dfrac{D_{11}}{D_{22}}}, & \text{then } m = 1 \\[2ex]
1.41 \sqrt[4]{\dfrac{D_{11}}{D_{22}}} < \phi < 2.45 \sqrt[4]{\dfrac{D_{11}}{D_{22}}}, & \text{then } m = 2, \\[2ex]
2.45 \sqrt[4]{\dfrac{D_{11}}{D_{22}}} < \phi < 3.46 \sqrt[4]{\dfrac{D_{11}}{D_{22}}}, & \text{then } m = 3
\end{cases}
\tag{2.63}
$$

and so forth.

3) For a given aspect ratio $\phi$, therefore, (2.60) can be used to determine the value of $m$ which holds at buckling; the use of this integer $m$ in (2.55) then produces the corresponding critical load $\lambda_{cr}$ for the given $\phi$.
In general, by virtue of (2.59), we may write that

$$
\lambda_{cr} = \frac{\pi^2 \sqrt{D_{11} \cdot D_{22}}}{b^2 h} k
$$

where

$$
k \equiv k_m = \sqrt{\frac{D_{11}}{D_{22}}} \left(\frac{m}{\phi}\right)^2 + \frac{2(D_{12} + 2D_{66})}{\sqrt{D_{11}D_{22}}} + \sqrt{\frac{D_{22}}{D_{11}}} \left(\frac{\phi}{m}\right)^2
\tag{2.64}
$$

In Fig. 2.21 (and Table 2.1) we show a typical graph of $k$ against $\phi$ for the case of compression along the external fibers while in Fig. 2.22 (and Table 2.2) we show the analogous results for compression across the external fibers; in the first case, $D_{11} > D_{22}$, while in the latter case, $D_{11} < D_{22}$. The figures referenced above clearly show that the number of half-waves in the $x$-direction is much larger for a given aspect ratio $\phi$ when compression is across the external fibers than when it is along them. In Fig. 2.23 we show the graph of $k$ against $\bar{\beta} = \phi \cdot \sqrt[4]{D_{22}/D_{11}}$ (which [43] defines to be an "effective" aspect ratio); the parameter $\alpha$, which varies along each family of curves, is

$$
\alpha = \frac{D_{12} + 2D_{66}}{\sqrt{D_{11}D_{22}}}
\tag{2.65}
$$

**(b) The Edges $y = 0$, $y = b$, $0 \le x \le a$, are Clamped**
In this case, along $y = 0$, $y = b, 0 \le x \le a$, the relevant boundary conditions consist of the second pair or conditions in (1.130a) and the second pair of conditions in (1.130b). The trial eigenfunction

$$
w = f(y) \sin \frac{m \pi x}{a}
\tag{2.66}
$$

satisfies the conditions of simple support along the loaded edges at $x = 0$ and $x = a$, for $0 \leq y \leq b$. If we substitute (2.66) into (2.51), we find that $f(y)$ must satisfy

$$
f'''(y) - 2 \left(\frac{m\pi}{a}\right)^2 [D_{12} + 2D_{66}] f''(y)
$$

$$
+ \left[ D_{11} \left(\frac{m\pi}{a}\right)^4 - \lambda h \left(\frac{m\pi}{a}\right)^2 \right] f(y) = 0
$$

(2.67)

Extracting the characteristic algebraic equation from (2.67), it is easily determined that the buckling mode must have the form

$$
w = (A \cosh k_1 y + B \sinh k_1 y + C \cos k_2 y
$$

$$
+ D \sin k_2 y) \sin \frac{m\pi x}{a}
$$

(2.68)

where the $k_1, k_2$, are determined by the roots of the characteristic equation associated with (2.67), i.e., by the roots of

$$
D_{22} r^4 - 2 \left(\frac{m\pi}{a}\right)^2 [D_{12} + 2D_{66}] r^2 + D_{11} \left(\frac{m\pi}{a}\right)^4 - \lambda h \left(\frac{m\pi}{a}\right)^2 = 0 \quad (2.69)
$$

while $A, B, C, D$ are determined by the conditions on $w$ along the edges $y = 0$, $y = b$, $0 \leq x \leq a$, i.e., by the condition that the plate is clamped along these edges; the equations generated in this manner for $A, \ldots, D$ then yield, in the usual way, the following transcendental algebraic equation for the values of $\lambda$ at which initial buckling might occur:

$$
\tanh k_1 b \tan k_2 b = \frac{2k_1 k_2}{k_1^2 - k_2^2} \left( 1 - \frac{1}{\cosh k_1 b \cos k_2 b} \right) \qquad (2.70)
$$

where, of course, the $k_i$, $i = 1, 2$, are, in view of (2.69), functions of $D_{11}, D_{12}, D_{22}, D_{66}, m, h$, and $\lambda$. For the compressed plate problem discussed in this subcase, Lekhnitskii [51] provides an approximate solution by choosing, in (2.66), an $f$ of the form

$$
f(y) = A_n \left( 1 - \cos \frac{2n\pi y}{b} \right), \qquad (2.71)
$$

and equating the work of the compressive edge forces with the potential energy acquired during bending; it is determined that $\lambda$ must satisfy

$$\lambda \equiv \lambda_{mn} = \frac{\pi^2\sqrt{D_{11}D_{22}}}{b^2h}\left[\sqrt{\frac{D_{11}}{D_{22}}}\left(\frac{m}{\phi}\right)^2\right.$$

$$+\frac{8}{3}n^2\frac{[D_{12}+2D_{66}]}{\sqrt{D_{11}D_{22}}}+\frac{16}{3}n^4\sqrt{\frac{D_{22}}{D_{11}}}\left(\frac{\phi}{m}\right)^2\right] \tag{2.72}$$

As was the case with (2.54), the smallest value of $\lambda$ occurs at $n = 1$; therefore, $\lambda_{cr}$ is to be determined from

$$\lambda = \lambda_{m1} = \frac{\pi^2\sqrt{D_{11}D_{22}}}{b^2h}\left[\sqrt{\frac{D_{11}}{D_{22}}}\left(\frac{m}{\phi}\right)^2+\frac{8}{3}\frac{[D_{12}+2D_{66}]}{\sqrt{D_{11}D_{22}}}\right.$$

$$+\frac{16}{3}\sqrt{\frac{D_{22}}{D_{11}}}\left(\frac{\phi}{m}\right)^2\right] \tag{2.73}$$

The results for (2.73) are now quite analogous to those for the case of compression in the $x$-direction with four simply supported edges:

1) If the aspect ratio $\phi$ satisfies $\phi = .658m'\sqrt[4]{\dfrac{D_{11}}{D_{22}}}$, for some integer $m'$, then at criticality, $m = m'$ and

$$\lambda_{cr} = \frac{8\pi^2\sqrt{D_{11}D_{22}}}{3b^2h}\left(\sqrt{3}+\frac{[D_{12}+2D_{66}]}{\sqrt{D_{11}D_{22}}}\right) \tag{2.74}$$

which may be compared with (2.56)

2) If

$$\phi = .658\sqrt{m(m+1)}\sqrt[4]{\frac{D_{11}}{D_{22}}},$$

with $m$ an integer, then for the same value of $\lambda_{cr}$, buckling modes with either $m$ or $m+1$ half-sine waves may occur in the $x$-direction:

$$w = A_1\left(1-\cos\frac{2\pi y}{b}\right)\sin\frac{m\pi x}{a} \tag{2.75a}$$

or

$$w = A_1\left(1-\cos\frac{2\pi y}{b}\right)\sin\frac{(m+1)\pi x}{a} \tag{2.75b}$$

3) For a value of the aspect ratio $\phi$ which does not satisfy either the conditions in 1 or 2, above, the value of $m$ may be determined by looking at the transitions governed by the relation in 2 (above) i.e., if

$$
\begin{cases}
0 < \phi < .931 \sqrt[4]{\dfrac{D_{11}}{D_{22}}}, & \text{then } m = 1 \\[2ex]
.931 \sqrt[4]{\dfrac{D_{11}}{D_{22}}} < \phi < 1.61 \sqrt[4]{\dfrac{D_{11}}{D_{22}}}, & \text{then } m = 2 \\[2ex]
1.61 \sqrt[4]{\dfrac{D_{11}}{D_{22}}} < \phi < 2.28 \sqrt[4]{\dfrac{D_{11}}{D_{22}}}, & \text{then } m = 3
\end{cases}
\tag{2.76}
$$

and so forth; once $m$ has been determined, the corresponding value of $\lambda_{cr}$ is then given by (2.73). In Fig. 2.23 we also show the graph of

$$
k \equiv k_m = \sqrt{\frac{D_{11}}{D_{22}}} \left(\frac{m}{\phi}\right)^2 + \frac{8}{3} \frac{[D_{12} + 2D_{66}]}{\sqrt{D_{11}D_{22}}}
$$
$$
+ \frac{16}{3} \sqrt{\frac{D_{22}}{D_{11}}} \left(\frac{\phi}{m}\right)^2,
\tag{2.77}
$$

against $\bar{\beta} = \phi \sqrt[4]{D_{22}/D_{11}}$, for various values of $\alpha$, as defined by (2.65), when the loaded ends of the plate at $x = 0$, $x = a$ are simply supported and the ends at $y = 0$, $y = b$ are clamped.

**(ii) Compression in the $x$-Direction with the Loaded Edges Clamped.**
The relevant differential equation is still (2.51); the boundary conditions along $x = 0$ and $x = a$, $0 \le y \le b$, are given by the first pair of conditions in each of (1.130a,b).

**(a) The Edges $y = 0$, $y = b$, $0 \le x \le a$ are Simply Supported.**
In this subcase $w = 0$ along $y = 0, y = b, 0 \le x \le a$, and the second pair of conditions in (1.148) hold as well. Analysis, similar to that presented in the case where the loaded edges were simply supported, again produces a relation of the form

$$
\lambda_{cr} = k \frac{\pi^2}{b^2 h} \sqrt{D_{11} D_{22}}
\tag{2.78}
$$

for an appropriately defined buckling coefficient $k = k_m$. The graph in Fig. 2.24 depicts $k$ vs. $\beta$, again for varying values of the parameter $\alpha$.

**(b) The Edges $y = 0$, $y = b$, $0 \le x \le a$, Are Clamped**

In this subcase, we have the second pair of conditions, in each of (1.130a) and (1.130b), holding along $y = 0$ and $y = b$, for $0 \le x \le a$. A relation of the type (2.78) again applies with the results displayed on the graph in Fig. 2.24.

**(iii) Compression in both the $x$ and $y$ Directions** The situation here corresponds to that depicted in Fig. 2.9; instead of equation (2.20), which governs initial buckling for this loading condition, in the isotropic case, we have ($\lambda = -\sigma_{xx}|_{x=0,a}, \xi = -\sigma_{yy}|_{y=0,b}$):

$$
D_{11}\frac{\partial^4 w}{\partial x^4} + 2[D_{12} + 2D_{66}]\frac{\partial^4 w}{\partial x^2 \partial y^2} + D_{22}\frac{\partial^4 w}{\partial y^4}
$$

$$
+ h\left(\lambda \frac{\partial^2 w}{\partial x^2} + \zeta \frac{\partial^2 w}{\partial y^2}\right) = 0
$$

(2.79)

We consider only the case where the edges of the plate are simply supported on all four sides so that $w = 0$ for $x = 0$, $x = a$, $0 \le y \le b$, $w = 0$, for $y = 0$, $y = b$, $0 \le x \le a$, and all four conditions in (1.148) apply as well. Because of the condition of simply supported edges, we again look for eigenfunctions of the form (2.52), in which case, substitution into (2.79) yields

$$
\lambda\left(\frac{m}{a}\right)^2 + \xi\left(\frac{n}{b}\right)^2 = \frac{\pi^2}{h}\left[D_{11}\left(\frac{m}{a}\right)^4 + \right.
$$

$$
\left. 2[D_{12} + 2D_{66}]\left(\frac{mn}{ab}\right)^2 + D_{22}\left(\frac{n}{b}\right)^4\right]
$$

(2.80)

Two subcases of (2.79) naturally suggest themselves which correspond to $\xi = \eta\lambda$, $\eta = $ const., i.e., to $\sigma_{yy} = \eta\sigma_{xx}$ along the edges of the plate (with $\eta < 0$, if $\sigma_{yy}$, along the edges at $y = 0$, $y = b$, is tensile). In this case, $\lambda_{cr}$ is to be determined from the relation

$$
\lambda \equiv \lambda_{mn} = \frac{\pi^2\sqrt{D_{11}D_{22}}}{b^2 h}\left\{\left[\sqrt{\frac{D_{11}}{D_{22}}}\left(\frac{m}{\phi}\right)^2\right.\right.
$$

$$
+ \frac{2[D_{12} + 2D_{66}]}{\sqrt{D_{11}D_{22}}}\cdot n^2
$$

(2.81)

$$
\left.\left. + \sqrt{\frac{D_{22}}{D_{11}}}\left(\frac{\phi}{m}\right)^2 n^4\right] \middle/ 1 + \eta\left(\frac{\phi}{m}\right)^2 n^2\right\}
$$

and the problem again consists of seeking values of the integers $m$ and $n$ which give the smallest value of $\lambda$ and, thus, $\lambda_{cr}$; numerical analysis yields the following [51] :

(a) The Plate is Square ($\phi = 1$), and Under Uniform Compression on all Sides, i.e., $\eta = 1$ so that $\lambda = \xi$.

In this case, we have $m = 1$, $n = 2$ and, (approximately),

$$\lambda_{cr} = 2.23 \frac{\pi^2 \sqrt{D_{11}D_{22}}}{b^2 h} \tag{2.82}$$

when $D_{11} > D_{22}$ ($x$-axis directed along the external fibers). The plate, upon initial buckling, exhibits one-half sine waves in the direction of the $y$-axis.

(b) The Plate is Square ($\phi = 1$), is Compressed along Two Edges, and is under Tension Along the Other Two, i.e. $\eta = -1$ so that $\lambda = -\xi$.

In this case, for compression in the $x$-direction (along the external fibers) and tension in the $y$-direction, we have, $m = 2$, $n = 1$, and, (approximately),

$$\lambda_{cr} = 19.67 \frac{\pi^2 \sqrt{D_{11}D_{22}}}{b^2 h} \tag{2.83}$$

while for compression in the $y$-direction (across the fibers), and tension in the $x$-direction, $m = 2$, $n = 1$, and, (approximately),

$$\lambda_{cr} = 3.72 \frac{\pi^2 \sqrt{D_{11}D_{22}}}{b^2 h} \tag{2.84}$$

Thus, the plate is more stable in the first case, represented by (2.83), as opposed to the second case, represented by (2.84) since the critical $\lambda$ in (2.83) is more than five times the size of the critical $\lambda$ in (2.84). Other subcases of this particular problem which have been treated in Lekhnitskii [51] include the following:

(c) The Plate is Simply Supported along all Four Edges, with a Compressive Stress $\sigma_{xx} = -\lambda$ Uniformly Distributed along the Edges $x = 0$, $x = a$, $0 \le y \le b$, and a Constant Tension given by $\sigma_{yy} = +\bar{\xi}$ Uniformly Distributed along the Edges $y = 0$, $y = b$, $0 \le x \le a$.

Setting $\xi = -\bar{\xi}$ in (2.80), we find that

$$
\lambda = \frac{\pi^2 \sqrt{D_{11} D_{22}}}{b^2 h} \left[ \sqrt{\frac{D_{11}}{D_{22}}} \left( \frac{m}{\phi} \right)^2 + 2 \frac{[D_{12} + 2D_{66}]}{\sqrt{D_{11} D_{22}}} n^2 \right.
$$
$$
\left. \sqrt{\frac{D_{22}}{D_{11}}} \left( \frac{\phi}{m} \right)^2 n^4 + \frac{\bar{\xi} h b^2}{\pi^2 \sqrt{D_{11} D_{22}}} \left( \frac{\phi}{m} \right)^2 n^2 \right] \tag{2.85}
$$

In (2.85), the smallest value of $\lambda$ is obtained when $n = 1$, with one-half sine wave in the direction of the applied tension; an analysis of (2.85) produces the following results:

1) If the aspect ratio satisfies the condition

$$
\phi = m' \Big/ \sqrt[4]{\frac{D_{22}}{D_{11}} + \frac{\bar{\xi} b^2 h}{\pi^2 D_{11}}} \tag{2.86}
$$

for some integer $m'$, then

$$
\lambda_{cr} = \frac{2\pi^2 \sqrt{D_{11} D_{22}}}{b^2 h} \left( \sqrt{1 + \frac{\bar{\xi} b^2 h}{\pi^2 D_{22}}} + \frac{(D_{12} + 2D_{66})}{\sqrt{D_{11} D_{22}}} \right) \tag{2.87}
$$

From (2.86) it follows that an additional tension (load) along the edges $y = 0$, $y = b$ raises the critical load and, thus, improves the stability of the plate. We also note that, by virtue of (2.86), for a given aspect ratio $\phi$, $m'$ may be large if either $D_{11}$ is small or $D_{22}$ is large.

2) The aspect ratio at which a transition from $m$ half sine waves to $m + 1$ half sine waves (in the $x$-direction) takes place is

$$
\phi \equiv \phi_m = \frac{\sqrt{m(m+1)}}{\sqrt[4]{\frac{D_{22}}{D_{11}} + \frac{\bar{\xi} b^2 h}{\pi^2 D_{11}}}} \tag{2.88}
$$

**(iv) Compressed in the $x$-Direction by Linearly Varying Edge forces**

We assume that all four edges of the plate are simply supported; the applicable figure in this case is the same as the one for the corresponding case of an isotropic plate, i.e., Fig. 2.13, and as in the isotropic case we

have, along the edges $x = 0$, $x = a$, $0 \leq y \leq b$, $\sigma_{xx} = \lambda \left(1 - c\frac{y}{b}\right)$.
Using an energy method, Lekhnitskii [51] shows that

$$
\lambda = \frac{1}{\int_0^a \int_0^b \left(1 - \frac{c}{b}y\right) \left(\frac{\partial w}{\partial x}\right)^2 dxdy} \left[\int_0^a \int_0^b \left\{ D_{11} \left(\frac{\partial^2 w}{\partial x}\right)^2 \right. \right.
$$

$$
\left. \left. +2D_{11}\nu_{21}\frac{\partial^2 w}{\partial x^2}\frac{\partial^2 w}{\partial y^2} + D_{22} \left(\frac{\partial^2 w}{\partial y^2}\right)^2 + 2[D_{12} + 2D_{33}] \left(\frac{\partial^2 w}{\partial x \partial y}\right)^2 \right\} dxdy \right.
$$

$$
\tag{2.89}
$$

The most general form for an initial buckling mode satisfying the conditions of simple support on all four edges is

$$
w = \sum_{m=1}^{\infty} \sum_{n=1}^{\infty} A_{mn} \sin \frac{m\pi x}{a} \sin \frac{n\pi y}{b}
$$

a first approximation to which may be taken to be

$$
w \equiv w_m = A_{m1} \sin \frac{m\pi x}{a} \sin \frac{\pi y}{b} \tag{2.90}
$$

Using (2.90), we obtain from (2.89),

$$
\lambda_{cr} = \frac{\pi^2 \sqrt{D_{11} D_{22}}}{b^2 h(1 - c/2)} \left[ \sqrt{\frac{D_{11}}{D_{22}}} \left(\frac{m}{\phi}\right)^2 \right.
$$

$$
\left. + \frac{2(D_{12} + 2D_{66})}{\sqrt{D_{11} D_{22}}} + \sqrt{\frac{D_{22}}{D_{11}}} \left(\frac{\phi}{m}\right)^2 \right] \tag{2.91}
$$

We note that (2.91) is valid only for small $c$ and, in fact, loses its meaning for $c = 2$, which corresponds to the case of pure bending. The calculations may be found in Lekhnitskii [51] for the case in which a second approximation to $w$ of the form

$$
w = \left(A_{m1} \sin \frac{\pi y}{b} + A_{m2} \sin \frac{2\pi y}{b}\right) \sin \frac{m\pi x}{a} \tag{2.92}
$$

is employed in (2.89), but we will not pursue them in detail here; some of the results, however, are well worth noting, as follows: If we set

$$
\begin{cases} a_{m1} = \sqrt{\frac{D_{11}}{D_{22}}} \left(\frac{m}{\phi}\right)^2 + \frac{2(D_{12} + 2D_{66})}{\sqrt{D_{11} D_{22}}} + \sqrt{\frac{D_{22}}{D_{11}}} \left(\frac{\phi}{m}\right)^2 \\ \\ a_{m2} = \sqrt{\frac{D_{11}}{D_{22}}} \left(\frac{m}{\phi}\right)^2 + \frac{8(D_{12} + 2D_{66})}{\sqrt{D_{11} D_{22}}} + 16\sqrt{\frac{D_{22}}{D_{11}}} \left(\frac{\phi}{m}\right)^2 \end{cases} \tag{2.93}
$$

then the energy method yields

$$\lambda = \frac{\pi^2\sqrt{D_{11}D_{22}}}{b^2 h} \left[ \frac{\frac{1}{2}\left(1 - \frac{1}{2}c\right)(a_{m1} + a_{m2})}{\left(1 - \frac{1}{2}c\right)^2 - \left(\frac{16c}{9\pi^2}\right)^2} \right.$$

$$\left. \pm \frac{\sqrt{\frac{1}{4}\left(1 - \frac{1}{2}c\right)^2 (a_{m1} - a_{m2})^2 + \left(\frac{16c}{9\pi^2}\right)a_{m1}a_{m2}}}{\left(1 - \frac{1}{2}c\right)^2 - \left(\frac{16c}{9\pi^2}\right)^2} \right]$$

(2.94)

As in many of the earlier cases we have considered, the procedure now would be to look for the integer $m$, which corresponds to a minimum value for $\lambda$, as given by (2.94); then $\lambda = \lambda_m \equiv \lambda_{cr}$ for the second approximation based on (2.92). For the specific case in which $c = 2$, so that the applied edge forces are reduced to a moment $M$, thus producing pure bending, the critical moment is

$$M_{cr} = \frac{\pi^2\sqrt{D_{11}D_{22}}}{6} \cdot k \tag{2.95}$$

with

$$k = \frac{9\pi^2}{32}\sqrt{a_{m1}a_{m2}} \tag{2.96}$$

and the critical load is then

$$\lambda_{cr} = \frac{\pi^2\sqrt{D_{11}D_{22}}}{b^2 h} k \tag{2.97}$$

An analysis of $k$, as defined by (2.93), (2.96) shows that ([51])

1) If $\phi = \frac{\sqrt{2}}{2}m'\sqrt[4]{\frac{D_{11}}{D_{22}}}$, with $m'$ an integer, then the plate buck-

les in $m = m'$ half-sine waves in the $x$-direction and $k \approx 11.1$ $\left(1.25\right.$

$\left. +\frac{2[D_{12} + 2D_{66}]}{\sqrt{D_{11}D_{22}}}\right)$ in (2.97).

2) A transition from $m$ to $m + 1$ half-sine waves takes place when $\phi = \sqrt{\frac{m(m+1)}{2}}\sqrt[4]{\frac{D_{11}}{D_{22}}}$, $m = 1, 2, 3, \ldots$; as a consequence, the following

results on the number of half-sine waves in the $x$-direction, which occur upon initial buckling, hold:

$$\begin{cases} 0 < \phi < \sqrt[4]{\dfrac{D_{11}}{D_{22}}}, & \text{then } m = 1 \\[3ex] \sqrt[4]{\dfrac{D_{11}}{D_{22}}} < \phi < 1.73\sqrt[4]{\dfrac{D_{11}}{D_{22}}}, & \text{then } m = 2 \\[3ex] 1.73\sqrt[4]{\dfrac{D_{11}}{D_{22}}} < \phi < 2.45\sqrt[4]{\dfrac{D_{11}}{D_{22}}}, & \text{then } m = 3 \end{cases} \qquad (2.98)$$

and so forth. Using these results, for a given value of the aspect ratio $\phi$ we may determine the number $m$ of half-sine waves in the $x$-direction into which the plate buckles and then, from (2.93), (2.96), and (2.97), the corresponding critical load. It may again be demonstrated that, as was the case with compression of the simply supported, rectilinearly orthotropic plate, (along the edges $x = 0$, $x = a$, $0 \le y \le b$) if $D_{11} < D_{22}$ a large number of half-sine waves occur upon initial buckling as compared with the case in which $D_{11} > D_{22}$ (loading parallel to the external fibers of the plate as opposed to loading perpendicular to the external fibers). For plywood plates deflected by a pure bending moment, we indicate in Table 2.3 the values of $k$ and $m$ (relative to (2.96), (2.97)) for various values of the aspect ratio $\phi$, both in the case where $D_{11} > D_{22}$ and in the case where $D_{11} < D_{22}$.

## (v) Rectangular Orthotropic Plates Subjected to Shearing Loads

The situation is again depicted, as in the isotropic case, in Fig. 2.16. Taking, once again, $\zeta = \sigma_{xy}$ (along the four edges of the plate), the partial differential equation which mediates the initial buckling of the plate is

$$D_{11}\frac{\partial^4 w}{\partial x^4} + 2(D_{12} + 2D_{66})\frac{\partial^4 w}{\partial x^2 \partial y^2}$$
$$+ D_{22}\frac{\partial^4 w}{\partial y^2} + 2\zeta h\frac{\partial^2 w}{\partial x^2 \partial y^2} = 0 \qquad (2.99)$$

The principal directions of the plate are, of course, taken to be parallel to the edges of the plate, all of which are simply supported. Lekhnitskii [51] employs the energy method to obtain a solution of the initial buckling problem in this case. The work done by the edge forces is, once again, given by (2.40b), as in the isotropic case, and equating this to the potential energy of bending yields

$$\zeta = \int_0^a \int_0^b \Psi \, dx dy \bigg/ 2h \int_0^a \int_0^b \frac{\partial w}{\partial x} \frac{\partial w}{\partial y} dx dy \qquad (2.100)$$

with

$$\Psi(x,y) \equiv D_{11} \left( \frac{\partial^2 w}{\partial x^2} \right)^2 + 2D_{11}\nu_{21} \frac{\partial^2 w}{\partial x^2} \frac{\partial^2 w}{\partial y^2}$$

$$+ D_{22} \left( \frac{\partial^2 w}{\partial y^2} \right)^2 + 4(D_{12} + 2D_{12}) \left( \frac{\partial^2 w}{\partial x \partial y} \right)^2 \qquad (2.101)$$

The initial buckling mode is again sought in the form (2.38), which automatically satisfies the simple support conditions along the edges; the rest of the solution now proceeds as in the isotropic case. It is noted in [51] that, with the external fibers of the plate parallel to the long sides of the plate, good accuracy is achieved with five terms in the expression (2.38) for the initial buckling mode, i.e., with

$$w = \left( A_{11} \sin \frac{\pi x}{a} + A_{31} \sin \frac{3\pi x}{a} \right) \sin \frac{\pi y}{b}$$

$$+ A_{22} \sin \frac{2\pi x}{a} \sin \frac{2\pi y}{b} \qquad (2.102)$$

$$+ \left( A_{13} \sin \frac{\pi x}{a} + A_{33} \sin \frac{3\pi x}{a} \right) \sin \frac{3\pi y}{b},$$

if the plate is square ($\phi = 1$); other approximations (for the aspect ratios $\phi = 2$ and $\phi = 3$) are also reported in Lekhnitskii [51]. In each of the cases referred to above, the critical (shear) buckling load assumes the form

$$\zeta_{cr} = 10^4 \left( \frac{h}{b} \right)^2 k \qquad (2.103)$$

where the buckling coefficient $k$, which depends on $\phi$, is tabulated in Table 2.4.

### Remarks : Shearing of a Very Narrow or Infinite Orthotropic Strip

A special case of the problem treated above occurs when the ratio of the sides is large, i.e., $\phi > 4$; in this case, as Lekhnitskii [51] points out, the effect of the short sides is negligible and the plate may be considered to be an infinite strip, as depicted below in Fig. 2.25. The rigorous solution of this problem is due to Seydel [84]–[86] who considered the cases where the principal directions are both parallel and perpendicular to the edges of the plate for the case of both simply supported as well as clamped edges; he found approximate solutions for both infinite and finite rectangular plates for arbitrary values of the $D_{ij}$. The results of Seydel's investigations for infinite plates were presented by Bergmann [87] in the following form, for the case of simply supported edges:

$$\zeta_{cr} = \frac{\sqrt{2D_{22}(D_{12} + 2D_{66})}}{(b/2)^2 h} \left[ 8.3 + 1.525 \frac{D_{11}D_{22}}{(D_{12} + 2D_{66})^2} \right.$$
$$\left. - .493 \frac{D_{11}^2 D_{22}^2}{(D_{12} + 2D_{33})^4} \right] \tag{2.104}$$

when

$$D_{11}D_{22} < (D_{12} + 2D_{66})^2 \tag{2.105}$$

and

$$\zeta_{cr} = \frac{\sqrt[4]{D_{11}D_{22}^3}}{(b/2)^2 h} \left[ 8.125 + 5.64 \sqrt{\frac{(D_{12} + 2D_{66})^2}{D_{11}D_{22}}} \right.$$
$$\left. - .6 \frac{(D_{12} + 2D_{33})^2}{D_{11}D_{22}} \right] \tag{2.106}$$

when

$$D_{11}D_{22} > (D_{12} + 2D_{66})^2 \tag{2.107}$$

For the finite, long, rectangular, orthotropic plate under shear along its edges, Seydel, op. cit., seeks a solution of (2.99) of the form

$$w = f(y) \exp\left(\frac{2\eta}{b} ix\right) \tag{2.108}$$

in which case $f(y)$ must satisfy

$$D_{22} f'''' - 2(D_{12} + 2D_{66}) \left(\frac{2\eta}{b}\right)^2 f''$$
$$+ i\zeta h \cdot \frac{2\eta}{b} f' + D_{11} \left(\frac{2\eta}{B}\right)^4 f = 0 \tag{2.109}$$

with associated characteristic equation

$$D_{22} r^4 - 2(D_{12} + 2D_{66}) \left(\frac{2\eta}{b}\right)^2 r^2 + i\zeta h \cdot \frac{2\eta}{b} r + D_{11} \left(\frac{2\eta}{b}\right)^4 = 0 \tag{2.110}$$

An analysis of (2.10) produces an expression for the initial buckling mode of the form

$$w = \exp\left(\frac{2\eta}{b}ix\right)\left(c_1 e^{2\beta_1 iy/b} + c_2 e^{2\beta_2 iy/b}\right.$$

$$\left. + c_3 e^{2\beta_3 iy/b} + c_4 e^{2\beta_4 iy/b}\right) \tag{2.111}$$

with the $\beta_i = \beta_i(\zeta)$ (which may be complex) given by the solutions of (2.110). The boundary conditions corresponding to either clamped edges, i.e., (1.130a,b), or simply supported edges, i.e. (1.130a), (1.148), are now imposed upon (2.111) to obtain the relevant algebraic equation for $\zeta$ which determines $\zeta_{cr}$. As noted in [51] the plate, upon buckling, exhibits a series of waves which make an angle with the edges of the plate; the parameter $\eta$ characterizes the wavelengths. Based on the analysis described above, Seydel [84] introduces the plate parameter $\theta$ by

$$\frac{1}{\theta} \equiv (D_{12} + 2D_{66})\Big/\sqrt{D_{11}D_{22}} \equiv \alpha \tag{2.112}$$

and shows that, for $\theta \geq 1$, the critical shear stress for the finite (but narrow) rectangular orthotropic plate has the form

$$\zeta_{cr} = \tilde{K}\pi^2 \sqrt[4]{D_{11}D_{22}^3}/b^2 h \tag{2.113}$$

In Fig. 2.26, taken from Seydel op. cit., $\tilde{K}$, the buckling coefficient, is plotted against $1/\bar{\beta}$ where $\bar{\beta}$ is the effective aspect ratio, $\bar{\beta} = \phi\sqrt[4]{D_{22}/D_{11}}$, for various values of the plate parameter $\theta(\geq 1)$, as given by (2.112); for the plot shown in Fig. 2.26, all the edges of the plate are simply supported. If the plate parameter $\theta$ satisfies $\theta < 1$, instead of $\theta \geq 1$, we have, in lieu of (2.113), an expression of the form

$$\zeta_{cr} = \tilde{K}\pi^2 \sqrt{D_{22}(D_{12} + 2D_{66})}/b^2 h \tag{2.114}$$

for the critical shear stress at which the plate buckles; another (similar) description of the results for this case is depicted in Fig. 2.27.

**(vi) Rectangular Orthotropic Plates Subject to a Combined Loading: Shear and Compression in the $x$-Direction.**
The relevant situation in this case is, once again, depicted in Fig. II.16 (except that the plate now exhibits rectilinear orthotropy); the equation

which governs initial buckling is

$$D_{11}\frac{\partial^2 w}{\partial x^4} + 2(D_{12} + 2D_{66})\frac{\partial^4 w}{\partial x^2 \partial y^2} + D_{22}\frac{\partial^4 w}{\partial y^4}$$

$$+\lambda h\frac{\partial^2 w}{\partial x^2} + 2\zeta h\frac{\partial^2 w}{\partial w \partial y} = 0 \qquad (2.115)$$

and the plate is assumed to be simply supported along all four edges. According to Lekhnitskii [51], a solution of the initial buckling problem is known only for the case of an infinite strip, i.e. aspect ratio $\phi = \infty$; The method consists of seeking an initial buckling mode in the form

$$w = A\sin\frac{\pi y}{b}\sin\frac{\pi}{s}(x - y\tan\psi) \qquad (2.116)$$

which represents a surface consisting of waves inclined to the $y$-axis at angle $\psi$ and having wavelength $s$ in the direction of the $x$-axis. Substituting (2.116) into (2.115), multiplying the resulting equation by $\sin\frac{\pi y}{b}\sin\frac{\pi}{s}(x - y\tan\psi)$, and then integrating over the domain $0 \le x \le s$, $0 \le y \le b$, produces the result

$$\lambda = \frac{\pi^2\sqrt{D_{11}D_{22}}}{b^2 h}\left[\sqrt{\frac{D_{11}}{D_{22}}}\gamma + \frac{2(D_{12} + 2D_{66})}{\sqrt{D_{11}D_{22}}}(\alpha^2\gamma + 1)\right.$$

$$\left. +\sqrt{\frac{D_{22}}{D_{11}}}\left(\alpha^4\gamma + 6x^2 + \frac{1}{\gamma}\right)\right] - 2\alpha\zeta h \qquad (2.117)$$

where

$$\gamma = (b/s)^2 \quad , \quad \alpha = \tan\psi \qquad (2.118)$$

Holding $\zeta$ constant, one now seeks a minimum value for $\lambda$ as a function of $\alpha$ and $\gamma$ by requiring that $\partial\lambda/\partial\alpha = 0$, $\partial\lambda/\partial\gamma = 0$; this leads to the relations

$$\begin{cases} \zeta = \frac{2\pi^2\sqrt{D_{11}D_{22}}}{b^2 h}\cdot\alpha\left[\frac{(D_{12} + 2D_{66})}{\sqrt{D_{11}D_{22}}}\gamma + \sqrt{\frac{D_{22}}{D_{11}}}(3 + \alpha^2\gamma)\right] \\[2em] \gamma = \sqrt{\frac{D_{22}}{D_{11} + 2(D_{12} + 2D_{66})\alpha^2 + D_{22}\alpha^4}} \end{cases} \qquad (2.119)$$

Substituting the value of $\gamma$ from (2.119) into (2.117), as well as into the expression for $\zeta$ in (2.119), yields the following results:

$$\lambda = \frac{2\pi^2\sqrt{D_{11}D_{22}}}{b^2 h}\left(\frac{2[D_{12}+2D_{66}]}{\sqrt{D_{11}D_{22}}} + 3\alpha^2\sqrt{\frac{D_{22}}{D_{11}}}\right.$$

$$\left. +\sqrt{1+\frac{2(D_{12}+2D_{33})}{D_{11}}\alpha^2 + \frac{D_{22}}{D_{11}}\alpha^4}\right) - 2\alpha\zeta h \qquad (2.120)$$

$$\zeta = \frac{2\pi^2\sqrt{D_{11}D_{22}}}{b^2 h}\cdot\alpha\left(3\sqrt{\frac{D_{22}}{D_{11}}}\right.$$

$$\left. +\frac{[D_{12}+2D_{66}]+\alpha^2 D_{22}}{\sqrt{1+\frac{2(D_{12}+2D_{66})}{D_{11}}\alpha^2 + \frac{D_{22}}{D_{11}}\alpha^4}}\right) \qquad (2.121)$$

Equations (2.120), (2.121) yield a relation between $\lambda$ and $\zeta$ in parametric form; as the parameter $\alpha$ is varied from 0 to $\infty$ [$\psi$ varies from 0 radians to $\pi/2$ radians ] we may plot this relation with the values of $\lambda$ as the abscissa and the values of $\zeta$ as the ordinate (see Fig. 2.28); the graph which results is a parabolic curve of the form

$$\frac{\lambda}{\bar{\lambda}_{cr}} + \frac{\zeta^2}{\bar{\zeta}_{cr}^2} = 1 \qquad (2.122)$$

so that the intercept on the abscissa, $\bar{\lambda}_{cr}$, is the critical compressive load (actually, the critical stress $(\sigma_{xx})_{cr}$) in the absence of shear forces, while the intercept on the ordinate axis, $\bar{\zeta}_{cr}$, is the critical shearing load (actually, the critical shear stress $(\sigma_{xy})_{cr}$) when the compressive forces are absent. As $\bar{\lambda}_{cr}$ and $\bar{\zeta}_{cr}$ are known from our earlier work, it is always possible to use (2.122) to find the critical compressive load for a given shearing load or vice versa.

For plates in which the stiffnesses in two directions differ by a large amount (i.e., plywood), Lekhnitskii [51] notes the following approximate formula for the case of simply supported edges

$$\lambda_{cr} + \frac{\zeta_{cr}^2}{\frac{2\pi^2}{b^2 h}\left(3D_{22}+[D_{12}+2D_{66}]\sqrt{\frac{D_{22}}{D_{11}}}\right)} \qquad (2.123)$$

$$= \frac{2\pi^2}{b^2 h}\left(\sqrt{D_{11}D_{22}}+\{D_{12}+2D_{66}\}\right)$$

and the following result for the case of a clamped strip:

$$\lambda_{cr} + \frac{\zeta_{cr}^2}{\frac{12\pi^2}{b^2 h}\left(3D_{22} + \{D_{12} + 2D_{66}\}\sqrt{\frac{D_{22}}{D_{11}}}\right)}$$
$$= \frac{15\pi^2}{b^2 h}\left(\sqrt{D_{11}D_{22}} + \{D_{12} + 2D_{66}\}\right) \qquad (2.124)$$

### 2.1.3 Postbuckling Behavior of Isotropic and Rectilinear Orthotropic Linear Elastic Rectangular Plates

Up to this point, we have considered only the initial buckling problem for linear elastic, isotropic, and rectilinearly orthotropic, rectangular plates; in sub-sections 2.1.1 and 2.1.2, the critical stress was determined for a wide range of loading and boundary conditions. However, rectangular plates can support stresses which are often significantly higher than the critical stress at which buckling occurs and still remain stable in the new buckled state because the out-of-plane deflections are accompanied by stretching of the middle surface of the plate. In §1.2.1, we formulated the von Karman equations which govern the large deflections of both isotropic and (rectilinearly) orthotropic linear elastic rectangular plates; in terms of the out-of-plane deflection $w$, and the net airy stress function $\Phi$, these equations are (1.53a,b) for the isotropic plate (with $t \equiv 0$ if there is no applied normal load) and (1.68), (1.69) for the rectilinearly orthotropic plate. For the (isotropic) plate, with a rigid boundary, we have noted that the in-plane boundary conditions are specified by the vanishing of the displacements $u, v$, rather than by specification of the middle surface forces, and that it may, therefore, be advantageous to express the large deflection equations for an initially flat plate of constant thickness in terms of the displacements $u, v, w$; these displacement equations appear in §1.2.1 as (1.59a,b,c) where we must set $t = 0$ in (1.59a) if there is no normal applied force.

In one form or another, all calculations dealing with postbuckling behavior are based on some expansion (e.g., a perturbation series expansion) of the variables $w$ and $\Phi$ (or $u, v, w$) about the in-plane equilibrium stress state; these expansions may be justified by use of either the implicit function theorem or some version of the Lyapunov–Schmidt reduction. Of course, one may also do a direct numerical analysis of the von Karman equations for the type of geometry and material behavior in question. In this subsection, we will content ourselves with applying, in a straightforward manner, some standard perturbation expansions for rectangular plates; however, in §3.1 we will go into some detail, with respect to an application of the Lyapunov–Schmidt reduction, when we consider the postbuckling behavior of compressed, linearly elastic, isotropic circular plates.

One of the first to look carefully at the postbuckling behavior of rectangular plates appears to have been Stein [61], and [88]–[93], who applied perturbation expansions to study the behavior of simply supported plates which are subjected to various combinations of compressive forces in the plane of the plate; in this case, the relevant equations for the isotropic plate may be taken as (1.59a,b,c), with $t \equiv 0$, in which case we must expand the displacements $u, v, w$ about the point of buckling in powers of a suitable parameter; there is some freedom in the choice of this expansion parameter, as is pointed out by Stein [61] who, for the problem of uniaxial compression of an isotropic rectangular plate, chooses the parameter $\epsilon = \{(\lambda - \lambda_{cr})/\lambda_{cr}\}^{1/2}$. The parameter $\epsilon$ will, indeed, be the expansion parameter of choice in much of what follows, because it is known that, immediately after buckling, the deflection increases in proportion to $\epsilon$; in our first discussion of postbuckling behavior, below, we will, however, employ (1.59a,b,c) in lieu of (1.53a,b), and expand the displacements $u, v, w$ in powers of the deflection $\delta$ at some fixed point $x_0, y_0$ in the plate, i.e.

$$
\begin{cases}
w = w_1(x,y)\delta + w_3(x,y)\delta^3 + \ldots \\
u = u_1(x,y) + u_2(x,y)\delta^2 + \ldots \\
v = v_0(x,y) + v_2(x,y)\delta^2 + \ldots
\end{cases}
\tag{2.125}
$$

Because a change in the sign of $w$ does not affect the displacements in the plane of the plate, only odd powers of $\delta$ appear in the expansion of $w$ and even powers in the expansions for $u$ and $v$. The functions $u_0$, $v_0$ represent the in-plane displacements of the plate at the onset of buckling. Each of the functions $w_{2k+1}$, $u_{2k}$, $v_{2k}$, $k = 0, 1, 2, \ldots$ must satisfy the relevant boundary conditions associated with the buckling problem. If we substitute (2.125) into (1.59a,b,c), and equate those terms which are independent of the perturbation parameter $\delta$, we recover the equations of plane stress expressed in terms of the displacements $u_0$, $v_0$:

$$
\begin{cases}
\dfrac{\partial}{\partial x}\left(\dfrac{\partial u_0}{\partial x} + \dfrac{\partial v_0}{\partial y}\right) + \left(\dfrac{1-\nu}{1+\nu}\right)\nabla^2 u_0 = 0 \\[4mm]
\dfrac{\partial}{\partial y}\left(\dfrac{\partial u_0}{\partial x} + \dfrac{\partial v_0}{\partial y}\right) + \left(\dfrac{1-\nu}{1+\nu}\right)\nabla^2 v_0 = 0
\end{cases}
\tag{2.126}
$$

Next, equating all terms of order $\delta$ reproduces the small-deflection or initial buckling equation, namely,

$$\frac{h^2}{12} \nabla^4 w_1 = \frac{\partial^2 w_1}{\partial x^2} \left( \frac{\partial u_0}{\partial x} + \nu \frac{\partial v_0}{\partial y} \right) + \frac{\partial^2 w_1}{\partial y^2} \left( \frac{\partial v_0}{\partial y} + \nu \frac{\partial u_0}{\partial x} \right)$$

$$+ (1-\nu) \frac{\partial^2 w_1}{\partial x \partial y} \left( \frac{\partial u_0}{\partial y} + \frac{\partial v_0}{\partial x} \right) \tag{2.127}$$

Each of the series of equations generated is then solved subject to the appropriate boundary conditions, and the results are cycled into the successive sets of equations which result from looking at increasing powers of $\delta$. As a specific example, consider the postbuckling behavior of a square isotropic plate simply supported along its edges $x = 0$, $x = a$, $0 \le y \le a$, and $y = 0$, $y = a$, $0 \le x \le a$, and subjected to a compressive load which causes the edges at $x = 0$, $x = a$ to approach each other by a fixed amount, while the distance between the edges at $y = 0$, $y = a$ remains constant. The boundary conditions are

$$\begin{cases} \dfrac{\partial u}{\partial y} = \dfrac{\partial v}{\partial x} = w = \dfrac{\partial^2 w}{\partial x^2} = 0, & \text{along } x = 0, \ x = a \\[3mm] \dfrac{\partial u}{\partial x} = v = w = \dfrac{\partial^2 w}{\partial y^2} = 0, & \text{along } y = 0, \ y = a \end{cases} \tag{2.128}$$

and have been chosen to illustrate the technique because they yield particularly simple results. The parameter $\delta$ is taken to be the deflection at the center of the plate, i.e., at $\left( \dfrac{a}{2}, \dfrac{a}{2} \right)$; while we have already seen that, for high enough values of the compression, two or more buckles can occur in the direction of the compression, we will confine our attention here to plates exhibiting a single initial buckle.

The solution of (2.126) which satisfies (2.128) is

$$u_0 = -k_0 x, \quad v_0 = 0 \tag{2.129}$$

with the constant $k_0$ still to be determined. Substitution of (2.129) into (2.127) yields the small-deflection equation, i.e., the equation for the initial buckling mode

$$\frac{h^2}{12} \nabla^4 w_1 = -k_0 \left( \frac{\partial^2 w_1}{\partial x^2} + \nu \frac{\partial^2 w_1}{\partial y^2} \right) \tag{2.130}$$

whose solution, subject to (2.128), determines not only $w_1$ but $k_0$ as well:

$$\begin{cases} w_1(x,y) = \sin \dfrac{\pi x}{a} \sin \dfrac{\pi y}{a} \\ k_0 = \pi^2 h^2 / 3(1+\nu)a^2 \end{cases} \tag{2.131}$$

By employing (2.131) in the equations obtained from substituting (2.125) into (1.59a,b,c), and equating powers of order $\delta^2$, we obtain as the equations for $u_2(x,y)$ and $v_2(x,y)$

$$a \begin{cases} \dfrac{\partial^2 u_2}{\partial x^2} + \dfrac{1}{2}(1-\nu)\dfrac{\partial^2 u_2}{\partial y^2} + \dfrac{1}{2}(1+\nu)\dfrac{\partial^2 v_2}{\partial x \partial y} \\ -\dfrac{1}{4}\left(\dfrac{\pi}{a}\right)^3 \left\{ 1 - \nu - 2\cos\dfrac{2\pi y}{a} \right\} \sin\dfrac{2\pi x}{a} = 0 \end{cases} \tag{2.132a}$$

and

$$\begin{cases} \dfrac{\partial^2 v_2}{\partial y^2} + \dfrac{1}{2}(1-\nu)\dfrac{\partial^2 v_2}{\partial x^2} + \dfrac{1}{2}(1+\nu)\dfrac{\partial^2 u_2}{\partial x \partial y} \\ -\dfrac{1}{4}\left(\dfrac{\pi}{a}\right)^3 \left\{ 1 - \nu - 2\cos\dfrac{2\pi x}{a} \right\} \sin\dfrac{2\pi y}{a} = 0 \end{cases} \tag{2.132b}$$

the solution of which, subject to the boundary data in (2.128), is

$$\begin{cases} u_2 = -k_2 x - \dfrac{\pi}{16a}\left(1 - \nu - \cos\dfrac{2\pi y}{a}\right)\sin\dfrac{2\pi x}{a} \\ v_2 = -\dfrac{\pi}{16a}\left(1 - \nu - \cos\dfrac{2\pi x}{a}\right)\sin\dfrac{2\pi y}{a} \end{cases} \tag{2.133}$$

with $k_2$ an (at this point) unknown constant. The equation for $w_3$ is now obtained by equating all terms of order $\delta^3$ in the equation resulting from the substitution of (2.125) into (1.59a,b,c) and then employing (2.133); we find that

$$\dfrac{h^2}{12}\nabla^4 w_3 + \dfrac{\pi^2 h^2}{3(1+\nu)a^2}\left(\dfrac{\partial^2 w_3}{\partial x^2} + \nu\dfrac{\partial^2 w_3}{\partial y^2}\right)$$

$$= \dfrac{\pi^2(1+\nu)}{a^2}\left\{k_2 - \dfrac{\pi^2}{8a^2}(3-\nu)\right\}\sin\dfrac{\pi x}{a}\sin\dfrac{\pi y}{a} \tag{2.134}$$

$$+\dfrac{\pi^4(1-\nu^2)}{16a^4}\left(\sin\dfrac{\pi x}{a}\sin\dfrac{3\pi y}{a} + \sin\dfrac{3\pi x}{a}\sin\dfrac{\pi y}{a}\right)$$

the solution of which, subject to (2.128), yields

$$w_3 = (A + B)\sin\frac{\pi x}{a}\sin\frac{\pi y}{a} + A\sin\frac{\pi x}{a}\sin\frac{3\pi y}{a}$$

$$+ B\sin\frac{3\pi x}{a}\sin\frac{\pi y}{a}$$

(2.135)

with

$$\begin{cases} A = 3(1-\nu)(1+\nu)^2/16(24+25\nu-9\nu^2)h^2 \\ B = 3(1-\nu)(1+\nu)^2/16(16+25\nu-\nu^2)h^2 \end{cases}$$

(2.136)

and

$$k_2 = \frac{\pi^2}{8a^2}(3-\nu)$$

(2.137)

The cycle of operations described above now continues so as to first determine $u_4$, $v_4$ and, then, $w_5$, etc. If $\delta u_{cr}$ represents the amount by which the edges at $x = 0$, $x = a$ approach each other upon initial buckling, then it is possible to relate the amount by which the loaded edges approach each other, beyond that point, to the magnitude of the center deflection $\delta$, namely,

$$\delta u - \delta u_{cr} = \frac{\pi^2(3-\nu)}{\delta a}\delta^2 + \mathcal{O}(\delta^2)$$

(2.138)

Of more practical importance is the relation between $\delta u - \delta u_{cr}$ and $P - P_{cr}$, where $P = h \int_0^a \sigma_{xx}dy$ denotes the total applied compressive load. However,

$$P = -\frac{Eh}{1-\nu^2}\int_0^a \left\{\frac{\partial u}{\partial x} + \nu\frac{\partial v}{\partial y} + \frac{1}{2}\left(\frac{\partial w}{\partial x}\right)^2 + \frac{1}{2}\nu\left(\frac{\partial w}{\partial y}\right)^2\right\}dy \quad (2.139)$$

so by (2.125), and the perturbation results described above, we find that

$$P - P_{cr} = \frac{\pi^2 Eh}{4(1+\nu)a}\delta^2 + \mathcal{O}(\delta^4)$$

(2.140)

Combining (2.140) with (2.138) then yields

$$\frac{P - P_{cr}}{\delta u - \delta u_{cr}} = \frac{2Eh}{(1+\nu)(3-\nu)} + \mathcal{O}(\delta^2)$$

(2.141)

As $(P-P_{cr})/(\delta u - \delta u_{cr})$ is the stiffness of the plate, immediately after the onset of buckling, and $Eh/(1-\nu^2)$ is the plate stiffness prior to buckling, the ratio of the stiffnesses is (approximately) dependent only on the Poisson ratio $\nu$ and may be computed to be $2(1-\nu)/(3-\nu)$. A graph of the postbuckling behavior for the square isotropic plate discussed above is shown in Fig. 2.29, in which $P/P_{cr}$ is plotted against $\delta_u/\delta_{u_{cr}}$; another postbuckling graph of $P/P_{cr}$, against the deflection $\delta$ of the center of the plate, when the plate buckles (initially, i.e., (2.131)) into one-half sine wave in the $x$-direction may be obtained by employing (2.140), i.e.,

$$P/P_{cr} = 1 + \frac{\pi^2 E}{4P_{cr}(1+\nu)a} \cdot \delta^2 + \mathcal{O}(\delta^4) \qquad (2.142)$$

In fact, for a rectangular, isotropic, linearly elastic plate, simply supported on all four edges and subjected to compressive loading along the edges at $x = 0$ and $x = a$, $0 \le y \le b$, such a postbuckling graph is shown in Fig. 2.30, which indicates that significant postbuckling loads can be carried when the deflection to thickness ratio $\delta/h$ increases over the range $\delta/h = 1$ to $\delta/h = 3$. Load versus deflection curves similar to Fig. 2.30 have been produced by (among others) Yamaki [94] for a number of conditions of different edge support in the case of an isotropic linearly elastic plate subject to end compression; several of these results for a square plate of edge length $a$ are shown in Fig. 2.31. For rectangular plates with edge lengths $a$ and $b = \frac{1}{2}a$ we show, in Fig. 2.32 some results of Walker [95] for the case of compression under a linearly varying end load when all four edges are simply supported. Examples of other types of postbuckling curves which apply in the case of isotropic rectangular plates may be gleaned from our discussion, below, on the postbuckling behavior of rectilinearly orthotropic rectangular plates.

The equations which govern the postbuckling behavior of rectilinearly orthotropic, linearly elastic, rectangular plates are the von Karman equations (1.68) and (1.69), which may also be reformulated as a system of equations for the displacements $u, v, w$. While there have been many studies of the postbuckling behavior of rectangular orthotropic plates, the paper of Chandra and Raju [96] has the advantage that it follows a direct perturbation expansion for the displacement fields and covers a wide range of loading and edge support conditions. Thus, in lieu of using (1.68), (1.69) directly, the work in [96] employs the equilibrium equations (1.42), the first of the von Karman equations (1.68), as well as the inverse of the strain-membrane force equations, i.e.,

$$N_{xx} = \frac{\bar{E}E_{11}h}{E - \nu_{12}^2}(\epsilon_{xx} + \nu_{12}\epsilon_{yy}) \qquad (2.143a)$$

$$N_{yy} = \frac{\bar{E}E_{11}h}{\bar{E} - \nu_{12}^2}(\bar{E}\epsilon_{yy} + \nu_{12}\epsilon_{xx}) \tag{2.143b}$$

and

$$N_{xy} = G_{12}h\gamma_{xy} \tag{2.143c}$$

For our purposes, the strain-displacement relations may be taken in the form (1.35), with $\zeta \equiv 0$, since integration over the thickness of the plate in the range $-h/2 < \zeta < h/2$ has the effect of eliminating the last term in each of the expressions in (1.35). The procedure in [96] is to assume a perturbation expansion for $u, v, w$ of the form (compare with (2.125))

$$\begin{cases} u = u_0 + \eta^2 u_2 + \eta^4 u_4 + \cdots \\ v = v_0 + \eta^2 v_2 + \eta^4 v_4 + \cdots \\ w = \eta w_1 + \eta^3 w_3 + \eta^5 w_5 + \cdots \end{cases} \tag{2.144}$$

where $\eta^2 = (P - P_{cr})/P_{cr}$ with $P$ the total compressive load in the $x$-direction and $P_{cr}$ the critical load. The perturbation expansions (2.144) are now substituted into the strain-displacement relations (1.35), with $\zeta = 0$, so as to generate perturbation expansions for the strain components $\epsilon_{xx}$, $\epsilon_{yy}$, and $\gamma_{xy}$; these are, in turn, substituted into (2.143a,b,c) so as to generate perturbation expansions for the averaged stresses $N_{xx}$, $N_{yy}$, and $N_{xy}$. Finally, the perturbation expansions which result for $N_{xx}$, $N_{yy}$, and $N_{xy}$ are substituted into the equilibrium equations (1.42), and the first von Karman equation (1.68), so as to generate an infinite set of algebraic equations for the coefficients $u_{2k}$, $v_{2k}$, and $w_{2k+1}$, $k = 0, 1, \ldots$ For the averaged stresses, it is possible to show that one obtains expansions of the form

$$\begin{cases} N_{xx} = \sum_{n=0,2,4\ldots}^{\infty} N_{xn}\eta^n + \sum_{n=1,3,5\ldots}^{\infty} \sum_{m=1,3,5\ldots}^{\infty} N_{xnm}\eta^{m+n} \\ N_{yy} = \sum_{n=0,2,4\ldots}^{\infty} N_{yn}\eta^n + \sum_{n=1,3,5\ldots}^{\infty} \sum_{m=1,3,5\ldots}^{\infty} N_{ynm}\eta^{m+n} \\ N_{xy} = \sum_{n=0,2,4\ldots}^{\infty} N_{xyn}\eta^n + \sum_{n=1,3,5\ldots}^{\infty} \sum_{m=1,3,5\ldots}^{\infty} N_{xymn}\eta^{m+n} \end{cases} \tag{2.145}$$

with

$$N_{xn} = \frac{\bar{E}E_{11}h}{\bar{E} - \nu_{12}^2}(u_{n,x} + \nu_{12}v_{n,y}), \tag{2.146}$$

$$N_{xyn} = Gh(u_{n,y} + v_{n,x}) \qquad (2.147)$$

and analogous expressions for the other coefficients $N_{xnm}$, $N_{yn}$, $N_{ynm}$ and $N_{xynm}$. Substitution of the expansions in (2.145) into (1.42) and (1.68) then yields a recursive collection of sets of three equations each for the $u_{2k}$, $v_{2k}$, $w_{2k+1}$, the first two of which are

$$
\begin{cases}
N_{x0,x} + N_{xy0,y} = 0 \\[2mm]
N_{y0,y} + N_{xy0,x} = 0 \\[2mm]
\mathcal{L}(D)w_1 - (N_{x0}w_{1,xx} + N_{y0}w_{1,yy} + 2N_{xy0}w_{1,xy}) = 0
\end{cases}
\qquad (2.148)
$$

where $\mathcal{L}(D) = D_{11}\dfrac{\partial^4}{\partial x^4} + 2(D_{12} + 2D_{66})\dfrac{\partial^4}{\partial x^2 \partial y^2} + D_{22}\dfrac{\partial^4}{\partial y^4}$, and

$$
\begin{cases}
N_{x2,x} + N_{xy2,y} = -(N_{x11,x} + N_{xy11,y}) \\[2mm]
N_{y2,y} + N_{xy2,x} = -(N_{y11,y} + N_{xy11,x}) \\[2mm]
\mathcal{L}(D)w_3 - (N_{x0}w_{3,xx} + N_{y0}w_{3,yy} + 2N_{xy0}w_{3,xy}) \\[2mm]
\qquad = (N_{x2} + N_{x11})w_{1,xx} + (N_{y2} + N_{y11})w_{1,yy} \\[2mm]
\qquad + 2(N_{xy2} + N_{xy11})w_{1,xy}
\end{cases}
\qquad (2.149)
$$

The set of boundary conditions considered in [96] include the conditions of simple support relative to $w$ along the four edges of the rectangular plate as well as the following:

(i) Constant in-plane displacements:

$$
\begin{cases}
u_{,y}(0, y) = u_{,y}(a, y) = 0, & 0 \le y \le b \\[2mm]
v_{,x}(x, 0) = v_{,x}(x, b) = 0, & 0 \le x \le a
\end{cases}
\qquad (2.150)
$$

(ii) zero shear stress:

$$
\begin{cases}
v_{,x}(0, y) = v_{,x}(a, y) = 0, & 0 \le y \le b \\[2mm]
u_{,y}(x, 0) = y_{,y}(x, b) = 0, & 0 \le y \le b
\end{cases}
\qquad (2.151)
$$

It is also assumed that the plate is subject to a uniform compressive load of magnitude $P$ in the $x$-direction so that, in the usual manner, on the loaded edges, $\displaystyle\int_0^b N_{xx}\Big|_{x=0} dy = \int_0^b N_{xx}\Big|_{x=a} dy = -P$, while on the non-loaded edges, $\int_0^a N_{yy}\big|_{y=0}\,dx = \int_0^a N_{yy}\big|_{y=b}\,dx = 0$; if into these load conditions on the edges we substitute the perturbation expansions for $N_{xx}$ and $N_{yy}$ from (2.45) and, also, replace $P$ by $P_{cr} + \eta^2 P_{cr}$, then we obtain a sequence of boundary conditions for the coefficients $N_{xn}$, $N_{xnm}$, ..., $N_{xynm}$ of the form

$$
\begin{cases}
\int_0^b N_{x0}\bigg|_{x=0,a}\,dy + P_{cr} = 0 \\[2mm]
\int_0^b (N_{x2} + N_{x11})\bigg|_{x=0,a}\,dy + P_{cr} = 0 \\[2mm]
\int_0^b (N_{xy} + 2N_{x13})\bigg|_{x=0,a}\,dy = 0
\end{cases}
\qquad (2.152)
$$

along the loaded edges at $x = 0$, $x = a$, while along the non-loaded edges at $y = 0$, $y = b$

$$
\begin{cases}
\int_0^a N_{y0}\bigg|_{y=0,b}\,dx = 0 \\[2mm]
\int_0^a (N_{y2} + N_{y11})\bigg|_{y=0,b}\,dx = 0 \\[2mm]
\int_0^a (N_{y4} + 2N_{y13})\bigg|_{y=0,b}\,dx = 0
\end{cases}
\qquad (2.153)
$$

We note that solutions of the first two equations in (2.148) provide the prebuckling stresses in the rectangular, orthotropic plate; these equations may be put in the form

$$
\begin{cases}
A u_{0,xx} + G_{12} h u_{0,yy} + B v_{0,xy} = 0 \\[2mm]
\bar{E} A v_{0,yy} + G_{12} h v_{0,xx} + B u_{0,xy} = 0
\end{cases}
\qquad (2.154)
$$

where

$$
A = \frac{\bar{E} E_{11} h}{\bar{E} - \nu_{12}^2}, \quad B = \nu_{12} A + G_{12} h \qquad (2.155)
$$

Solutions of (2.154) satisfying the boundary conditions (2.150), 2.151), and the first relation in (2.152) are given in [96] as

$$\begin{cases} u_0 = -\dfrac{P_{cr}}{hbE_{11}}\left(x - \dfrac{a}{2}\right) \\[2ex] v_1 = \nu_{12}\dfrac{P_{cr}}{hbE_{11}}\cdot\dfrac{1}{E}\cdot\left(y - \dfrac{b}{2}\right) \end{cases} \tag{2.156a}$$

$$N_{x0} = -\dfrac{P_{cr}}{b}, \quad N_{y0} = N_{xy0} = 0 \tag{2.156b}$$

The use of (2.156b) then permits us to write the last equation in (2.148) in the form

$$\mathcal{L}(D)w_1 + \left(\dfrac{P_{cr}}{b}\right)w_{1,xx} = 0 \tag{2.157}$$

and imposition of the simple support condition along the four edges of the plate again leads to an initial buckling mode of the form

$$w_1(x,y) = A_1 \sin\dfrac{m\pi x}{a} \sin\dfrac{n\pi y}{b} \tag{2.158}$$

As in our previous discussion of the initial buckling problem, for a simply supported, orthotropic, rectangular plate which is compressed in the $x$-direction, we substitute (2.158) into (2.157) so as to determine that

$$P_{cr} = \dfrac{b}{(m\pi/a)^2}\left[D_{11}\left(\dfrac{m\pi}{a}\right)^4 + 2(D_{12} + 2D_{66})\left(\dfrac{m\pi}{a}\right)^2\left(\dfrac{n\pi}{b}\right)^2 \right.$$
$$\left. +D_{22}\left(\dfrac{n\pi}{b}\right)^4\right] \tag{2.159}$$

For a uniform distribution of stress along the edges at $x = 0$, $x = a$,

$$P = h\int_0^b \sigma_{xx}\Big|_{x=0,\ a}\, dy = \lambda h b \text{ so that } P_{cr} = hb\lambda_{cr} \text{ and (2.159) may be}$$

directly related to our earlier result in terms of the aspect ratio $\phi = a/b$. The lowest buckling load is, as previously discussed, determined by the choice of $m$ and $n$ in (2.159) for a given aspect ratio $\phi$ and the value of $A_1$ is then determined from the subsequent postbuckling analysis, as follows: First we note that several of the terms in the first pair of equations in (2.149) involve the expression (2.158) for $w_1$; in general

$$N_{xnm} = \dfrac{\bar{E}E_{11}h}{2\left(\bar{E} - \nu_{12}^2\right)}\left(w_{m,x}w_{n,x} + \nu_{12}w_{m,y}w_{n,y}\right)$$

$$N_{ynm} = \frac{\bar{E}E_{11}h}{2\left(\bar{E} - \nu_{12}^2\right)}\left(\bar{E}w_{n,y}w_{m,y} + \nu_{12}w_{n,x}w_{m,x}\right)$$

and

$$N_{xymn} = G_{12}h(w_{m,x}w_{n,y})$$

Thus, if we rewrite the first pair of equations in (2.149) in terms of $u_2$ and $v_2$, and use (2.158), we find that $Au_{2,xx} + G_{12}hu_{2,yy} + Bv_{2,xy}$

$$= -\left(\frac{m\pi}{a}\right)\frac{AA_1^2}{4}\left[-\left(\frac{m\pi}{a}\right)^2 + \nu_{12}\left(\frac{n\pi}{b}\right)^2\right]\sin\frac{2m\pi x}{a}$$

$$-\left(\frac{m\pi}{a}\right)\frac{A_1^2}{4}\left[A\left(\frac{m\pi}{a}\right)^2 + (B + G_{12}h)\left(\frac{n\pi}{b}\right)^2\right] \qquad (2.160)$$

$$\times \sin\frac{2m\pi x}{a}\cos\frac{2n\pi y}{b}$$

and

$$\bar{E}Av_{2,yy} + G_{12}hv_{2,xx} + Bu_{2,xy}$$

$$= -\left(\frac{n\pi}{b}\right)\frac{AA_1^2}{4}\left[-\bar{E}\left(\frac{n\pi}{b}\right)^2 + \nu_{12}\left(\frac{m\pi}{a}\right)^2\right]\sin\frac{2n\pi y}{b} \qquad (2.161)$$

$$-\left(\frac{n\pi}{b}\right)\frac{A_1^2}{4}\left[\bar{E}A\left(\frac{n\pi}{b}\right)^2 + (B + G_{12}h)\left(\frac{m\pi}{a}\right)^2\right]$$

$$\times \sin\frac{2n\pi y}{b}\cos\frac{2m\pi x}{a}$$

where $A, B$ are given by (2.155). The solutions of (2.160), (2.161), subject to (2.150), (2.151), and the second condition in (2.152), are

$$u_2 = -\left[\frac{P_{cr}}{E_{11}hb} + \frac{A_1^2}{8}\left(\frac{m\pi}{a}\right)^2\right]\left(x - \frac{a}{2}\right) \qquad (2.162)$$

$$-\frac{A_1^2}{16}\left\{\frac{(m\pi/a)^2 - \nu_{12}(n\pi/b)^2}{(m\pi/a)}\sin\frac{2m\pi x}{a}\right.$$

$$\left. -\frac{m\pi}{a}\sin\frac{2m\pi x}{a}\cos\frac{2n\pi y}{b}\right\}$$

and

$$v_2 = -\left[\frac{P_{cr}}{E_{11}hb}\cdot\frac{\nu_{12}}{\bar{E}} - \frac{A_1^2}{8}\left(\frac{n\pi}{b}\right)^2\right](y - b/2) \qquad (2.163)$$

$$-\frac{A_1^2}{16} \left\{ \frac{\bar{E}(n\pi/b)^2 - \nu_{12}(m\pi/a)^2}{\bar{E}(n\pi/b)} \sin\frac{2n\pi y}{b} \right.$$

$$\left. -\left(\frac{n\pi}{b}\right)\cos\frac{2m\pi x}{a}\sin\frac{2n\pi y}{b} \right\}$$

Finally, employing the expressions in (2.162) and (2.163) for $u_2$ and $v_2$, respectively, and the expression (2.158) for $w_1$, in the last equation in (2.149), we obtain as the equation for $w_3$

$$\mathcal{L}(D)w_3 + \frac{P_{cr}}{b}w_{3,xx} \tag{2.164}$$

$$= A_1 \left\{ \frac{P_{cr}}{b}\left(\frac{m\pi}{a}\right)^2 - \frac{E_{11}hA_1^2}{16}\left[\left(\frac{m\pi}{a}\right)^4 \right.\right.$$

$$\left.\left. +\bar{E}\left(\frac{n\pi}{b}\right)^4\right]\right\}\sin\frac{m\pi x}{a}\sin\frac{n\pi y}{b}$$

$$+\frac{E_{11}hA_1^3}{16}\left(\frac{m\pi}{a}\right)^4\sin\frac{m\pi x}{a}\sin\frac{3n\pi y}{b}$$

$$+\frac{E_{11}hA_1^3}{16}\bar{E}\left(\frac{n\pi}{b}\right)^4\sin\frac{n\pi y}{b}\sin\frac{3m\pi x}{a}$$

It is not difficult to prove that a solution of (2.164), satisfying the conditions of simple support along the four edges of the plate, does not exist unless the coefficient in (2.164) of $\sin\dfrac{m\pi x}{a}\sin\dfrac{n\pi y}{b}$ vanishes, in which case

$$A_1^2 = \frac{16}{E_{11}h}\cdot\frac{P_{cr}}{b}\left(\frac{m\pi}{a}\right)^2\left[\left(\frac{m\pi}{a}\right)^4 + \bar{E}\left(\frac{n\pi}{b}\right)^4\right]^{-1} \tag{2.165}$$

and

$$w_3(x,y) = A_3\sin\frac{m\pi x}{a}\sin\frac{n\pi y}{b}$$

$$+ A_{31}\sin\frac{m\pi x}{a}\sin\frac{3n\pi y}{b} + A_{32}\sin\frac{3m\pi x}{a}\sin\frac{n\pi y}{b} \tag{2.166}$$

with

$$
\begin{aligned}
A_{31} = {} & \frac{E_{11}hA_1^3}{16}\left(\frac{m\pi}{a}\right)^4\left[D_{11}\left(\frac{m\pi}{a}\right)^4\right. \\
& + 2(D_{12}+2D_{66})\left(\frac{m\pi}{a}\right)^2\left(\frac{3n\pi}{b}\right)^2 \\
& \left.+ D_{22}\left(\frac{3n\pi}{b}\right)^4 - \frac{P_{cr}}{b}\left(\frac{m\pi}{a}\right)^2\right]^{-1}
\end{aligned}
\tag{2.167a}
$$

and

$$
\begin{aligned}
A_{32} = {} & \frac{E_{11}hA_1^3}{16}\left(\frac{n\pi}{b}\right)^4\left[D_{11}\left(\frac{3m\pi}{a}\right)^4\right. \\
& + 2(D_{12}+2D_{66})\left(\frac{3m\pi}{a}\right)^2\left(\frac{n\pi}{b}\right)^2 \\
& \left.+D_{22}\left(\frac{n\pi}{b}\right)^4 - \frac{P_{cr}}{b}\left(\frac{3m\pi}{a}\right)^2\right]^{-1}
\end{aligned}
\tag{2.167b}
$$

The cycle is now repeated so as to compute $u_4$ and $v_4$ in (2.144) and, then, $A_3$ in (2.166); using these results, the perturbation expansions in (2.144) are determined up to terms of order $\eta^4$ in the in-plane displacements $u, v$ and up to the term of order $\eta^3$ in the deflection $w$. If we define the total end shortening $\triangle$ to be $\triangle \equiv u(0,y) - u(a,y)$, $0 \le y \le b$, (assumed to be independent of $y$ for a uniform compressive loading) then it may be shown [96] that

$$
\begin{aligned}
\frac{\triangle}{a} \cdot \frac{3}{\pi^2} \cdot \frac{(\bar{E}-\nu_{12}^2)\,b^2}{h} = {} & \frac{Pb\bar{E}}{4D_{11}\pi^2} + \frac{\mu^2\Lambda^2}{2} \\
& + \Lambda^4\left[\bar{A}_3\left(\frac{1-\nu_{12}^2}{\bar{E}-\nu_{12}^2}\cdot\Lambda^2 + \nu_{12}\left(\frac{\bar{E}-1}{\bar{E}-\nu_{12}^2}\right)n^2\right)\right. \\
& \left.+\frac{\bar{E}-1}{\bar{E}-\nu_{12}^2}\cdot 6\nu_{12}n^2\bar{A}_{31}\right]
\end{aligned}
\tag{2.168}
$$

where $\Lambda = m/\phi$, $\mu^2 = \dfrac{3}{4}(\bar{E}-\nu_{12}^2)\dfrac{\eta^2 A_1^2}{h^2}$, and

$$
\bar{A}_3 = \frac{3}{2}\frac{\bar{A}_{31}\left\{\Lambda^4 + \left[8\bar{E}^2\left(\bar{E}-1\right)/\left(\bar{E}-\nu_{12}^2\right)\right]n^4\right\} + \bar{E}\bar{A}_{32}n^4}{\Lambda^4 + \bar{E}n^4 + \left[2\bar{E}\left(\bar{E}-1\right)/\left(\bar{E}-\nu_{12}^2\right)\right]n^2\left(\bar{E}h^2 - \nu_{12}\Lambda^2\right)}
\tag{2.169}
$$

with

$$\bar{A}_{31} = \frac{\Lambda^4 \bar{E}}{16} \left[ \left( \nu_{12}\bar{E} + 2\frac{G_{12}}{E_{11}} \left( \bar{E} - \nu_{12}^2 \right) \right) \Lambda^2 n^2 + 5\bar{E}^2 n^4 \right]^{-1} \quad (2.170a)$$

and

$$\bar{A}_{32} = \frac{\bar{E} n^4}{8 \left( 9\Lambda^4 - \bar{E} n^4 \right)} \quad (2.170b)$$

An effective width $b_e$ is also defined in [96] by

$$\left( \frac{b_e}{a} \right) \triangle = P/h E_{11} \quad (2.171)$$

and, as a consequence of (2.168), may be expressed in the form

$$\frac{b_e}{b} = \bar{E} \frac{Pb}{4D_{11}\pi^2} \left\{ \frac{Pb\bar{E}}{4D_{11}\pi^2} + \frac{\mu^2 \Lambda^2}{2} \right.$$

$$+ \mu^4 \left[ \bar{A}_3 \left( \frac{1 - \nu_{12}^2}{\bar{E} - \nu_{12}^2} \Lambda^2 + \nu_{12} \left\{ \frac{\bar{E} - 1}{\bar{E} - \nu_{12}^2} \right\} n^2 \right) \right.$$

$$\left. \left. + \left\{ \frac{\bar{E} - 1}{\bar{E} - \nu_{12}^2} \right\} 6\nu_{12} n^2 \bar{A}_{31} \right] \right\}^{-1}$$

Based on the analysis in [96], some of which we have described above, both initial buckling and postbuckling results are presented, in the form of various graphs, for several different rectilinearly orthotropic rectangular plates (i.e. plates $A$–$N$) whose properties are displayed in Table 2.5. The results of the initial buckling analysis are presented in Figs. 2.33 through 2.37 for plates $C, F$ and $I$, as well as for the isotropic plate with $\bar{E} = 1$. As we have noted previously, the transition from $m$ to $m+1$ half-sine waves in the $x$-direction occurs when $\phi = \sqrt{m(m+1)}/\bar{E}^{1/4}$. Among the results, therefore, which can be gleaned from the five graphs referenced above are the following: For a relatively high stiffness ratio $\bar{E}$ of around 13, the transition from a single to a double half-sine wave in the $x$-direction takes place at an aspect ratio $\phi$ of .735, while the corresponding aspect ratios are 1.19, 1.68, 2.72, and $\sqrt{2}$ for (approximate) $\bar{E}$ values of 1.97, .51, .07, and 1.0, respectively. In Fig. 2.38 we show a plot, from [96], of the variation of the buckling load parameter $P_{cr} b/4\pi^2 D_{11}$ against $\Lambda = m/\phi$; the graph indicates that the buckling load parameter increases with $\Lambda$, monotonically, for the isotropic plate, and plates $A$–$F$, but for plates $L, K, J, I, H$, and $G$ the buckling parameter first decreases, and then increases again, as $\Lambda$ is increased. Finally, in Figs. 2.39 through 2.45, we show some of the (postbuckling) load-shortening curves from [96]; from the graphs it may be deduced that postbuckling strength in the case of rectangular orthotropic plates depends on the ratios $\bar{E}$, $G_{12}/E_{11}$, $\nu_{12}$, and the parameter $\Lambda = m/\phi$, where $m$ is the number of half-sine waves in the $x$-direction which are present in the buckling mode. In particular, plates $G$–$E$ (for which $\bar{E} > 1$) have a higher postbuckling strength as compared with plates $A$–$F$ (for which $\bar{E} < 1$). For $\Lambda = 1$, plates $E, F, I, J, K$, and $L$ possess distinct postbuckling strengths, whereas the postbuckling behavior of plates $A$–$D$ is the same as that of the isotropic plate; however, as $\Lambda$ increases, the postbuckling strengths (load-shortening curves) of plates $A$–$F$ are very different from that of the isotropic case (i.e., Fig. 2.41). The graph in Fig. 2.43 is a plot of non-dimensional effective width against non-dimensional load $P/P_{cr}$ for several of the cases in Table 2.5. The last of this series of graphs, i.e., Fig. 2.45, displays some results from [96] for the isotropic plate that show the change in the nature of the postbuckling curves (plotted as non-dimensionalized load against non-dimensionalized end shortening) as $\Lambda$ is varied; the authors [96] indicate that the results displayed here are very close to those obtained earlier by Stein [93] for postbuckling of an isotropic plate.

## 2.2 Plates with Nonlinear Elastic Behavior: The Johnson-Urbanik Generalization of the von Karman Equations

The only generalization of the von Karman equations in the literature which has been formulated *directly* to deal with nonlinear elastic, rectilinearly orthotropic behavior appears to be that of Johnson and Urbanik [73], [74]; in these papers a theory of thin plates is developed (which is physically, as well as kinematically, nonlinear) and used to characterize elastic material behavior for arbitrary stretching and bending deformations. The theory is based on the use of an effective strain measure; it begins by simplifying the equations of three-dimensional nonlinear elasticity through use of a scaling argument, and then proceeds by partially integrating the resulting equations to obtain two-dimensional plate equations. The results in [73], [74] cover only the case of initial buckling, i.e., an equation analogous to the second von Karman equation (1.69) is not derived; also, the extant results are limited only to applications to orthotropic, nonlinear elastic rectangular plates.

The basic equations of three-dimensional nonlinear elasticity are the equilibrium equations

$$\frac{\partial P_{ij}}{\partial x_j} = 0, \quad i = 1, 2, 3, \tag{2.172}$$

where the Piola-Kirchoff stress tensor $P_{ij}$ is related to the Cauchy stress tensor $\sigma_{ij}$ by,

$$P_{ij} = \left(\delta_{jk} + \frac{\partial u_j}{\partial x_k}\right)\sigma_{ik} \tag{2.173}$$

The constitutive relations are given by

$$\sigma_{ij} = \frac{\partial H(\boldsymbol{E})}{\partial E_{ij}} \tag{2.174}$$

with the strain-displacement relations

$$E_{ij} = \frac{1}{2}\left(\frac{\partial u_i}{\partial x_j} + \frac{\partial u_j}{\partial x_i} + \frac{\partial u_k}{\partial x_i}\frac{\partial u_k}{\partial x_j}\right) \tag{2.175}$$

In (2.172) - (2.175), $x_1 = x$, $x_2 = y$, $x_3 = z$, the $u_i$ are the displacement components and $H$ is the strain-energy function. In general, the $u_i$ in (2.175) and, thus, the $E_{ij}$ may depend on $z$. In [73] the authors take the middle surface of the undeformed plate to be given by $x_3 = 0$, the thickness to be $h$, and introduce a lateral length scale $L$, so that for a thin plate $\epsilon = h/L$ is a small parameter. Scaling the coordinates, displacement components, and strain components as in the von Karman theory (i.e., $x_1, x_2 = \mathcal{O}(L)$, $x_3 = \mathcal{O}(h)$, $u_1, u_2 = \mathcal{O}(\epsilon h)$, $u_3 = \mathcal{O}(h)$, $E_{11}, E_{22}, E_{12}, E_{33} = \mathcal{O}(\epsilon^2)$, and $E_{13}, E_{23} = \mathcal{O}(\epsilon^3)$) it turns out that the lowest order terms in the strain-displacement relations (2.175) can be written in the form

$$\begin{cases} E_{11} = \epsilon_1 - K_1 x_3 \\ E_{22} = \epsilon_2 - K_2 x_3 \\ E_{12} = \epsilon_{12} - K_{12} x_3 \end{cases} \tag{2.176}$$

and a *direct* correlation with (1.35) may be made as follows: $E_{11} = \epsilon_{xx}$, $E_{22} = \epsilon_{yy}$, $E_{12} = \epsilon_{xy}$, $x_3 = \zeta$, $K_1 = w_{,xx}$, $K_2 = w_{,yy}$, $K_{12} = w_{,xy}$, while

$$\begin{cases} \epsilon_1 = u_{,x} + \dfrac{1}{2}w_{,x}^2 \\[2mm] \epsilon_2 = v_{,y} + \dfrac{1}{2}w_{,y}^2 \\[2mm] \epsilon_{12} = \dfrac{1}{2}(u_{,y} + v_{,x} + w_{,x}w_{,y}) \end{cases} \qquad (2.177)$$

are the middle surface strain components. The approximate relation between the displacements $u_i$, $i = 1, 2, 3$, in general, and the $u, v, w$ in (2.177), which are functions of $x, y$ only, has been given before as (1.30), i.e.

$$\begin{cases} u_3 = w(x, y) \\[2mm] u_1 = u(x, y) - z\dfrac{\partial w}{\partial x} \\[2mm] u_2 = v(x, y) - z\dfrac{\partial w}{\partial y} \end{cases} \qquad (2.178)$$

We now give below a brief alternative derivation of the moment- curvature relations (deduced in [73]) which are to be substituted into the same equilibrium equation, i.e. (1.45), which applies in the case of linear elastic response; in using (1.45) we will, for now, set $W \equiv 0$, and $t \equiv 0$. Thus, we must relate $M_{11} = M_x$, $M_{12} = M_{xy}$, and $M_{22} = M_y$ to $K_1, K_2$, and $K_3$ just as we did in (1.39) and (1.41), for the linear elastic isotropic case, and in (1.67a,b,c) for the linear elastic rectilinearly orthotropic case.

From (2.174) we have ($\sigma_{11} \equiv \sigma_{xx}$) :

$$\sigma_{11} \simeq \frac{\partial H(\boldsymbol{E})}{\partial E_{11}} \qquad (2.179)$$

If we set $\tilde{H}(\epsilon_1, \epsilon_2, \epsilon_3) = H(\boldsymbol{E})\Big|_{x_3=0}$, then

$$H(\boldsymbol{E}) \simeq \tilde{H}(\epsilon_1, \epsilon_2, \epsilon_3) + \frac{\partial H}{\partial x_3}\Big|_0 x_3 + \frac{1}{2}\frac{\partial^2 H}{\partial x_3^2}\Big|_0 x_3^2 \qquad (2.180)$$

where the zero subscripts denote evaluation at $x_3 = 0$. Then,

$$\left.\frac{\partial H}{\partial x_3}\right|_0 = \left.\frac{\partial H}{\partial E_{11}}\right|_{x_3=0} \cdot (-K_1) + \left.\frac{\partial H}{\partial E_{22}}\right|_{x_3=0} \cdot (-K_2)$$

$$+ \left.\frac{\partial H}{\partial E_{12}}\right|_{x_3=0} \cdot (-K_{12}) \tag{2.181}$$

$$\simeq -\frac{\partial \tilde{H}}{\partial \epsilon_1} K_1 - \frac{\partial \tilde{H}}{\partial \epsilon_2} K_2 - \frac{\partial \tilde{H}}{\partial \epsilon_{12}} K_{12}$$

On the other hand, for $|x_3|$ small we have, by virtue of (2.179)

$$\sigma_{11} \simeq \frac{\partial \tilde{H}}{\partial \epsilon_1} + \frac{\partial}{\partial \epsilon_1}\left(\left.\frac{\partial H}{\partial x_3}\right|_0\right) x_3 + \frac{1}{2}\frac{\partial}{\partial \epsilon_1}\left(\left.\frac{\partial^2 H}{\partial x_3^2}\right|_0\right) x_3^2 \tag{2.182}$$

in which case

$$M_x \equiv \int_{-h/2}^{h/2} \sigma_{xx}\zeta d\zeta$$

$$\simeq \int_{-h/2}^{h/2} \frac{\partial \tilde{H}}{\partial \epsilon_1}\zeta d\zeta + \int_{-h/2}^{h/2} \frac{\partial}{\partial \epsilon_1}\left(\left.\frac{\partial H}{\partial x_3}\right|_0\right)\zeta^2 d\zeta$$

$$+ \frac{1}{2}\int_{-h/2}^{h/2} \frac{\partial}{\partial \epsilon_1}\left(\left.\frac{\partial^2 H}{\partial x_3^2}\right|_0\right)\zeta^3 d\zeta \tag{2.183}$$

$$\equiv \int_{-h/2}^{h/2} \frac{\partial}{\partial \epsilon_1}\left(\left.\frac{\partial H}{\partial x_3}\right|_0\right)\zeta^2 d\zeta$$

Finally, if we use (2.181) in (2.183), we have

$$M_x = \left(-\kappa_1 \frac{\partial^2 \tilde{H}}{\partial \epsilon_1^2} - \kappa_2 \frac{\partial^2 \tilde{H}}{\partial \epsilon_1 \partial \epsilon_2} - \kappa_{12}\frac{\partial^2 \tilde{H}}{\partial \epsilon_1 \partial \epsilon_{12}}\right)\int_{-h/2}^{h/2}\zeta^2 d\zeta$$

or

$$M_x = -\frac{h^3}{12}\left(\kappa_1 \tilde{H}_{11} + \kappa_2 \tilde{H}_{12} + \kappa_{12}\tilde{H}_{13}\right) \tag{2.184}$$

where the $\tilde{H}_{ij}$ are the bending stiffness moduli, $\tilde{H}_{11} = \partial^2 \tilde{H}/\partial \epsilon_1^2$, $\tilde{H}_{12} = \partial^2 \tilde{H}/\partial \epsilon_1 \partial \epsilon_2$, $\tilde{H}_{13} = \partial^2 \tilde{H}/\partial \epsilon_1 \partial \epsilon_{12}$, etc. The other two moment-curvature relations are

$$M_y = -\frac{h^3}{12}\left(\kappa_1 \tilde{H}_{12} + \kappa_2 \tilde{H}_{22} + \kappa_3 \tilde{H}_{23}\right) \tag{2.185}$$

$$M_{xy} = -\frac{h^3}{24}\left(\kappa_1 \tilde{H}_{13} + \kappa_2 \tilde{H}_{23} + \kappa_{12}\tilde{H}_{33}\right) \qquad (2.186)$$

As noted in [73] , linear isotropic plate theory corresponds to the (approximate) strain-energy

$$\tilde{H} = \frac{E}{2(1-\nu^2)}(\epsilon_1 + \epsilon_2)^2 - \frac{E}{1+\nu}(\epsilon_1\epsilon_2 - \epsilon_{12}^2) \qquad (2.187)$$

while linear orthotropic plate theory corresponds to

$$\tilde{H} = \frac{E_1}{2(1-\nu_{12}\nu_{21})}(\epsilon_1^2 + \nu_{21}\epsilon_1\epsilon_2)$$

$$+\frac{E_2}{2(1-\nu_{12}\nu_{21})}(\epsilon_2^2 + \nu_{12}\epsilon_1\epsilon_2) \qquad (2.188)$$

$$+2G_{12}\epsilon_{12}^2$$

For a nonlinear orthotropic elastic material it has been shown (e.g., Green and Adkins [97]) that

$$\tilde{H} = \tilde{H}(\epsilon_1, \epsilon_2, \epsilon_{12}^2) \qquad (2.189)$$

We also note that, for the constitutive theory at hand, it follows directly that the averaged stresses are

$$\begin{cases} N_x = h\dfrac{\partial \tilde{H}}{\partial \epsilon_1}, \quad N_y = h\dfrac{\partial \tilde{H}}{\partial \epsilon_2} \\[2mm] N_{xy} = N_{yx} = \dfrac{1}{2}h\dfrac{\partial \tilde{H}}{\partial \epsilon_{12}} \end{cases} \qquad (2.190)$$

Now, as noted by Johnson and Urbanik [73], because $\tilde{H}$ is a function of the three middle surface strains $\epsilon_1, \epsilon_2$, and $\epsilon_{12}$, in order to completely determine $\tilde{H}$ one would need a complete set of data on biaxial stretching and shear of a rectangular plate; because such data was not available at the time of the writing of [73], the analysis in [73] proceeds by noting that the linear orthotropic relation (2.188) can be written in the equivalent form

$$\tilde{H} = \frac{\nu_{21}E_1}{2(1-\nu_{12}\nu_{21})}e \qquad (2.191)$$

where $e$ is given by

$$
\begin{cases}
e = \nu_{21}^{-1}\epsilon_1^2 + \nu_{12}^{-1}\epsilon_2^2 + 2\epsilon_1\epsilon_2 + c\epsilon_{12}^2 \\[2mm]
c = \dfrac{4(1 - \nu_{12}\nu_{21})}{\nu_{12}E_2}G_{12}
\end{cases}
\tag{2.192}
$$

The quantity $\sqrt{e}$ is an effective strain measure and, in [73] and [74], an approximate theory of nonlinear elastic orthotropic behavior is formulated in analogy with (2.191), by assuming that $\tilde{H} = \tilde{H}(e)$; the functional form of $\tilde{H}(e)$ which is appropriate, e.g., for paper, must be determined by experiment. With the assumption that $H = \tilde{H}(e)$, the average stresses in (2.190) take the form

$$
\begin{cases}
N_x = 2h\tilde{H}'(e)(\nu_{21}^{-1}\epsilon_1 + \epsilon_2) \\[2mm]
N_y = 2h\tilde{H}'(e)(\nu_{12}^{-1}\epsilon_2 + \epsilon_1) \\[2mm]
N_{xy} = N_{yx} = ch\tilde{H}'(e)\epsilon_{12}
\end{cases}
\tag{2.193}
$$

and explicit expressions for the $\tilde{H}_{ij}$ in (2.185) - (2.187), in terms of $\tilde{H}'(e)$ and $\tilde{H}''(e)$, may be computed as well. For uniaxial extension in the $y$ direction, $N_x = 0$, which implies that $\epsilon_1 = -\nu_{21}\epsilon_2$; also, in that particular situation, it follows from (2.192), (2.193), that $e = (\nu_{12}^{-1} - \nu_{21})\epsilon_2^2$ and, as $\epsilon_{12} = 0$, $N_{xy} = N_{yx} = 0$, while $N_y = 2h(\nu_{12}^{-1} - \nu_{21})\tilde{H}'(e)\epsilon_2$. Explicit forms for the $\tilde{H}_{ij}$, for this case, are also computed in [73], i.e.

$$
\begin{cases}
\tilde{H}_{11} = 2\nu_{21}^{-1}\tilde{H}'(e) \\[2mm]
\tilde{H}_{22} = 4(\nu_{12}^{-1} - \nu_{21})^2\tilde{H}''(e)\epsilon_2^2 + 2\nu_{12}^{-1}\tilde{H}'(e) \\[2mm]
\tilde{H}_{12} = 2\tilde{H}'(e), \quad \tilde{H}_{33} = 2c\tilde{H}'(e) \\[2mm]
\tilde{H}_{13} = \tilde{H}_{23} = 0
\end{cases}
\tag{2.194}
$$

and analogous results hold for uniaxial extension in the $x$ direction.

Several sets of comparison (experimental) data for paper are reported in [74] which appear to conform to the following form of the strain energy function $\tilde{H}$:

$$
\tilde{H}(e) = \left(\frac{c_1^2}{c_2}\right) \ln \cosh \left(\frac{c_2}{c_1}\sqrt{\frac{\nu_{12}e}{1 - \nu_{12}\nu_{21}}}\right)
\tag{2.195}
$$

In (2.195) the parameters $c_1$ and $c_2$ must be determined through an experimental data fit. For uniaxial deformation in the $x$-direction, (2.195) leads to

$$\sigma_{xx}(\epsilon_1) = c_1 \sqrt{\frac{\nu_{12}}{\nu_{21}}} \tanh\left(\frac{c_2}{c_1}\sqrt{\frac{\nu_{12}}{\nu_{21}}}\epsilon_1\right) \tag{2.196}$$

while for uniaxial deformation in the $y$-direction one obtains

$$\sigma_{yy}(\epsilon_2) = c_1 \tanh(c_2\epsilon_2/c_1) \tag{2.197}$$

Finally, if we employ the form (2.195) of the strain energy function $\tilde{H}(e)$, in (2.194), then for compression in the $y$-direction of the plate one finds that the bending stiffness moduli are given by

$$
\begin{cases}
\tilde{H}_{11} = \dfrac{c_1\nu_{12}/\nu_{21}}{1 - \nu_{12}\nu_{21}} \cdot \dfrac{\tanh(c_2\epsilon_2/c_1)}{\epsilon_2} \\[2ex]
\tilde{H}_{22} = \dfrac{c_2}{\cosh^2(c_2\epsilon_2/c_1)} + \dfrac{\nu_{12}\nu_{21}c_1}{1 - \nu_{12}\nu_{21}} \cdot \dfrac{\tanh(c_2\epsilon_2 K_1)}{\epsilon_2} \\[2ex]
\tilde{H}_{12} = \dfrac{c_1\nu_{12}}{1 - \nu_{12}\nu_{21}} \cdot \dfrac{\tanh(c_2\epsilon_2 K_1)}{\epsilon_2} \\[2ex]
\tilde{H}_{33} = c\tilde{H}_{12} \\[2ex]
\tilde{H}_{13} = \tilde{H}_{23} = 0
\end{cases}
\tag{2.198}
$$

with $c$ defined as in (2.192). For situations in which the shear modulus $G_{12}$ is not known Panc [98] and others, have suggested the formula

$$G_{12} = \frac{\sqrt{E_1 E_2}}{2(1 + \sqrt{\nu_{12}\nu_{21}})} = \frac{\sqrt{\nu_{12}/\nu_{21}}E_2}{2(1 + \sqrt{\nu_{12}\nu_{21}})} \tag{2.199}$$

in which case $c$ may be expressed in the form

$$c = \frac{2\left(1 - \sqrt{\nu_{12}\nu_{21}}\right)}{\sqrt{\nu_{12}\nu_{21}}} \tag{2.200}$$

As an example of the kind of initial buckling problem that can be handled within the context of the Johnson-Urbanik theory, consider a rectangular, orthotropic, nonlinearly elastic plate which is in a state of uniform axial compression in the $y$-direction, assumed e.g., to be the cross-machine direction (CD) of a rectangular piece of paper; then $N_x =$

$N_{xy} = 0$, while the second relation in (2.193), coupled with $\epsilon_1 = -\nu_{21}\epsilon_2$, and (2.195), yield

$$N_y = -N_0 = -hc_1 \tanh(c_2\epsilon_2/c_1) \tag{2.201}$$

Using (1.45), and (2.184) - (2.186), and assuming that the middle surface strain field is homogeneous, so that the $\tilde{H}_{ij}$ in (2.198) are independent of $x$ and $y$, we obtain as the equation governing the initial buckling mode

$$\tilde{H}_{11}\frac{\partial^4 w}{\partial x^4} + \tilde{H}_{22}\frac{\partial^4 w}{\partial y^4} + \left(2\tilde{H}_{12} + \tilde{H}_{33}\right)\frac{\partial^4 w}{\partial x^2 \partial y^2} + \frac{12}{h^3}N_0\frac{\partial^2 w}{\partial y^2} = 0 \tag{2.202}$$

In [73] solutions of (2.202) are sought in the form

$$w(x,y) = Ae^{i\lambda y + \alpha x} \tag{2.203}$$

Substitution of (2.203) into (2.202) leads to a fourth-degree equation for $\alpha(\lambda)$, with four roots of the form $\alpha, -\alpha, i\beta$, and $-i\beta$, which we will not delineate here, but which satisfy

$$\alpha^2 - \beta^2 = \frac{\left(2\tilde{H}_{12} + \tilde{H}_{33}\right)\lambda^2}{\tilde{H}_{11}} \tag{2.204}$$

These roots lead to solutions (even in $x$) of the type

$$w = [a_1 \cosh \alpha x + a_2 \cos \beta x]\,e^{i\lambda y} \tag{2.205}$$

and to solutions (odd in $x$) of the form

$$w = [a_3 \sinh \alpha x + a_4 \sin \beta x]\,e^{i\lambda y} \tag{2.206}$$

It is noted in [73] that the even solutions lead to lower buckling loads than the odd solutions; the imposition of support conditions along the edges that are parallel to the $x$-axis, in an (assumed) long plate, then yields transcendental characteristic equations for the buckling strain as a function of the wave number $\lambda$ and the critical buckling strain is obtained by minimizing the buckling strain with respect to $\lambda$. If we introduce the dimensionless (buckling) strain $\hat{\epsilon} = c_2\epsilon_2/c_1$, and the dimensionless stiffness coefficient $\mathcal{S} = (c_2/c_1)(h/2a)^2 \sqrt{\nu_{12}/\nu_{21}}$, where $a$ is the half-width of the plate in the $x$ direction, then Fig. 2.46 depicts a plot of $\hat{\epsilon}$ as a function of $\mathcal{S}$ for three mean Poisson's ratios $\nu$ $(= \sqrt{\nu_{12}\nu_{21}})$ and two different edge support conditions along the non-loaded edges at $x = \pm a$; similarly, Fig. 2.47 is a plot of the corresponding normalized buckling stress $\hat{\sigma}$ against $\mathcal{S}$ for the same three mean Poisson's ratios and the same two-edge support conditions along $x = \pm a$. The ratio of the clamped edge buckling stress to the simply supported buckling stress is graphed in Fig. 2.48 as a function of $\mathcal{S}$ for the same three values of the mean Poisson ration $\nu$. Finally, in Fig. 2.49 we depict, for the same type of edge supports along $x = \pm a$ and the same mean Poisson's ratios, the predicted variation with $\mathcal{S}$ of the ratio of the buckling stress given by the nonlinear elastic plate theory of Johnson-Urbanik to the buckling stress for a linear elastic, rectilinearly orthotropic plate, with the same $E_1, E_2, \nu_{12}, \nu_{21}$, and $G_{12}$.

---

## 2.3   Plates which Exhibit Elastic-Plastic or Viscoelastic Behavior

For an elastic plate which is perfectly flat, buckling occurs at a critical value $\lambda_c$ of the elastic stress; loads associated with stresses larger than $\lambda_c$ can then be supported (postbuckling theory), as the lateral deflections grow, because of stretching of the middle surface of the plate. Eventually, a maximum load associated with a value $\lambda_{\max}$ of the applied stress field is reached, at which point the increase of deflection with load is limited due to plastic breakdown of the plate. For most cases involving the postbuckling behavior of rectangular plates, it is usually true that $\lambda_{\max} > \lambda_c$; however, it is possible for breakdown to occur when the plate is still flat and in a state of plane stress (before $\lambda_c$ is reached). If this happens, the plate will buckle and collapse at a lower value of the applied stress than $\lambda_c$, i.e., $\lambda_{\max} < \lambda_c$ and plastic buckling occurs. As noted in [17] the development of plastic buckling theory for plates parallels that of the theory for plastic buckling of struts. Considère [99] and von Karman [100] considered the idea of replacing the Young's modulus $E$ in the elastic buckling formula for struts by a reduced modulus $\bar{E}$; as noted in Hutchinson [17], Shanley [101] showed that for plastic buckling of struts, within the context of a $J_2$ flow theory, the correct value of the reduced modulus was the tangent modulus $E_t$ (see Fig. 2.50), which in a $J_2$ plastic flow theory, e.g., is a function of $J_2$ and is defined, in the usual way by $\dot{\sigma} = E_t \dot{\epsilon}$ for a uniaxial tensile history. The other modulus of importance, which occurs in theories of plastic buckling, is associated with $J_2$ deformation theories of plasticity, namely, the secant modulus $E_s$ which is defined by $E_s = \sigma/\epsilon$ for uniaxial tension tests. In addition to replacing (for an isotropic plate) the Young's modulus $E$ by a reduced modulus $\bar{E}$, the Poisson's ratio $\nu$ is usually also replaced by a modified ratio $\bar{\nu}$, although this is not uniformly done in all theories of plastic buckling of plates. Specific forms for $\bar{E}$ and $\bar{\nu}$ which apply to the plastic buckling of a clamped circular plate, subject to a uniform radial stress, are delineated in Hutchinson [17] and will be discussed in Chapter 3.3. We now briefly discuss some of the existing results for plastic buckling of (initially) isotropic elastic rectangular plates, both within the context of a $J_2$ deformation theory — which some authors believe should be favored for engineering prediction purposes over incremental flow theories with smooth yield surfaces — as well as within the context of a $J_2$ flow theory. In addition, we take note of the work of Stowell [102], who, using a total strain theory in which $\sigma = E_s \epsilon$ for a material loaded uniaxially, deduces, in the plastic range, the relations

$$\begin{cases} \epsilon_{xx} = \dfrac{1}{E_s}\left(\sigma_{xx} - \dfrac{1}{2}\sigma_{yy}\right) \\[2mm] \epsilon_{yy} = \dfrac{1}{E_s}\left(\sigma_{yy} - \dfrac{1}{2}\sigma_{xx}\right) \\[2mm] \gamma_{xy} = \dfrac{3}{E_s}\left(\sigma_{xy}\right) \end{cases} \tag{2.207}$$

The moment-curvature relations computed in [102] have the form

$$\begin{cases} M_x = -K'\left[\left(\dfrac{1}{4} + \dfrac{3E_t}{4E_s}\right)w_{,xx} + \dfrac{1}{2}w_{,yy}\right] \\[2mm] M_y = -K'\left[w_{,yy} + \dfrac{1}{2}w_{,xx}\right] \\[2mm] M_{xy} = -\dfrac{1}{2}K'w_{,xy} \end{cases} \tag{2.208}$$

with

$$K' = E_s h^3/12(1 - \nu^2) \equiv E_s h^3/9 \tag{2.209}$$

and the corresponding equilibrium equation governing initial buckling of a plate compressed in the $x$-direction is

$$\left(\dfrac{1}{4} + \dfrac{3E_t}{4E_s}\right)\dfrac{\partial^4 w}{\partial x^4} + 2\dfrac{\partial^4 w}{\partial x^2 \partial y^2} + \dfrac{\partial^4 w}{\partial y^4} = -\dfrac{\lambda h}{K'}\dfrac{\partial^2 w}{\partial x^2} \tag{2.210}$$

As in the case of elastic buckling, (2.210) must now be solved subject to known support conditions along the edges. For a long plate whose unloaded edges are simply supported Stowell [102] reports the critical plastic stress for one transverse half-wave at buckling as

$$\lambda_c^P = \frac{\pi^2 E_s}{9} \left(\frac{h}{b}\right)^2 \left[\left(\frac{1}{4} + \frac{3}{4}\frac{E_t}{E_s}\right)\frac{1}{\phi^2} + \phi^2 + 2\right] \tag{2.211}$$

where $\phi = a/b$ is, once again, the aspect ratio of the plate. It is noted in [102] that the formula in (2.211) is similar to the one obtained for elastic buckling of a compressed plate which is simply supported along its unloaded edges, except that $\nu$ is replaced by $1/2$, $E$ by $E_s$, and a factor multiplying the term $1/\phi^2$ has been introduced. In fact, as indicated in [102], the ratio of the minimum plastic buckling stress $(\lambda_c^P)_{\min}$ to the minimum elastic buckling stress $(\lambda_c^e)_{\min}$, in this case, is given by

$$\eta = \frac{(\lambda_c^P)_{\min}}{(\lambda_c^e)_{\min}} = \frac{4E_s}{3E} \left(\frac{1}{2} + \frac{1}{2}\sqrt{\frac{1}{4} + \frac{3E_t}{4E_s}}\right)(1 - \nu^2) \tag{2.212}$$

where $\eta$ is called the *plasticity reduction factor*. From (2.211) it follows that (initially) isotropic rectangular plates will buckle plastically with a reduced wave length, as opposed to the case of elastic buckling, because the smallest value of the critical stress occurs for

$$\phi = \sqrt[4]{\frac{1}{4} + \frac{3E_t}{4E_s}} \tag{2.213}$$

and $E_t < E_s$. Besides the condition of simple support along the edges at $y = 0$, $y = b$, $0 \le x \le a$, which has been discussed above, for end compression of a rectangular (initially isotropic) plate, the following plasticity reduction factors have been reported in Bulson [32] and are obtained by solving (2.210) subject to the boundary conditions indicated:

(i) $y = 0$ simply supported, $y = b$ free,

$$\eta = \frac{4E_s}{3E}(1 - \nu^2) \tag{2.214}$$

(ii) $y = 0$ clamped, $y = b$ free,

$$\eta = \frac{4E_s}{3E}\left(\frac{1}{3} + \frac{2}{3}\sqrt{\frac{1}{4} + \frac{3E_t}{4E_s}}\right)(1 - \nu^2) \tag{2.215}$$

(iii) $y = 0$, $y = b$ clamped,

$$\eta = \frac{4E_s}{3E} \left( .352 + .648\sqrt{\frac{1}{4} + \frac{3E_t}{4E_s}} \right) (1 - \nu^2) \qquad (2.216)$$

Much of the more recent work in plastic buckling, i.e., Hutchinson [17], Tvergaard [103], Hutchinson [45], [46], and Needleman and Tvergaard [104] is based on a small strain theory of plasticity in which the relation between the stress rates and strain rates is assumed to be of the form

$$\dot{\sigma}_{ij} = L_{ijk\ell}\dot{\epsilon}_{kl} \qquad (2.217)$$

with the $L_{ijk\ell}$ "instantaneous" moduli which depend, in general, on the stress history; it is further assumed that there are two branches of the tensor $L_{ijk\ell}$, depending on whether loading or unloading occurs. For each of the two basic phenomenological plasticity theories alluded to above (flow versus deformation) the plastic deformation depends only on the $J_2$ invariant

$$J_2 = \frac{1}{2}S_{ij}S_{ij} \simeq 1/3 \left( \sigma_{11}^2 + \sigma_{22}^2 + 3\sigma_{12}^2 - \sigma_{11}\sigma_{22} \right) \qquad (2.218)$$

where $S_{ij} = \sigma_{ij} - \frac{1}{3}\sigma_{kk}\delta_{ij}$ is the stress deviator. The instantaneous moduli for $J_2$ flow theory with isotropic elastic properties, are given, e.g., in Hutchinson [17] as

$$L_{ijk\ell} = \frac{E}{1+\nu} \left\{ \frac{1}{2} (\delta_{ik}\delta_{j\ell} + \delta_{i\ell}\delta_{jk}) \right.$$
$$\left. + \frac{\nu}{1-2\nu}\delta_{ij}\delta_{kl} - \frac{\mathcal{F}\delta_{ij}S_{kl}}{1+\nu+2\mathcal{F}J_2} \right\} \qquad (2.219)$$

where

$$\mathcal{F} \equiv 0, \quad \text{for } \dot{J}_2 < 0 \text{ or } J_2 < (J_2)_{\max},$$
$$\text{i.e., in the elastic range or for elastic unloading}$$

and

$$\mathcal{F} = \frac{3}{4J_2}\left(\frac{E}{E_t} - 1\right), \quad \text{for } J_2 = (J_2)_{\text{max}} \text{ and } \dot{J}_2 \geq 0,$$

$$\text{i.e., for loading in the plastic range}$$

For $J_2$ deformation theory, with unloading incorporated,

$$L_{ijk\ell} = \frac{E}{1 + \nu + \mathcal{G}}\left\{\frac{1}{2}\left(\delta_{ik}\delta_{j\ell} + \delta_{i\ell}\delta_{jk}\right)\right.$$

$$\left. + \frac{3\nu + \mathcal{G}}{3(1 - 2\nu)}\delta_{ij}\delta_{k\ell} - \frac{\mathcal{G}'S_{ij}S_{k\ell}}{1 + \nu + \mathcal{G} + 2\mathcal{G}'J_2}\right\} \tag{2.220}$$

where $\mathcal{G}' = \dfrac{d\mathcal{G}}{dJ_2}$; for unloading $\mathcal{G} = \mathcal{G}' = 0$, while for loading, $\mathcal{G} = 3\left(\dfrac{E}{E_s} - 1\right)$. The relations (2.219), (2.220), however, were proposed in conjunction with the problem of buckling of a complete spherical shell under external pressure. For the specific problem of buckling of square elastic-plastic plates under axial compression, Needleman and Tvergaard [104] assume (2.217), with the elastic tensor of moduli $L_{ijk\ell}^e$ isotropic, i.e.,

$$L_{ijk\ell}^e = \frac{E}{1 + \nu}\left[\frac{1}{2}\left(\delta_{ik}\delta_{j\ell} + \delta_{i\ell}\delta_{jk}\right) + \frac{\nu}{1 - 2\nu}\delta_{ij}\delta_{k\ell}\right] \tag{2.221}$$

and then restricting themselves to a $J_2$ incremental flow theory take

$$L_{ijk\ell} = L_{ijk\ell}^e - \mathcal{H}\frac{S_{ij}S_{k\ell}}{\sigma_e^2} \tag{2.222}$$

where

$$\sigma_e = \sqrt{\frac{3}{2}S_{ij}S_{ij}} \tag{2.223}$$

and

$$\mathcal{H} = \begin{cases} 0, & \text{if } \sigma_e < \mathcal{Y} \text{ or } \dot{\sigma}_e < 0 \\[2ex] \dfrac{3}{2} \cdot \dfrac{E}{1+\nu}\left\{\dfrac{1 - E_t/E}{\dfrac{2}{3}(1+\nu)\dfrac{E_t}{E} + 1 - (E_t/E)}\right\}, & \text{if } \sigma_e = \mathcal{Y} \text{ and } \dot{\sigma}_e \geq 0 \end{cases} \tag{2.224}$$

In (2.224), $\mathcal{Y}$ is the flow (yield) stress which is taken to be the greater of the maximum value of $\sigma_e$ (over the stress history) and the (initial) yield stress $\sigma_y$ associated with uniaxial stress-strain behavior; the uniaxial stress-strain behavior chosen is that of a piecewise power law with yield stress $\sigma_y$ and a continuous tangent modulus, i.e.

$$\frac{\epsilon}{\epsilon_y} = \begin{cases} \sigma/\sigma_y, & \text{for } \sigma < \sigma_y \\ \dfrac{1}{n}\left(\dfrac{\sigma}{\sigma_y}\right)^n + (1 - \dfrac{1}{n}), & \text{for } \sigma \geq \sigma_y \end{cases} \tag{2.225}$$

The parameter $n$ in (2.225) is the strain hardening exponent and $\epsilon_y = \sigma_y/E$.

In [104] it is assumed that the plate is square (of side $a$), is simply supported on all four edges, and is uniformly compressed in the $x$-direction by a uniformly distributed load with $N_x = -\lambda h$ along $x = 0$, $x = a$, for $0 \leq y \leq a$; bifurcation from the associated plane stress solution associated with this loading first becomes possible when $\lambda = \lambda_c$, where

$$\lambda_c = \frac{E}{12}\left(\frac{\pi h}{a}\right)^2 \left[\frac{2}{1+\nu} + \frac{9 + \dfrac{E_t}{E}(8\nu - 1)}{(5 - 4\nu) - \dfrac{E_t}{E}(1 - 2\nu)^2)}\right] \tag{2.226}$$

while the corresponding eigenmode (displacement) turns out to be

$$w_c(x, y) = \pm h \sin\frac{\pi x}{a}\sin\frac{\pi y}{a} \tag{2.227}$$

A perturbation analysis, similar to those we have used previously, is then employed to analyze the postbuckling behavior in the elastic range, i.e., for $\lambda_c < \sigma_y$; the initial postbuckling behavior parallels that for postbuckling of a simply supported, linearly elastic, isotropic square plate compressed in the $x$-direction. For $\lambda_c > \sigma_y$, the initial postbuckling behavior is determined by using Hutchinson's asymptotic theory of postbuckling in the plastic range [46]; we summarize the results of the analysis as follows: with the amplitude of the eigenmode (2.227) denoted by $\xi$, so that $w(x, y) = \xi w_c(x, y) + \ldots$, the applied stress varies as $\lambda = \lambda_c + \lambda_1\xi + \ldots$. With the sign of $w_c$ chosen so that $\xi$ is positive it may be shown [45] that

$$\lambda_1 = \left(\frac{\Pi h}{a}\right)^2 E_t \frac{(1 + \nu)}{(5 - 4\nu) - \frac{E_t}{E}(1 - 2\nu)^2} \tag{2.228}$$

and the initial slope $\lambda_1$ of the $\lambda(\xi)$ curve is uniquely determined by the condition that plastic loading takes place through the plate except at one point where neutral loading occurs; in accordance with the original concept of Shanley [101], the slope $\lambda_1 > 0$, so that buckling takes place under increasing load. After buckling, a region of elastic unloading spreads out from the point of neutral loading and the next term in the postbuckling expansion for $\lambda$ takes the penetration of this region of elastic unloading into account and turns out to be proportional to $\xi^{4/3}$ (e.g., Hutchinson [17]):

$$\lambda = \lambda_c + \lambda_1 \xi + \lambda_2 \xi^{4/3} + \ldots \tag{2.229}$$

In all cases considered, the authors of [104] determine that $\lambda_2 < 0$ and, thus, the first three terms of (2.229) can be used to estimate the value of the maximum stress $\lambda_{\max}$ that can be supported by the plate. In [104] one also finds an analysis of the imperfection sensitivity of compressed, simply supported, elastic-plastic square plates which ignore elastic unloading; we will not go into any details of the imperfection-sensitivity analysis here but do note the following: attention in [104] is confined to imperfections which have the shape of the buckling mode (2.227) with the amplitude of the imperfection denoted by $\bar{\xi}$. For $\lambda < \lambda_c$, a regular perturbation analysis shows that the lowest order effect of the imperfection on the magnitude $\xi$, of the out-of-plane deflection of the plate, is given by

$$\xi = \bar{\xi}(1 - \lambda/\lambda_c)^{-\Psi} p(\lambda), \tag{2.230}$$

where $\Psi \geq 1$ is a (positive) constant, whose structure we do not delineate here, while $p(\lambda)$ is a function of $\lambda$ that is finite at $\lambda = \lambda_c$. For an elastic material (even nonlinear elastic) $\Psi = 1$. In a neighborhood of $\lambda = \lambda_c$, a singular perturbation analysis produces

$$\lambda = \lambda_c + \lambda_2 \xi^2 + c \bar{\xi}^{(2\Psi+1/\gamma\Psi)} \xi^{-(1/\Psi)} \tag{2.231}$$

We note that $c$ and $\gamma$ are undetermined by the analysis that produces (2.231) but are, subsequently, determined by matching the solutions (2.230) and (2.231), i.e., $\gamma = 2\Psi + 1$ and $c = -\lambda_c(p(\lambda_c))^{1/\Psi}$. If $\lambda_2 < 0$ then the maximum support stress $\lambda_{\max}$ which is predicted by (2.231) is given by

$$\frac{\lambda_{\max}}{\lambda_c} = 1 - \mu(\xi)^{2/(2\Psi+1)} \tag{2.232}$$

with

$$\mu = -\frac{\lambda_2}{\lambda_c}(2\Psi+1)\left(-\frac{2\Psi\lambda_2}{\lambda_c}\right)^{-2\Psi/(2\Psi+1)} \cdot p(\lambda_c)^{2/(2\Psi+1)}. \qquad (2.233)$$

The results of the analysis presented in [104] for three different cases are depicted in Figs. 2.51, 2.52, and 2.53.

Finally, we turn our attention to the analysis of the initial and post-buckling behavior of rectangular viscoelastic plates; a formulation of the buckling equations which are suitable for dealing with this problem may be found in the work of Brilla [106] although only the stability problems, both linearized and nonlinear are dealt with. The buckling of circular viscoelastic plates has been treated in [140] and will be discussed in the next chapter. The governing equations for the large deflection theory of viscoelastic plates considered in [106] are of two types, the first of which is of the form

$$\frac{h^3}{12}K_{ijk\ell}(D)w_{,ijk\ell} \qquad (2.234)$$
$$= K(D)(q + h\epsilon_{ik}\epsilon_{j\ell}w_{,ij}\,\Phi_{,k\ell})$$

with $i, j, k, \ell = 1, 2$, and

$$\epsilon_{im}\epsilon_{jn}\epsilon_{\ell s}L_{mnrs}(D)\Phi_{,ijk\ell} \qquad (2.235)$$
$$= -\frac{1}{2}\epsilon_{ik}\epsilon_{j\ell}L(D)w_{,ij}\,w_{,k\ell}$$

where

$$\begin{cases} K_{ijk\ell}(D) = \displaystyle\sum_{\gamma=0}^{r} K_{ijk\ell}^{(\gamma)}D^\gamma \\[2mm] L(D) = \displaystyle\sum_{\gamma=0}^{s} L_\gamma D^\gamma \end{cases} \qquad (2.236)$$

are polynomial operators in $D = \dfrac{\partial}{\partial t}$, the $L_{ijk\ell}(D)$, and $K(D)$, have analogous representations, and

$$K(D)\left[K_{ijk\ell}(D)\right]^{-1} = L(D)^{-1}L_{ijk\ell}(D) \qquad (2.237)$$

The operators $K(D), L(D), K_{ijk\ell}(D)$, and $L_{ijk\ell}(D)$ are the differential operators of linear viscoelasticity, $w$ and $\Phi$ are, of course, the transverse plate deflection and Airy function, respectively, $q$ is the transverse plate loading, $h$ the plate thickness, and $\epsilon_{ij}$ the standard alternating tensor. By $w_{,i}$ we mean $\dfrac{\partial}{\partial x_i}w$, etc., and we employ the usual convention of summing over repeated indices. The coefficients $K_{ijk\ell}$ are assumed to possess the symmetries given by

$$K_{ijk\ell}^{(\delta)} = K_{jik\ell}^{(\delta)} = K_{ij\ell k}^{(\delta)} = K_{\ell kij}^{(\delta)} \qquad (2.238)$$

and satisfy

$$K_{ijk\ell}^{(\delta)}(D)\epsilon_{ij}\epsilon_{k\ell} \geq 0 \qquad (2.239)$$

with equality holding only for $\epsilon=0$. The polynomials $K_{ijk\ell}$ and $L$ are assumed to have only negative real roots. The boundary conditions considered in [106] are either of the form

$$w = \frac{\partial w}{\partial n} = 0, \quad \text{on } \partial\Omega, \qquad (2.240)$$

$\partial\Omega$ being the boundary of the domain $\Omega$ occupied by the thin plate (in which (2.234), (2.235) are to hold), or

$$w = 0, \ K_{ijk\ell}(D)w_{,ij}\,\nu_{kn}\nu_{\ell n} = 0, \ \text{on } \partial\Omega \qquad (2.241)$$

where the $\nu_{kn} = \cos(x_k, \boldsymbol{n})$ are the direction cosines of the (unit) outward normal $\boldsymbol{n}$ to $\partial\Omega$. Also, with respect to the Airy function $\Phi$, we have either

$$\frac{\partial\Phi}{\partial n} = \frac{\partial^3\Phi}{\partial n^3} = 0, \quad \text{on } \partial\Omega \qquad (2.242)$$

or

$$\frac{\partial^2\Phi}{\partial n^2} = \frac{\partial^2\Phi}{\partial s\partial n} = 0, \ \text{on } \partial\Omega \qquad (2.243)$$

where $\dfrac{\partial}{\partial n}, \dfrac{\partial}{\partial s}$ denote, respectively, the normal and tangential derivatives on $\partial\Omega$. Finally, if the differential operator $L_{ijk\ell}(D)$ is of order $k$, then the initial conditions for the problem assume, in general, the form

$$\begin{cases} \dfrac{\partial^\gamma w}{\partial t^\gamma}\bigg|_{t=0} = w_\gamma \ (\gamma = 0, 1, \dots, k-1) \\[4mm] \dfrac{\partial^\gamma \Phi}{\partial t^\gamma}\bigg|_{t=0} = 0 \ (\gamma = 0, 1, \dots, k-1) \end{cases} \qquad (2.244)$$

A second set of governing equations considered in [106] are those appropriate for the large deflection theory of viscoelastic plates of Boltzmann type, namely,

$$
\frac{h^3}{12} \int_0^t G_{ijk\ell}(t-\tau) \frac{\partial}{\partial \tau} w_{,ijk\ell}(\tau) d\tau
$$

$$
= q + h\epsilon_{ik}\epsilon_{j\ell} w_{,ij} \, \Phi_{,k\ell}
$$

(2.245)

and

$$
\int_0^t \epsilon_{im}\epsilon_{jn}\epsilon_{kr}\epsilon_{\ell s} J_{mnrs}(t-\tau) \frac{\partial}{\partial t} \Phi_{,ijk\ell}(\tau) d\tau
$$

$$
= -\frac{1}{2}\epsilon_{ik}\epsilon_{j\ell} w_{,ij} \, w_{,k\ell}
$$

(2.246)

The boundary and initial conditions associated with the system (2.245), (2.246) are of the same type as those delineated, above, for the system (2.234), (2.235).

We begin our discussion of the analysis in [106] by detailing the description of the linearized stability problem, i.e., the problem of initial buckling. Since we are dealing now with a stability problem for a time-dependent process, we must consider perturbations from an equilibrium state; in the usual manner, we neglect the nonlinear terms in (2.234) and (2.235) when we address the linearized stability problem. We assume that, as a consequence of the loading and boundary conditions, the plate is subjected to an in-plane stress distribution of the form $\lambda\sigma_{ij}^{\circ}$. With $N_{ij}^{\circ} = -h\sigma_{ij}^{\circ}$, we have as a consequence of (2.234), with $q \equiv 0$, the following equation governing the initial buckling of the plate within the context of the (linear) viscoelasticity model of differential type:

$$
\frac{h^3}{12} K_{ijk\ell}(D) w_{,ijk\ell} + \lambda N_{ij}^{\circ} w_{,ij} = 0
$$

(2.247)

Equation (2.247) is subject to boundary conditions of the type (2.240) or (2.241) and initial conditions

$$
w|_{t=0} = w_0, \quad \left.\frac{\partial^\gamma w}{\partial t^\gamma}\right|_{t=0} = 0, \quad \gamma = 1, 2, \cdots, r-1
$$

(2.248)

If we seek a solution of (2.247) in the form

$$
w(x, y, t) = e^{\mu t} u(x, y)
$$

(2.249)

then it is easily seen that $u(x, y)$ must satisfy the equation

$$
\frac{h^3}{12} \sum_{\gamma=0}^{r} K_{ijk\ell}^{(\gamma)} \mu^\gamma u_{,ijk\ell} + \lambda \sum_{\gamma=0}^{s} K_\gamma \mu^\gamma N_{ij}^{\circ} u_{,ij} = 0
$$

(2.250)

A nontrivial solution $u(x, y)$ of (2.250) exists only for values $\mu = \mu_n$, the generalized eigenvalues of the linearized problem. Specifically, if we assume that $r = s$, then (2.250) becomes

$$\sum_{\gamma=0}^{r} \mu^{\gamma} \left[ \frac{h^3}{12} K_{ijk\ell}^{(\gamma)} u_{,ijk\ell} + \lambda K_{\gamma} N_{ij}^{\circ} u_{,ij} \right] = 0 \qquad (2.251)$$

If $\phi_n$ is an eigenfunction corresponding to the generalized eigenvalue $\mu_n$ then, with

$$(\phi, \Psi) = \int_{\Omega} \phi(x, y) \Psi(x, y) dx dy, \text{ we have}$$

$$\sum_{\gamma=0}^{r} \mu_n^{\gamma} \left[ \frac{h^3}{12} K_{ijk\ell}^{(\gamma)} (\phi_{n,ij}, \phi_{n,k\ell}) - \lambda K_{\gamma} N_{ij}^{\circ} (\phi_{n,i}, \phi_{n,j}) \right] = 0 \qquad (2.252)$$

Using the positive definiteness assumption (2.239), relative to the $K_{ijk\ell}^{(\gamma)}$, and choosing the $N_{ij}^{\circ}$ such that

$$\sum_{\gamma=0}^{r} \mu^{\gamma} K_{\gamma} N_{ij}^{\circ} (\phi_{,i}, \phi_{,j}) > 0 \qquad (2.253)$$

it follows that, for sufficiently small $\lambda$, there exists $\kappa > 0$ such that

$$\sum_{\gamma=0}^{r} \mu^{\gamma} \left[ \frac{h^3}{12} K_{ijk\ell}^{(\gamma)} (\phi_{,ij}, \phi_{,k\ell}) - \lambda K_{\gamma} N_{ij}^{\circ} (\phi_{,i}, \phi_{,k}) \right] \geq \kappa \|\phi\|^2 \qquad (2.254)$$

where $\|\phi\|^2 = (\phi, \phi)$.

The polynomial on the left-hand side of (2.254) has positive coefficients for $\lambda$ sufficiently small and is a monotonically increasing function for $\mu > 0$; this, in turn, implies that the roots of this polynomial are either negative or have negative real parts; for a real viscoelastic material the roots must, in fact, all be negative. If, in addition, the roots are simple, then the solution of (2.247) can be expressed in the form

$$w(x, y, t) = \sum_{n=1}^{\infty} \sum_{k=1}^{r} A_{nk} w_{on} \phi_n(x, y) e^{-\mu_{nk} t} \qquad (2.255)$$

where the (real) roots of (2.252) have been denoted by $-\mu_{nk}$ and $w_{on} = (w_0, \phi_n)$. It can be shown that the coefficients $A_{nk}$ in the expansion (2.255) are given by [106]:

$$A_{nk} = \frac{\mu_{n1} \mu_{n2} \cdots \mu_{n(k-1)} \mu_{n(k+1)} \cdots \mu_{nr}}{(\mu_{nk} - \mu_{n1}) \cdots (\mu_{nk} - \mu_{n(k-1)})(\mu_{nk} - \mu_{n(k+1)}) \cdots (\mu_{nk} - \mu_{nr})} \qquad (2.256)$$

With all the $\mu_{nk} > 0$ the solution is stable, while if at least one $\mu_{nk} < 0$ then the solution is unstable. As the left-hand side of (2.252) is a continuous, increasing function of $\lambda$, when $\lambda$ increases critical values of $\lambda$ are reached at which the roots of (2.252) successively change their signs; to determine these critical roots the polynomial equation (2.252), which can be written in the form

$$\sum_{\gamma=0}^{r} \mu_n^\gamma A_{n\gamma}(\lambda) = 0 \tag{2.257}$$

must be analyzed. It is shown in [106] that changes in the signs of the roots $\mu_{nk}$, which satisfy

$$\prod_{\ell=1}^{r} \mu_{n\ell}(\lambda) = A_{nr}^{-1}(\lambda) A_{no}(\lambda) \tag{2.258}$$

occur at those values of $\lambda$ which satisfy the equations

$$\frac{h^3}{12} K_{ijk\ell}^{(0)}(\phi_{n,ij}, \phi_{n,k\ell}) - \lambda K_0 N_y^0(\phi_{n,i}, \phi_{n,j}) = 0 \tag{2.259}$$

and

$$\frac{h^3}{12} K_{ijk\ell}^{(r)}(\phi_{n,ij}, \phi_{n,k\ell}) - \lambda K_r N_{ij}^0(\phi_{n,i}, \phi_{n,j}) = 0 \tag{2.260}$$

If the Laplace transform is applied to (2.247), we find that the values of $\lambda$, in question, are eigenvalues of the system

$$\frac{h^3}{12} K_{ijk\ell}^{(0)} w_{,ijk\ell}(\infty) + \lambda N_{ij}^0 K_0 w_{,ij}(\infty) = 0 \tag{2.261}$$

and

$$\frac{h^3}{12} K_{ijk\ell}^{(r)} w_{,ijk\ell}(0) + \lambda N_{ij}^0 K_r w_{,ij}(0) = 0 \tag{2.262}$$

Whenever $\lambda$ is an eigenvalue of (2.261), one of the roots of (2.252) is equal to zero and, as $\lambda$ increases above this value, one of the $\mu_{nk}$ becomes negative; when $\lambda$ is an eigenvalue of (2.262) one of the roots of (2.252) must be infinite. In [106], eigenvalues of (2.261) are called critical values for infinite critical time and are denoted by $\lambda_{cr}$. For $\lambda < \min \lambda_{cr}$ each of the $\mu_{nk} > 0$ and the basic solution of (2.47) is stable. For $\lambda = \min \lambda_{cr}$ we have neutral stability, while for $\lambda > \min \lambda_{cr}$ at least one of the $\mu_{nk} < 0$ and the solution is said to be unstable (with infinite critical time). On the other hand, eigenvalues of (2.262) are said to be critical values of instant instability, or critical values for finite critical time, and are denoted as $\lambda_{cr}^0$. When $\lambda$ achieves the value min $\lambda_{cr}^0$ the viscoelastic plate becomes (instantly) unstable.

The results delineated above hold for viscoelastic plates of "differential type," i.e., those which are governed by the large deflection equations (2.234), (2.235); for the case of viscoelastic plates of the "Boltzmann type," namely, those for which large deflection relations of the form (2.245), (2.246) are valid, the critical values are shown in [106] to correspond to the eigenvalues of the system

$$\begin{cases} G_{ijk\ell}(\infty)w_{,ijk\ell}(\infty) + \lambda N_{ij}^0 w_{,ij}(\infty) = 0 \\ G_{ijk\ell}(0)w_{,ijk\ell}(0) + \lambda N_{ij}^0 w_{,ij}(0) = 0 \end{cases} \quad (2.263)$$

Besides treating the linearized stability problem in [106], Brilla also considers the nonlinear stability problem, albeit only for an isotropic, viscoelastic plate of differential type for which the (generalized) von Karman equations have the form

$$K(1 + \alpha D)\,\triangle^2 w = h(1 + \beta D)(\lambda\,[\Phi_0, w] + \left[\tilde{\Phi}, w\right]) \quad (2.264)$$

and

$$(1 + \beta D)\,\triangle^2\,\tilde{\Phi} = -\frac{1}{2}E(1 + \alpha D)[w, w] \quad (2.265)$$

where $\Phi_0(x, y)$ satisfies $\triangle^2\Phi_0 = 0$, with boundary conditions that correspond to the given loading condition(s) on the edge(s), $K$ is the plate bending stiffness, $E$ the Young modulus, and $\alpha, \beta$ constants which serve to characterize the viscoelastic behavior of the plate. For $\alpha = \beta = 0$, (2.264), (2.265) reduce to the usual set of von Karman equations for a linearly elastic, isotropic plate, in Cartesian coordinates, with $\Phi = \lambda\Phi_0 + \tilde{\Phi}$ the (total) Airy function. Using the boundary conditions (2.240), or (2.241), and the initial conditions

$$w(x, y, 0) = w_0(x, y), \quad \tilde{\Phi}(x, y, 0) = 0, \quad (2.266)$$

it is demonstrated in [106] that critical values with infinite critical time exist; the corresponding stationary solutions are given by the equations

$$K\,\triangle^2 w = h\left(\lambda\,[\Phi_0, w] + \left[\tilde{\Phi}, w\right]\right) \quad (2.267)$$

and

$$\triangle^2\tilde{\Phi} = -\frac{1}{2}E[w, w] \quad (2.268)$$

The result referenced above may be deduced as follows: we rewrite equations (2.264), (2.265) in the form

$$K \left[ \frac{\alpha}{\beta} \Delta^2 w + \left(1 - \frac{\alpha}{\beta}\right) \frac{1}{\beta} \int_0^t (\Delta^2 w) \, e^{-\frac{1}{\beta}(t-\tau)} d\tau \right.$$

$$\left. - \frac{\alpha}{\beta} \Delta^2 w_0 \cdot e^{-\frac{1}{\beta}t} \right] \tag{2.269}$$

$$= h \left( \lambda \, [\Phi_0, w] + [\tilde{\Phi}, w] - \lambda \, [\Phi_0, w] \, e^{-\frac{1}{\beta}t} \right)$$

and

$$\frac{\beta}{\alpha} \Delta^2 \tilde{\Phi} + \left(1 - \frac{\beta}{\alpha}\right) \frac{1}{\alpha} \int_0^t (\Delta^2 \tilde{\Phi}) \, e^{-\frac{1}{\alpha}(t-\tau)} d\tau$$

$$= -\frac{1}{2} E \left( [w, w] - [w_0, w_0] \, e^{-\frac{1}{\alpha}t} \right) \tag{2.270}$$

Taking the scalar product of (2.269) with $w$, and then using $[w, w]$ as given by (2.270), it is possible to show that

$$K \left\{ \frac{\alpha}{\beta} (\Delta w, \Delta w) + \left(1 - \frac{\alpha}{\beta}\right) \frac{1}{\beta} \left( \Delta w, \int_0^t (\Delta w) \, e^{-\frac{1}{\beta}(t-\tau)} d\tau \right) \right\}$$

$$- h\lambda \, (\Phi_0, [w, w]) + \frac{2h}{E} \left\{ \frac{\beta}{\alpha} \left( \Delta \tilde{\Phi}, \Delta \tilde{\Phi} \right) \right.$$

$$\left. + \left(1 - \frac{\beta}{\alpha}\right) \frac{1}{\alpha} \left( \Delta \tilde{\Phi}, \int_0^t (\Delta \tilde{\Phi}) \, e^{-\frac{1}{\alpha}(t-\tau)} d\tau \right) \right\}$$

$$= K \frac{\alpha}{\beta} (\Delta w_0, \Delta w_0) \, e^{-\frac{1}{\beta}t} - h\lambda \, (\Phi_0, [w_0, w]) \, e^{-\frac{1}{\beta}t} - h \left( \tilde{\Phi}, [w_0, w_0] \right) e^{-\frac{1}{\alpha}t} \tag{2.271}$$

from which, with the assumption of finite total energy, we obtain, in the limit, as $t \to \infty$

$$K \, (\Delta w, \Delta w) - h\lambda \, (\Phi_0, [w, w]) + \frac{2}{E} h \left( \Delta \tilde{\Phi}, \Delta \tilde{\Phi} \right) = 0 \tag{2.272}$$

Because $(\Delta \tilde{\Phi}, \Delta \tilde{\Phi}) \geq 0$, a nonzero solution of (2.272) exists if and only if $\lambda$ is larger than the critical values (with infinite critical time) of the linearized problem, which are also critical values of the non-linear problem (2.267), (2.268). If $w_0 = 0$, the right-hand side of (2.271) is equal to zero and, setting $t = 0$, we then find that

$$K\alpha \, (\Delta w, \Delta w) - h\lambda\beta \, (\Phi_0, [w, w]) + \frac{2\beta^2}{E\alpha} h \left( \Delta \tilde{\Phi}, \Delta \tilde{\Phi} \right) = 0 \tag{2.273}$$

Also, the corresponding reduced (generalized) von Karman equations (which are obtained from (2.269), (2.270)), are

$$
\begin{cases}
K\alpha \, \triangle^2 \, w = \beta h \left( \lambda \left[ \Phi_0, w \right] + \left[ \tilde{\Phi}, w \right] \right) \\
\beta \, \triangle^2 \, \tilde{\Phi} = -\dfrac{1}{2} E\alpha \left[ w, w \right]
\end{cases}
\tag{2.274}
$$

It is not difficult to show that (2.273), and, thus, also the system (2.274) can have a non-zero solution if and only if

$$
K\alpha \left( \triangle w, \triangle w \right) - h\beta\lambda \left( \Phi_0, [w, w] \right) < 0
\tag{2.275}
$$

which leads to the following conclusion: the critical values for zero critical time, which are the eigenvalues of (2.274), are critical values for zero critical time of the linearized problem. No numerical computations for the critical values or the corresponding eigenfunctions (buckling modes) have been carried out in [106]; there exists no postbuckling analysis either, so that these are likely subjects for future research work in this area.

---

## 2.4 Comparisons of Initial and Postbuckling Behavior of Rectangular Plates

In this section, we will make use of the analysis (and corresponding graphs) from sections 2.1, 2.2, and 2.3 to describe some of the major differences which occur in both the initial buckling and postbuckling behavior of rectangular plates as one varies the boundary conditions imposed along the edges of the plate, the type of material symmetry assumed, and the nature of the constitutive response.

### 2.4.1 Variations with Respect to Boundary Conditions

In Figs. 2.1–2.8 we display some of the results which apply with respect to buckling in the first mode ($m = 1$, i.e., one-half sine-wave in the $x$-direction) for an isotropic, linearly elastic, rectangular plate which is compressed in the $x$-direction; the support conditions which apply in these various figures are, respectively, those which conform to simple support along all four edges, simple support along the loaded edges but clamped along the longitudinal edges, simple support along the loaded edges but clamped along one longitudinal edge and simply supported along the other, simply supported along the loaded edges but free along one longitudinal edge with the other edge clamped, simply supported along the loaded edges with one longitudinal edge simply supported and the other free, clamped along the loaded edges with the two longitudinal edges simply supported, and clamped along all four edges. In addition, Fig. 2.2 depicts the initial buckling behavior for the compressed isotropic plate, simply supported on all four edges as the aspect ratio increases into the range where buckling into $m = 2, 3$, and 4 half-sine waves in the $x$-direction occurs, while Fig. 2.7 shows the plot in Fig. 2.2 (up to $\phi = 3$) for the simply supported case as well as the variation in the initial buckling pattern for the plate which is clamped along the loaded edges but simply supported along the longitudinal edges. Figure 2.8 compares (up to an aspect ratio of $\phi = 3$) the variation in the initial buckling pattern for plates which are clamped along all four edges with those which are simply supported along the loaded edges and clamped along the longitudinal edges.

One way to compare the results with respect to buckling into the lowest mode ($m = 1$) in Figs. 2.1–2.8 is to fix an aspect ratio, say $\phi = 1$ (the square plate) and compare the corresponding values of the buckling parameter $\tilde{K}$, which is proportional to the buckling stress $\lambda_{cr}$; these results are displayed in Table 2.6 where it is clear that, as one reduces the number of clamped edges and shifts, first to an increasing number of simply supported edges and then to cases where one of the edges is free, the critical buckling load decreases. Another way to display the results in Figs. 2.1–2.8, is to look at the variation in the minimum buckling parameter value and the (approximate) corresponding value of the aspect ratio $\phi$ at which it occurs. These results are displayed in Table 2.7; again, as the number of clamped edges decreases in favor of simply supported edges and then free edges, the minimal buckling stress decreases. One other comparison which follows from Figs. 2.7 and 2.8 relates to the way in which variations in the support conditions along the edges of the plate affect the transition from buckling, initially, into one-half sine wave in the $x$-direction to buckling into two-half sine waves in the $x$-direction; these comparative results are summarized in Table 2.8.

Figures 2.11 and 2.12 may be compared so as to gauge how varying the support conditions from simply supported along all four edges to simply supported along the edges $x = 0, x = a, 0 \leq y \leq b$, and clamped along the edges $y = 0$, $y = 0$, $0 \leq x \leq a$, affects the variation with $\phi$ of the normalized loading $\lambda/\lambda_{cr}$, in the $x$-direction, for a rectangular isotropic plate compressed in both directions; the parameter along the curves is the compressive stress $\xi$ normalized by $\lambda_{cr}$. Here $\lambda_{cr}$ denotes the critical value of the compressive stress in the $x$-direction for the given support conditions in the absence of the compressive stress $\xi$; among the results that may be gleaned from Figs. 2.11 and 2.12 are the following: fix the value of $\xi/\lambda_{cr}$, say, $\xi/\lambda_{cr} = 1.0$; then for $\phi = 1$ (square plate), $\lambda/\lambda_{cr} < 1$ for the plate which is simply supported along the edges at $x = 0$, $x = a$, $0 \leq y \leq b$ and clamped along the edges at $y = 0$, $y = b$, $0 \leq x \leq a$, while $\lambda/\lambda_{cr} > 1$ (in fact, approximately 1.8) for the plate which is simply supported along all four edges. Also, in the first case, the plate buckles into one-half wave in the $x$-direction, while in the second case the plate buckles into two half-waves in the $x$-direction.

For an isotropic, linearly elastic, rectangular plate which is simply supported on all four sides, Fig. 2.14 gives a very clear picture of how the buckling parameter $\tilde{K}$ varies with the aspect ratio $\phi$ as the loading along the edges $x = 0$, $x = a$, $0 \leq y \leq b$, varies according to $\sigma_{xx}(y) = \sigma_o(1 - \frac{c}{b}y)$, $c = 0$ corresponding to pure compression along these edges; the graphs also depict the transitions from buckling into one-half wave ($m = 1$) in the $x$-direction to buckling into two-half waves ($m = 2$) in the $x$-direction. For example, with $m = 1$ in each case, the *minimum buckling load increases*, and the *corresponding aspect ratio $\phi$ decreases* as $c$ increases monotonically from $c = 0$ to $c = 2$.

Figure 2.19 clearly displays the comparative results for shear buckling of a linearly elastic isotropic rectangular plate for four different kinds of edge support conditions; all the graphs of $\tilde{K}$ against $\phi$ are monotonically decreasing and as the graphs do not intersect for $\phi > 1$ it is sufficient to display, as we have done in Table 2.9, the results for just one aspect ratio, say $\phi = 2$. Once again, we see an increase in the buckling load as we move from edges which are all simply supported to edges which are pairwise simply supported and clamped to the case where all edges are clamped.

For linearly elastic rectangular plates which are orthotropic Figs. 2.23 and 2.24 display the variation of the buckling constant $k$ (for a plate which is compressed in the $x$-direction) with the effective aspect ratio $\bar{\beta} = \phi \sqrt[4]{\frac{D_{22}}{D_{11}}}$. The parameter $\alpha$ along the curves is defined in (2.65), while $k$ is related to $\lambda_{cr}$ by

$$k = \frac{b^2 h}{\pi^2 \sqrt{D_{11} \cdot D_{22}}} \lambda_{cr} \qquad (2.276)$$

For a given value of $\alpha$, say $\alpha = 1.0$, and a given value of $\phi$, say, $\phi = 1$ (with $D_{11}, D_{22}$ and $D_{66}$ fixed) one may use Figs. 2.23 and 2.24 to gauge the effect of varying the support conditions along the edges, as, e.g., in Table 2.10.

As in the isotropic case, $\lambda_{cr}$ increases as we move from edges which are all simply supported to edges which are pairwise simply supported and clamped to edges which are all clamped.

A postbuckling figure which very clearly displays the influence of varying the support conditions along the edges of an isotropic linearly elastic (square) plate which is compressed in the $x$-direction is Fig. 2.31, which is taken from Yamaki [94]. From Fig. 2.31, it is clear that of the four support conditions considered, the steepest of the load versus deflection curves associated with postbuckling behavior occurs for the case of a plate simply supported along its loaded edges and clamped along the other edges, while the next steepest load deflection curve is associated with the case of a plate clamped along all four edges.

For the case of end compression of a nonlinear elastic rectangular plate which is simply supported along its loaded edges, Figs. 2.46 and 2.47 display, respectively, the effect of going from clamped longitudinal edges to simply supported longitudinal edges on the normalized buckling strain and normalized buckling stress as functions of the dimensionless plate stiffness, for several different values of the mean Poisson ratio. Within the context of the Johnson-Urbanik theory [73], [74] for buckling of nonlinear elastic rectangular plates, Fig. 2.49 depicts the way in which the variation of the ratio of the nonlinear elastic buckling stress to the linear elastic buckling stress varies with the dimensionless stiffness $S$ as one transitions (for three different values of the mean Poisson ratio $\nu$) from clamped longitudinal edges to simply supported longitudinal edges in a plate which is simply supported along its loaded edges.

## 2.4.2 Variations with Respect to Symmetry

For the case of end compression of a rectangular linearly elastic plate which is simply supported along all four edges, we consider, in this subsection, the variation in both initial buckling and postbuckling behavior as one varies the material symmetry properties of the plate; by varying the material symmetry properties of the plate we mean not only the process of going from isotropic to (rectilinearly) orthotropic but also the process of going from one degree of orthotropy, as measured, e.g., by the ratio $\bar{E} = E_{22}/E_{11}$, to another.

The most revealing of the graphs presented in [96] that are related to initial buckling is depicted in Fig. 2.38; this figure depicts the variation in buckling load, as a function of $\Lambda = m/\phi$, as one varies the orthotropic behavior of the plate among cases A-L in Table 2.5. The curves in Fig. 2.38 turn out to be ordered from bottom to top $(F, E, D, C, B, A)$, isotropic, $(G, H, I, J, K, L)$ according to increasing $\bar{E} = E_{22}/E_{11}$; for fixed $m$ and $\phi$ the buckling load increases as the ratio of the Young's moduli $E_{22}$ to $E_{11}$ increases. In Figs. 2.33 through 2.37, we show the initial buckling graphs of critical stress versus aspect ratio $a/b$ as $\bar{E}$ increases through the values $\bar{E} = .072, .51, 1.0$ (isotropic), 1.97, and 13.73, respectively; these graphs again illustrate the increase in buckling load with aspect ratio for a wide range of buckling wavelengths in the direction of compression; perhaps, more importantly, they depict the aspect ratios at which a transition from $m = 1$ to $m = 2$ half-waves occurs. From Table 2.11 and the graphs in Figs. 2.33 through 2.37, it should be clear that, as $\bar{E}$ increases, the transition from $m = 1$ to $m = 2$ half-waves in the direction of compression of the plate comes at smaller and smaller values of the aspect ratio.

Variations in postbuckling behavior for linearly elastic rectangular orthotropic plates may be gauged from the graphs in Figs. 2.39 through 2.42, which depict normalized load versus normalized shortening (or contraction) curves for a simply supported plate in compression; in Figs. 2.39–2.41, $\Lambda = m/\phi$ varies over the values $1.0, 1.33$, and $2.0$, respectively, while, in Fig. 2.42, the results are for an infinitely long plate. The values of $\bar{E}$ for the plates depicted are $L$ (13.73), $K$ (7.6), $J$ (3.22), $I$ (1.97), $A$ (.83), $B$ (.70), $C$ (.506), $D$ (.303), $E$ (.131), and $F$ (.072). For the case $\Lambda = 1$ one cannot distinguish the postbuckling behavior for plates $A, B, C, D$, and an isotropic plate ($\bar{E} = 1.0$); otherwise, it is clear from the graphs in the first three figures in this group that the steepness of the postbuckling curves decrease as the value of $\bar{E} = E_{22}/E_{11}$ decreases, i.e., *for a fixed load, beyond the critical load, the normalized shortening increases as the value of $\bar{E}$ decreases.* For any given plate, the postbuckling curves, for the corresponding values of $\bar{E}$, exhibit a decrease in steepness as the value of $\Lambda$ increases. Note that, for the infinitely long plate, it is clear from Fig. 2.42 that the steepness of the load-shortening curves decreases as the value of $\bar{E}$, for the plate, increases; thus for the compressed, simply supported, infinitely long plate, and a given load beyond the critical load, the shortening (contraction) of the plate increases as the ratio $\bar{E}$ of $E_{22}$ to $E_{11}$ increases. In Fig. 2.43 we depict the monotonic decreases in the effective widths of simply supported, compressed rectangular plates as the load increases beyond the critical load; we note that the steepness of the curves (or the rate of decrease of effective width with increasing load) increases as the value of the ratio $\bar{E}(= E_{22}/E_{11})$ decreases. For a given load, beyond the critical load, a smaller effective width is obtained for smaller values of $\bar{E}$.

### 2.4.3    Variations with Respect to Constitutive Response

This is the area, related to comparative features of both initial and postbuckling response, in which the least amount of work appears to have been done. In Fig. 2.49 (which is based on the Johnson-Urbanik generalization of the von Karman equations for nonlinear elastic response of rectangular plates) we depict the ratio of the nonlinear elastic buckling stress $\sigma_c^N$, to the linear elastic buckling stress $\sigma_c^L$, as a function of stiffness of the plate for two types of support (clamped and simply supported) along the long (longitudinal) edges, and three different values of the mean Poisson ratio $\nu = \sqrt{\nu_{12}\nu_{21}}$; the plate is compressed in the $x$-direction and is simply supported along its loaded edges. Clearly, for all the cases shown, and any fixed value of the stiffness of the plate, $\sigma_c^N/\sigma_c L < 1$, which means that the nonlinear elastic plate will always buckle at a lower critical load than the corresponding linear elastic plate with the same type of support conditions and mean Poisson ratio. For both clamped and simply supported longitudinal edges, the ratio of $\sigma_c^N$ to $\sigma_c^L$ is smaller for increasing values of $\nu$ for a fixed stiffness, and for increasing values of the stiffness for a fixed $\nu$. No results for postbuckling within the context of the Johnson-Urbanik theory currently exist, even for rectangular plates, which would allow for a comparative study of the postbuckling behavior of linear and nonlinear elastic plates.

As has been indicated in the previous section, very few results appear to exist relative to the postbuckling behavior of rectangular viscoelastic plates. For a compressed, initially isotropic, long rectangular plate which is simply supported along its unloaded edges, the critical stress at which plastic buckling occurs is given by (2.21), i.e.

$$\lambda_c^P = \frac{\pi^2 E_s}{9}\left(\frac{h}{b}\right)^2\left[\left(\frac{1}{4} + \frac{3}{4}\frac{E_t}{E_s}\right)\frac{1}{\phi^2} + \phi^2 + 2\right]$$

where $\phi = a/b$ is, again, the aspect ratio of the plate while $E_t$ and $E_s$ are, respectively the tangent and secant moduli associated with the elastic-plastic behavior of the plate. To compare the critical stress levels required to initiate plastic and (linear) elastic buckling, we have the relation (2.212) for the plasticity reduction factor $\eta$, which holds in the case where the plate is simply supported along its unloaded edges, namely,

$$\eta \equiv \frac{(\lambda_c^P)_{\min}}{(\lambda_c^e)_{\min}} = \frac{4E_s}{3E} \left( \frac{1}{2} + \frac{1}{2}\sqrt{\frac{1}{4} + \frac{3E_t}{4E_s}} \right) (1 - \nu^2).$$

The reduction in critical stress levels for plastic, as opposed to (linear) elastic, buckling are given in (2.214), (2.215), and (2.216) for three other types of support conditions along the unloaded edges. Graphs of comparative results for the postbuckling behavior of elastic and plastic rectangular plates do not appear to be available, although it should be clear from the graphs in Figs. 2.51–2.53 that, even in the case of a perfect elastic-plastic plate, the load versus deflection curves are concave-down as opposed to the concave-up (convex) behavior of such postbuckling curves in the elastic case (e.g., Fig. 2.31). Beyond a maximum stress level, the postbuckling curves for the elastic-plastic case all exhibit a decrease (along the load-deflection curves) as opposed to the monotonically increasing behavior exhibited in the elastic range, where the concept of a maximum stress does not enter the picture.

The scope of this chapter has not been intended to be exhaustive. For other work in this extensive area the reader may wish to consult the following references:

(i) For some of the research on specific buckling problems for rectangular plates: [107]–[122].

(ii) For fundamental work on buckling of rectangular plates and the von Karman Equations: [123]–[129].

(iii) For additional work on plastic buckling of rectangular plates: [130]–[134].

(iv) For work on the buckling behavior of rectangular plates with specific reference to the buckling of paper sheets: [135]–[138].

Papers related to the buckling and postbuckling behavior of rectangular plates which involve either imperfection buckling or secondary buckling will be referenced in Chapter 5.

## 2.5 Initial Buckling/Postbuckling Figures, Graphs, and Tables: Rectangular Plates

**Table 2.1** Values of Coefficient $k$ for Different Values of $\phi$ for Plywood Compressed in a Principal Direction $(D_{11} > D_{22})$ [51] (From Lekhnitskii, S.G., *Anisotropic Plates*, Gordon and Breach Pub. Co., N.Y., 1968 and OPA, N.V. With permission)

| $\phi$ | 0.5 | 1 | 1.86 | 2 | 2.63 | 3 | $\infty$ |
|---|---|---|---|---|---|---|---|
| $k$ | 14.75 | 4.53 | 2.76 | 2.79 | 3.27 | 2.96 | 2.76 |
| $m$ | 1 | 1 | 1 | 1 | 1-2 | 2 | - |

**Table 2.2** Values of Coefficient $k$ for Different Values of $\phi$ for Plywood Compressed in a Principal Direction $(D_{11} < D_{22})$ [51] (From Lekhnitskii, S.G., *Anisotropic Plates*, Gordon and Breach Pub. Co., N.Y., 1968 and OPA, N.V. With permission)

| $\phi$ | 0.5 | 0.54 | 0.76 | 1 | 1.31 | 1.62 | 1.86 | 2 | $\infty$ |
|---|---|---|---|---|---|---|---|---|---|
| $k$ | 2.79 | 2.76 | 3.27 | 2.79 | 2.93 | 2.76 | 2.85 | 2.79 | 276 |
| $m$ | 1 | 1 | 1-2 | 2 | 2-3 | 3 | 3-4 | 4 | - |

**Table 2.3** Values of Coefficient $k$ for Plywood Plates Deflected by Bending Moment [51] (From Lekhnitskii, S.G., *Anisotropic Plates*, Gordon and Breach Pub. Co., N.Y., 1968 and OPA, N.V. With permission)

| $\phi$ | $D_{11} > D_{22}$ | | $D_{11} < D_{22}$ | |
|---|---|---|---|---|
| | $k$ | $m$ | $k$ | $m$ |
| 0.5 | 45.5 | 1 | 19.8 | 1 |
| 1 | 19.8 | 1 | 18.5 | 3 |
| 2 | 19.8 | 2 | 18.2 | 5 |
| 3 | 18.5 | 2 | 18.2 | 8 |
| $\infty$ | 18.1 | - | 18.1 | - |

**Table 2.4** Values of Coefficient $k$ for a Plywood Plate Under Shear Forces [51] (From Lekhnitskii, S.G., *Anisotropic Plates*, Gordon and Breach Pub. Co., N.Y., 1968 and OPA, N.V. With permission)

| | $\phi = 1$ | 1.2 | 1.4 | 1.6 | 1.8 | 2 | 2.5 | 3 | $\infty$ |
|---|---|---|---|---|---|---|---|---|---|
| External fibers parallel to the short sides | 38.39 | - | - | - | - | 35.06 | - | 33.66 | 32.15 |
| External fibers parallel to the long sides | 38.39 | 27.39 | 21.28 | 17.63 | 15.35 | 13.87 | 11.98 | 11.29 | 9.28 |

Table 2.5 Elastic Constants for the Various Cases of Composite Plates A-N [96]   (Reprinted from the *Int. J. Mech. Sci.*, 15, Chandra, R. and Raju, B., Postbuckling Analysis of Rectangular Orthotropic Plates, 81–97, 1973, with permission from Elsevier Science)

| Designation of different cases | $E_{11}$ | $E_{22}$ | $G_{12}^*$ | $\bar{E}=E_{22}/E_{11}$ | $\dfrac{G_{12}}{E_{11}}$ | $\nu_{21}$ | $\nu_{12}$ |
|---|---|---|---|---|---|---|---|
| A | 1.887 | 1.5772 | 0.6123 | 0.8358 | 0.3245 | 0.3 | 0.25074 |
| B | 3.761 | 2.648 | 1.061 | 0.7041 | 0.2823 | 0.3 | 0.21123 |
| C | 7.068 | 3.578 | 1.409 | 0.5064 | 0.1994 | 0.3 | 0.15192 |
| D | 14.130 | 4.282 | 1.714 | 0.3031 | 0.1213 | 0.3 | 0.09093 |
| E | 36.390 | 4.788 | 1.959 | 0.1315 | 0.0538 | 0.3 | 0.03945 |
| F | 69.660 | 5.071 | 2.062 | 0.0728 | 0.0296 | 0.3 | 0.02184 |
| G | 1.5772 | 1.87 | 0.6123 | 1.1964 | 0.3882 | 0.25074 | 0.3 |
| H | 1.648 | 3.761 | 1.061 | 1.4202 | 0.4009 | 0.21123 | 0.3 |
| I | 3.578 | 7.068 | 1.409 | 1.9747 | 0.3937 | 0.15192 | 0.3 |
| J | 4.282 | 14.13 | 1.714 | 3.2292 | 0.4002 | 0.09093 | 0.3 |
| K | 4.788 | 36.39 | 1.959 | 7.6045 | 0.4091 | 0.03945 | 0.3 |
| L | 5.071 | 69.66 | 2.062 | 13.7362 | 0.4065 | 0.02184 | 0.3 |
| M | 2.6 | 7.5 | 2.1 | 2.89 | 0.424 | 0.08675 | 0.25 |
| N | 7.5 | 2.6 | 1.1 | 0.347 | 0.147 | 0.25 | 0.08673 |

**Table 2.6**  Buckling Parameter $\tilde{K}(\sim \lambda_{cr})$ for a Linearly Elastic Isotropic Rectangular Plate Compressed in the x-Direction; the Aspect Ratio is $\phi = 1$. Based on the Data Presented in [32].

| Loaded Edges | Longitudinal Edges | $\tilde{K}$ |
|---|---|---|
| clamped | clamped | 10.2 |
| S.S. | clamped | 10.0 |
| clamped | S.S. | 6.9 |
| S.S. | clamped/S.S. | 5.7 |
| S.S. | S.S. | 4.0 |
| S.S. | clamped/free | 1.7 |
| S.S. | S.S./free | 1.3 |

**Table 2.7**  Minimum Values of $\tilde{K}(\sim \lambda_{cr})$ for a Linearly Elastic Isotropic Rectangular Plate, Compressed in the $x$-Direction; the Corresponding Value of $\phi = 1$. Based on the Data Presented in [32].

| Loaded Edges | Longitudinal Edges | $\tilde{K}$ | $\phi$ |
|---|---|---|---|
| clamped | clamped | 9.80 | 1.20 |
| S.S. | clamped | 6.97 | .66 |
| S.S. | clamped/S.S. | 6.41 | .80 |
| clamped | S.S. | 5.30 | 1.70 |
| S.S. | S.S. | 4.00 | 1.00 |
| S.S. | clamped/free | 1.25 | 1.64 |
| S.S. | S.S./free | | $\begin{cases} \tilde{K} \text{ decreases as} \\ \phi \text{ increases} \end{cases}$ |

**Table 2.8** Transition from $m = 1$ to $m = 2$ Half-Sine Waves in the $x$-Direction for Initial Buckling of a Compressed, Linearly Elastic Isotropic Rectangular Plate. Based on the Data Presented in [32].

| Loaded Edges | Longitudinal Edges | $\tilde{K}$ | $\phi(m = 1 \rightarrow m = 2)$ |
|---|---|---|---|
| clamped | clamped | 9.9 | 1.2 |
| S.S. | clamped | 8.1 | .9 |
| clamped | S.S. | 5.2 | 1.7 |
| S.S. | S.S. | 4.5 | 1.4 |

**Table 2.9** Variation of $\tilde{K}(\sim \lambda_{cr})$ for an Isotropic Linearly Elastic Rectangular Plate with Aspect Ratio $\phi = 2$, Under Shear Loading. Based on the Data Presented in [32].

| Short Edges $x = 0, x = a$ | Long Edges $y = 0, y = b$ | $\tilde{K}$ |
|---|---|---|
| S.S. | S.S. | 6.3 |
| clamped | S.S. | 6.7 |
| S.S. | clamped | 10.2 |
| clamped | clamped | 10.5 |

**Table 2.10** Variation of $k (\sim \lambda_{cr})$ for Compression of an Orthotropic Rectangular Linearly Elastic Plate ($\alpha = 1, \phi = 1$). Based on the Data Presented in [43].

| Loaded Edges | Longitudinal Edges | $k$ |
|:---:|:---:|:---|
| S.S. | S.S. | 4.0 $(m = 1)$ |
| clamped | S.S. | 6.8 $(m = 1)$ |
| S.S. | clamped | 7.8 $(m = 2)$ |
| clamped | clamped | 10.1 $(m = 1)$ |

**Table 2.11** Variation in the Transition from $m = 1$ to $m = 2$ Half-Waves, with $\bar{E} = E_{22}/E_{11}$, the Plate Compressed in the $x$-Direction and Simply Supported on all Sides. Based on the Data Presented in [43].

| $\bar{E}$ | $(\phi)$Transition $m = 1$ to $m = 2$ | Normalized Critical Stress (approximate) |
|:---:|:---:|:---:|
| .072 | 2.72 | .8 |
| .51 | 1.68 | 2.7 |
| 1.0 | 1.41 $(\sqrt{2})$ | 4.6 |
| 1.97 | 1.19 | 5.6 |
| 13.73 | .735 | 11.4 |

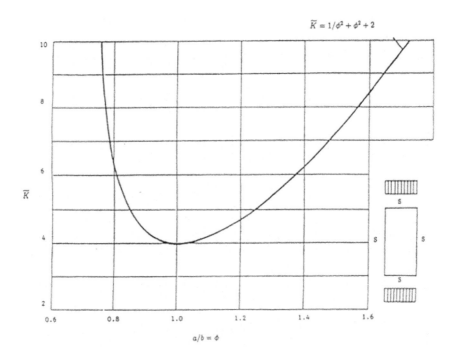

**FIGURE 2.1**
Buckling of simply supported plates [32]. (From Bulson, P.S.,
*The Stability of Flat Plates*, American Elsevier, N.Y., 1969)

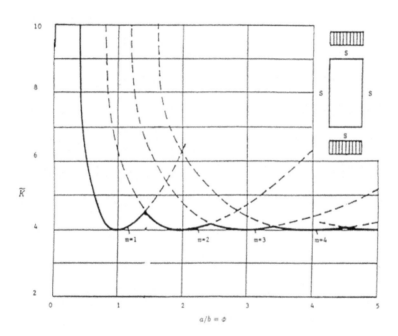

**FIGURE 2.2**
$\bar{K} \sim \phi$ curves for long plates [32]. (From Bulson, P.S., *The Stability of Flat Plates*, American Elsevier, N.Y., 1969)

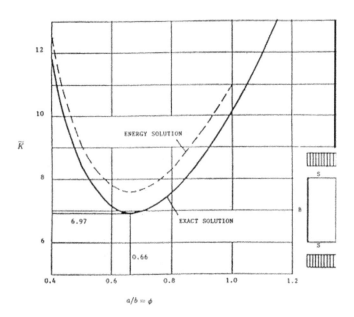

**FIGURE 2.3**
Buckling of plates with built-in longitudinal edges [32]. (From Bulson, P.S., *The Stability of Flat Plates*, American Elsevier, N.Y., 1969)

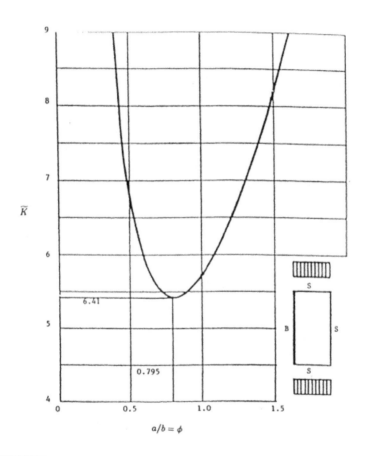

FIGURE 2.4
One longitudinal edge built-in, one simply supported [32].
(From Bulson, P.S., *The Stability of Flat Plates*, American
Elsevier, N.Y., 1969)

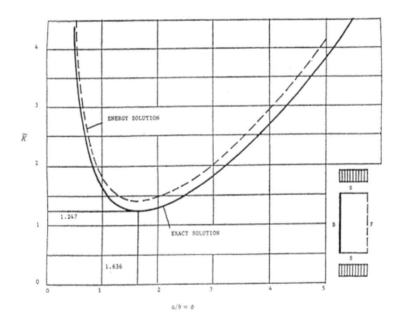

**FIGURE 2.5**
One longitudinal edge built-in, one free [32]. (From Bulson,
P.S., *The Stability of Flat Plates*, American Elsevier, N.Y.,
1969)

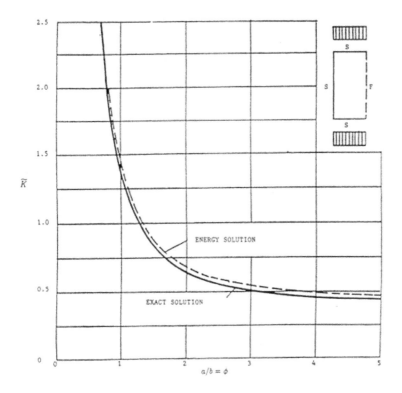

**FIGURE 2.6**
One longitudinal edge simply supported, one free [32]. (From
Bulson, P.S., *The Stability of Flat Plates*, American Elsevier,
N.Y., 1969)

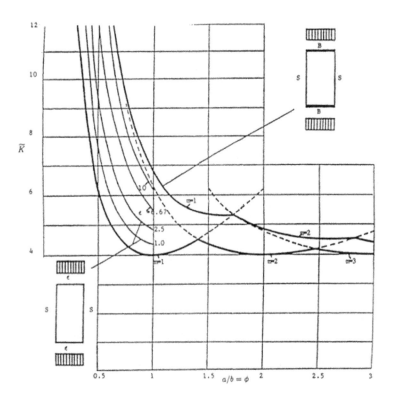

**FIGURE 2.7**
Built-in loaded edges. (Adopted, in modified form, from [75].)

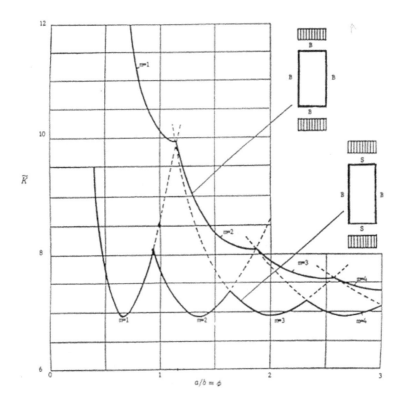

**FIGURE 2.8**
All edges built in. (Adopted, in modified form, from [76].)

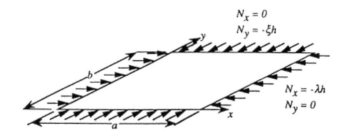

**FIGURE 2.9**
Rectangular plate compressed in both directions.

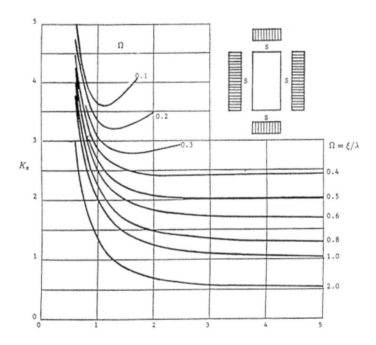

**FIGURE 2.10**
$K_x$ for simply supported edges [32]. (From Bulson, P.S., *The Stability of Flat Plates*, American Elsevier, N.Y., 1969)

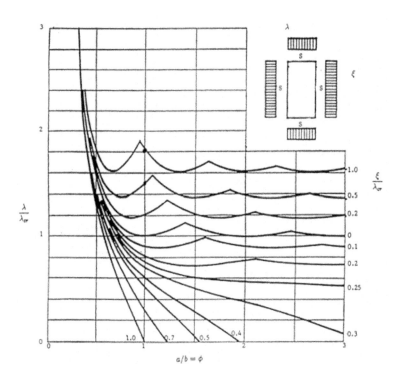

**FIGURE 2.11**
Relationship between $\lambda$ and $\xi$ [78].

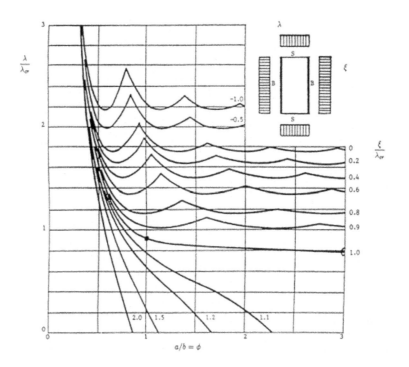

**FIGURE 2.12**
Relationship between $\lambda$ and $\xi$ [78].

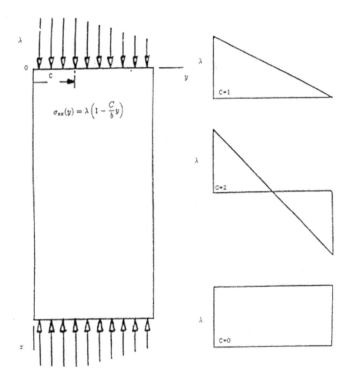

**FIGURE 2.13**
Linearly varying edge forces [32]. (From Bulson, P.S., *The Stability of Flat Plates*, American Elsevier, N.Y., 1969)

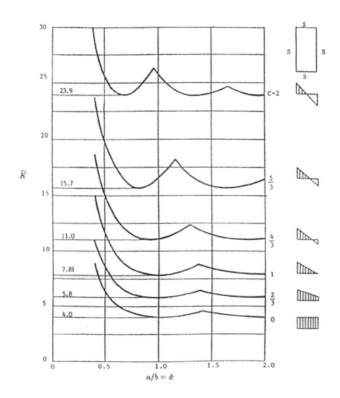

**FIGURE 2.14**
Simply supported plates with linearly varying edge forces [32].
(From Bulson, P.S., *The Stability of Flat Plates*, American
Elsevier, N.Y., 1969)

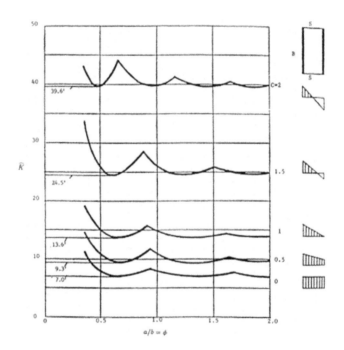

**FIGURE 2.15**
Longitudinal edges built-in. (Adopted, in modified form, from
[79].)

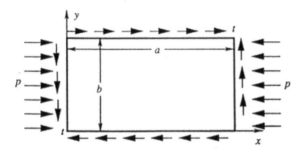

**FIGURE 2.16**
Plate subjected to a combined loading.

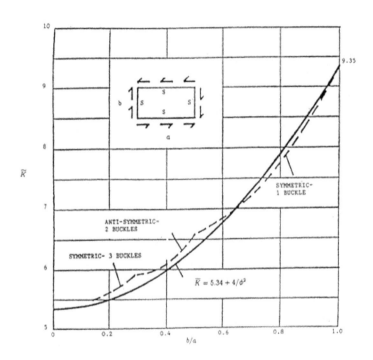

**FIGURE 2.17**
Simply supported plates in shear. (Adopted, in modified form, from [80].)

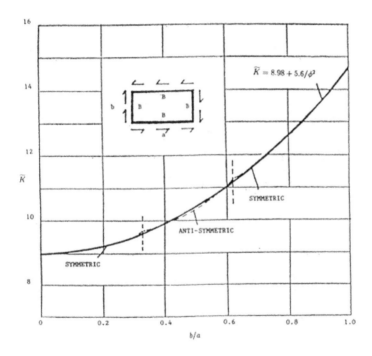

**FIGURE 2.18**
Built-in plates in shear. (Adopted, in modified form, from [81].)

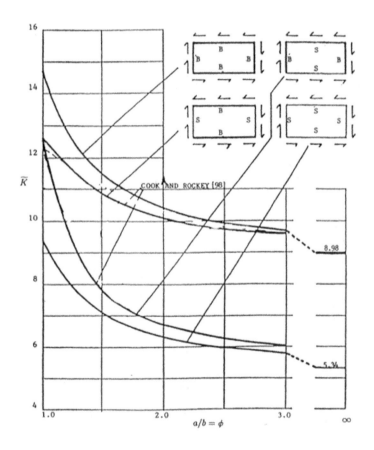

**FIGURE 2.19**
Shear buckling coefficient for various edge conditions [32].
(From Bulson, P.S., *The Stability of Flat Plates*, American
Elsevier, N.Y., 1969)

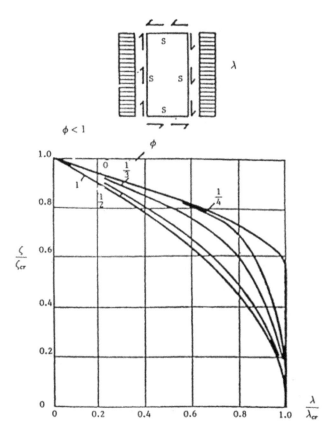

**FIGURE 2.20**
Shear and compression, restrained edges, $\phi < 1$. (Adopted, in modified form, from [83].)

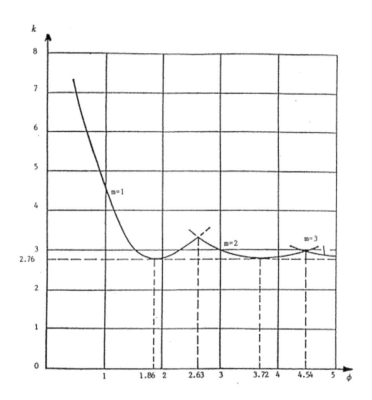

FIGURE 2.21
Compression along the external fibers [51]. (From Lekhnitskii,
S.G., *Anisotropic Plates*, Gordon and Breach Pub. Co., N.Y.,
1968 and OPA, N.V. With permission)

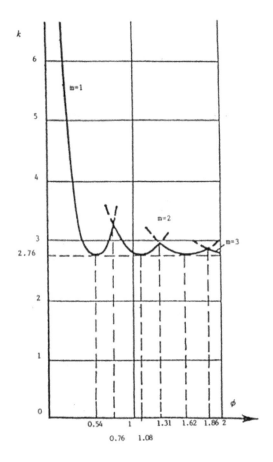

**FIGURE 2.22**
Compression across the external fibers [51]. (From Lekhnitskii,
S.G., *Anisotropic Plates*, Gordon and Breach Pub. Co., N.Y.,
1968 and OPA, N.V. With permission)

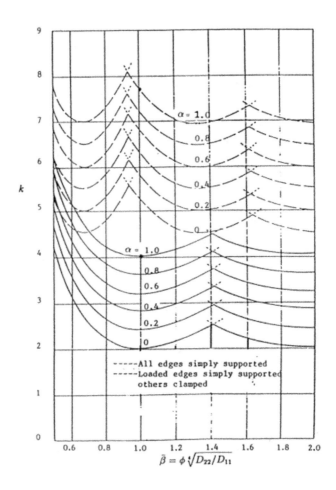

**FIGURE 2.23**
Coefficients of plate buckling in compression, I [43]. (From Hoff, N.J. and Stavsky, Y., "Mechanics of Composite Structures", in *Composite Engineering Laminates*, A.G.H. Dietz, ed., MIT Press, Cambridge, MA., 1969, 5-59, with permission.)

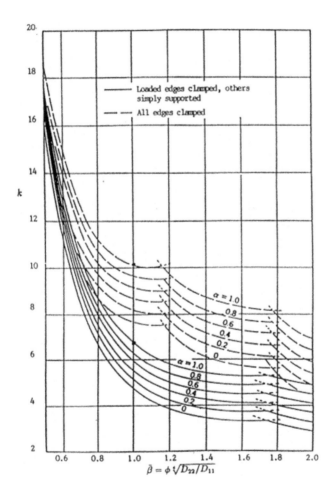

FIGURE 2.24
Coefficients of plate buckling in compression, II [43]. (From Hoff, N.J. and Stavsky, Y., "Mechanics of Composite Structures", in *Composite Engineering Laminates*, A.G.H. Dietz, ed., MIT Press, Cambridge, MA., 1969, 5-59, with permission.)

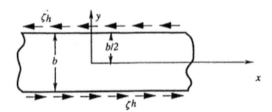

**FIGURE 2.25**
Shearing of a very narrow or infinite orthotropic strip.

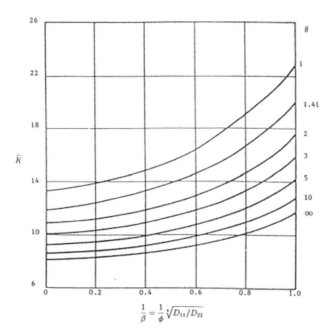

**FIGURE 2.26**
Coefficients of plate buckling in shear. (Adopted, in modified
form, from [84].)

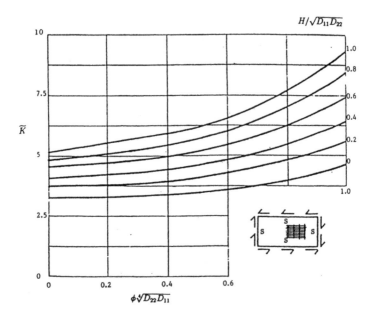

**FIGURE 2.27**
Orthotropic plates in shear. (Adopted, in modified form, from [84].)

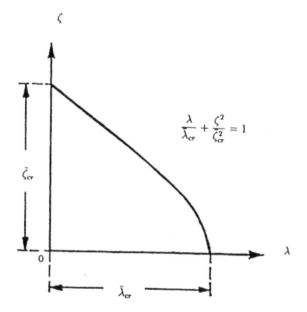

**FIGURE 2.28**
Orthotropic plates subjected to a combination of shear and compressive loading [51]. (From Lekhnitskii, S.G., *Anisotropic Plates*, Gordon and Breach Pub. Co., N.Y., 1968 and OPA, N.V. With permission)

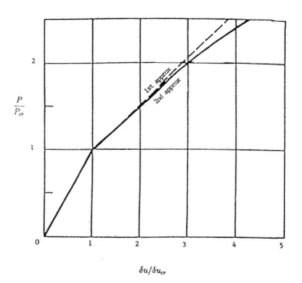

**FIGURE 2.29**
Postbuckling behavior for a square isotropic plate. (Adopted, in modified form, from [61].)

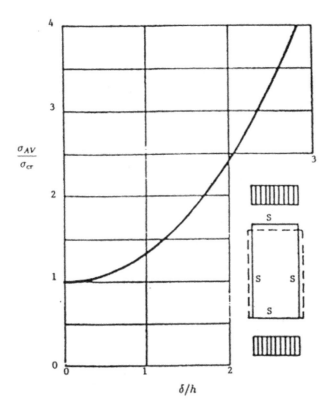

**FIGURE 2.30**
Postbuckling deflections of rectangular plate in end compression [32]. (From Bulson, P.S., *The Stability of Flat Plates*, American Elsevier, N.Y., 1969)

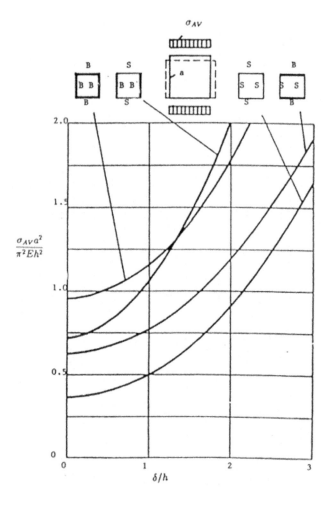

**FIGURE 2.31**
Postbuckling deflections. (Adopted, in modified form, from [94].)

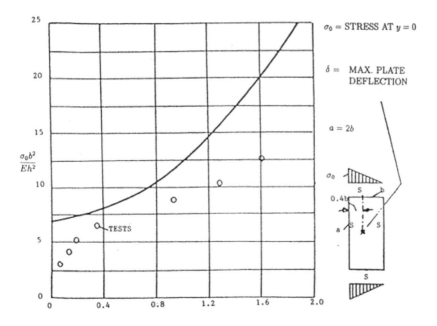

**FIGURE 2.32**
Postbuckling deflections of plates under a linearly varying end load.
(Reprinted with permission of The Random House Group Limited©
A.H. Chilver, Editor.)

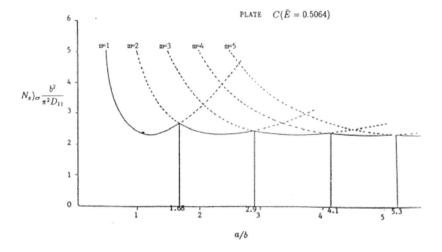

**FIGURE 2.33**
Critical stress versus $a/b$ ratio for plate $C(\bar{E} = 0.51)$ [96]. (Reprinted from the *Int. J. Mech. Sci.*, 15, Chandra, R. and Raju, B., Postbuckling Analysis of Rectangular Orthotropic Plates, 81-97, 1973, with permission from Elsevier Science)

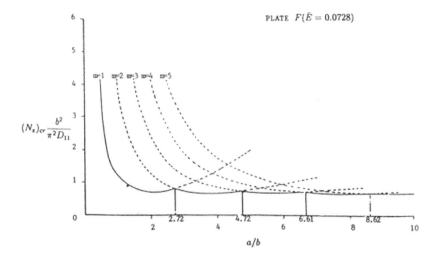

**FIGURE 2.34**
Critical stress versus $a/b$ ratio for plate $F(\bar{E} = 0.072)$ [96]. (Reprinted from the *Int. J. Mech. Sci.*, 15, Chandra, R. and Raju, B., Postbuckling Analysis of Rectangular Orthotropic Plates, 81-97, 1973, with permission from Elsevier Science)

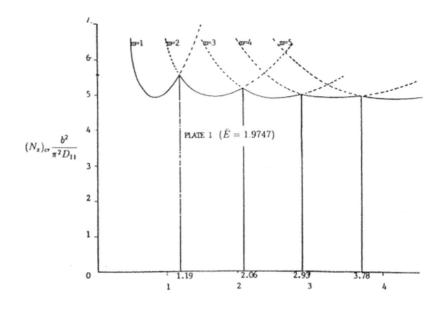

**FIGURE 2.35**
Critical stress versus $a/b$ ratio for plate $I(\bar{E} = 1.97)$ [96] .
(Reprinted from the *Int. J. Mech. Sci.*, 15, Chandra, R. and
Raju, B., Postbuckling Analysis of Rectangular Orthotropic
Plates, 81-97, 1973, with permission from Elsevier Science)

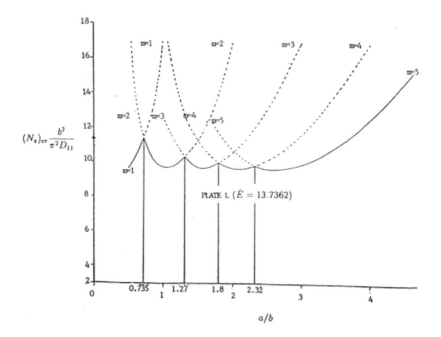

**FIGURE 2.36**
Critical stress versus $a/b$ ratio for plate $L(\bar{E} = 13.73)$ [96].
(Reprinted from the *Int. J. Mech. Sci.*, 15, Chandra, R. and
Raju, B., Postbuckling Analysis of Rectangular Orthotropic
Plates, 81-97, 1973, with permission from Elsevier Science)

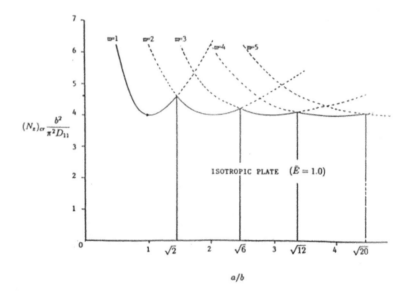

**FIGURE 2.37**
Critical stress versus $a/b$ ratio for the isotropic plate ($\bar{E} = 1.00$) [96]. (Reprinted from the *Int. J. Mech. Sci.*, 15, Chandra, R. and Raju, B., Postbuckling Analysis of Rectangular Orthotropic Plates, 81-97, 1973, with permission from Elsevier Science)

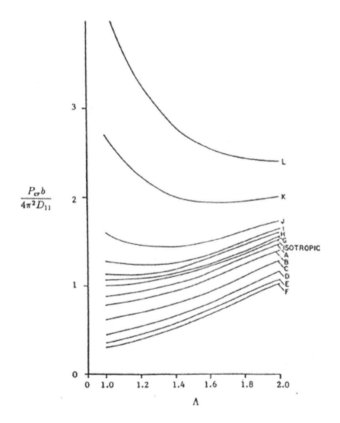

**FIGURE 2.38**
Variation of the buckling load parameter with $\Lambda$ for various cases, $\Lambda = m/\phi$ [96]. (Reprinted from the *Int. J. Mech. Sci.*, 15, Chandra, R. and Raju, B., Postbuckling Analysis of Rectangular Orthotropic Plates, 81-97, 1973, with permission from Elsevier Science)

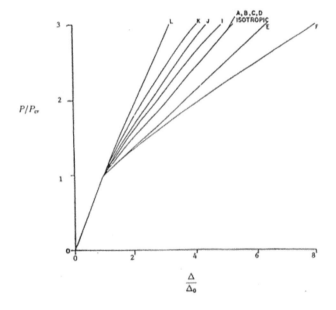

**FIGURE 2.39**
Load-shortening curves for the case $\Lambda = 1.0$, $\Lambda = m/\phi$ [96]. (Reprinted from the *Int. J. Mech. Sci.*, 15, Chandra, R. and Raju, B., Postbuckling Analysis of Rectangular Orthotropic Plates, 81-97, 1973, with permission from Elsevier Science)

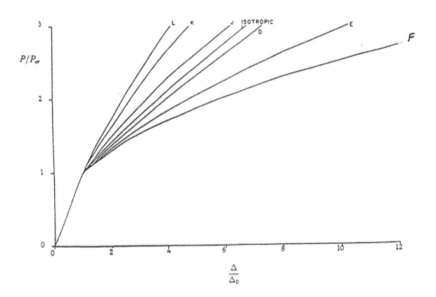

**FIGURE 2.40**
Load-shortening curves for the case $\Lambda = 1.33$, $\Lambda = m/\phi$ [96]. (Reprinted from the *Int. J. Mech. Sci.*, 15, Chandra, R. and Raju, B., Postbuckling Analysis of Rectangular Orthotropic Plates, 81-97, 1973, with permission from Elsevier Science)

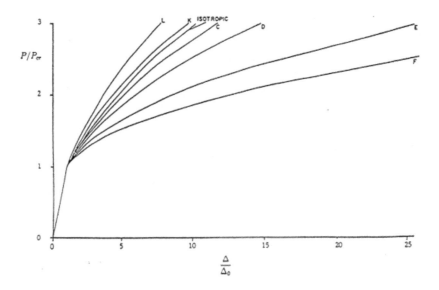

**FIGURE 2.41**
Load-shortening curves for the case $\Lambda = 2.0$, $\Lambda = m/\phi$ [96]. (Reprinted from the *Int. J. Mech. Sci.*, 15, Chandra, R. and Raju, B., Postbuckling Analysis of Rectangular Orthotropic Plates, 81-97, 1973, with permission from Elsevier Science)

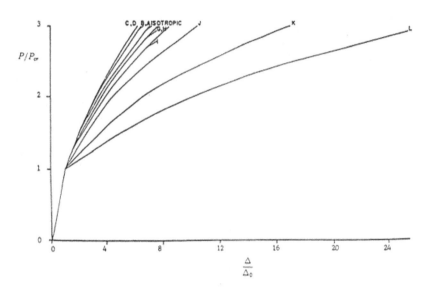

**FIGURE 2.42**
Load-shortening curves for the infinitely long plate [96].
(Reprinted from the *Int. J. Mech. Sci.*, 15, Chandra, R. and
Raju, B., Postbuckling Analysis of Rectangular Orthotropic
Plates, 81-97, 1973, with permission from Elsevier Science)

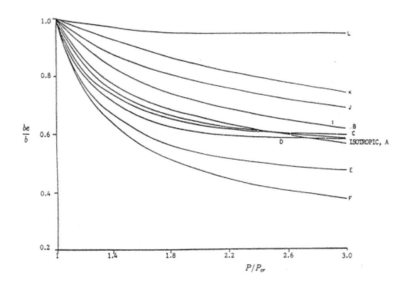

**FIGURE 2.43**
Effective width versus load curves for the case $\Lambda = 1.0$, $\Lambda = m/\phi$ [96]. (Reprinted from the *Int. J. Mech. Sci.*, 15, Chandra, R. and Raju, B., Postbuckling Analysis of Rectangular Orthotropic Plates, 81-97, 1973, with permission from Elsevier Science)

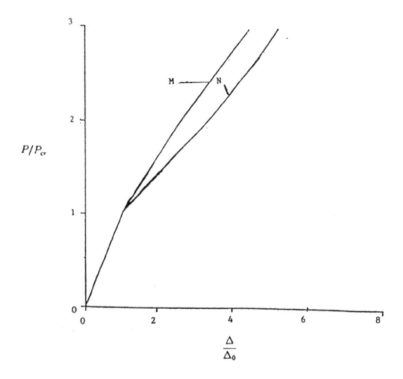

**FIGURE 2.44**
Comparison of load-shortening curves [96]. (Reprinted from
the *Int. J. Mech. Sci.*, 15, Chandra, R. and Raju, B., Postbuckling Analysis of Rectangular Orthotropic Plates, 81-97,
1973, with permission from Elsevier Science)

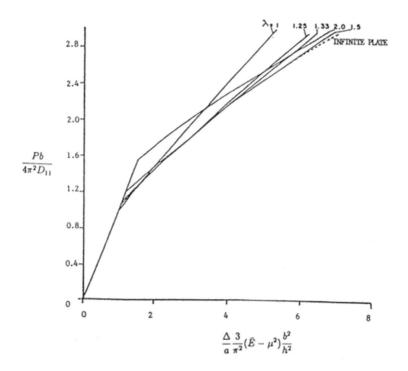

**FIGURE 2.45**
Load-shortening curves for different values of $\lambda$ for the isotropic case [96]. (Reprinted from the *Int. J. Mech. Sci.*, 15, Chandra, R. and Raju, B., Postbuckling Analysis of Rectangular Orthotropic Plates, 81-97, 1973, with permission from Elsevier Science)

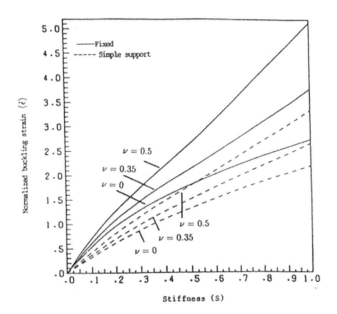

**FIGURE 2.46**
Plots of $\hat{\epsilon}$ as a function of $S$ for three mean Poisson's ratios and two edge-support conditions [74]. (From Johnson, M. and Urbanik, T.J., *Wood and Fiber Science*, 19, 135, 1987, with permission)

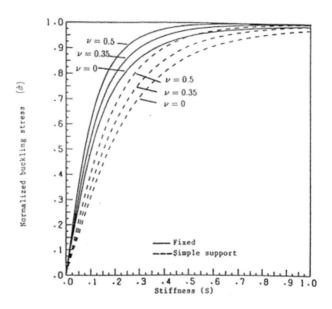

**FIGURE 2.47**
Plots of $\hat{\sigma}$ as a function of $S$ for three mean Poisson's ratios and two edge-support conditions [74]. (From Johnson, M. and Urbanik, T.J., *Wood and Fiber Science*, 19, 135, 1987, with permission)

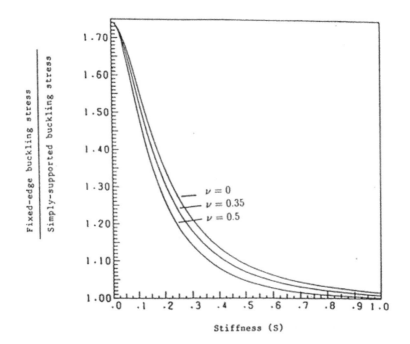

**FIGURE 2.48**
Ratio of fixed-edge buckling stress to simply-supported buckling stress as a function for $S$ for three mean Poisson's ratios
[74]. (From Johnson, M. and Urbanik, T.J., *Wood and Fiber
Science*, 19, 135, 1987, with permission)

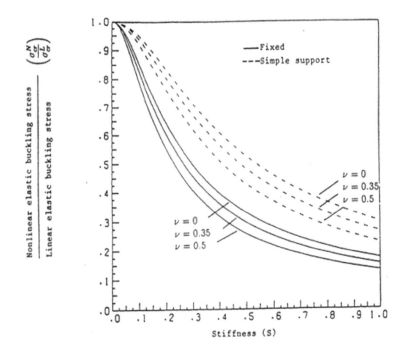

**FIGURE 2.49**
Ratio of nonlinear elastic buckling stress to linear elastic buckling stress as a function of $S$ for three mean Poisson's ratios and two edge-support conditions [74]. (From Johnson, M. and Urbanik, T.J., *Wood and Fiber Science*, 19, 135, 1987, with permission)

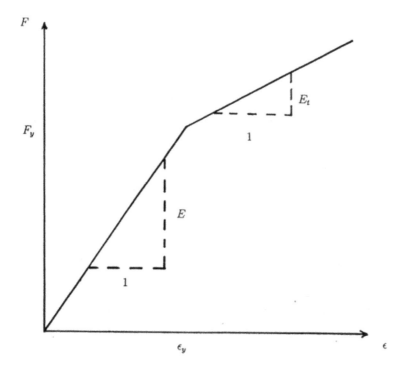

**FIGURE 2.50**
Young's modulus and tangent modulus for an elastic-plastic
material.

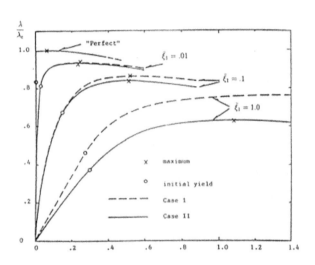

**FIGURE 2.51**
Load versus buckling mode displacement for square plate
$(\sigma_y/E = 0.00337,\ h/a = 0.035,\ n = 10, \nu = 0.3)$ [104].
(Reprinted from the *Int. J. Solids and Structures*, 12,
Needleman, A. and Tvergaard, V., An Analysis of the Imper-
fection Sensitivity of Square Elastic-Plastic Plates under Axial
Compression, 185-201, 1976, with permission from Elsevier
Science)

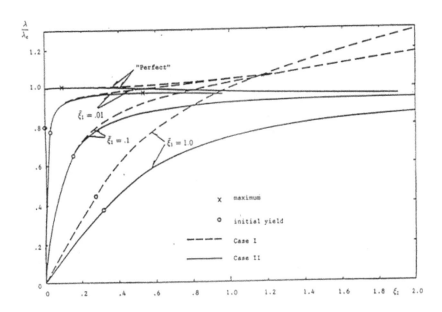

**FIGURE 2.52**
Load versus buckling mode displacement for a Square plate $(\sigma_y/E = 0.00337,\ h/a = 0.035, n = 3, \nu = 0.3)$ [104]. (Reprinted from the *Int. J. Solids and Structures*, 12, Needleman, A. and Tvergaard, V., An Analysis of the Imperfection Sensitivity of Square Elastic-Plastic Plates under Axial Compression, 185-201, 1976, with permission from Elsevier Science)

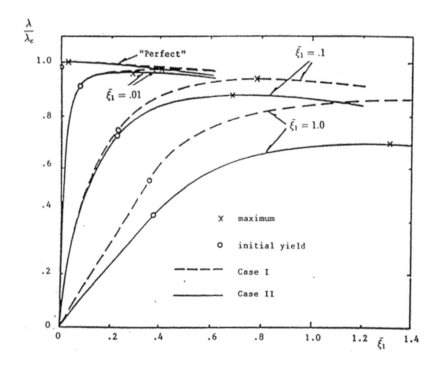

**FIGURE 2.53**
Load versus buckling mode displacement for a Square
plate $(\sigma_y/E = 0.00337,\ h/a - 0.031,\ n = 10, \nu = 0.3)$ [104].
(Reprinted from the *Int. J. Solids and Structures*, 12, Needleman, A. and Tvergaard, V., An Analysis of the Imperfection
Sensitivity of Square Elastic-Plastic Plates under Axial Compression, 185-201, 1976, with permission from Elsevier Science)

CHAPTER 3

# Initial and Postbuckling Behavior of Thin Circular Plates

In this chapter, we review the initial and postbuckling behavior of perfect, thin circular plates; the most prevalent types of support and loading conditions on the boundary of circular plates will be considered. We begin our study of the results for circular plate buckling by assuming the plate to exhibit linear elastic response; three subcases then arise, namely, plates exhibiting isotropic material symmetry, cylindrically orthotropic symmetry, or rectilinear orthotropic symmetry. The case of a circular plate exhibiting rectilinear orthotropic symmetry presents particular difficulties because of the mismatch between the geometry of the plate and the material symmetry reflected in the constitutive response of the plate.

Once the linear elastic case has been treated, we turn to circular plates exhibiting nonlinear elastic, viscoelastic, or elastic-plastic response. While very general treatments of the local and global buckling behavior of thin, nonlinear elastic, circular plates take into account both shearing and flexure of the plate (i.e., Antman [21], [139] exist in the literature), a model similar to the Johnson-Urbanik theory ([73], [74]) for nonlinear elastic rectangular plates, in which one obtains a generalization of the classical von Karman equations, does not appear to have been formulated, although extending the Johnson-Urbanik formulation to circular or annular geometries should be a straightforward if tedious task. For viscoelastic response, the results of Brilla [140] exist, and for the case of elastic-plastic circular plates, there is a well-developed literature, e.g., Needleman [141], Hutchinson [17], [45], [46], Tvergaard [103], and Onat and Drucker [142]. As was the case in Chapter 2 for rectangular plates, we compare, in §3.3, the initial and postbuckling behavior of circular plates as one varies the boundary (support and loading) conditions along the edge of the plate, the type of material symmetry exhibited by the plate, and the constitutive response of the plate.

## 3.1 Plates with Linear Elastic Behavior

Circular plates exhibiting linear elastic behavior may be classified according to whether the plate, which is assumed to be thin and without initial imperfections, possesses isotropic, cylindrically orthotropic, or rectilinearly orthotropic material symmetry.

### 3.1.1 Isotropic Symmetry: Initial Buckling and Postbuckling Behavior

We assume that the plate occupies the region $0 \leq r \leq a$, $r = \sqrt{x^2 + y^2}$, in the $x, y$ plane. The full set of von Karman equations for this problem consists of (1.78), (1.79) in the general situation for which the out-of-plane displacement (deflection) $w = w(r, \theta)$; for the special case of a deflection $w$ exhibiting radial symmetry, these equations reduce to (1.80), (1.81). The function $\Phi = \Phi(r, \theta)$ in (1.79) is, of course, the Airy stress function. Initial buckling in the non-axially symmetric case is governed by (1.78) with $\Phi = \Phi_0(r, \theta)$ being the Airy function describing the in-plane prebuckling stress distribution resulting from the applied loads and support conditions which are in force along the edge of the plate, i.e., $\Phi_0$ satisfies (1.79) with $w \equiv 0$; for the radially symmetric case initial buckling is governed by (1.80) with $\Phi = \Phi_0(r)$. We recall ((1.74)) that $N_r = \frac{1}{r}\Phi_{,r} + \frac{1}{r^2}\Phi_{,\theta\theta}$, $N_\theta = \Phi_{,rr}$, $N_{r\theta} = \frac{1}{r^2}\Phi_{,\theta} - \frac{1}{r}\Phi_{,r\theta}$ so that, e.g., (1.80), for the initial buckling problem, may be written in the form

$$
\begin{aligned}
K[w_{,rrrr} + \frac{2}{r}w_{,rrr} - \frac{1}{r^2}w_{,rr} + \frac{1}{r^3}w_{,r}] \\
= N_r^0 w_{,rr} + \frac{1}{r}N_\theta^0 w_{,r}
\end{aligned}
\tag{3.1}
$$

where $N_r^0 = \frac{1}{r}(\Phi_0)_{,r}$, and $N_\theta^0 = (\Phi_0)_{,rr}$. An alternative form for the full set of von Karman equations, in the case of axially symmetric deformations, may be obtained by multiplying both (1.80) and (1.81) by $r$ and integrating; this yields the system

$$
Kr\frac{d}{dr}\frac{1}{r}\frac{d}{dr}r\frac{dw}{dr} = \frac{d\Phi}{dr}\frac{dw}{dr} + c_1
\tag{3.2}
$$

and

$$
\frac{1}{Eh}r\frac{d}{dr}\frac{1}{r}\frac{d}{dr}r\frac{d\Phi}{dr} = \frac{1}{2}\left(\frac{dw}{dr}\right)^2 + c_2
\tag{3.3}
$$

At $r = 0$, the left-hand sides of (3.2), (3.3) vanish and, as a consequence of the radial symmetry of the solutions and the physical requirement of regularity, $\dfrac{dw}{dr}\Big|_{r=0} = 0$. Thus, $c_1 = c_2 = 0$ in (3.2), (3.3). If we expand the left-hand sides of (3.2), (3.3) we then have, in lieu of (1.80), (1.81), the equations

$$K(w_{,rrr} + \frac{1}{r}w_{,rr} - \frac{1}{r^2}w_{,r}) = \frac{1}{r}\Phi_{,r}w_{,r} \tag{3.4}$$

and

$$\frac{1}{Eh}(\Phi_{,rrr} + \frac{1}{r}\Phi_{,rr} - \frac{1}{r^2}\Phi_{,r}) = \frac{1}{2r}(w_{,r})^2 \tag{3.5}$$

With $\Phi = \Phi_0(r)$ in (3.4), the initial buckling equation for an isotropic circular plate is now obtained in the form

$$K(w_{,rrr} + \frac{1}{r}w_{,rr} - \frac{1}{r^2}w_{,r}) = N_r^0 w_{,r} \tag{3.6}$$

If we let $\varphi$ be the angle between the central axis of the plate and the normal to the deflected surface $w = w(r)$ at any point, then elementary geometry establishes that $\varphi = w_{,r}$ so that, e.g., (3.6) may be rewritten in terms of $\varphi$ as

$$\varphi_{,rr} + \frac{1}{r}\varphi_{,r} - \left(\frac{N_r^0}{K} + \frac{1}{r^2}\right)\varphi = 0 \tag{3.7}$$

One other form of the von Karman equations for axisymmetric buckling of an isotropic, thin, linearly elastic circular plate which is prevalent in the literature, especially in the work of Friedrichs and Stoker [72], [143], is obtained by introducing the functions

$$\begin{cases} p = \frac{1}{r}\Phi_{,r}(\equiv N_r) \\ q = -\frac{a}{r}w_{,r} \end{cases} \tag{3.8}$$

the constant $\eta = \sqrt{K}/a$, and the linear operator

$$G \equiv \frac{a^2}{r}\frac{d}{dr}\frac{1}{r}\frac{d}{dr}r^2 = a^2\left(\frac{d^2}{dr^2} + \frac{3}{r}\frac{d}{dr}\right) \tag{3.9}$$

In terms of $p, q, \eta$, and $G$ the von Karman equations (3.4), (3.5) become

$$\eta^2 Gq - pq = 0 \tag{3.10a}$$

$$\frac{1}{Eh} \cdot Gp = \frac{1}{2}q^2 \qquad (3.10b)$$

**Remarks:** In the work of Friedrichs and Stoker, op.cit., the first equation in the set (3.10) (with the definitions (3.8), (3.9)) has the form $\eta^2 Gq + pq = 0$; the difference in sign appears as a result of defining the Airy function so that, in the radially symmetric case, $N_r = \frac{1}{r}\Phi_{,r}$ instead of $N_r = -\frac{1}{r}\Phi_{,r}$. If we had defined the Airy function so that $N_r = -\frac{1}{r}\Phi_{,r}$, for axially symmetric deformations, then in lieu of (3.4) we would have

$$K\left(w_{,rrr} + \frac{1}{r}w_{,rr} - \frac{1}{r^2}w_{,r}\right) = -\frac{1}{r}\Phi_{,r}w_{,r} \qquad (\overline{3.4})$$

and, with the definitions of $p$, $q$, and $G$ in (3.8), (3.9), the first equation in (3.10) would read

$$\eta^2 Gq + pq = 0 \qquad (\overline{3.10a})$$

while the equilibrium equation in (3.7) would become

$$\varphi_{,rr} + \frac{1}{r}\varphi_{,r} + \left(\frac{N_r^0}{K} - \frac{1}{r^2}\right)\varphi = 0 \qquad (\overline{3.7})$$

with $\varphi = \dfrac{dw}{dr}$.

**Remarks:** If $\Phi(r)$ and $w(r)$ are to have continuous fourth derivatives, so that (1.80), (1.81) may be interpreted in the classical sense, then the boundary conditions which must be satisfied at the center of the plate are shown in [72], [143] to be

$$\left.\frac{dp}{dr}\right|_{r=0} = 0, \quad \left.\frac{dq}{dr}\right|_{r=0} = 0 \qquad (3.11)$$

**Remarks:** For circular, isotropic, linearly elastic plates the two basic support conditions that we will consider along the edge at $r = a$ are of the clamped and simply supported kinds; in the first case, we have, as a consequence of (1.139)

$$w(a, \theta) = w_{,r}(a, \theta) = 0, \ 0 \le \theta \le 2\pi, \qquad (3.12)$$

with the obvious simplifications for the axially symmetric situation, while for the second case the conditions are, as a direct consequence of (1.140),

$$
\begin{cases}
w(a,\theta) = 0 \\
\left[ w_{,rr} + \nu \left( \dfrac{1}{r}\, w_{,r} + \dfrac{1}{r^2}\, w_{,\theta\theta} \right) \right] \Bigg|_{r=a} = 0
\end{cases}
\tag{3.13}
$$

for $0 < \theta \le 2\pi$, with the obvious simplifications for the axially symmetric case.

**Remarks:** For the case where one considers only rotationally symmetric buckled states of an isotropic, linearly elastic, circular plate, which is simply supported along its edge at $r = a$, and is subject to a uniform compressive thrust $N_r = \lambda h$, per unit length, along $r = a$, which lies in the plane of the plate, several fairly general results have appeared, most notably in [72], [144], and [145]; in the first paper it was shown there is one pair of buckled states with no internal node for all values of the thrust parameter $\lambda$ greater than the first eigenvalue (predicted by the linear buckling theory) and that no other buckled states exist when $\lambda$ is less than or equal to the second eigenvalue. In the second of the papers referenced above, i.e., [140], it was proven that for every positive integer $n$, a pair of axially symmetric buckled states with $n$–1 internal nodes exists when the thrust $\lambda$ is slightly larger than $\lambda_n$, the $n$th eigenvalue of the linear buckling problem. In [145], Wolkowisky uses the Schauder fixed point theorem to prove that the pairs of buckled states established in [140] continue to exist for all $\lambda > \lambda_n$; in fact, Wolkowisky [145] proves that for all $\lambda > \lambda_n$ there exist $n$ pairs of nontrivial solutions of (3.10a,b); the members of each pair differ only in the sign of $q$. The $q$ in the first pair has no internal nodes, the $q$ in the second has one internal pair, and so on up to the $q$ in the $n$th pair which has $n$–1 internal nodes.

We begin our delineation of the results on initial buckling of a linearly elastic, isotropic, circular plate subject to the action of a radial compressive force $N_r = -\lambda h$ along its edge at $r = a$, by looking at rotationally symmetric buckling into the first mode for a plate with a clamped edge. In this case, the equation governing initial buckling is (3.7), with $\varphi = w'(r)$ and $N_r^0 = -\lambda h$, and the boundary conditions are (see (3.11)) $w(a) = w'(a) = 0$. If (3.7) is multiplied through by $r^2$ we obtain

$$
r^2 \frac{d^2\varphi}{dr^2} + r\frac{d\varphi}{dr} + \left( \frac{N_r^0 r^2}{K} - 1 \right)\varphi = 0
\tag{3.14}
$$

which is Bessel's differential equation. It can be shown (see, e.g., Timoshenko and Gere [64]) that the solution of (3.14) subject to the clamped boundary conditions $w(a) = w'(a) = 0$ leads to a critical value of the compressive stress $\lambda$, which is given by

$$\lambda_{cr} = \frac{14.68K}{a^2 h} \tag{3.15}$$

The result in (3.15) will also follow, later on, from a perturbation expansion approach to the postbuckling analysis for this case; the clamped plate in edge compression is depicted in Fig. 3.1.

The next case is that of the isotropic circular plate which is subject to a uniform radially compressive force and is simply supported along its edge at $r = a$; we consider axially symmetric buckling into the first mode so that $w'(0) = 0$ and $w(a) = 0$, as well as $w''(a) + \frac{\nu}{a} w'(a) = 0$. The case of a simply supported circular plate is more amenable to postbuckling analysis than that of a clamped plate and has received far more attention in the literature. Using Bessel functions (see, once again, Timoshenko and Gere [64]), it may be shown that

$$\lambda_{cr} = \frac{4.20K}{a^2 h} \tag{3.16}$$

a result which will also follow from the postbuckling analysis of this problem.

A more general problem than that considered in either of the two preceeding cases concerns the case of elastically restrained edges, from which one may deduce (i.e., Reismann [146]) a critical stress for any degree of restraint along the edge at $r = a$ between clamped and simply supported. If one restricts consideration to buckling only into the first axially symmetric mode, then a coefficient $\mu$ of restraint may be introduced into the support condition along $r = a$ in such a way that $\mu = 0$ corresponds to the simply supported case, while $\mu = \infty$ corresponds to a clamped edge; specifically, we have, in all situations, $w(a) = 0$, and

$$\frac{\mu \phi(a)}{a} = -\left( \frac{d\phi}{dr} + \frac{\nu}{r} \phi \right) |_{r=a} \tag{3.17}$$

Physically, (3.17) says that the inclination of the middle surface of the plate with respect to the plane of the undeflected plate is proportional to the radial edge moment. We still maintain the condition $w'(0) = 0$ at the center of the plate. By imposing the conditions $w(a) = 0$, $w'(0) = 0$, and (3.17), on the general solution of (3.14)—which will be discussed, in detail during the course of the postbuckling analysis—it may be determined that

$$\lambda_{cr} = \frac{kK}{a^2 h} \tag{3.18}$$

where the relationship between the buckling parameter $k$ and $\mu$ is indicated in Fig. 3.2, which is taken from Reismann's work [146]; it is clear from this figure that as $\mu \to 0$, $k \to 14.68$ and as $\mu \to \infty$, $k \to 4.20$, in agreement with (3.15) and (3.16), respectively, for the clamped and simply supported conditions along the edge at $r = a$.

Higher order (non-axially symmetric) buckling modes, which are associated with the uniform compressive buckling of linearly elastic, isotropic, circular plates of radius $a$, are governed by the linear partial differential equation (1.78) with $N_r = N_r^0 \equiv -\dfrac{1}{r}\Phi_{,r}^0$, $\Phi_{,\theta}^0 = \Phi_{,r\theta}^0 = \Phi_{,\theta\theta}^0 = 0$, where $\Phi^0$ is, once again, the Airy function corresponding to the prebuckled in-plane stress distribution; under this condition, of a uniform, radially compressive, force $N_r = -\lambda h$ per unit length, applied to the edge of the plate at $r = a$, (1.78), with $N_r = N_r^0 \equiv -\dfrac{1}{r}\Phi_{,r}^0$, may be rewritten in the form

$$\left( \frac{\partial^2}{\partial r^2} + \frac{1}{r}\frac{\partial}{\partial r} + \frac{1}{r^2}\frac{\partial^2}{\partial \theta^2} \right) \left( \frac{\partial^2}{\partial r^2} + \frac{1}{r}\frac{\partial}{\partial r} + \frac{1}{r^2}\frac{\partial^2}{\partial \theta^2} + \frac{N_r^0}{K} \right) w(r,\theta) = 0$$

(3.19)

The usual procedure at this point (which we will substantially elaborate upon when we consider the problem of initial buckling of isotropic annual plates) is to look for solutions of (3.19) of the form

$$w(r,\theta) = \sum_{n=1}^{\infty} A_n(r) \sin n\theta \tag{3.20}$$

which will correspond to the existence, in the buckled state, of nodal diametric lines in the plate as indicated in Fig. 3.3. By substituting (3.20) into (3.19), it may be shown that the $A_n(r), n = 1, ..,$ must be Bessel functions and by imposing the boundary (support) conditions, along $r = a$, two linear homogeneous algebraic equations result, with the vanishing of the determinant of the matrix of coefficients of these equations determining the possible corresponding buckled modes.

For a plate which is elastically restrained against rotation at the edge $r = a$, of the circular plate, we have $w(a,\theta) = 0$, $0 < \theta \le 2\pi$, as well as the condition which generalizes (3.17) to the non-axially symmetric case, namely,

$$\left[ \frac{\partial^2 w}{\partial r^2} + \left( \frac{\nu}{r} + \frac{\mu}{a} \right) \frac{\partial w}{\partial r} + \frac{\nu}{r^2} \frac{\partial^2 w}{\partial \theta^2} \right]_{r=a} = 0 \tag{3.21}$$

One now makes the additional assumption that the smallest value of $\lambda_{cr}$ occurs when the middle surface of the plate forms one full wavelength in the circumferential direction with a nodal line along the diameter as in Fig. 3.3. For such a case we obtain, therefore, $n = 1$, and a numerical solution giving the buckling parameter

$$k = \frac{a^2 h}{K} \lambda_{cr} \qquad (3.22)$$

as a function of $\mu a / K$ is shown as the upper curve in Fig. 3.4, where the lower curve again depicts the same function for the axially symmetric first mode.

We now turn our attention to the postbuckling analysis for a uniformly (radially) compressed linearly elastic, isotropic circular plate which is either clamped or simply supported along the edge at $r = a$; the earliest analysis of this problem which yielded substantive results appears to be in the work of Friedrichs and Stoker [72], [143] which treats the simply supported case; further work on the case where the boundary at $r = a$ is simply supported has been carried out by Sherbourne [147], while the work of Friedrichs and Stoker, op. cit, has been summarized in the monograph by Stoker [148]. The methods employed by Friedrichs and Stoker, op. cit, for the simply supported case have been extended by Bodner [149] to treat the case where the edge at $r = a$ is clamped. All the work referenced above deals with the von Karman equations for this problem written in terms of $\Phi$ and $w$; attention is also confined, with respect to the problem of buckling beyond the critical load, to that situation where the plate buckles, initially, into an axisymmetric mode. Beyond reviewing the results of Friedrichs and Stoker, [72], [143], and Sherbourne [147] (for a simply supported edge), and those of Bodner [149] (for a clamped edge), we will also discuss, for both types of support conditions along $r = a$, results which follow, for axially symmetric initial buckling, from carrying out the postbuckling analysis for the von Karman equations in terms of the in-plane ($u = u(r)$) and out-of-plane ($w = w(r)$) displacements instead of in terms of $\Phi$ and $w$.

We begin with the work of Friedrichs and Stoker [72], [143], [148] for a simply supported edge; to bring our results completely into line with their work we will take $N_r = -\frac{1}{r} \Phi_r$, replace $G$ in (3.9) by $\tilde{G} = G/Eh$, and $\eta^2 = K/a^2$ by $\tilde{\eta}^2 = Eh\eta^2$, in which case we have, in place of (3.10a,b), the equations (we drop the ˜):

$$\eta^2 G q + p q = 0 \qquad (3.23a)$$

$$G p = \frac{1}{2} q^2 \qquad (3.23b)$$

Here $p = \dfrac{1}{r}\Phi_{,r}$ is the compressive radial membrane stress, and $q = -\dfrac{a}{r}\phi \equiv -\dfrac{a}{r}\dfrac{dw}{dr}$, so that $\dfrac{qr}{a}$ is the slope of the deflected plate. At the center of the plate at $r = 0$, $p(r)$ and $q(r)$ must satisfy the conditions in (3.11). At the edge of the plate, at $r = a$, we have the boundary conditions

$$p = \frac{1}{r}\Phi_{,r}\Big|_{r=a} = -\,N_r|_{r=a} = \bar{p}(= -\lambda h) \tag{3.24}$$

where $\bar{p}$ is the applied compressive pressure (force per unit length) at the edge of the plate, and

$$\mathcal{B}_\nu q|_{r=a} \equiv r\frac{dq}{dr} + (1+\nu)q\Big|_{r=a} = 0 \tag{3.25}$$

which implies the vanishing of the bending moment at the edge of the plate.

**Remarks:** As $q = -\dfrac{a}{r}w_{,r}$, $q' = -\dfrac{a}{r}w_{,rr} + \dfrac{a}{r^2}w_{,r}$ so

$$r\frac{dq}{dr} + (1+\nu)q = (-aw_{,rr} + \frac{a}{r}w_{,r}) - (1+\nu)\frac{aw_{,r}}{q}$$

and (3.25) is, thus, a direct consequence of the condition of simple support, (1.140), with $w_{,\theta} = 0$ and $R_2 = a$.

**Remarks:** If one has determined $p$ and $q$, then all other physical quantities of interest may be easily computed, e.g., the circumferential membrane stress is

$$p_c = \mathcal{B}_0 p \equiv rp'(r) + p(r), \tag{3.26}$$

while the normal deflection $w$ is

$$w(r) = \frac{1}{a}\int_r^a q(r)r\,dr \tag{3.27}$$

The basis for the postbuckling analysis of a linearly elastic, isotropic, circular plate subject to a uniform radial compressive loading along its edge at $r = a$, and simply supported along that edge consists, therefore, of (3.23a,b), (3.11), and (3.24), (3.25). The trivial solution of this boundary-value problem, i.e., $q \equiv 0, p \equiv \bar{p}$ (const.) corresponds to $p_c = \bar{p}$, $w \equiv 0$, and, thus, represents a state of pure hydrostatic compression of the plate. For $\bar{p}$, sufficiently large buckling occurs (at $\bar{p} = \bar{p}_0 \equiv \lambda_{cr} \cdot h$); it is natural to employ a perturbation expansion approach to describe the postbuckling behavior of the plate and to develop

$p$ and $q$ with respect to some convenient parameter $\epsilon$ which vanishes at the onset of buckling, i.e.,

$$\begin{cases} \bar{p} = \bar{p}_0 + \epsilon^2 \bar{p}_2 + \epsilon^4 \bar{p}_4 + \cdots \\ p = p_0 + \epsilon^2 p_2 + \epsilon^4 p_4 + \cdots \\ q = \epsilon q_1 + \epsilon^3 q_3 + \epsilon^5 q_5 + \cdots \end{cases} \tag{3.28}$$

The expansions in (3.28) may be justified by appealing to general results in bifurcation theory ( [10], [12], or [26]). The perturbation expansion approach to postbuckling consists of substituting the expressions in (3.28) into (3.23a,b), (3.11), (3.24), (3.25), and equating like powers of $\epsilon$ so as to generate a sequence of differential equations for $p_0, q_1, p_2, q_3, \cdots$ together with a corresponding sequence of boundary conditions; the first several members of this sequence of boundary value problems is indicated in the Table 3.1.

The general solution of the first equation in the sequence, $Gp_0 = 0$, is $p_0 = c_1 r^{-2} + c_2$ and the boundary conditions at $r = 0$ and $r = a$ yield $c_1 = 0$ and $c_2 = \bar{p}_0$. Thus, the only solution to problem $L_0$ is $p_0 \equiv \bar{p}_0$, where $\bar{p}_0$ is the critical (compressive) buckling load which is determined by the solution of the initial buckling problem. By (3.16) we have $\bar{p}_0 = 4.2K/a^2$ and we will now indicate, in some detail, how this result arises. As $p_0$ is constant, problem $L_1$, delineated in Table 3.1 (being a homogeneous linear differential equation with associated homogeneous boundary conditions) has only the solution $q_1 \equiv 0$ unless $p_0 \equiv \bar{p}_0$ is an eigenvalue; the eigenvalue problem for determining $\bar{p}_0$ is the same one which arises in the linear (initial) buckling theory. We now set

$$u = \frac{\lambda_0 r}{a}, \quad \left( \lambda_0^2 = \frac{\bar{p}_0}{\eta^2} \right), \quad Q_1 = q_1 \frac{r}{a} \tag{3.29}$$

and the differential equation in $L_1$ becomes

$$u^2 \frac{dQ_1}{du} + u \frac{dQ_1}{du} + (u^2 - 1)Q_1 = 0 \tag{3.30}$$

Equation (3.30) is the Bessel equation of the first order and has the general solution

$$Q_1(u) = c_1 J_1(u) + c_2 Y_1(u) \tag{3.31}$$

where $J_1(u)$ and $Y_1(u)$ are, respectively, the Bessel functions of the first and second kinds of the first order. Because $Y_1$ is not bounded in a neighborhood of $u = 0$, the conditions in the physical problem require that $c_2 = 0$, and the remaining boundary conditions then imply that

$$c_1 \lambda_0 \left( \frac{dJ_1}{du} + \frac{\nu}{u} J_1 \right) \bigg|_{u=\lambda_0} = 0 \qquad (3.32)$$

and

$$c_1 a^{-1} \lambda_0^2 \left( \frac{1}{u} \frac{dJ_1}{du} - \frac{1}{u^2} J_1 \right) \bigg|_{u=0} = 0 \qquad (3.33)$$

As $J_1'(u) = J_0(u) - \frac{1}{u} J_1(u)$, where $J_0$ is the Bessel function of zero order, except for the trivial solution in which $c_1 = 0$, the boundary condition (3.32) is satisfied only if $\lambda_0$ is a root of the transcendental equation

$$\lambda_0 J_0(\lambda_0) - (1 - \nu) J_1(\lambda_0) = 0 \qquad (3.34)$$

while the second boundary condition, i.e., (3.33) is automatically satisfied. For each allowable value of the Poisson ratio $\nu$, (3.34) has infinitely many solutions; the smallest value of $\lambda_0$ which satisfies (3.34) then corresponds to the smallest value of $\bar{p}_0$ which satisfies problem $L_1$ and, therefore, characterizes the onset of buckling for the plate. For $\nu = 0.318$, Friedrichs and Stoker [143] note that the smallest root of (3.34) is $\lambda_0 = 2.06$ and that moderate variations in $\nu$ have little effect on the solution, e.g., if $\nu = 0.3$ then the smallest root of (3.34) is $\lambda_0 = 2.05$. Thus, in this case, buckling will begin when the compressive edge pressure is $\bar{p} = \bar{p}_0 = p_0 = 4.2436\eta^2$. Once $\bar{p}_0$ has been computed, it may be determined that $q_1(r)$, the first (axisymmetric) buckling mode, has the form

$$q_1(r) = c \frac{r}{a} J_1 \left( \lambda_0 \frac{r}{a} \right) \qquad (3.35)$$

where the amplitude of the mode $q_1$ has not yet been uniquely determined because $p_2, q_3, ...$, depend on $q_1$. Using an analogous approach, the problems $L_2, L_3, ...$ can be solved, in turn, for $p_2, q_3, p_4, ...$, etc., however, these problems are no longer homogeneous and it is noted in [148] that the solutions are not expressible (in closed form) in terms of known functions. For $\lambda$ satisfying $\lambda_0 \leq \lambda \leq \lambda_1$, where $\lambda^2 = \bar{p}^2 / \eta^2$, and $\lambda_1$ is the second eigenvalue of the linear problem $L_1$, there are, at most, two buckled states of the circular plate which differ only in the direction of buckling (i.e., up or down).

Because problems $L_2, L_3, ...$, cannot be solved in closed form, numerical techniques are employed to generate the successive coefficients $p_2, q_3, p_4$, etc., in the postbuckling expansions (3.28); it has also been standard practice in the analysis of this problem to replace $\lambda$ by $\Lambda =$

$\bar{p}/\bar{p}_0 \equiv \lambda^2/\lambda_0^2$ as the basic buckling parameter and to choose $\epsilon = \sqrt{\Lambda - 1}$ so that the onset of buckling is characterized by $\Lambda = 1$ and $\epsilon = 0$. In this section, we will present the results of a calculation based on the Lyapunov-Schmidt reduction ([10], [12]), which shows that $\epsilon$ in (3.28) should have the form

$$\epsilon = \left( \frac{\int_0^a q_1^2 dr}{\int_0^a q_1 a_3 dr} \right)^{\frac{1}{2}} \sqrt{\Lambda - 1} \tag{3.36}$$

and which results in an expansion for $q$ of the form

$$q \simeq \left( \sqrt{\frac{\int_0^a q_1^2 dr}{\bar{p}_0 \int_0^a q_1 q_3 dr}} \, \bar{\lambda}^{\frac{1}{2}} q_1 + \cdots \right) \tag{3.37}$$

with $\bar{\lambda} = \bar{p} - \bar{p}_0$. For various values of $\Lambda = \bar{p}/\bar{p}_0$, the results for radial and circumferential membrane stresses and radial bending stress which follow from the work of Friedrichs and Stoker [72], [143] are displayed, respectively, in Figs. 3.5, 3.6, and 3.7; the corresponding postbuckling graphs for the slope and deflection of the (uniformly) radially compressed circular plate are depicted in Figs. 3.8 and 3.9. It must be emphasized that the results shown are for first-order (axisymmetric) buckling corresponding to the lowest eigenvalue of the linear (initial) buckling theory; there are many indications that this first buckled state becomes unstable for sufficiently large $\Lambda$ and that a second buckling occurs which, subsequently, also becomes unstable, e.g., the second-order buckling, which corresponds to the second eigenvalue of the linear problem $L_1$, appears in the work of Friedrichs and Stoker [143] for $\Lambda$ in the range $\Lambda > 5$. The issue of secondary buckling into non-axisymmetric modes ("wrinkling") for compressed, linearly elastic, isotropic, circular plates will be discussed in detail in § 5.2.

A slightly different analysis of the postbuckling behavior of a simply supported uniformly (radially) compressed, elastic, isotropic plate has been given by Sherbourne [147] who again recovers the result (3.16) for the critical stress at which the plate will first buckle into an axisymmetric mode; for large elastic deformations, the compressive stress $\lambda = \sigma_{rr}|_{r=a}$ then increases as the central deflection $\delta = w(0)$ grows. Sherbourne

[147] shows that the relationship between $\lambda/\lambda_{cr}$ and $\delta/h$ takes the form

$$\frac{\lambda}{\lambda_{cr}} = 1 + 0.241 \left(\frac{\delta}{h}\right)^2 \tag{3.38}$$

whose graph is depicted in Fig. 3.10. A remarkable feature of the relationship (3.38) is that it is independent of the radius-to-thickness ratio, $a/h$, of the plate; just as was the case for rectangular isotropic plates, significant postbuckling loads may be carried by an elastic, isotropic, circular plate as the central deflection to plate thickness ratio increases in the range from $\delta/h = 1$ to $\delta/h = 3$, with ultimate failure of the plate being associated with the development of plastic strains.

For the clamped, linearly elastic, isotropic plate under uniform radial compression, the critical compressive edge stress is given by (3.15), and, as $\lambda = \sigma_{rr}|_{r=a}$ is increased beyond this value, the plate develops significant membrane stresses. Bodner [149], employing the methods developed by Friedrichs and Stoker, op. cit., has compared the radial membrane stresses developed in the cases of simply supported and clamped edges; the results of his analysis are displayed in Fig. 3.11, which compares $\lambda/\lambda_{cr}$ with $p(0)/\lambda$ where $p(0) = \frac{1}{r}\Phi_{,r}\Big|_{r=0}$ is the radial membrane stress at the center of the plate. The radial membrane stress changes from compression to tension when $\lambda/\lambda_{cr} = 1.57$, for simply supported plates, and 1.98 for clamped plates. Also, as $\frac{\lambda}{\lambda_{cr}} \to \infty$, $p(0)/\lambda \to -0.47$, for simply supported plates, and $-0.13$ for clamped plates.

As we have previously indicated, the postbuckling analysis for the elastic, isotropic, circular plate may be framed in terms of the deflection and in-plane displacement of the middle surface of the plate instead of in terms of the deflection and the Airy stress function. Referring to Fig. 3.12 and assuming an axisymmetric deformation of the plate, we denote the in-plane displacement of the plate by $u = u(r)$ and, once again, use $\bar{p}$ for the applied compressive pressure (force per unit length) at the edge of the plate, $\bar{p} = -\lambda h$; then, prior to the onset of buckling, we have

$$w(r) = 0, \ u(r) \equiv u^o(r) = -\frac{\bar{p}(1-\nu)}{Eh}r \tag{3.39}$$

and, if we set $v(r) = u(r) - u^o(r)$, then in terms of $w, v$, the von Karman equations assume the form

$$w_{,rrr} + \frac{1}{r}w_{,rr} - \frac{1}{r^2}w_{,r} - \frac{12}{h^2}\left(v_{,r} + \frac{1}{2}w_{,r}^2 + v\frac{v}{r}\right)w_{,r}$$

$$+ \left(\frac{\bar{p}}{K}\right)w_{,r} = 0 \qquad (3.40a)$$

and

$$v_{,rr} + \frac{1}{r}v_{,r} - \frac{1}{r^2}v + w_{,r}w_{,rr} + \frac{(1-v)}{2r}w_{,r}^2 = 0 \qquad (3.40b)$$

For a simply supported plate we have, furthermore, the boundary conditions

$$\begin{cases} w(a) = w_{,rr} + v\frac{w_{,r}}{r}\bigg|_{r=a} = 0 \\[2mm] v_{,r} + \frac{1}{2}w_{,r}^2 + v\frac{v}{r}\bigg|_{r=a} = 0 \\[2mm] \dot{w}(0) = v(0) = 0, \end{cases} \qquad (3.41a)$$

while, for a clamped plate, these conditions read

$$\begin{cases} w(a) = w_{,r}|_{r=a} = 0 \\[2mm] v_{,r} + \frac{1}{2}w_{,r}^2 + v\frac{v}{r}\bigg|_{r=a} = 0 \\[2mm] \dot{w}(0) = v(0) = 0, \end{cases} \qquad (3.41b)$$

In the spirit of the perturbation analysis employed by Friedrichs and Stoker [143], we seek solutions of (3.40a,b) in the form

$$\begin{cases} w(r) = \epsilon w_1(r) + \epsilon^2 w_2(r) + \epsilon^3 w_3(r) + \cdots \\[2mm] v(r) = \epsilon v_1(r) + \epsilon^2 v_2(r) + \epsilon^3 v_3(r) + \cdots \\[2mm] \bar{p} = \bar{p}_0 + \epsilon\bar{p}_1 + \frac{\epsilon^2}{2}\bar{p}_2 + \cdots \end{cases} \qquad (3.42)$$

with $\epsilon$ once again the independent expansion parameter which regulates progress along the post-buckling path. Substituting the expansions (3.42) into the equilibrium equations (3.40a,b) we generate a sequence of equations (all linear) for the coefficients $w_i$ and $v_i$, the first three of which in each case are (out-of plane equations):

$$w_{1,rrr} + \frac{1}{r}w_{1,rr} - \frac{1}{r^2}w_{1,r} + \left(\frac{\bar{p}_0}{K}\right)w_{1,r} = 0 \qquad (3.43a)$$

$$w_{2,rrr} + \frac{1}{r}w_{2,rr} - \frac{1}{r^2}w_{2,r} - \left(\frac{12}{h^2}\right)\left(w_{1,r}v_{1,r} + \nu\frac{w_{1,r}v_1}{r}\right)$$
$$+ \left(\frac{\bar{p}_0}{K}\right)w_{2,r} + \left(\frac{\bar{p}_1}{K}\right)(1 - \nu^2)w_{1,r} = 0 \tag{3.43b}$$

$$w_{3,rrr} + \frac{1}{r}w_{3,rr} - \frac{1}{r^2}w_{3,r} - \left(\frac{12}{h^2}\right)\{w_{1,r}v_{2,r}$$
$$+ \nu\frac{w_{1,r}v_2}{r} + \frac{1}{2}w_{1,r}^3 + w_{2,r}v_{1,r} + \nu\frac{w_{2,r}v_1}{r}\}$$
$$+ \frac{1}{K}\left\{\bar{p}_0 w_{3,r} + \bar{p}_1 w_{2,r} + \frac{1}{2}\bar{p}_2 w_{1,r}\right\} = 0 \tag{3.43c}$$

and (in-plane equations):

$$v_{1,rr} + \frac{1}{r}v_{1,r} - \frac{1}{r^2}v_1 = 0 \tag{3.44a}$$

$$v_{2,rr} + \frac{1}{r}v_{2,r} - \frac{1}{r^2}v_2 + w_{1,r}w_{1,rr} + \frac{1}{2r}(1 - v)w_{1,r}^2 = 0 \tag{3.44b}$$

$$v_{3,rr} + \frac{1}{r}v_{3,r} - \frac{1}{r^2}v_3 + w_{1,r}w_{2,rr}$$
$$+ w_{2,r}w_{1,rr} + (1 - \nu)\frac{w_{1,r}w_{2,r}}{r} = 0 \tag{3.44c}$$

For the simply supported circular plate, the conditions (3.41a) imply that

$$\begin{cases} \dot{w}_i(0) = v_i(0) = 0, \quad \text{all } i \\ w_i(a) = 0, \ w_{i,rr} + \frac{\nu}{r}w_{i,r}\Big|_{r=a} = 0, \quad \text{all } i \end{cases} \tag{3.45a}$$

$$\begin{cases} v_{1,r} + \frac{\nu}{r}v_1\Big|_{r=a} = 0 \\ v_{2,r} + \frac{1}{2}w_{1,r}^2 + \frac{\nu}{r}v_2\Big|_{r=a} = 0 \\ v_{3,r} + w_{1,r}w_{2,r} + \frac{\nu}{r}v_3\Big|_{r=a} = 0 \end{cases} \tag{3.45b}$$

The first out-of-plane equation has the solution

$$w_{1,r} = A_1 J_1(\gamma r) + B_1 Y_1(\gamma r) \tag{3.46}$$

with $A_1, B_1$ constants of integration, $\gamma = \sqrt{\bar{p}_0/K}, \bar{p}_0$ the critical load for buckling of the plate, and $J_1$ and $Y_1$ the Bessel functions of the first order. Regularity at $r = 0$ requires that $B_1 = 0$. If we set the parameter, $\epsilon = \delta \equiv w(0)$, the central deflection of the plate, we find that $w_1(0) = 1, w_j(0) = 0, j \neq 1$; using this result, together with $w_1(a) = 0$, we obtain

$$w_{1,r} = \gamma J_1(\gamma r)/(J_0(\gamma a) - 1) \tag{3.47}$$

and

$$w_{1,rr} = \frac{\gamma^2}{J_0(\gamma a) - 1} \left\{ J_0(\gamma r) - \frac{J_1(\gamma r)}{\gamma r} \right\} \tag{3.48}$$

with $J_0$, once again, the Bessel function of the first kind of zero order. Using the second condition of simple support along the edge at $r = a$, i.e., $w_{1,rr} + \dfrac{\nu}{r} w_{1,r} \Big|_{r=a} = 0$, we then obtain the relation

$$\gamma a J_0(\gamma a) - (1 - \nu) J_1(\gamma a) = 0 \tag{3.49}$$

which is, of course, the same as (3.34) with $\lambda_0 = \gamma a$.

The solution of (3.49), for $\nu = 0.3$, yields $\gamma a = 2.049$ and, thus, the result previously indicated, i.e., (3.16), or $\bar{p}_0 = 4.2K/a^2$; the corresponding buckling mode, which is in line with our earlier result (3.35), is

$$w_1(r) = \frac{J_0(\gamma a) - J_0(\gamma r)}{J_0(\gamma a) - 1} \tag{3.50}$$

The first in-plane and second out-of-plane equations yield the anticipated results

$$v_1(r) = w_2(r) = \bar{p}_1 = 0 \tag{3.51}$$

and the second in-plane equation yields

$$v_2(r) = \frac{\gamma}{4[J_0(\gamma a) - 1]^2} \{ (1 + \nu) J_0(\gamma r) J_1(\gamma r)$$

$$+ \nu \gamma r \left[ J_0^2(\gamma a) - J_0^2(\gamma r) + J_1^2(\gamma a) - J_1^2(\gamma r) \right] \tag{3.52}$$

$$- \gamma r J_1^2(\gamma a) \}$$

We may now write the third-order equation for the lateral deflection in the form

$$w_{3,rrr} + \frac{1}{r} w_{3,rr} - \frac{1}{r^2} w_{3,r} + \left( \frac{\bar{p}_0}{K} \right) w_{3,r}$$

$$= \frac{12}{h^2} \left[ v_{2,r} + \frac{1}{2} w_{1,r}^2 + \nu \frac{v_2}{r} \right] w_{1,r} - \left( \frac{\bar{p}_2}{K} \right) w_{1,r} \tag{3.53}$$

so that the right-hand side of (3.53) is known except for $\bar{p}_2$. If (3.53) is multiplied by $rw_{1,r}$ and integrated with respect to $r$, from $r = 0$ to $r = a$, and the boundary conditions at $r = 0$ and $r = a$ are used, we find that

$$\bar{p}_2 = 2.269K/a^2h^2 \tag{3.54}$$

so that the leading terms in the postbuckling expansion for the load yield

$$\bar{p}/\bar{p}_0 = 1 + 0.27 \left(\frac{\delta}{h}\right)^2, \quad \delta = w(0) \tag{3.55}$$

which is in line with Sherbourne's [147] result (3.38). The corresponding analysis for the clamped (plate) boundary conditions yields the equation

$$\frac{\gamma J_1(\gamma r)}{J_0(\gamma a) - 1} = 0, \quad \gamma = \sqrt{\bar{p}_0/K} \tag{3.56}$$

whose solution again produces (3.15), or $\bar{p}_0 = 14.68K/a^2$, and the buckling mode (3.50) with $\gamma$ now satisfying $\gamma a = 3.832$. We also find that, for the case of clamped boundary conditions, in lieu of (3.54) we have

$$\bar{p}_2 = 6.017K/a^2h^2 \tag{3.57}$$

and, thus, the postbuckling expansion

$$\bar{p}/\bar{p}_0 = 1 + 0.2049 \left(\frac{\delta}{h}\right)^2, \quad \delta = w(0) \tag{3.58}$$

The results of the postbuckling analyses for linearly elastic, isotropic, clamped and simply supported plates that are subject to a uniform radial compression along (the edge at) $r = a$ are depicted in Fig. 3.13; graphs $A$ and $C$ are based on the perturbation expansion analysis of the equations for the displacements $w(r), u(r)$ and the solution for the simply supported case is compared with that which follows from the work of Friedrichs and Stoker [143] in the same case.

The buckling modes $w_1(r)$ are shown in Fig. 3.14 for both the clamped and simply supported cases.

We now consider an application of the Lyapunov-Schmidt reduction scheme to the problem under consideration. We begin with equations (3.23a,b), which we may write in the form

$$\Phi \equiv \left(G_p - \frac{1}{2}q^2, \eta^2 Gq + pq\right) = \mathbf{0} \tag{3.59}$$

Setting $\phi = (p, q)$, we define the mapping $\boldsymbol{G} : \phi \to \boldsymbol{\Phi}$ by

$$
\begin{cases}
\Phi_1 = G_1(p, q) \equiv G_p - \dfrac{1}{2}q^2 \\[2mm]
\Phi_2 = G_2(p, q) \equiv \eta^2 Gq + pq
\end{cases}
\tag{3.60}
$$

and we note that $\boldsymbol{G}(\bar{p}, 0) = 0$. Computing the Frechét derivative $\boldsymbol{A} = \boldsymbol{G}_\phi(\bar{p}, 0)$ we have, first,

$$
\boldsymbol{G}_\phi(p', q')[(p, q)] = (Gp - q'q, \eta^2 Gq + p'q + q'p)
\tag{3.61}
$$

and, then,

$$
\boldsymbol{G}_\phi(\bar{p}, 0)[(p, q)] \equiv \boldsymbol{A}(p, q) = \left(Gp, \eta^2 Gq + \bar{p}q\right)
\tag{3.62}
$$

thus, if $\mathcal{N}(\boldsymbol{A})$ denotes the null space of $\boldsymbol{A}$, we have $(p, q) \in \mathcal{N}(\boldsymbol{A})$ if and only if $Gp = 0$ and $\eta^2 Gq + \bar{p}q = 0$. Therefore,

$$
\mathcal{N}(\boldsymbol{A}) = \mathrm{span}\ (\bar{p}_0, q_1)
\tag{3.63}
$$

We write

$$
\begin{cases}
p = \bar{p} + \tilde{p} \\
q = \epsilon q_1 + \tilde{q}
\end{cases}
\tag{3.64}
$$

where

$$
\begin{cases}
\tilde{p} = \epsilon^2 p_2 + \cdots \\
\tilde{q} = \epsilon^3 q_3 + \cdots
\end{cases}
\tag{3.65}
$$

and, as $\tilde{q}$ belongs (via the Lyapunov-Schmidt decomposition) to $\mathcal{N}(\boldsymbol{A})^\perp$, the orthogonal complement of $\mathcal{N}(\boldsymbol{A})$,

$$
\int_0^a \tilde{q}(r)q_1(r)dr = 0
\tag{3.66}
$$

We define the projection operators

$$
\boldsymbol{P}(p, q) = (\bar{p}, \epsilon q_1), \quad (\boldsymbol{I} - \boldsymbol{P})(p, q) = (\tilde{p}, \tilde{q})
\tag{3.67}
$$

The von Karman equations for this problem have the form $\boldsymbol{G}\phi = \boldsymbol{0}$ and, thus, employing the projection $\boldsymbol{P}$ ( onto $\mathcal{N}(\boldsymbol{A})$) we have $\boldsymbol{P}\boldsymbol{G}\phi = \boldsymbol{0}$ or

$$
\left\langle \left(Gp - \frac{1}{2}q^2,\ \eta^2 Gq + pq\right), (\bar{p}, q_1)\right\rangle = 0
\tag{3.68}
$$

where $< (p, q), (p', q') > = \int_0^a (pp' + qq')dr$. Thus,

$$\bar{p} \int_0^a (Gp - \frac{1}{2}q^2)dr + \int_0^a (\eta^2 Gq + pq) q_1 dr = 0 \qquad (3.69)$$

However,

$$\int_0^a (Gp - \frac{1}{2}q^2)dr$$
$$= \int_0^a G(\bar{p} + \tilde{p})dr - \frac{1}{2}\int_0^a (\epsilon q_1 + \tilde{q})^2 dr$$
$$= \int_0^a G\tilde{p}dr - \frac{1}{2}\int_0^a (\epsilon^2 q_1^2 + 2\epsilon q_1 \tilde{q} + \tilde{q}^2)dr \qquad (3.70)$$
$$= \epsilon^2 \int_0^a Gp_2 dr - \frac{1}{2}\epsilon^2 \int_0^a q_1^2 dr - \frac{1}{2}\int_0^a \tilde{q}^2 dr$$

as $\int_0^a q_1 \tilde{q} dr = 0$. Using the fact that $Gp_2 = \frac{1}{2}q_1^2$ we have, from (3.70),

$$\int_0^a (Gp - \frac{1}{2}q^2)dr = -\frac{1}{2}\int_0^a \tilde{q}^2 dr$$
$$= -\frac{1}{2}\int_0^a (\epsilon^3 q_3 + \cdots)^2 dr \qquad (3.71)$$
$$= O(\epsilon^6)$$

Next, we need to compute $\int_0^a (\eta^2 Gq + pq)q_1 dr$. We begin with

$$\eta^2 Gq + pq = \eta^2 Gq + \bar{p}q + \tilde{p}q$$
$$= \eta^2 G(\epsilon q_1 + \tilde{q}) + \bar{p}q + \tilde{p}q$$
$$= \epsilon \eta^2 Gq_1 + \eta^2 G\tilde{q} + \bar{p}q + \tilde{p}q$$

But

$$\eta^2 Gq_1 = -\bar{p}_0 q_1$$

so

$$\eta^2 Gq + pq = -\epsilon \bar{p}_0 q_1 + \eta^2 G\tilde{q} + \bar{p}q + \tilde{p}q$$
$$= -\epsilon \bar{p}_0 q_1 + \eta^2 G\tilde{q} + \bar{p}(\epsilon q_1 + \tilde{q}) + \tilde{p}q$$
$$= \epsilon \lambda q_1 + \eta^2 G\tilde{q} + \bar{p}\tilde{q} + \tilde{p}q$$

where, once again, we have set $\lambda = \bar{p} - \bar{p}_0$. Thus,

$$
\begin{aligned}
0 &= \int_0^a (\eta^2 Gq + pq) q_1 \, dr \\
&= \epsilon\lambda \int_0^a q_1^2 \, dr + \eta^2 \int_0^a q_1 G \tilde{q} \, dr + \bar{p} \int_0^a \tilde{q} q_1 \, dr \\
&\quad + \int_0^a \tilde{p} q q_1 \, dr \\
&= \epsilon\lambda \int_0^a q_1^2 \, dr + \eta^2 \int_0^a q_1 G(\epsilon^3 q_3) \, dr \\
&\quad + \int_0^k (\epsilon^2 p_2)(\epsilon q_1 + \epsilon^3 q_3 + \cdots) q_1 \, dr + O(\epsilon^7)
\end{aligned}
$$

where we have used the fact that $\int_0^a \tilde{q} q_1 \, dr = 0$. Therefore,

$$
\begin{aligned}
0 &= \epsilon\lambda \int_0^a q_1^2 \, dr + \epsilon^3 \eta^2 \int_0^a (Gq_3) q_1 \, dr \\
&\quad + \epsilon^3 \int_0^a p_2 q_1^2 \, dr + \epsilon^5 \int_0^a p_2(q_3 + \epsilon^2 q_5 + \cdots) \, dr + O(\epsilon^7)
\end{aligned}
\tag{3.72}
$$

or

$$
\begin{aligned}
0 &= \epsilon\lambda \int_0^a q_1^2 \, dr + \epsilon^3 \int_0^a (-\bar{p}_0 q_3 - p_2 q_1) q_1 \, dr \\
&\quad + \epsilon^3 \int_0^a p_2 q_1^2 \, dr + O(\epsilon^5)
\end{aligned}
\tag{3.73}
$$

as

$$
\eta^2 Gq_3 + \bar{p}_0 q_3 + p_2 q_1 = 0,
\tag{3.74}
$$

Therefore,

$$
\begin{aligned}
\int_0^a (\eta^2 Gq + pq) q_1 \, dr &= \epsilon\lambda \int_0^a q_1^2 \, dr - \bar{p}_0 \epsilon^3 \int_0^a q_3 q_1 \, dr + O(\epsilon^5) \\
&= 0
\end{aligned}
\tag{3.75}
$$

Hence,

$$
\begin{cases}
\lambda A - \bar{p}_0 \epsilon^2 B = 0 \\[2mm]
A = \int_0^a q_1^2 \, dr, \quad B = \int_0^a q_3 a_1 \, dr
\end{cases}
\tag{3.76}
$$

or

$$\epsilon = \left( \frac{\int_0^a q_1^2 \, dr}{\int_0^a q_1 q_3 \, dr} \right)^{\frac{1}{2}} \sqrt{\frac{\bar{p}}{p_0}} - 1 \tag{3.77}$$

Therefore,

$$q = \epsilon q_1 + (\epsilon^3 q_3 + \cdots) \tag{3.78}$$

where $\epsilon$ is given by (3.77) so that, finally,

$$q \equiv q(\lambda) = \left( \sqrt{\frac{\int_0^a q_1^2 \, dr}{\bar{p}_0 \int_0^a q_1 q_3 \, dr}} \right) \lambda^{1/2} q_1 + \cdots \tag{3.79}$$

## 3.1.2 Cylindrically Orthotropic Symmetry: Initial Buckling and Postbuckling Behavior

We again assume that the plate occupies the region $0 \le r \le a$, $r = \sqrt{x^2 + y^2}$ in the $x, y$ plane. In this case, the full set of von Karman equations consists of (1.98) and (1.99) for the general situation in which the transverse deflection $w = w(r, \theta)$; for the case of a cylindrically orthotropic plate undergoing an axisymmetric deformation, the relevant equations are (1.101) and (1.102), where $\Delta_r^2, \beta$, and $F(w, \Phi)$ are given by (1.103). Initial buckling of the plate is governed by (1.98), with $\Phi = \Phi_0(r, \theta)$ the Airy function which describes the in-plane prebuckling stress distribution that arises as a consequence of the applied loads and support conditions along the edge of the plate at $r = a$; this equation may be rewritten in the form

$$D_r\left(w_{,rrrr} + \frac{2}{r}w_{,rrr}\right) + D_\theta\left(-\frac{1}{r^2}w_{,rr} + \frac{1}{r^3}w_{,r}\right.$$
$$\left. + \frac{2}{r^4}w_{,\theta\theta} + \frac{1}{r^4}w_{,\theta\theta\theta\theta}\right)$$
$$+ 2D_{r\theta}\left(\frac{1}{r^2}w_{,rr\theta\theta} - \frac{1}{r^3}w_{,r\theta\theta} + \frac{1}{r^4}w_{,\theta\theta}\right)$$
$$= \sigma_{rr}^o h w_{,rr} + \sigma_{\theta\theta}^o h\left(\frac{1}{r}w_{,r} + \frac{1}{r^2}w_{,\theta\theta}\right) \tag{3.80}$$

where $D_r, D_\theta$, and $D_{r\theta}$ are given by (1.95), (1.96), and (1.97) and

$$\begin{cases} \sigma_{rr}^o \cdot h \equiv N_r^o = \dfrac{1}{r}\Phi_{0,r} + \dfrac{1}{r^2}\Phi_{0,\theta\theta} \\[2mm] \sigma_{\theta\theta}^o \cdot h \equiv N_\theta = \Phi_{0,rr} \end{cases} \tag{3.81}$$

so that $\sigma_{rr}^o, \sigma_{\theta\theta}^o$ are the non-zero components of the in-plane (prebuckling) stress tensor in polar coordinates.

The initial buckling problem governed by (3.80) was analyzed by Woinowsky-Krieger [150] and a more general solution which includes the results in [148] was then presented by Mossakowski [151]. We assume that the deflection $w(r,\theta)$ can be expressed in the form

$$w(r,\theta) = w_0(r) + \sum_{n=1}^{\infty} w_r(r)\cos n\theta \tag{3.82}$$

which results in replacing (3.80) by an infinite system of ordinary differential equations. The solutions in Mossakowski [151] are then given in terms of various approximate formulas for $\lambda_{cr} = (\sigma_{rr}|_{r=a})_{cr}$ and the following cases have been considered:

**(i) The Edge at $r = a$ is Clamped**
In this case, the boundary conditions reduce to (1.139), where $i = 1$ and $R_1 = a$; at $r = 0$ the assumptions take either the form (1.144) or the form (1.145).

**(a) The Deformation is Axisymmetric, i.e., $w_{,\theta} = 0$.**
For this subcase, Mossakowski [151] shows that

$$\lambda_{cr} = \frac{D_r \left(1 + \sqrt{\beta}\right)^2}{a^2 h} \left[1.2024 + \frac{1.4882\sqrt{\beta}}{1 + \sqrt{\beta}} - \frac{0.1228\beta}{(1 + \sqrt{\beta})^2}\right]^2 \tag{3.83}$$

with $\beta$ the orthotropy ratio given by (1.103). For $\beta = 1$ (the isotropic circular plate) we recover our earlier result $\lambda_{cr} = 14.68 D_r / a^2 h$ (i.e., (3.15)). If we define a buckling parameter $k$ by $k = \lambda_{cr}\dfrac{a^2 h}{D_r}$ then the relation of the buckling load to the orthotropy ratio is depicted in Fig. 3.15.

**(b) The Deformation is Antisymmetric, i.e., $w_{,\theta} \neq 0$.**
In this subcase, it is shown in [151] that

$$\lambda_{cr} = \frac{D_r(1 + \sqrt{\beta})^2}{a^2 h} \left[1.2341 + 1.3931\eta + 0.0594\eta^2\right]^2 \tag{3.84}$$

where

$$\eta = 1 + \frac{2\sqrt{\beta}(\chi - 1)}{(1 + \sqrt{\beta})^2} \qquad (3.85)$$

and

$$\chi = \frac{D_{r\theta}}{\sqrt{D_r D_\theta}} \qquad (3.86)$$

with the result applying to the geometries that are depicted in Fig. 3.16 i.e., to the case of a circular plate clamped along its edge and having a hinged support at its center and to the case of a circular plate clamped along its edge with hinged support along one diameter.

**(ii) The Edge at $r = a$ is Simply Supported.**
In this case, the boundary conditions reduce to $w(a, \theta) = 0$, $0 < \theta \le 2\pi$, and the first condition in (1.153), as well as one of the sets of conditions, referenced above, at $r = 0$.

**(a) The Deformation is Axisymmetric, i.e., $w_{,\theta} = 0$**
For this subcase, it is no longer possible to obtain values for $\lambda_{cr}$ in the form (3.83) or (3.84); however, if one expresses the buckling parameter $k$ as

$$k = (1 + \beta)\Psi^2 \qquad (3.87)$$

then (Mossakowski [151]) a graphical representation of the solution may be given. For various values of $(1 - \sqrt{\beta})/(1 + \sqrt{\beta})$, these curves, which depict $\Psi$ as a function of $(1 - \nu_\theta)/(1 + \sqrt{\beta})$, are shown in Fig. 3.17.

**(b) The Deformation is Antisymmetric, i.e., $w_{,\theta} \ne 0$**
With the buckling parameter expressed in the form (3.87), one finds (Mossakowski [151]) curves relating $\Psi$ and $\eta - \left( \dfrac{1 - \nu_\theta}{1 + \sqrt{\beta}} \right)$ where $\eta$ is defined by (3.85), (3.86); these graphs are presented in Fig. 3.18.

We now turn to the problem of postbuckling for a (uniformly) radially compressed, linearly elastic, cylindrically orthotropic plate which is either clamped or simply supported along the edge at $r = a$. Many of the postbuckling results for circular, cylindrically orthotropic plates will appear as "special cases" of the corresponding results for annular plates in the next section; indeed, for a cylindrically orthotropic, annular plate occupying the region $b \le r \le a$, we will have the opportunity in the next section to compare the postbuckling results for the case $b \ne 0$ and different aspect ratios $a/b$.

One of the earliest works to treat the postbuckling analysis of cylindrically orthotropic, circular plates appears to be the paper [152] of

Iwinski and Nowinski; other notable contributions include the work of Pandalai and Patel [153] and C.L. Huang [154]. In the work which we present below, we follow the analysis in the survey paper of Sherbourne and Pandey [155]; this comprehensive work contains a comparison of the postbuckling behavior of circular and annular plates exhibiting cylindrically orthotropic symmetry. The authors [155] also make note of the fact that, with the exception of a handful of studies, most of the postbuckling analyses that have been carried out for plates with a circular geometry and cylindrically orthotropic symmetry, have been effected for annular as opposed to circular plates; they note that "this somewhat surprising neglect of postbuckling analysis for circular orthotropic plates may not be coincidental as the problem is characterized by singularities due to orthotropy and axisymmetry of deformations thus contributing to mathematical complexity." One study of postbuckling analysis which the authors [155] do point to specifically is the finite-element analysis of Raju and Rao [156], which highlights one of the difficulties inherent in the subject: it is shown in [156] that for the cylindrically orthotropic, circular plate, an increase in the orthotropy ratio increases the linear buckling load but decreases the postbuckling stiffness, i.e., the load-carrying capacity of the plate.

The plane stress solution for a circular orthotropic plate appears, e.g., in Lekhnitskii [51] and Pandalai and Patel [153]; the plate is assumed to be subjected to a uniform compressive stress $\lambda = \sigma_{rr}|_{r=a}$. For the isotropic circular plate, the prebuckling, in-plane stress distribution is characterized by a uniform compressive stress distribution throughout the entire plate; however, for the case of a cylindrically orthotropic, circular plate, the in-plane prebuckling stress state varies along the direction of the radial coordinate according to

$$
\begin{cases}
\sigma_{rr} = -\left(\sigma_{rr}|_{r=a}\right)\left(\frac{r}{a}\right)^{\sqrt{\beta}-1} \\[2mm]
\sigma_{\theta\theta} = -\left(\sigma_{rr}|_{r=a}\right)\cdot\sqrt{\beta}\left(\frac{r}{a}\right)^{\sqrt{\beta}-1} \\[2mm]
\sigma_{r\theta} = 0
\end{cases}
\tag{3.88}
$$

so that the qualitative nature of the prebuckling stress distribution is governed by the orthotropy ratio. Thus, for $\beta > 1$, stresses decrease as we approach the center of the plate, with the opposite situation holding for $\beta < 1$ ( and the stresses becoming infinite at the center of the plate); stress singularities, therefore, appear at $r = 0$, whenever $\beta < 1$, and, as the authors [155] explicitly point out, "this peculiar feature of the

plane stress solution is tacitly recognized in the literature since most
solutions for circular and annular plates are limited to cases of $k > 1$.
The only exception appears to be the study by Turvey and Drinali [157]
where numerical results are reported for cases including $\beta < 1$ using the
dynamic relaxation method.

**Remarks:** As has been pointed out in [155], G.F. Carrier [158] argued
that it is pointless to speak of material orthotropy at the center of a
circular plate and that such a concept is simply a mathematical ab-
straction. In practical terms, cylindrical orthotropy does not exist at
$r = 0$ because of the coincidence there of the radial and circumferential
directions.

The scope of the nonlinear analysis which is presented in [155] is limited to axisymmetric deformations and proceeds as follows: we first define, in the usual way, the in-plane strains by

$$e_{rr} = u'(r) + \frac{1}{2}w'^2(r), e_{\theta\theta} = \frac{1}{4}u(r) \tag{3.89}$$

where $u(r), w(r)$ have the same interpretation as for the case of the
isotropic circular plate undergoing axisymmetric deformations. As the
plate is assumed to be subjected to a uniform radial compression applied
along its edge at $r = a$, $\sigma_{r\theta} = 0$ in (1.91) and, thus, the inverted version
of these constitutive relations assumes the form

$$\begin{pmatrix} \sigma_{rr} \\ \sigma_{\theta\theta} \end{pmatrix} = \frac{E_r}{(1 - \nu_\theta \nu_r)} \begin{bmatrix} 1 & \nu_\theta \\ \nu_\theta & \beta \end{bmatrix} \begin{pmatrix} e_{rr} \\ e_{\theta\theta} \end{pmatrix} \tag{3.90}$$

with $\beta = \dfrac{E_\theta}{E_r} = \dfrac{\nu_\theta}{\nu_r}$, while the moment-curvature relation is

$$\begin{pmatrix} M_r \\ M_\theta \end{pmatrix} = \frac{E_r h^3}{(1 - \nu_\theta \nu_r)} \begin{bmatrix} 1 & \nu_\theta \\ \nu_\theta & \beta \end{bmatrix} \begin{pmatrix} \kappa_r \\ \kappa_\theta \end{pmatrix} \tag{3.91}$$

with the middle-surface curvatures $\kappa_r, \kappa_\theta$ given by (1.86) for the case
of radially symmetric deformations of the plate. In terms of the displacements $u$ and $w$, the equilibrium equations for the cylindrically orthotropic, linearly elastic, circular plate assume the form

$$\frac{d^2 u}{dr^2} = -\frac{1}{r}\frac{du}{dr} - \frac{(1 - \nu_\theta)}{2r}\left(\frac{dw}{dr}\right)^2 + \frac{\beta u}{r^2} - \frac{dw}{dr}\frac{d^2 w}{dr^2} \tag{3.92}$$

and

$$\frac{d^3w}{dr^3} = -\frac{1}{r}\frac{d^2w}{dr^2} + \frac{\beta}{r^2}\frac{dw}{dr} + \frac{12}{h^2}\frac{dw}{dr}\left[\frac{du}{dr} + \frac{1}{2}\left(\frac{dw}{dr}\right)^2 + \nu_\theta \frac{u}{r}\right] \tag{3.93}$$

If we introduce the nondimensional variables

$$\omega = \frac{w}{h}, \ \rho = \frac{r}{h}, \ v = \frac{u}{h} \tag{3.94}$$

then (3.92) and (3.93) assume the respective forms

$$\rho \frac{d^2v}{d\rho^2} = -\frac{dv}{d\rho} - \frac{(1-\nu_\theta)}{2}\left(\frac{d\omega}{d\rho}\right)^2 + \frac{\beta v}{\rho} - \rho \frac{d\omega}{d\rho}\frac{d^2\omega}{d\rho^2} \tag{3.95}$$

and

$$\rho \frac{d^3\omega}{d\rho^3} = -\frac{d^2\omega}{d\rho^2} + \frac{\beta}{\rho}\frac{d\omega}{d\rho} + 12\rho\frac{d\omega}{d\rho}\left[\frac{dv}{d\rho} + \frac{1}{2}\left(\frac{d\omega}{d\rho}\right)^2 + \nu_\theta \frac{v}{\rho}\right] \tag{3.96}$$

while the symmetry of the deformation with respect to the pole at $\rho = 0$ requires that

$$v(0) = \left.\frac{dv}{d\rho}\right|_{\rho=0} = 0 \tag{3.97}$$

The equations (3.95), (3.96) provide the basis for a complete characterization of the postbuckling behavior of the cylindrically orthotropic plate once boundary data is specified at $r = a$, i.e., along the edge of the plate.

To eliminate singularities at $\rho = 0$ in (3.95), (3.96), Sherbourne and Pandey [155] introduce the new variables $x, U$, and $W$ by

$$x = \rho^{\sqrt{\beta}+1}, \ v = \rho^{\sqrt{\beta}+1} \cdot U, \ \omega = W \tag{3.98}$$

in which case, with a prime denoting differentiation with respect to $x$, the equilibrium equations (3.95), (3.96) become

$$U'' = -\frac{(3\sqrt{\beta}+1)}{(\sqrt{\beta}+1)} \cdot \frac{U'}{x} - (1 - \nu_\theta + 2\sqrt{\beta})\frac{W'^2}{2x} \tag{3.99}$$
$$-(\sqrt{\beta}+1)W'W''$$

and

$$W''' = -\frac{(3\sqrt{\beta}+1)}{(\sqrt{\beta}+1)} \cdot \frac{W''}{x} + \frac{12(\sqrt{\beta}+\nu_\theta)}{(\sqrt{\beta}+1)^2} \cdot \frac{UW'}{x} \tag{3.100}$$
$$+6W'\left[\frac{2U'}{(\sqrt{\beta}+1)} + W'^2\right]$$

As for the initial conditions which are to be associated with the transformed equilibrium equations, we require, following Sherbourne [147] and Huang and Sandman [159], that

$$U''(0) < \infty, \ W'''(0) < \infty \tag{3.101}$$

which implies that the deformed shape of the circular plate remains smooth at the center of the plate (at $x = 0$); in order that (3.101) hold, it follows, from (3.99), (3.100), that the coefficients of those terms involving $1/x$ in these equations must vanish, which, in turn, implies that

$$U'(0) = \frac{(\nu_\theta - 2\sqrt{\beta} - 1)(\sqrt{\beta} + 1)}{2(3\sqrt{\beta} + 1)} \cdot W'^2(0) \tag{3.102}$$

and

$$W''(0) = \frac{12(\sqrt{\beta} + \nu_\theta)}{(3\sqrt{\beta} + 1)(\sqrt{\beta} + 1)} U(0) W'(0) \tag{3.103}$$

Expanding $U(x)$ and $W(x)$ in Maclaurin series, in a neighborhood of $x = 0$, substituting the results in (3.99), (3.100), and then setting $x = 0$, one obtains for $U''(0)$ and $W'''(0)$ the relations

$$U''(0) = -U(0)W'^2(0) \left[ \frac{6(4\sqrt{\beta} + 3 - \nu_\theta)(\sqrt{\beta} + \nu_\theta)}{(4\sqrt{\beta} + 2)(3\sqrt{\beta} + 1)} \right] \tag{3.104}$$

$$
\begin{aligned}
(4\sqrt{\beta} + 2)W'''(0) &= \frac{12(2\sqrt{\beta} + \nu_\theta + 1)}{(\sqrt{\beta} + 1)} U'(0)W'(0) \\
&+ \frac{12(\sqrt{\beta} + \nu_\theta)}{(\sqrt{\beta} + 1)} \Omega(0)W''(0) + 6(\sqrt{\beta} + 1)W'(0)W'^2(0)
\end{aligned} \tag{3.105}
$$

If we introduce the variables $Y_i$, $i = 1, \cdots, 5$,

$$Y_1 = W, \ Y_2 = U, \ Y_3 = W', \ Y_4 = U', \ Y_5 = W'' \tag{3.106}$$

the axisymmetric postbuckling behavior of the cylindrically orthotropic circular plate may be described by the first-order system

$$Y_1' = Y_3, \ Y_2' = Y_4, \ Y_3' = Y_5, \ Y_4' = U'', \ Y_5' = W''' \tag{3.107}$$

where $U'', W'''$ may be written in terms of the $Y_i$'s by using (3.99), (3.100); a Runge-Kutta method for the system (3.107) may then be employed in which the initial values $U(0), W'(0)$ are assigned, while the initial values of $U'(0)$, $U''(0), W''(0)$, and $W'''(0)$ are evaluated using (3.102)–(3.105). A key step in the procedure consists of examining the boundary condition along the edge of the plate at $r = a$; if this boundary condition is not satisfied, then $W'(0)$ is systematically varied, keeping $U(0)$ constant, and the system of ODEs is numerically integrated until the prescribed boundary condition is met.

We want to elaborate upon some of the results which follow from numerical solutions of the system (3.107); however, it is first worthwhile to append some remarks to the discussion of initial buckling for the cylindrically orthotropic circular plate that was presented earlier.

**Remarks:** In [160] Swamidas and Kunukkasseril have shown that the buckling equation for a clamped, circular, cylindrically orthotropic plate has the form

$$J_\eta(\sigma) = 0; \quad \eta = \frac{2\sqrt{\beta}}{(\sqrt{\beta}+1)}, \quad \sigma = \frac{2\sqrt{\lambda_{cr}}}{(\sqrt{\beta}+1)}, \tag{3.108}$$

with $J_\eta$ the Bessel function of the first kind of order $\eta$, while for simply supported plates, the critical buckling loads are obtained from the solutions of the equation

$$(\sqrt{\beta}+\nu_\theta)J_\eta(\sigma) - \left(\frac{\sqrt{\beta}+1}{2}\right)\sigma J_{\eta+1}(\sigma) = 0 \tag{3.109}$$

with the argument $\sigma$ as given in (3.108). The buckling loads which are obtained by solving (3.108) and (3.109) for various values of the orthotropy ratio $\beta$, including the isotropic case $\beta = 1$, are displayed in Table 3.2.

The postbuckling behavior of clamped and simply supported cylindrically orthotropic circular plates has been presented, in detail, in [155], for a plate with nondimensional radius $a = 50$, four orthotropy ratios of $\beta = 1$ (isotropic case), 2, 5, and 8, and a (major) Poisson ratio of $\nu_\theta = 0.3$; these results have also been compared, in [155] with those which were obtained earlier by Raju and Rao [156] using a finite element method. For the clamped circular plate with $\beta = 2$, the postbuckling curves are presented in Fig. 3.19 and remarkable agreement between the two approaches is evident; "present" on this figure refers to the work of Sherbourne and Pandey [155]. The postbuckling variations of transverse deflection $w$ and radial shortening $u$, with respect to the applied compressive radial loading, is displayed in Figures 3.20a and 3.20b; in these graphs the applied load has been normalized by the buckling load of a similar, isotropic circular plate and curves are shown for orthotropy ratios of $\beta = 1, 2, 5$, and 8. The plate is assumed to be clamped in all the cases considered. From the graphs referenced above, it is clear that plates with higher orthotropy ratios $\beta$ not only have proportionately higher buckling loads, but also tend to be stiffer than plates with lower orthotropy ratios; such results are further confirmed by the results displayed in Table 3.3 which, for both the clamped and the simply supported cases, shows that the maximum central plate deflection at a load ratio of $\lambda/\lambda_{cr} = 3$ decreases, consistently, as the orthotropy ratio is increased from 1 to 8. For the same load ratio of $\lambda/\lambda_{cr} = 3$, the radial distribution of stresses and moments is presented in Figures 3.21a,b and 3.22a,b for the case of a clamped plate. As noted in [155], the most notable result here is that the membrane stresses become tensile in a small region near the plate center and change abruptly to compressive stresses. In comparison with isotropic plates, the variation of the bending moments in orthotropic plates is remarkably different, because, in the isotropic case, the bending moments assume a finite value at $r = 0$ and then exhibit a mild variation with $r$, $0 \leq r \leq a$; however, the moments in orthotropic plates are always zero at the origin $(r = 0)$ but are then marked by a highly nonlinear variation in $r$. In general, both membrane and bending stresses are increasing as the degree of orthotropy increases.

**Remarks:** In Figures 3.21a,b and 3.22a,b we have set, in accordance with the notation in [155]

$$\beta_r = \frac{\sigma_{rr}}{f_r}, \ \beta_\theta = \frac{\sigma_{\theta\theta}}{f_r}, \ \gamma_r = \frac{M_r}{m_r}, \ \gamma_\theta = \frac{M_\theta}{m_r} \tag{3.110}$$

where

$$f_r = E_r/(1 - \nu_\theta\nu_r) \tag{3.111}$$

and

$$m_r = E_r h^2/12(1 - \nu_\theta\nu_r) \tag{3.112}$$

For simply supported plates, the qualitative nature of the load versus deflection and load versus radial shortening curves are similar to those for clamped cylindrically orthotropic circular plates. Increases in the orthotropy ratio result in increases in the postbuckling stiffness and a reduced transverse deflection. Also, the zone of the normalized tensile radial stresses which surrounds the origin $r = 0$, i.e. $\beta_r = \sigma_{rr}/f_r$, $f_r$, as given by (3.111), is much larger than that for clamped plates; this is depicted in Fig. 3.23a. In Fig. 3.23b we depict the corresponding normalized circumferential stresses for the simply supported case. Both Figs. 3.23a and 3.23b correspond to a load ratio of $\lambda/\lambda_{cr} = 3$. Finally, contrary to the case of an isotropic plate, a peak in the tensile stress $\beta_r$

is observed at some distance away from the center of the plate; this fact is quite evident from Fig. 3.23b.

**Remarks:** Although we have not yet considered the initial or postbuckling behavior of cylindrically orthotropic annular plates, this is a good place to make note of some of the differences of a qualitative nature which have been observed in the literature. As noted in [155], Huang [154] has reported an increase in the maximum transverse deflection of annular, cylindrically orthotropic plates (with a free inner edge and an outer edge that is either clamped or simply supported) as the orthotropy ratio increases, while the opposite trend is observed in the solutions for cylindrically orthotropic circular plates, which are given in [155]. It is natural to ask whether one can recover the circular plate solutions from the solutions obtained for annular plates by letting the hole size in the annular case shrink to zero; this does not appear to be mathematically feasible. At the free inner edge of an annular plate the boundary conditions are $\beta_r = \gamma_r = 0$, and these lead to solutions which yield finite values for the circumferential stress and moment at the free edge. However, in formulating the problem for the cylindrically orthotropic circular plate, we impose the boundary conditions

$$\beta_r = \beta_\theta = \gamma_r = \gamma_\theta = 0, \quad \text{at } r = 0 \qquad (3.113)$$

If we let the hole in an annular plate shrink to zero, the conditions $\beta_\theta = \gamma_\theta = 0$, at $r = 0$, cannot be enforced, and this precludes being able to deduce results for the circular plate from the corresponding results for an annular plate.

**Remarks:** While we will discuss variations in both initial and postbuckling behavior for circular plates in § 3.3, as one varies the boundary conditions, material symmetry, and constitutive response, some simple and important observations based directly on the work in [155] are worth making here. First of all, for the cylindrically orthotropic circular plate, the buckling load and postbuckling stiffness are both consistently improved by increasing the orthotropy ratio, $\beta = E_\theta/E_r$, i.e. the ratio of the circumferential to the radial elastic modulus. The circumferential direction is the principal load-carrying direction in circular plates so an increase in the elastic modulus $E_\theta$ can be expected to lead to an increase in the prebuckling stiffness and, as a consequence, to an increase in the buckling load. As we transition to the regime in which postbuckling of the cylindrically orthotropic, circular plate occurs, an increase in the orthotropy ratio $\beta$ produces an intensification in the zone of tensile stresses about the center of the plate at $r = 0$ which leads to stretching and flattening of the interior of the plate and, thus, to an improvement in the overall plate stiffness. Therefore, both prebuckling and postbuckling behavior of a cylindrically orthotropic circular plate improve with increases in $\beta = E_\theta / E_r$.

## 3.1.3 Rectilinear Orthotropic Symmetry: Initial Buckling and Postbuckling Behavior

As was indicated in Chapter 1, the only substantive work which has been done, to date, on the initial or postbuckling behavior of linearly elastic, rectilinearly orthotropic thin plates possessing a circular geometry, occurs in the thesis of Coffin [71], in which the prebuckling stress distributions and initial buckling loads are computed for various cases associated with a rectilinearly orthotropic annular plate; no postbuckling analysis for this situation appears in [71] nor does the problem of initial buckling seem to have been addressed for the circular, linearly elastic plate which exhibits rectilinearly orthotropic material symmetry.

The first von Karman equation for a linearly elastic, rectilinearly orthotropic, thin circular plate is precisely the same as the corresponding partial differential equation for the case of an annular plate which was employed in [71]; thus, the first von Karman equation which applies in the present situation is (1.109), where $[w, \Phi]$ is given by (1.74) and (1.75b), while the $\tilde{D}_j$, $j = 1, 2, 3, 6$ and the coefficients $\tilde{D}_{12}, \tilde{D}_{16}$, and $\tilde{D}_{26}$ are given by (1.108) and (1.110). An indication of how one would derive the second of the von Karman equations (in terms of the Airy stress function) was given in Chapter 1; in lieu of that approach we may complement (1.109) with the two in-plane equations which govern the behavior of the displacements $u_r(r, \theta)$ and $u_\theta(r, \theta)$. To this end, we set $\sigma_{rz} = \sigma_{\theta z} = 0$, $F_r = F_\theta = 0$ in the set of in-plane equilibrium equations (I.72a) and then substitute, into the resulting pair of equations, the inverted form of the constitutive relations (1.112). Finally, replacing $e_{rr}, e_{\theta\theta}$, and $\gamma_{r\theta}$ by their equivalent expressions, as given by (1.71), we obtain the following system of equations for $u = u_r(r, \theta)$ and $v = u_\theta(r, \theta)$

$$I_{1,1}\frac{\partial^2 u}{\partial r^2} + I_{1,2}\frac{\partial u}{\partial r} + I_{1,3}u + I_{1,4}\frac{\partial^2 u}{\partial \theta^2} + I_{1,5}\frac{\partial u}{\partial \theta} + I_{1,6}\frac{\partial^2 u}{\partial r \partial \theta} +$$
$$J_{1,1}\frac{\partial^2 v}{\partial r^2} + J_{1,2}\frac{\partial v}{\partial r} + J_{1,3}v + J_{1,4}\frac{\partial^2 v}{\partial \theta^2} + J_{1,5}\frac{\partial v}{\partial \theta} + J_{1,6}\frac{\partial^2 v}{\partial r \partial \theta} = 0$$

$$(3.114)$$

$$I_{2,1}\frac{\partial^2 u}{\partial r^2} + I_{2,2}\frac{\partial u}{\partial r} + I_{2,3}u + I_{2,4}\frac{\partial^2 u}{\partial\theta^2} + I_{2,5}\frac{\partial u}{\partial\theta} + I_{2,6}\frac{\partial^2 u}{\partial r\partial\theta} +$$

$$J_{2,1}\frac{\partial^2 v}{\partial r^2} + J_{2,2}\frac{\partial v}{\partial r} + J_{2,3}v + J_{2,4}\frac{\partial^2 v}{\partial\theta^2} + J_{2,5}\frac{\partial v}{\partial\theta} + J_{2,6}\frac{\partial^2 v}{\partial r\partial\theta} = 0$$

$$(3.115)$$

In the "in-plane" equations (3.114), (3.115) for $u = u_r(r,\theta)$ and $v = u_\theta(r,\theta)$ the coefficients, modulo a factor of $h$ (the plate thickness) are given by

$$\frac{1}{r}I_{1,1} = \frac{a}{2}\cos 2\theta + \frac{b}{8}\cos 4\theta + c \tag{3.116a}$$

$$I_{1,2} = -\frac{3}{8}b\cos 4\theta + c \tag{3.116b}$$

$$-rI_{1,3} = -\frac{3}{8}b\cos 4\theta + c \tag{3.116c}$$

$$rI_{1,4} = -\frac{1}{8}b\cos 4\theta + \frac{1}{8}b \tag{3.116d}$$

$$rI_{1,5} = \frac{1}{2}b\sin 4\theta \tag{3.116e}$$

$$I_{1,6} = -\frac{1}{2}a\sin 2\theta - \frac{1}{4}b\sin 4\theta \tag{3.116f}$$

$$\frac{1}{r}I_{2,1} = -\frac{1}{4}a\sin 2\theta - \frac{1}{8}b\sin 4\theta \tag{3.116g}$$

$$I_{2,2} = -\frac{3}{4}a\sin 2\theta + \frac{3}{8}b\sin 4\theta \tag{3.116h}$$

$$rI_{2,3} = \frac{3}{4}a\sin 2\theta - \frac{3}{8}b\sin 4\theta \tag{3.116i}$$

$$rI_{2,4} = -\frac{1}{4}a\sin 2\theta + \frac{1}{8}b\sin 4\theta \tag{3.116j}$$

$$rI_{2,5} = -a\cos 2\theta + \frac{1}{2}b\cos 4\theta + \frac{1}{2}b + (D_3 + G_{12}) \qquad (3.116\text{k})$$

$$I_{2,6} = -\frac{1}{4}b\cos 4\theta + \frac{1}{4}b + (D_3 - G_{12}) \qquad (3.116\text{l})$$

and

$$\frac{1}{r}J_{1,1} = -\frac{1}{4}a\sin 2\theta - \frac{1}{8}b\sin 4\theta \qquad (3.117\text{a})$$

$$J_{1,2} = \frac{1}{4}a\sin 2\theta + \frac{3}{8}b\sin 4\theta \qquad (3.117\text{b})$$

$$rJ_{1,3} = -\frac{1}{4}a\sin 2\theta - \frac{3}{8}b\sin 4\theta \qquad (3.117\text{c})$$

$$rJ_{1,4} = -\frac{1}{4}a\sin 2\theta + \frac{1}{8}b\sin 4\theta \qquad (3.117\text{d})$$

$$rJ_{1,5} = \frac{1}{2}b\cos 4\theta - \frac{1}{2}b - (D_3 + G_{12}) \qquad (3.117\text{e})$$

$$J_{1,6} = -\frac{1}{4}b\cos 4\theta + \frac{1}{4}b + (D_3 - G_{12}) \qquad (3.117\text{f})$$

$$\frac{1}{r}J_{2,1} = \frac{1}{8}b\cos 4\theta - \frac{1}{8}b + G_{12} \qquad (3.117\text{g})$$

$$J_{2,2} = -\frac{1}{2}a\cos 2\theta + \frac{3}{8}b\cos 4\theta + \frac{1}{8}b + G_{12} \qquad (3.117\text{h})$$

$$rJ_{2,3} = \frac{1}{2}a\cos 2\theta - \frac{3}{8}b\cos 4\theta - \frac{1}{8}b - G_{12} \qquad (3.117\text{i})$$

$$rJ_{2,4} = -\frac{1}{2}a\cos 2\theta + \frac{1}{8}b\cos 4\theta + c \qquad (3.117\text{j})$$

$$\frac{r}{4}J_{2,5} = -\frac{1}{8}b\sin 4\theta + \frac{1}{4}a\sin 2\theta \qquad (3.117\text{k})$$

$$J_{2,6} = -\frac{1}{2}a\sin 2\theta + \frac{1}{4}b\sin 4\theta \qquad (3.117l)$$

where the parameters $a, b, c$ are given by

$$\begin{cases} a = D_1 - D_2 \\ b = (D_1 + D_2) - 2D_3 \\ c = \frac{3}{8}(D_1 + D_2) + \frac{1}{4}D_3 \end{cases} \qquad (3.118)$$

with

$$D_1 = \frac{E_1}{1 - \nu_{12}\nu_{21}}, \quad D_2 = \frac{E_2}{1 - \nu_{12}\nu_{21}}, \quad D_3 = \frac{E_1\nu_{12}}{1 - \nu_{12}\nu_{21}} + 2G_{12} \quad (3.119)$$

Along the edge of the plate, at $r = a$, we prescribe, in general, the radial and tangential components $p_r(\theta)$ and $p_\theta(\theta)$, respectively, of the applied traction where, as in Chapter 1

$$\begin{cases} p_r(\theta) = h\left[\sigma_{xx}\cos^2\theta + \sigma_{yy}\sin\theta\cos\theta + \sigma_{xy}\sin 2\theta\right]\big|_{r=a} \\ p_\theta(\theta) = h\left[(\sigma_{yy} - \sigma_{xx})\sin\theta\cos\theta - \sigma_{xy}\cos 2\theta\right]\big|_{r=a} \end{cases} \qquad (3.120)$$

i.e.

$$p_r(\theta) = h\sigma_{rr}\big|_{r=a}, \quad p_\theta(\theta) = h\sigma_{r\theta}\big|_{r=a} \qquad (3.121)$$

and regularity conditions must, in addition, be prescribed at the center of the plate, i.e., at $r = 0$, with respect to both the Airy function $\Phi$ and the transverse deflection $w$. To the best of the author's knowledge, the system of equations (3.114), (3.115) and the associated structure (3.116a)–(3.120) for the coefficients, have never appeared explicitly in the literature on plate buckling.

---

## 3.2 Plates which Exhibit Nonlinear Elastic, Viscoelastic, or Elastic-Plastic Behavior

The literature on the initial and postbuckling behavior of nonlinear elastic, viscoelastic, and elastic-plastic thin circular plates is much sparser than that which exists for linear elastic circular plates. The largest body of work on the buckling behavior of nonlinearly elastic circular plates can be attributed to Antman [21], [22], and [139]; the approach in this work is very general, allowing for a broad class of nonlinear elastic constitutive behavior and treating both isotropic and anisotropic material, but is limited to axisymmetric deformations. The theory promulugated in [21], [22], and [139] is geometrically exact and takes into account the fact that, in general, a thin circular plate may suffer flexure, mid-surface extension, and shear; the point of view taken in these papers considers the plate to be a shell with a flat natural configuration. As Antman points out in [22], "there exists a voluminous and contentious literature on the derivation of various approximate theories of shells, both linear and nonlinear, such as the von Kármán theory. Most of these theories were originally devised by imposing ad hoc truncations on the equations describing the deformation," (e.g., the truncation which takes us from (1.33) to (1.34)). The author [22] continues by noting that "some of these (approximate) theories can now be obtained as formal asymptotic limits of the equations of the three-dimensional theory as a thickness parameter goes to zero, provided that the data, in the form of applied forces and boundary conditions, have special scalings in terms of the thickness." An outline of the approach taken (Antman, op. cit.) in the analysis of the buckling of nonlinear elastic plates will be presented as an appendix to this subsection. An alternative to the exact geometrical theory, which is outlined in the appendix to this subsection, would be an approximate theory leading to some suitable modification of the von Karman equations for circular plates with either isotropic, cylindrically orthotropic, or rectilinearly orthotropic symmetry; a logical candidate for such an approximate theory would be an extension of the Johnson-Urbanik theory detailed in Section 2.2. Even for rectangular plates, however, only equations governing the initial buckling of a nonlinear elastic plate (e.g. 2.202) are available so that, as a first exercise, the second of the von Karman equations would have to be derived and then the entire system would have to be transformed into polar coordinates; alternatively, the analysis in Chapter 2.2 could be emulated for the case of a circular geometry, thereby yielding approximate relations for $M_r, M_\theta, M_{r\theta}$, etc., directly in terms of a polar coordinate representation.

The development of a framework to treat the initial and postbuckling behavior of viscoelastic plates may be found in the recent work of Brilla [106] and [140]; the analysis in [106], which is suitable for application to rectangular plates, has been described in Section 2.3. In many respects, the formulation of the governing viscoelastic equations is much easier than the nonlinear elastic case because of the assumed linearity of the constitutive response although, of course, memory effects must now be taken into consideration, as is clear from the discussion in Section 2.2. The general system of governing equations for viscoelastic buckling presented in [106] has been specialized in [140] to treat the problems of initial buckling and postbuckling for thin, circular, viscoelastic plates; these equations assume the form (recall that $D \equiv \dfrac{\partial}{\partial t}$) :

$$(1 + \alpha D)\frac{\partial}{\partial r}\left(r^3 \frac{\partial w'(r,t)}{\partial r}\right)$$
$$= (1 + \beta D)\, r^3 w'(r,t)\left[\Phi'(r,t) - \lambda\right] \tag{3.122}$$

$$(1 + \beta D)\frac{\partial}{\partial r}\left(r^3 \frac{\partial \Phi'(r,t)}{\partial r}\right)$$
$$= -(1 + \alpha D)r^3 w'^2(r,t) \tag{3.123}$$

for $0 < r < 1$, $0 \le t \le T$, $T < \infty$, where $w'$ is the spatial derivative of the transverse displacement of the plate, $\Phi'$ is the spatial derivative of the Airy stress function, $\lambda$ is, in the usual way, the positive parameter of proportionality for the given uniform radial loading along the edge of the circular plate (located at $r = 1$) and $\alpha$, $\beta$ are positive "viscous" parameters, where we assume that $\alpha < \beta$. Associated with the generalized von Karman equations (3.122), (3.123) are the boundary conditions

$$w'(0,t) < \infty, \ \ \Phi'(0,t) < \infty, \ \ 0 \le t \le T \tag{3.124}$$

$$w'(1,t) = 0, \ \ \Phi'(1,t) = 0, \ \ 0 \le t \le T \tag{3.125}$$

and the initial conditions

$$w'(r,0) = w_0'(r), \ \ \Phi'(r,0) = 0, \ \ 0 < r < 1 \tag{3.126}$$

We introduce the transformation

$$\frac{1}{\beta}\int_0^t f(\tau)K(t - \tau)d\tau = (1 + \beta D)^{-1}f(t), \tag{3.127}$$

with the kernel $K$ having the form,

$$K(t - \tau) = exp\left\{-\frac{1}{\beta}(t - \tau)\right\}, \tag{3.128}$$

and then integrate both (3.122) and (3.123) twice, with respect to the spatial variable $r$, so as to obtain an integral operator formulation of the problem (3.122) - (3.123), i.e.,

$$\begin{cases} Z_\lambda\left[w'(r,t)\right] = S\left[w'(r,t)\right] \\ \Phi'(r,t) = R\left[w'(r,t)\right] \end{cases} \qquad (3.129)$$

where

$$Z_\lambda[w'(r,t)] = w'(r,t) - \lambda\frac{\beta}{\alpha}\int_r^1 \frac{1}{\xi^3}\int_0^\xi \mu^3 w'(\mu,t)d\mu d\xi$$

$$+ \int_r^1 \frac{1}{\xi^3}\int_0^\xi \mu^3 w'(\mu,t)\int_\mu^1 \frac{1}{\gamma^3}\int_0^\gamma \psi^3 w'^2(\psi,t)d\psi d\gamma d\mu d\xi$$

and

$$S\left[w'(r,t)\right] = \frac{1}{\alpha}\left(\frac{\alpha}{\beta} - 1\right)\int_0^t K(t-\tau)\{w(r,t)$$

$$+ \int_r^1 \frac{1}{\xi^3}\int_0^\xi \mu^3 w'(\mu,t)\int_\mu^1 \frac{1}{\psi^3}\int_0^\gamma \psi^3 w'^2(\psi,\tau)\,d\psi d\gamma d\mu d\xi\}\,d\tau$$

$$R[w'(r,t)] = \frac{\alpha}{\beta}\int_r^1 \frac{1}{\xi^3}\int_0^\xi \mu^3 w'^2(\mu,t)d\mu d\xi$$

$$-\frac{1}{\beta}\left(\frac{\alpha}{\beta} - 1\right)\int_0^t \int_r^1 \frac{1}{\xi^3}\int_0^\xi \mu^3 w'^2(\mu,\tau)K(t-\tau)d\mu d\xi d\tau$$

The equations in (3.129) are uncoupled because $w'$ may be determined independently of $\Phi'$; thus it is sufficient to consider only the first equation in the set (3.129). We begin with the linearized (stability) problem, which is governed by the initial boundary value problem

$$(1+\alpha D)\frac{\partial}{\partial r}\left(r^3\frac{\partial w'(r,t)}{\partial r}\right) + \lambda(1+\beta D)r^3 w'(r,t) = 0 \qquad (3.130)$$

$$\begin{cases} w(0,t) = 0, \ w(1,t) = 0, \ 0 \le t \le T \\ w(r,0) = w_0'(r), \ 0 < r < 1 \end{cases} \qquad (3.131)$$

In terms of the integral operator formulation introduced above, the problem (3.130), (3.131) may be written in the form

$$w'(r,t) = \lambda\frac{\beta}{\alpha}\int_r^1 \frac{1}{\xi^3}\int_0^\xi \mu^3 w'(\mu,t)d\mu d\xi$$

$$+\frac{1}{\alpha}\left(\frac{\alpha}{\beta} - 1\right)\int_0^t w'(r,\tau)K(t-\tau)d\tau \qquad (3.132)$$

It may be shown that the eigenvalues of (3.132) which correspond to the time $t = 0$ are simple and constitute a sequence of discrete numbers $\lambda_n^o \to \infty$, as $n \to \infty$, while the eigenvalues of (3.132) which correspond to time $t = \infty$ are also simple and form a sequence $\lambda_n^\infty \to \infty$, as $n \to \infty$. Furthermore, for each $n = 1, 2, \cdots$

$$\lambda_n^o = \frac{\alpha}{\beta} \lambda_n^\infty \qquad (3.133)$$

For $\lambda \neq \lambda_n^o$ the linearized problem (3.132) may be solved in the form

$$w'(r, t) = \sum_{n=1}^{\infty} c_n \phi_n(r) exp\left[-\frac{1}{\beta}\left(\frac{\lambda_n^\infty - \lambda}{\lambda_n^o - \lambda}\right)t\right] \qquad (3.134)$$

where the $\phi_n(r)$ are the eigenfunctions which correspond to the eigenvalues $\lambda_n^o$, and the $c_n$ are the (generalized) Fourier coefficients in the expansion of the initial transverse perturbation $w_0'(r)$ with respect to the $\phi_n(r)$. For $\lambda < \lambda_n^\infty$, it is clear from (3.134) that all transverse perturbations of the viscoelastic plate tend to zero as $t \to \infty$; for $\lambda_n^\infty < \lambda < \lambda_n^o$ we have an unstable situation, i.e., there exist perturbations which grow without bound as $t \to \infty$. Finally, for $\lambda = \lambda_n^o$, the plate is said to be instantly unstable.

In dealing with the postbuckling analysis of the circular viscoelastic plate within the context of the present model, we will consider the limiting cases of the first equation in the set (3.129) at times $t = 0$ and $t = \infty$, namely,

$$w'(r, 0) - \lambda \frac{\beta}{\alpha} \int_r^1 \frac{1}{\eta^3} \int_0^\eta \mu^3 w'(\mu, 0) d\mu d\eta$$
$$+ \int_r^1 \frac{1}{\eta^3} \int_0^\eta \mu^3 w'(\mu, 0) \int_\mu^1 \frac{1}{\gamma^3} \int_0^\gamma \psi^3 w'^2(\psi, 0) d\psi d\gamma d\mu d\eta$$
$$= 0$$

$$\qquad (3.135)$$

and

$$w'(r, \infty) - \lambda \int_r^1 \frac{1}{\xi^3} \int_0^\zeta \mu^3 w'(\mu, \infty) d\mu d\zeta$$
$$+ \int_r^1 \frac{1}{\zeta^3} \int_0^\zeta \mu^3 w'(\mu, \infty) \int_\mu^1 \frac{1}{\gamma^3} \int_0^\gamma \psi^3 w'^2(\psi, \infty) d\psi d\gamma d\mu d\zeta$$
$$= 0$$

$$\qquad (3.136)$$

These are stationary equations which correspond to the von Karman equations for an elastic plate. By employing results from the theory of monotone operators, as well as standard results on solutions of the von Karman equations for stationary (equilibrium) problems, Brilla [140] has been able to prove the following results concerning the solutions of the first equation in (3.129) in a neighborhood of the first critical point $\lambda_1 = \lambda_1^o$: Let $\lambda^* > \dfrac{\alpha}{\beta}\lambda_1$; then the first equation in the set (3.129) has

(i) only the trivial solution for $\lambda < \lambda_1$

(ii) exactly two symmetric nontrivial solutions, one positive and the other negative (in addition to the trivial solution, for $\lambda = \lambda_1$) which start at time $t = 0$ from zero, i.e., $w'(r,0) = 0$, $0 \le r \le 1$.

(iii) exactly two nontrivial solutions, one positive and the other negative (in addition to the trivial solution) for $\lambda_1 < \lambda < \min\{\lambda_c, \lambda_2\}$, where, $\lambda_1 < \lambda_c < \dfrac{\beta}{\alpha}\lambda^*$, which start at time $t = 0$ from nonzero values, i.e., $|w'(r,0)| > 0$, $0 \le r \le 1$.

(iv) The positive solutions are nondecreasing in the time variable and nonincreasing in the space variable, while the negative solutions are nonincreasing in the time variable and nondecreasing in the space variable.

In the analysis referenced above, $\lambda^*$ and $\lambda_c$ are such that if one constructs, for $\lambda > \lambda_1$, the sequence

$$w_i'(r,t) = Z_\lambda^{-1}\left\{S[w_{i-1}'(r,t)]\right\} \tag{3.137}$$

$i = 1, 2, \cdots$, with

$$w_0'(r,t) = Z_\lambda^{-1}\left\{S[w_d'(r,t)]\right\} \tag{3.138}$$

for $0 < r < 1$, $0 < t < T$, where $w_d(r,t) = w^+(r,0)$, $0 < t < T$, $w^+(r,0)$ being the only positive solution of (3.135), then for $\lambda_1 < \lambda < \lambda_c$ the operator $Z_\lambda^{-1}$ is monotone. As the operator $S$ is also monotone, in (3.137), for $0 < r < 1$, and $0 < t < T$, we have, with $\lambda_1 < \lambda < \left(\dfrac{\beta}{\alpha}\right)\lambda^*$,

$$0 \le w_d'(r,t) \le w_0'(r,t) \le \cdots \le w_i'(r,t) \le \cdots \tag{3.139}$$

with the sequence, defined by (3.137), bounded from above by the function $\bar{w}'(r,t) = w_+'(r,\infty)$, $0 < t < T$, where $w_+'(r,\infty)$ is the only positive solution of (3.136). By taking the limit as $i \to \infty$, in (3.137), one obtains a function $w(r,t)$, which satisfies the first equation in (3.129), for

$0 < r < 1$, $0 < t < T$, so that $\lambda_1 = \lambda_1^0$ is a bifurcation point (i.e. buckling load). The postbuckling behavior relative to branching from $\lambda_1 = \lambda_1^0$ is shown in Fig. 3.24 and our results may be summarized as follows: For $\lambda < \lambda_1$, only the trivial solution exists; however, in contrast to the situation for the stationary von Karman equations, where for $\lambda = \lambda_1$ only the trivial solution exists, for the generalized von Karman equations associated with buckling of a circular viscoelastic plate, of differential type, exactly two nontrivial solutions exist for $\lambda = \lambda_1$. One of these two solutions, for $\lambda = \lambda_1$, is increasing while the other is decreasing from the value $w = 0$, at time $t = 0$, to the values of $w$, which are given by the solutions of the associated limiting stationary equations at $t = \infty$. For $\lambda > \lambda_1$, there exist solutions starting from the values given by the solutions of the limiting stationary equations at time $t = 0$ which assume, as $t \to \infty$, values determined by the solutions of the limiting stationary equations as $t \to \infty$. A more detailed (numerical) analysis than the one carried out in [140] remains to be done, inasmuch as no numerical approximations are provided for the eigenvalues $\lambda_n^0$, or for the associated eigenfunctions $\phi_n(r)$, and the same remarks apply, of course, to the eigenvalues and associated eigenfunctions of (3.132) which are associated with time $t = \infty$.

We now turn our attention to the initial buckling and postbuckling behavior of elastic-plastic circular plates, a concise treatment of which may be found in the work of Needleman [141]. In [141], the author notes that the postbuckling behavior of elastic circular plates is characterized by (i) the stability of the initial postbuckling behavior, i.e., that immediately after buckling, the load increases, and (ii) imperfection insensitivity, in the sense that an imperfect plate can support loads above the buckling load; a primary focus of the work in [141] is the way in which this characteristic postbuckling behavior for elastic-plastic circular plates contrasts to that for elastic-plastic circular plates. In [141] the elastic-plastic circular plate is assumed to be subject to in-plane radial compressive loading, and the material of which the plate is constructed is taken to be strain hardening and characterized by a flow theory of plasticity with a smooth yield surface (in fact, the constitutive theory mirrors the one used to describe the buckling of elastic-plastic rectangular plates in Chapter 2.3). Both simply supported and clamped boundary conditions are considered, but attention is restricted to axisymmetric deformations of the plate. As in Chapter 2.3, an application is made of the asymptotic theory of Hutchinson [46] for postbuckling in the plastic range, which yields an approximate expression for the maximum load that can be supported by the plate as well as expressions for the buckling deflection and the size of the unloading region when the plate is at its maximum load. Postbuckling behavior for the elastic-plastic circular plate is also obtained in [141] numerically by a finite element calculation, and considerable attention is paid to the imperfection sensitivity of elastic-plastic circular plates (both when buckling of the perfect plate occurs in the elastic range as well as when buckling of the perfect plate occurs in the plastic range), although we will refine most of our discussion of imperfection sensitivity to Chapter 5.1.

We begin with the mathematical formulation of the buckling problem for an elastic-plastic circular plate. As in the earlier analyses of circular plates in this chapter, we take the circular plate to have radius $a$ and thickness $h$; the in-plane radial compressive loading results in a monotonically increasing compressive displacement $U$ at the outer edge of the plate which is assumed to be thin so that $h/a << 1$. With the radial in-plane and lateral displacements again given, respectively, by $u(r)$ and $w(r)$, the non-zero strain rates for an assumed axisymmetric deformation at a distance $z$ from the middle surface of the plate, are

$$
\begin{cases}
\dot{e}_{rr} = \dot{u}_{,r} + w_{,r}\dot{w}_{,r} - z\dot{w}_{,rr} \\
\dot{e}_{\theta\theta} = \dfrac{\dot{u}}{r} - \dfrac{z}{r}\dot{w}_{,r}
\end{cases}
\tag{3.140}
$$

where $(\dot{\ }) \equiv \dfrac{\partial}{\partial t}$. The stress rates are related to the strain rates in (3.140), by $\dot{\boldsymbol{\sigma}} = \boldsymbol{L}\dot{\boldsymbol{e}}$ where the tensor of moduli $\boldsymbol{L}$ possesses two branches, one of which corresponds to loading and the other corresponds to elastic unloading. The elastic part of $\boldsymbol{L}$, which we denote by $\boldsymbol{\mathcal{L}}$, is taken to be isotropic in [141] so that

$$
\mathcal{L}_{ijkl} = \frac{E}{1+\nu}\left[\frac{1}{2}(\delta_{ik}\delta_{jl} + \delta_{il}\delta_{jk}) + \frac{\nu}{1-2\nu}\delta_{ij}\delta_{kl}\right]
\tag{3.141}
$$

For the particular theory of plasticity employed in [141], a $J_2$ flow theory with isotropic hardening, the tensor of moduli $\boldsymbol{L}$ is given by

$$
L_{ijkl} = \mathcal{L}_{ijkl} - \frac{3}{2}\left(\frac{E}{1+\nu}\right)\frac{c\left(\dfrac{1}{E_t} - \dfrac{1}{E}\right)\delta_{ij}\delta_{kl}}{\left(\dfrac{2(1+\nu)}{3}\dfrac{}{E} + \dfrac{1}{E_t} - \dfrac{1}{E}\right)\sigma_e^2}
\tag{3.142}
$$

where

$$
\begin{cases}
\sigma_e = \left( \dfrac{3}{2} S_{ij} S_{ij} \right)^{\frac{1}{2}} \\[2mm]
S_{ij} = \sigma_{ij} - \dfrac{1}{3} \delta_{ij} \sigma_{kk}
\end{cases}
\tag{3.143}
$$

and

$$
c = \begin{cases}
0, \ \text{if } \sigma_e < (\sigma_e)_{\max} \text{ or} \\
\qquad \sigma_e = (\sigma_e)_{\max} \text{ and } \dot{\sigma}_e \leq 0 \\
1, \ \text{if } \sigma_e = (\sigma_e)_{\max} \text{ and } \dot{\sigma}_e \geq 0
\end{cases}
\tag{3.144}
$$

The tangent modulus, which we have depicted in Fig. 2.50, in connection with our discussion in Chapter 2.3, is the slope of the uniaxial stress-strain curve and is a function of $\sigma_e$. The uniaxial stress-strain behavior employed in [141] has a representation in terms of a modified piecewise power law of the form

$$
\frac{\epsilon}{\epsilon_y} = \begin{cases}
\dfrac{\sigma}{\sigma_y}, \ \sigma \leq \sigma_y \\[3mm]
\dfrac{1}{n} \left( \dfrac{\sigma}{\sigma_y} \right)^n + 1 - \dfrac{1}{n}, \ \sigma \geq \sigma_y
\end{cases}
\tag{3.145}
$$

where $n$ is the strain-hardening exponent, $\sigma_y$ is the yield stress, and $\epsilon_y = \sigma_y / E$; this representation of the uniaxial stress-strain behavior exhibits a sharp yield point and has a tangent modulus which is a continuous function of the stress at $\sigma_y$.

Because each surface which is parallel to the middle surface of the plate is in a state of (approximate) plane stress, only the in-plane stresses need to be considered in the constitutive theory; for this reason, the author [141] introduces a tensor $\boldsymbol{K}$ of in-plane moduli which are defined in terms of the three-dimensional moduli by

$$
K_{\alpha\beta\gamma\delta} = L_{\alpha\beta\gamma\delta} - \frac{L_{\alpha\beta 33} L_{\gamma\delta 33}}{L_{3333}}
\tag{3.146}
$$

with $\alpha, \beta, \gamma, \delta = 1, 2$. Both simply supported and clamped plates are considered in [141]; for the simply supported plate we have, at $r = a$,

$$
\begin{cases}
\dot{u}(a) = \dot{U}, \ \dot{w}(a) = 0 \\[3mm]
\displaystyle\int_{h/2}^{h/2} \dot{\sigma}_{rr}(a, \eta) \eta \, d\eta = 0
\end{cases}
\tag{3.147}
$$

while for a clamped plate

$$\dot{u}(a) = \dot{U}, \quad \dot{w}(a) = 0, \quad \dot{w}_{,r}(a) = 0 \tag{3.148}$$

The equations of (incremental) equilibrium, for the class of problems described above, are formulated in terms of a variational principle, i.e., among all incremental fields $\dot{u}(r)$ and $\dot{w}(r)$ which satisfy the kinematic boundary condition at $r = a$, the actual fields satisfy $\delta \mathcal{I} = 0$, where

$$\mathcal{I} = \frac{1}{2} \int_0^a \int_{-\frac{h}{2}}^{\frac{h}{2}} (\dot{\sigma}_{rr}\dot{\varepsilon}_{rr} + \dot{\sigma}_{\theta\theta}\dot{\varepsilon}_{\theta\theta} + \sigma_{rr}\dot{w}_{,r}^2)rdrd\zeta \tag{3.149}$$

with the stress increments related to the strain increments by the plane stress moduli (3.146); the variational principle described above serves as the basis in [141] for implementing a finite element solution of the buckling problem. We now delineate the initial buckling and postbuckling results which follow from the elastic-plastic model of the circular plate, beginning with the onset of buckling.

In both the elastic and plastic ranges, the prebuckling solution is that of uniform radial compression, so that in the plastic range, the plane stress moduli (3.146) at buckling correspond to an isotropic material. The critical stress for buckling in the plastic range is (see Chapter 2.3) given by the same expression as that in the elastic range except that, instead of $E$ and $\nu$, an effective Young's modulus $E_f$ and an effective Poisson ratio $\nu_f$ are employed. The critical compressive stress is given by

$$\sigma_c \equiv \lambda_{cr} = \frac{\tilde{\beta}^2}{12} \left( \frac{E_f}{1 - \nu_f^2} \right) \left( \frac{h}{a} \right)^2 \tag{3.150}$$

with

$$\frac{E}{E_f} = 1 + \frac{1}{4} \left( \frac{E}{E_t} - 1 \right) \tag{3.151}$$

and

$$\nu_f = \frac{E_f}{E} \left\{ \nu - \frac{1}{4} \left( \frac{E}{E_t} - 1 \right) \right\} \tag{3.152}$$

where, for a clamped plate, $\tilde{\beta}$ is the smallest non-zero root of $J_1(\tilde{\beta}) = 0$, while for a simply supported plate $\tilde{\beta}$ is the smallest non-zero root of

$$\tilde{\beta}J_0(\tilde{\beta}) = (1 - \nu_f)_c J_1(\tilde{\beta})$$

The functions $J_0, J_1$ are, once again, the Bessel functions of the first kind of orders zero and one, respectively, while the subscript $c$, above, denotes evaluation of the effective Poisson ratio $\nu_f$ at buckling. For both the simply supported and clamped boundary conditions, the buckling mode $w_c(r)$, associated with $\sigma_c$, as given by (3.150), has the form

$$w_c(r) = w_c(0) \frac{J_0(\tilde{\beta} r/a) - J_0(\tilde{\beta})}{1 - J_0(\tilde{\beta})} \qquad (3.153)$$

For the case of a clamped plate it may be computed that $\tilde{\beta} \simeq 3.832$.

In Fig. 3.25, we depict the buckling stress for both the clamped and simply supported cases, with $\nu = 1/3$, $n = 12$, as a function of the parameter $\tilde{\eta}$ where

$$\tilde{\eta} = \frac{\tilde{\beta}_E^2}{12(1 - \nu^2)\epsilon_y} \left(\frac{h}{a}\right)^2 \qquad (3.154)$$

with $\tilde{\beta}_E$ the value of $\tilde{\beta}$ when buckling occurs in the elastic range. For the simply supported plate $\tilde{\beta}_E = 2.069$, when $\nu = \frac{1}{3}$, while for the clamped plate $\tilde{\beta}_E = 3.832$. We also show, in Fig. 3.25, the uniaxial stress-strain curve for the modified piecewise power law (3.145) with strain-hardening exponent $n = 12$.

**Remarks:** It has been noted in [141] that, for a circular plate with $\nu = \frac{1}{3}$ and $n = 12$, the discrepancy between the buckling stresses predicted by $J_2$ flow theory and those predicted by a $J_2$ deformation theory is within one percent, provided $\tilde{\eta}$, as given by (3.154), satisfies $\tilde{\eta} \leq 2$; for larger values of $\tilde{\eta}$, e.g., for thicker plates, the discrepancy between the values of the predicted buckling stresses can be significant.

We now present those results of Needleman [141] which pertain to the initial postbuckling behavior of circular plates in the plastic range. The results presented are for both the clamped and simply supported situations and follow as a consequence of the asymptotic theory of Hutchinson [17]; in this analysis, the amplitude $\xi$ of the relevant eigenmode is employed as the expansion variable in the postbuckling perturbation analysis.

The average radial stress applied at the edge of the circular plate at bifurcation is given by

$$\sigma_{av} = \sigma_c + \sigma_1 \frac{\xi}{h} + \cdots \qquad (3.155)$$

where $\sigma_c$, the bifurcation stress, is given by (3.156) and

$$\sigma_{av} = -\frac{1}{h} \int_{-h/2}^{h/2} \sigma_{rr}(a, \eta) d\eta \equiv N_r \bigg|_{r=a} \qquad (3.156)$$

so that $\sigma_{av} > 0$ for compressive loading. In (3.155), the initial slope $\sigma_1$ is uniquely determined by the condition that plastic loading takes place throughout the plate except at one point where neutral loading occurs; this leads to

$$\frac{\sigma_1}{\sigma_c} = \frac{3(1+\nu_f)_c}{1 - J_0(\tilde{\beta})} \tag{3.157}$$

and $\sigma_1 > 0$ so that buckling takes place under increasing load.

The next term in the postbuckling expansion for the average applied compressive stress $\sigma_{av}$ turns out to be proportional to $\xi^{\frac{4}{3}}$, i.e.,

$$\sigma_{av} = \sigma_c + \sigma_1 \frac{\xi}{h} + \sigma_2 \left(\frac{\xi}{h}\right)^{\frac{4}{3}} + \cdots \tag{3.158}$$

and

$$\sigma_2 = -\frac{3}{4}\sigma_1 \left(\frac{\Delta}{1 - J_0(\tilde{\beta})}\right)^{\frac{1}{3}} \tag{3.159}$$

where $\Delta$ is given by

$$\Delta = 24\tilde{\beta}^2 N_c \left\{ I_1 + I_0 \left[ -\frac{\tilde{\beta}^2}{24} \left(\frac{h}{a}\right)^2 \left.\frac{dL_f}{d\sigma_c}\right|_c \right. \right.$$
$$\left. + \frac{3}{4}\left(\frac{1-\nu}{1+\nu}\right)(1-\nu_f)_c N_c^{-1} \right] \tag{3.160}$$
$$\left. -\frac{3}{2}\left(\frac{[1-J_0(\tilde{\beta})]J_1(\tilde{\beta})}{\tilde{\beta}}\right)^2 \left(\frac{1-\nu}{1+\nu}\right)(1-\nu_f)_c N_c^{-1} \right\}$$

with

$$I_0 = \int_2^1 x J_0^2(\tilde{\beta}x)dx, \quad I_1 = \int_0^1 x J_1^2(\tilde{\beta}x)dx \tag{3.161a}$$

and

$$L_f = \left.\frac{E_f}{1-\nu_f}\right|_c, \quad N_c = \left.\frac{L_f}{(E/(1-\nu)) - L_f}\right|_c \tag{3.161b}$$

The first three terms in the postbuckling expansion (3.158) are used, in [141], to estimate $\sigma_{max}$, the maximum applied stress that can be supported by the plate, as well as the value of the amplitude $\xi$ of the buckling mode at $\sigma_{max}$; the results obtained have the form

$$\frac{\sigma_{max}}{\sigma_c} = 1 + \frac{3}{4} \cdot \frac{(1+\nu_f)_c}{\Delta} \tag{3.162a}$$

and

$$\left(\frac{\xi}{h}\right)_{\text{max}} = \frac{1 - J_0(\tilde{\beta})}{\Delta} \qquad (3.162b)$$

Also, for both the clamped and simply supported circular plates, the point at which neutral loading occurs at buckling turns out to be $r = 0$, $z = \frac{1}{2}h$.

When $\xi$ increases from zero, the neutral loading surface (i.e., the surface which separates the regions of plastic loading and elastic unloading), spreads out from the point $r = 0$, $z = \frac{1}{2}h$. The equation of the neutral loading surface is

$$\frac{1}{2}\left(\frac{\xi}{\xi_{\text{max}}}\right)^{1/3} + \left(\frac{z}{h} - \frac{1}{2}\right) - \frac{\tilde{\beta}^2}{8}\left(\frac{r}{a}\right)^2 = 0 \qquad (3.163)$$

It is noted in [141] that, at maximum load, the greatest penetration of the neutral loading surface, through the thickness of the circular plate, occurs at $r = 0$, where the neutral loading surface has penetrated halfway through the thickness of the plate. Away from the axis of symmetry, at $r = 0$, the greatest penetration occurs on the surface $z = \frac{1}{2}h$. In Table 3.4, we display the results of the postbuckling analysis obtained in [141] for three different cases; among other things, Table 3.4 shows that the maximum support load, which is given by (3.162a), is only slightly above the buckling load for all three cases indicated and that, for the simply supported case, the analysis predicts that, at the maximum load, the plastic zone extends all the way out to the edge of the circular plate, while for the clamped plate $r_{\text{max}}/a$ is 0.52 at the maximum load irrespective of the amount of plastic strain prior to buckling.

In [141], Needleman also presents numerical results based on a finite-element analysis of the variational equation $\delta \mathcal{I} = 0$, where the functional $\mathcal{I}$ is given by (3.149); in this analysis an imperfection $w_0(r)$ is considered which is proportional to the buckling mode as given by (3.153). The imperfection mode in [141] is specified by $\hat{\xi}$, where $\hat{\xi} = w_0(0)$. The results given in [141], for a "perfect" plate, are actually those which are computed using the very small imperfection $\hat{\xi} = 10^{-4}h$. We will put off a complete discussion of the results obtained in [141], relative to the imperfection sensitivity of elastic-plastic circular plates, until Chapter 5. For now, however, we reproduce from [141], Figs. 3.26–3.29, which indicate, respectively, the curves of average applied stress vs. buckling deflection for two different cases of a simply supported circular plate, the curves governing buckling deflection versus applied displacement for a simply supported circular plate, and the curves governing average applied stress versus buckling deflection for a clamped circular plate; besides indicating the imperfection sensitivity for the various cases considered, all these graphs clearly depict the nature of the postbuckling behavior for the perfect elastic-plastic circular plate.

In Fig. 3.27, buckling of the perfect plate takes place just outside the elastic region; this plate is simply supported with $\sigma_c/\sigma_y = 1.02$, $\tilde{\eta} = 1.28$, $n = 12$, and $\nu = \frac{1}{3}$; for this case the initial slope $\sigma_1$, as computed from (3.157), is also shown on the graph (which is valid only for very small values of the buckling deflection because the maximum load, which is given by $\sigma_{max}/\sigma_c = 1.006$, is reached at $\xi = 0.008h$). The unloading region for the perfect plate of Fig. 3.27, at maximum load, is the region $1 < z < \frac{h}{2}$; beyond the maximum load point, the load decreases rapidly and then levels off at larger buckling deflections.

The graph in Fig. 3.28 depicts the curves of buckling deflection $\xi$ vs. applied edge displacement $U$ for the same values of the parameters, geometrical and physical, as those considered in the case depicted in Fig. 3.27; the applied displacement has been normalized by its value $U_c$ at buckling. For the perfect plate, the buckling deflection $\xi$ is zero until $U$ reaches the value $U_c$ then, at $U = U_c$, the deflection begins to grow (but very slowly). Near the maximum load point, $\xi$ grows at nearly constant edge displacement $U$.

In Fig. 3.29, we depict the average applied stress versus buckling deflection curves for a case in which the buckling occurs further into the plastic range. The circular plate for the case depicted in Fig. 3.29 is clamped with $\sigma_c/\sigma_y = 1.12$, $\tilde{\eta} = 2.0$, $\nu = \frac{1}{3}$, and $n = 12$; the initial slope $\sigma_1$, as computed from (3.157), is once again shown for the perfect plate but, as in the earlier case, this slope is only valid for very small values of the buckling deflection $\xi$. Also, the maximum load, which is given by $\sigma_{max}/\sigma_c = 1.018$, occurs at a value of $\xi = 0.065h$, which is larger than the corresponding value of $\xi$ in the previous example. The graph associated with the perfect plate in Fig. 3.29 shows that, when buckling occurs well into the plastic range, the load falls off much more gradually after the maximum load point is achieved than when $\sigma_c \approx \sigma_y$.

Finally, Fig. 3.30 depicts the development of the plastic zone and the unloading zone for the same case as that shown in Fig. 3.29. For the perfect plate, unloading starts at $r = 0$, $z = \frac{1}{2}h$. As the primary zone of unloading grows, a region near the clamped edge of the circular plate is subjected to unloading; however, this unloading region near the clamped edge does not have a significant effect on the overall deformation of the plate because this region is highly constrained by the condition that the edge is clamped. Once the maximum load point is reached, this zone near the clamped edge undergoes loading again. A fundamental conclusion, based on the analysis of perfect plates in [160], is that whenever buckling occurs in the plastic range of the circular plate, the maximum support stress is only slightly higher than the buckling stress; the asymptotic analysis of the postbuckling behavior for the perfect plate in the plastic range, as based on the theory first formulated by Hutchinson [17], [46], yields a good estimate of the maximum support stress and of the growth of the unloading zone in the plate after buckling.

## Appendix: A Geometrically Exact Theory for the Axisymmetric Deformation of Nonlinear Elastic Plates

In this appendix we will present some of the work of Antman [21], [22], and [139], in which a geometrically exact theory is formulated for the axisymmetric deformation of nonlinear elastic circular plates which may undergo flexure, mid-surface extension, and shear. The basic equations are actually constructed for the axisymmetric deformation of shells, a plate then being defined to be a shell with a flat natural configuration.

We begin by letting $\{i, j, k\}$ be a fixed orthonormal basis for the Euclidean 3-space and for each real (angle) $\phi$ we set

$$
\begin{cases}
e_1(\phi) = \cos \phi i + \sin \phi j \\[2mm]
e_2(\phi) = -\sin \phi i + \cos \phi j \\[2mm]
e_3 = k
\end{cases}
\tag{3.164}
$$

An axisymmetric configuration of the shell is determined by the pair of vector-valued functions

$$
r(s, \phi) = r(s)e_1(\phi) + z(s)k
$$
$$
b(s, \phi) = -\sin \theta(s)e_1(\theta) + \cos \theta(s)k
$$

$$
\tag{3.165}
$$

with $s_1 \leq s \leq s_2$, $0 \leq \phi \leq 2\pi$; we also set

$$
a(s, \phi) = e_2(\phi) \times b(s, \phi)
$$
$$
= \cos \theta(s)e_1(\phi) + \sin \theta(s)k
$$

$$
\tag{3.166}
$$

If we denote the values of geometrical variables in the reference configuration by a superscript $*$, then we have the following interpretation of the quantities appearing in (3.165), (3.166): the function $r$ defines an axisymmetric surface in Euclidean 3-space while $r^*$, usually chosen as the mid-surface of the shell, is the axisymmetric "reference surface." The variable $s$ is the arc length parameter along the curves $r^*(\cdot, \phi)$, for each fixed $\phi$, so that $|r^*_{,s}(s, \phi)| = 1$. Thus $r^*(s, \phi)$ identifies a typical material point on the reference surface, while $r(s, \phi)$ is the corresponding deformed position of that point. The vector $b(s, \phi)$ characterizes the deformed configuration of that material fiber in the shell whose reference configuration lies on the normal to the reference surface through the point $r^*(s, \phi)$. We now set

$$r_{,s}(s, \phi) \equiv \nu(s)a(s, \phi) + \eta(s)b(s, \phi) \qquad (3.167)$$

and

$$\tau(s) = \frac{r(s)}{r^*(s)}, \quad \sigma(s) = \frac{\sin \theta(s)}{r^*(s)}, \quad \mu(s) = \theta'(s) \qquad (3.168)$$

thus defining strains $\nu(s), \eta(s), \mu(s), \tau(s)$, and $\sigma(s)$. We note that $|r_{,s}| = \sqrt{\nu^2(s) + \eta^2(s)}$. The strain $\nu(s) = r_{,s} \cdot a$ is the local ratio of deformed area to reference area for the parallelogram with sides $r_{,s}$ and $b$; it is convenient to think of $\nu$ as measuring "stretching of the base curve $r^*$," while $\eta$ is a measure of "shear," i.e., of the reduction in the angle between $r^*_{,s}$ and $b^*$. The strain measure $\mu$ is related to changes in curvature which are associated with pure bending as opposed to stretching, while $\tau(s)$ measures stretching in the azimuthal direction and $\sigma(s)$ measures the amount of bending about $a$. The strain "vector" associated with axisymmetric deformations of the shell is, therefore,

$$S = (\nu, \eta, \tau, \sigma, \mu) \qquad (3.169)$$

Now, let $n_1(s_0, \phi)$ and $m_1(s_0, \phi)$ denote the resultant contact force and contact couple, per unit reference length of the circle $r^*(s_0, \phi)$, of radius $r^*(s_0)$, that are exerted across this section at $r^*(s_0, \phi)$, by the material with $s \geq s_0$ on the material with $s < s_0$. Let $n_2(s, \phi)$ and $m_2(s, \phi)$ denote the resultant contact force and contact couple per unit reference length of the curve $r^*(s, \phi)$ that are exerted across this section at $r^*(s, \phi)$ by the material with $\phi_0 \leq \phi \leq \phi_0 + \epsilon$ on the material with $\phi_0 - \epsilon \leq \phi < \phi_0$, where $\epsilon > 0$ is a small number. It may be shown that, in view of our limitation to axisymmetric deformations, the force and couple resultants have the form:

$$\begin{cases} \boldsymbol{n}_1(s,\phi) = N(s)\boldsymbol{a}(s,\phi) + H(s)\boldsymbol{b}(s,\phi) \\[2mm] \boldsymbol{n}_2(s,\phi) = T(s)\boldsymbol{e}_2(\phi) \\[2mm] \boldsymbol{m}_1(s,\phi) = -M(s)\boldsymbol{e}_2(\phi) \\[2mm] \boldsymbol{m}_2(s,\phi) = \Sigma(s)\boldsymbol{a}(s,\phi) \end{cases} \tag{3.170}$$

If we assume that the shell is subjected to an axisymmetric body force of intensity,

$$\boldsymbol{f}(s,\phi) = f_1(s)\boldsymbol{e}_1(\phi) + f_3(s)\boldsymbol{k} \tag{3.171}$$

and a body couple of intensity $-\ell(s)\boldsymbol{e}_2(s,\phi)$, per unit reference area of $\boldsymbol{r}^*$ at $(s,\phi)$, then by summing all the forces and moments on a segment which consists of all those material points with coordinates $(\xi,\psi)$, $s_1 \leq \xi \leq s$, $0 \leq \psi \leq \phi$, and subsequently differentiating the resulting integral balance laws with respect to $s$ and $\phi$, we obtain the equilibrium equations for the shell in the form:

$$\frac{\partial}{\partial s}\left[r^*(s)N(s)\boldsymbol{a}(s,\phi) + H(s)\boldsymbol{b}(s,\phi)\right]$$
$$-T(s)\boldsymbol{e}_1(\phi) + r^*(s)\boldsymbol{f}(s,\phi) = \boldsymbol{0} \tag{3.172a}$$

and

$$\frac{d}{ds}\left[r^*(s)M(s)\right] - \Sigma(s)\cos\theta(s)$$
$$+r^*(s)\left(\nu(s)H(s) - \eta(s)N(s)\right) \tag{3.172b}$$
$$+r^*(s)\ell(s) = 0$$

The shell is defined to be nonlinearly elastic if there are functions $\hat{T}, \hat{N}, \hat{H}, \hat{\Sigma},$ and $\hat{M}$, of $\boldsymbol{S}$ and $s$, such that

$$T(s) = \hat{T}(\boldsymbol{S}(s), s), \text{etc.} \tag{3.173}$$

It is required that the constitutive functions be sufficiently smooth and that they satisfy the condition that the Jacobian matrices

$$\frac{\partial(\hat{N}, \hat{H}, \hat{M})}{\partial(\nu, \eta, \mu)} \quad \text{and} \quad \frac{\partial(\hat{T}, \hat{\Sigma})}{\partial(\tau, \sigma)} \tag{3.174}$$

be positive definite; such conditions ensure that, e.g., an increase in the bending strain $\mu$ is associated with an increase in the bending couple

$M$. The constitutive functions must also satisfy certain compatible "co-ercivity" conditions (see [22], eqns. (1.15 a-d)) on which we will not elaborate here; such conditions are set down to guarantee, e.g., that if we consider a small part of the shell whose reference configuration is a square, with $\tau(s)$ measuring the stretch in the horizontal direction and $\nu(s)$ the stretch in the vertical direction, then if $\tau(s)$ becomes infinite, while $\nu(s)$ has a positive lower bound, the horizontal tension $T(s)$ must become infinite in order to maintain an equilibrium state.

**Remarks:** In the discussion which follows, it will be assumed that the natural (reference) state of the shell is stress-free, i.e.

$$\hat{T}(\boldsymbol{S}, s) = \hat{N}(\boldsymbol{S}, s) = \hat{H}(\boldsymbol{S}, s) = \hat{\Sigma}(\boldsymbol{S}, s) = \hat{M}(\boldsymbol{S}, s) = 0 \qquad (3.175)$$

for $\boldsymbol{S} = (1, 1, 0, \sigma^*, \theta^{*\prime})$

We now specialize the theory delineated above so as to consider the buckling of a nonlinear elastic, isotropic circular plate. We take the plate to have unit radius so that $s_1 = 0, s_2 = 1$, and

$$r^*(s) = s, \quad \theta^*(s) = 0 \qquad (3.176)$$

We also take $\boldsymbol{f} = \boldsymbol{0}, \ell = 0$ so that there is neither an applied force nor an applied couple acting on the plate. The limitation to axisymmetry requires that, at the center of the plate,

$$r(0) = 0, \quad \theta(0) = 0, \quad \eta(0) = 0 \qquad (3.177)$$

We assume that the edge of the plate, at $s = 1$, is constrained to be parallel to $\boldsymbol{k}$ so that $\theta(1) = 0$ and that a normal pressure of intensity $\lambda g(r(1))$ is applied at that edge so that

$$N(1) = -\lambda g(r(1)), \quad H(1) = 0 \qquad (3.178)$$

When the edge pressure has intensity $\lambda$ units of force per deformed length, $g(r) = r$, while, if it has intensity $\lambda$ units of force per reference length, $g(r) = 1$; in [22] the latter situation is assumed. The integral form of the balance of forces equation for the plate produces the relations

$$-sN(s) = \left[\lambda + \int_s^1 T(\xi)d\xi\right]\cos\theta(s) \qquad (3.179\text{a})$$

$$sH(s) = \left[\lambda + \int_s^1 T(\xi)d\xi\right]\sin\theta(s) \qquad (3.179\text{b})$$

and, upon differentiation, these yield the system

$$\frac{d}{ds}[sN(s)] = sH(s)\theta'(s) + T(s)\cos\theta(s) \qquad (3.180a)$$

$$-\frac{d}{ds}[sH(s)] = sN(s)\theta'(s) - T(s)\sin\theta(s) \qquad (3.180b)$$

which is equivalent to (3.172a) for the present situation.
By substituting from (3.179a,b) into (3.172b), we obtain

$$\frac{d}{ds}[sM(s)] - \Sigma(s)\cos\theta(s)$$
$$+ \left[\lambda + \int_s^1 T(\xi)d\xi\right](\nu(s)\sin\theta(s) + \eta(s)\cos\theta(s)) = 0 \qquad (3.180c)$$

Next, (3.185a,b) easily yield the equations

$$H(s)\cos\theta(s) = -N(s)\sin\theta(s), \quad (sN(s)/\cos\theta(s))' = T(s), \qquad (3.181)$$

(where $\prime = d/ds$), which are equivalent to (3.180a,b).
Before considering the nature of the unbuckled states of the nonlinear elastic circular plate, we note the material symmetry restrictions which must be satisfied; these are of two types. First of all, there are those conditions which ensure that deformed states of the plate occur as mirror images, i.e.,

$$\hat{T}, \hat{N}, \hat{\Sigma}, \hat{M} \text{ are even functions of } \eta,$$
$$\text{while } \hat{H} \text{ is an odd function of } \eta \qquad (3.182a)$$

$$\hat{T}, \hat{N}, \text{ and } \hat{H} \text{ are invariant under the}$$
$$\text{mapping } (\sigma, \mu) \to (-\sigma, -\mu) \qquad (3.182b)$$

and

$$\hat{\Sigma} \text{ and } \hat{M} \text{ change sign under the mapping}$$
$$(\sigma, \mu) \to (-\sigma, -\mu) \qquad (3.182c)$$

Second, it is assumed (in order to simplify the exposition in [22]) that the constitutive functions are independent of $s$ and satisfy the restricted isotropy conditions ("restricted" because $\eta \equiv 0$):

$$\begin{cases} \hat{N}(\tau, \nu, 0, \sigma, \mu) = \hat{T}(\nu, \tau, 0, \mu, \sigma) \\ \hat{M}(\tau, \nu, 0, \sigma, \mu) = \hat{\Sigma}(\tau, \nu, 0, \sigma, \mu) \end{cases} \qquad (3.183)$$

In [22], the relations (3.183) are said to define a plate which is transversely isotropic; in particular, (3.183) implies that $\hat{N}$ depends on $\nu$ in the same manner that $\hat{T}$ depends on $\tau$, etc. More generally, (3.183) implies that for axisymmetric deformations, the constitutive functions are invariant under rotations of $\dfrac{\pi}{2}$ radians, which take longitudinal fibers into azimuthal fibers and vice versa.

If we look for unbuckled states for which $\eta = \theta = 0$, it turns out that all the equilibrium equations are satisfied with the exception of (3.179a). The constitutive equations (3.173) are then substituted into (3.179a), and solutions are sought in the form

$$r(s) = r_0(s) = ks \tag{3.184}$$

with $k$ a constant that must be determined. Use of the isotropy condition results in the condition

$$\hat{T}(k, k, 0, 0, 0) = -\lambda \tag{3.185}$$

If we supplement the positive-definiteness of the Jacobian matrices in (3.174), and the coercivity conditions, with the assumptions that $\hat{T}(k, k, 0, 0, 0)$ decreases strictly from zero to $-\infty$, as $k$ decreases from one to zero, and that $\hat{T}(k, k, 0, 0, 0) > 0$, for $k > 1$, then for each $\lambda \geq 0$, (3.185) has a unique solution $k = k(\lambda)$.

In order to linearize (3.180c) and (3.173) about the unbuckled state defined by $r_0(s, \lambda) = k(\lambda)s$, we let $\hat{R}$ represent any of the constitutive functions in (3.173)—or a derivative of such a constitutive function—and we set

$$R^0(\lambda) = \hat{R}(k(\lambda), k(\lambda), 0, 0, 0) \tag{3.186}$$

The linearization is then given by the Bessel equation

$$(s\theta')' - s^{-1}\theta + \gamma(\lambda)^2 s\theta = 0 \tag{3.187}$$

with

$$
\begin{aligned}
\gamma(\lambda)^2 &= \frac{\lambda}{M_{,\mu}^0(\lambda)} \left[ k(\lambda) + \frac{\lambda}{H_{,\eta}^0(\lambda)} \right] \\
&= -\frac{N^0(\lambda)}{M_{,\mu}^0(\lambda)} \left[ k(\lambda) - \frac{N^0(\lambda)}{H_{,\eta}^0(\lambda)} \right]
\end{aligned}
\tag{3.188}
$$

The eigenvalues of (3.187), subject to $\theta(0) = \theta(1) = 0$, namely, $\lambda^0$, are the solutions of $J_1(\gamma(\lambda)) = 0$, where $J_1$ is the Bessel function of the first kind of order one and the corresponding eigenfunction is $J_1(\gamma(\lambda^0)s)$.

**Remarks**: Global bifurcation results for the above problem are obtained in [22] by transforming this problem as follows: we substitute the constitutive relations (3.173) into (3.180a,b,c) and carry out the indicated differentiations; the positive-definiteness of the first Jacobian matrix in (3.174) allows us to solve the resulting equations for $s\nu'$, $s\eta'$, and $s\mu'$ and an application of (3.165) then produces

$$
\begin{cases}
L_1\rho \equiv \dfrac{d}{ds}(s\rho') - \rho/s = f_1 \\[2mm]
L_2z \equiv \dfrac{d}{ds}(sz') = f_2 \\[2mm]
L_3\theta \equiv \dfrac{d}{ds}(s\theta') - (\theta/s) = f_3
\end{cases}
\tag{3.189}
$$

with $\rho \equiv r - r_0$ and the $f_i$, $i = 1, 2, 3$ involved functions containing all the geometric variables and the parameter $\lambda$. We also have the associated boundary conditions relative to the variables $\rho, z$, and $\theta$:

$$
\begin{cases}
\rho(0) = 0 \\
z(0) = z'(1) = 0 \\
\theta(0) = \theta(1) = 0
\end{cases}
\tag{3.190a}
$$

and

$$
N^0_{,\nu}\rho'(1) + N^0_{,\tau}\rho(1) = -\hat{N}(\boldsymbol{S}(1))
$$
$$
-\lambda + N^0_{,\nu}\rho'(1) + N^0_{,\tau}\rho(1)
\tag{3.190b}
$$

If we define $\boldsymbol{u} = (u_1, u_2, u_3)$ by

$$
\begin{cases}
(L_1\rho)(s) \equiv \sqrt{s}\,u_1(s) \\
(L_2z)(s) \equiv \sqrt{s}\,u_2(s) \\
(L_3\theta)(s) \equiv \sqrt{s}\,u_3(s)
\end{cases}
\tag{3.191}
$$

then these equations may be integrated, subject to the boundary conditions (3.190a,b), in order to express $\rho, z$, and $\theta$ as integral operators acting on $(u_1, u_2, u_3)$. It follows, in particular, from the work in [22] that

$$
\theta[u_3](s) = \int_0^1 K_3(s, \xi)\sqrt{\xi}\,u_3(\xi)d\xi
\tag{3.192}
$$

where $K_3$, which is the Green's function for $L_3$, subject to $\theta(0) = \theta(1) = 0$, is given by

$$K_3(s,\xi) = \begin{cases} \dfrac{1}{2}(s - \dfrac{1}{s})\xi, & \xi < s \\[2mm] \dfrac{1}{2}(\xi - \dfrac{1}{\xi})s, & s < \xi \end{cases} \tag{3.193}$$

If every expression in the $f_i$s is replaced by representations of the form (3.192), we obtain the $f_i$s in the form $f_i = f_i(\lambda, \boldsymbol{u})$; therefore, through the use of (3.189), and (3.191) the boundary value problem for buckling of the nonlinear elastic circular plate can be put in the form

$$u_i(s) = f_i(\lambda, \boldsymbol{u})(s)/\sqrt{s} \tag{3.194}$$

and the subsequent analysis in [22] establishes that the right-hand side of (3.194) defines a continuous and compact mapping (from $\mathcal{R} \times (C^0[0,1])^3$ to $(C^0[0,1])^3$). The compactness of the mapping referenced above permits the application of a global bifurcation theorem; a clear mathematical statement of this global theorem is beyond the level of the present work (the interested reader should consult [22]) but it may be of some interest to mention just one of the implications for the buckling problem under consideration. Thus, suppose that a solution of (3.194) generates, via (3.192), a $\theta(s)$ which possesses a double zero at $s = 0$. As $\rho$ and $z$ also have representations like (3.192), we may conclude that $\rho(0) = \rho'(0) = 0$ and $\eta(0) = 0$. It then follows from the definition of $\rho$ that

$$\hat{N}(\boldsymbol{S}(s)) \to N^0 = -\lambda, \quad \text{as } s \to 0 \tag{3.195}$$

**Remarks:** Within the context of the present theory, buckling problems for nonlinear elastic circular plates which do not exhibit transverse isotropy have also been considered; we sketch only a few of the details here. It may be shown that, even when (3.183) does not hold, if (3.182a,b,c) still apply, the governing equations, for $\theta = \eta = 0$, reduce to

$$\frac{d}{ds}\left[s\hat{N}(r(s)/s, r'(s))\right] = \hat{T}(r(s)/s, r'(s)) \tag{3.196}$$

Setting $\hat{N}(\tau, \nu) = \hat{N}(\tau, \nu, 0, 0, 0)$, etc., the boundary conditions are

$$r(0) = 0, \quad \hat{N}(r(1), r'(1)) = -\lambda \tag{3.197}$$

To avoid technical problems, the following conditions are imposed in [22]:

$$\begin{cases} \dfrac{\partial(\hat{T}, \hat{N})}{\partial(\tau, \nu)} \text{ is positive-definite} \\[2mm] \hat{T}_{,\nu} > 0, \ \hat{N}_{,\tau} > 0 \end{cases} \tag{3.198}$$

Now, let $\nu^{\#}(\tau, \nu)$ denote the unique solution of the equation $\hat{N}(\tau, \nu) = n$ and set $T^{\#}(\tau, n) \equiv \hat{T}(\tau, \nu^{\#}(\tau, n))$. Then the boundary-value problem (3.202), (3.203) is equivalent to

$$\frac{d}{ds}(s\tau) = \nu^{\#}(\tau, n) \tag{3.199a}$$

$$\frac{d}{ds}(sn) = T^{\#}(\tau, n) \tag{3.199b}$$

$$s\tau(s) \to 0, \text{ as } s \to 0 \tag{3.199c}$$

$$n(1) = -\lambda, \tau(s) > 0, \text{ for } s > 0 \tag{3.199d}$$

It has been proven in [22] (see also [161]) that, for $\lambda \geq 0$, (3.196), (3.197) has a unique solution $r_0(\cdot, \lambda)$ and, equivalently, (3.199a-d) has a unique solution $(\tau_0(\cdot, \lambda), N^0(\cdot, \lambda))$; to analyze the buckling problem, we set

$$
\begin{cases}
A_1(s, \lambda) \equiv \dfrac{T^0_{,\nu}(s, \lambda) - N^0_{,\tau}(s, \lambda)}{2\overset{\circ}{N}_{,\nu}(s, \lambda)} \\[2ex]
B_1(s, \lambda) \equiv \sqrt{A_1^2(s, \lambda) + \dfrac{T^0_{,\tau}(s, \lambda)}{N^0_{,\nu}(s, \lambda)}} \\[2ex]
\alpha_1(\lambda) \equiv A_1(0, \lambda) \\[1ex]
\beta_1(\lambda) \equiv B_1(0, \lambda)
\end{cases}
\tag{3.200}
$$

Then, the following result has been established in [22]: if $-\infty < N^0(0, \lambda) < 0$, the limits of $\tau_0(s, \lambda)$ and $\nu_0(s, \lambda)$, as $s \to 0$, exist and are equal and positive. Moreover, $\alpha_1(\lambda) + \beta_1(\lambda) \geq 1$ or, equivalently,

$$T^0_{,\tau}(0, \lambda) + T^0_{,\nu}(0, \lambda) \geq N^0_{,\tau}(0, \lambda) + N^0_{,\nu}(0, \lambda) \tag{3.201}$$

and there are numbers $A(\lambda), B(\lambda), C(\lambda)$, and $D(\lambda)$ such that

$$
\begin{Bmatrix} r_0(s, \lambda) \\ N^0(s, \lambda) \end{Bmatrix} = \begin{Bmatrix} s\tau_0(0, \lambda) \\ N^0(0, \lambda) \end{Bmatrix} + \begin{Bmatrix} A(\lambda) \\ B(\lambda) \end{Bmatrix} s^{\alpha_1(\lambda) + \beta_1(\lambda)}
$$
$$
+ \begin{Bmatrix} C(\lambda) \\ D(\lambda) \end{Bmatrix} \rho^{2[\alpha_1(\lambda) + \beta_1(\lambda)]} + \mathcal{O}(s^{2[\alpha_1(\lambda) + \beta_1(\lambda)]}),
\tag{3.202}
$$

as $s \to 0$

**Remarks**: For the aelotropic, nonlinear elastic circular plate, the linearized problem is of the form

$$\left[sM^0_{,\mu}(s,\lambda)\theta'\right]' - 2M'_{,\mu}(s,\lambda)A_3(s,\lambda)\theta'$$

$$M^0_{,\sigma}(s,\lambda)'\theta + M'_{,\mu}(s,\lambda)\left[A_3^2(s,\lambda) - B_3^2(s,\lambda)\right]\frac{\theta}{s} \qquad (3.203)$$

$$+sN^0(s,\lambda)\left[N^0(s,\lambda)/H^0_{,\eta}(s,\lambda) - r'_0(s,\lambda)\right]\theta = 0$$

where

$$
\begin{cases}
A_3(s,\lambda) \equiv \dfrac{\Sigma^0_{,\mu}(s,\lambda) - M^0_{,\sigma}(s,\lambda)}{2M^0_{,\mu}(s,\lambda)} \\[3mm]
B_3(s,\lambda) \equiv \sqrt{A_3^2(s,\lambda) + \dfrac{\Sigma^0_{,\sigma}(s,\lambda)}{M^0_{,\mu}(s,\lambda)}}
\end{cases}
\qquad (3.204)
$$

with

$$M^0_{,\mu}(s,\lambda) = M_{,\mu}(r_0(s,\lambda)/s,\ r'_0(s,\lambda),0,0,0), \qquad (3.205)$$

the value of $M_{,\mu}$ at the trivial state $(\tau_0(0,\lambda), N'(0,\lambda))$. As has been noted in [22], the analysis of (3.203) is a far more difficult task than the analysis of the linearized problem in the transversely isotropic case, i.e., (3.187). The expansions (3.202) may be used as the basis of an analysis to determine the nature of the eigenvalues and eigenfunctions of (3.203) subject to the boundary conditions $\theta(0) = \theta(1) = 0$.

---

## 3.3  Comparisons of Initial and Postbuckling Behavior of Circular Plates

In this section, we will review some of the differences that can be documented in the buckling and postbuckling behavior of circular plates as one varies the support conditions along the edge of the plate and the type of material symmetry assumed. Our work in this section is limited by the body of knowledge that exists at the present time with respect to buckling and postbuckling behavior of circular plates as opposed, say, to the equivalent body of knowledge for rectangular plates. For example, it is very difficult at this stage to talk about variations in behavior with respect to constitutive response; for the circular plate the nonlinear elastic equivalent of a Johnson-Urbanik type model has yet to be formulated and studied, while serious numerical studies of the geometrically exact model for nonlinear elastic buckling of a circular plate do not appear to have been carried out. The results we have presented for viscoelastic buckling refer, as they must, to a dynamic as opposed to a static problem, and the limiting cases for times $t = 0$ and $t \to \infty$ have not been subjected to serious numerical analysis. Finally, the models of plastic buckling we have presented are highly dependent on the type of flow rule assumed, on the nature of the hardening law, and even on whether the plate, assumed to be subject to a uniform applied radial stress at its edge, buckles in the elastic domain (below yield) or in the plastic domain (above yield). Even within the domain of the linear elastic circular plate, the case of the rectilinearly orthotropic circular plate appears to have not been studied at all. Thus, our discussion here is focused, primarily, on the variations in initial and postbuckling behavior for linearly elastic circular plates with cylindrically orthotropic symmetry (including isotropic response as a special case) as one varies the degree of orthotropy and the support conditions along the edge of the plate.

We begin with the problem of initial buckling of an isotropic circular plate in edge compression as depicted in Fig. 3.1 (in this figure, the plate is shown with a clamped edge). If we refer to Fig. 3.2 and recall that the parameter $\epsilon$ measures the degree of restraint of the edge of the plate at $r = a$, with $\epsilon = 0$ corresponding to a simply supported edge, while $\epsilon = \infty$ corresponds to a clamped edge, we see that as the restraint on the edge of the plate is relaxed, i.e., as we go from $\epsilon = \infty$ down to $\epsilon = 0 (1/\epsilon$ goes from zero to $\infty$) the buckling parameter $k$ decreases monotonically, $k = \lambda_{cr} \cdot a^2 h / K$, and appears to asymptote, as $\epsilon \to 0$, to the value 4.20, which holds for the simply supported situation. Buckling of the compressed isotropic circular plate into the next highest mode as depicted in Fig. 3.3, leads to the existence of nodal diametric lines in the plate. The variation of the buckling parameter $k$ for this second mode is shown in Fig. 3.4, where the behavior is compared with that for buckling in the first mode; clearly, for each fixed degree of restraint $\epsilon$, the value of $k$ ranges from (approximately) twice the corresponding value for the first mode case, when the edge is clamped, up to three times the corresponding value for the first mode case, when the edge of the plate is simply supported. As was the case for buckling into the first mode, for buckling in the second mode (for an isotropic circular plate), $k$ is monotonically increasing as we go from a simply supported edge to a clamped edge. The postbuckling behavior of the isotropic simply supported elastic circular plate, subject to uniform radial compression along its outer edge at $r = a$, is depicted in Figs. 3.9, 3.10, and 3.13; the last of these graphs depicts the normalized edge load vs. the normalized central deflection of the plate and is particularly clear with respect to the conclusion that the postbuckling curve, when presented in this fashion, is steeper for the plate with a simply supported edge than it is for a plate with a clamped edge. We have already made reference to the behavior depicted in Fig. 3.11, which follows from the work of Bodner [149] and which compares $\lambda/\lambda_{cr}$ with $\left. \frac{1}{r}\Phi_{,r}\right|_{r=0} \Big/ \lambda$ (the radial membrane stress at the center of the plate) for both the simply supported and clamped conditions at $r = a$; in both cases a change from compression to tension occurs, at $\lambda/\lambda_{cr} = 1.57$ for simply supported isotropic plates and at $\lambda/\lambda_{cr} = 1.98$ for clamped plates. The asymptotic value of $\left. \frac{1}{r}\Phi_{,r}\right|_{r=0}$, as $\lambda/\lambda_{cr} \to \infty$, is also higher for the clamped plate than the simply supported plate.

We now turn to the linearly elastic, cylindrically orthotropic circular plate, of which the isotropic circular plate is but a special case in which the orthotropy ratio $\beta \equiv E_\theta/E_r = 1$. For buckling into the first (axisymmetric) mode, Fig. 3.15 depicts the buckling parameter $k = \lambda_{cr}\dfrac{a^2 h}{D_r}$ as a function of $\sqrt{\beta}$ when the edge at $r = a$ is clamped; this graph corresponds to the relation (3.83); the corresponding (and far more complicated) initial buckling graph for buckling into the first (axisymmetric) mode when the edge at $r = a$ is simply supported is shown in Fig. 3.17 where, for various values of the parameter $(1-\sqrt{\beta})/(1+\sqrt{\beta})$, the variable $(1 - \nu_\theta)/(1 + \sqrt{\beta})$ has been graphed as a function of $\psi$, $\psi = \sqrt{(1 + \beta)k}$. For $\beta = 1$ ( the isotropic case) we obtain $1 - \nu_\theta$ as a function of $\sqrt{k}$ from Fig. 3.17. It is, of course, difficult to use Figs. 3.15 and 3.17 for comparative purposes; however, Table 3.2 clearly depicts the variation in the initial buckling load with the orthotropy ratio for both the case of a clamped and a simply supported cylindrically orthotropic plate; it is clear from this table that the buckling load increases as the orthotropy ratio $\beta$ increases, for both the clamped and the simply supported cases and that, for each fixed $\beta$, the buckling load in the clamped case is more than three times the buckling load for the simply supported case.

Two different types of postbuckling curves for a clamped, linear elastic, cylindrically orthotropic circular plate subjected to a uniform (radially) compressive load along its edge at $r = a$ are depicted in Fig. 3.20; these graphs show the postbuckling variations of transverse deflection and radial shortening, respectively, with respect to the applied compressive radial loading where the applied load has been normalized by the buckling load of a similar isotropic circular plate. The graphs in Fig. 3.20 correspond to orthotropy ratios of $\beta = 1, 4, 25$, and 64 and it follows, e.g., that the load vs. deflection curves become steeper as the orthotropy ratios increase, i.e., the plates with higher orthotropy ratios are stiffer with respect to their postbuckling response (in addition to having higher buckling loads). Figure 3.21 depicts the distributions of normalized radial and circumferential stresses for the clamped, cylindrically orthotropic circular plate while Fig. 3.22 depicts, for the same situation, the graphs of the normalized radial and circumferential bending moments; both membrane and bending stresses tend to increase as the degree $\beta$ of orthotropy increases. As we have indicated earlier, for the simply supported case, the qualitative nature of the load vs. deflection and load vs. radial shortening curves are similar to those for cylindrically orthotropic, clamped circular plates, e.g., increases in the orthotropy ratio result in an increase in the postbuckling stiffness and a reduction in the transverse deflection. One difference between the clamped and simply supported cases (see Fig. 3.23 is that the zone of normalized tensile radial stresses which surrounds the origin of the plate at $r = 0$, is much larger, for each orthotropy ratio $\beta$, for the simply supported cylindrically orthotropic circular plate than it is for the corresponding clamped plate. In Table 3.3, we indicate how the maximum displacement, i.e., $w(0)$, and the radial shortening of the circular plate, vary with increasing orthotropy ratio $\beta$ in both the clamped and simply supported cases; for both types of support the maximum deflection decreases and the radial shortening increases as $\beta$ increases.

**Remarks**: Although we have presented concrete analytical and graphical results for elastic-plastic plates, we will defer a discussion of comparative results for the initial and postbuckling behavior of such plates to Chapter 5, i.e., until we discuss the critical issue of imperfection sensitivity for such plates; such comparative results are already clearly present, however, in Figs. 3.26–3.29.

## 3.4 Initial Buckling/Postbuckling Figures, Graphs, and Tables: Circular Plates

**Table 3.1** Equations for Determining the Coefficients in the Postbuckling Expansions for a Circular Plate [148] (From Stoker, J.J., *Nonlinear Elasticity*, Gordon and Breach Publishers, N.Y., 1968. With permission)

|  | DE | BC, at $r = 0$ | BC, at $r = R$ |
|---|---|---|---|
| $L_0:$ | $Gp_0 = 0$ | $\dfrac{dp_0}{dr} = 0$ | $p_0 = \bar{p}_0$ |
| $L_1:$ | $\eta^2 Gq_1 + p_0 q_1 = 0$ | $\dfrac{dq_1}{dr} = 0$ | $B_\nu q_1 = 0$ |
| $L_2:$ | $Gp_2 = \dfrac{1}{2}q_1^2$ | $\dfrac{dp_2}{dr} = 0$ | $p_2 = \bar{p}_2$ |
| $L_3:$ | $\eta^2 Gq_3 + p_0 q_3 + p_2 q_1 = 0$ | $\dfrac{dq_3}{dr} = 0$ | $B_\nu q_3 = 0$ |
| etc. | ............. | ......... | ......... |

**Table 3.2  Linear Buckling Loads [160]**

| $\beta = E_\theta/E_r$ (1) | Clamped (2) | Simple (3) |
|---|---|---|
| 1 | 14.6819 | 4.1977 |
| 2 | 24.0349 | 7.1979 |
| 5 | 49.2942 | 15.5208 |
| 8 | 73.0251 | 23.4736 |

**Table 3.3** Maximum Displacement at Load $(\lambda/\lambda_{cr}) = 3$ [155] (From Sherbourne, A.N. and Panday, M.D., Postbuckling of Polar Orthotropic Circular Plates—Retrospective, *J. Eng. Mech. Div.*, ASCE, 118, 2087, 1992. Reproduced by permission of the publisher, ASCE.)

| $\beta$ | Clamped | | Simple | |
|---|---|---|---|---|
| | $w_0$ | $u_R$ | $w_0$ | $u_R$ |
| (1) | (2) | (3) | (4) | (5) |
| 1 | 3.4738 | 0.1733 | 2.84 | 0.0713 |
| 2 | 3.29 | 0.1723 | 2.7252 | 0.0954 |
| 5 | 3.1705 | 0.202 | 2.657 | 0.1327 |
| 8 | 3.14 | 0.23 | 2.6426 | 0.1583 |

**Table 3.4** Buckling Data for the Elastic-Plastic Circular Plate [141] (Reprinted from *Int. J. Mech. Sci.*, 17, Needleman, A., Postbifurcation behavior and imperfection sensitivity of elastic-plastic circular plates, 1–13, 1975, with permission from Elsevier Science.)

| $n$ | $v$ | $\rho$ | $\sigma_c/\sigma_y$ | Boundary condition * | $\beta$ | $\zeta_{max}/t$ | $\sigma_{max}/\sigma_c$ | $r_{max}/P_c$ |
|---|---|---|---|---|---|---|---|---|
| 12 | $\frac{1}{3}$ | 1.28 | 1.02 | S.S. | 1.999 | 0.002 | 1.002 | 1.0 |
| 12 | $\frac{1}{3}$ | 2.0 | 1.12 | cl. | 3.832 | 0.016 | 1.007 | 0.52 |
| 4 | $\frac{1}{3}$ | 2.0 | 1.33 | cl. | 3.832 | 0.019 | 1.010 | 0.52 |

* S.S., Simply supported; cl., clamped.

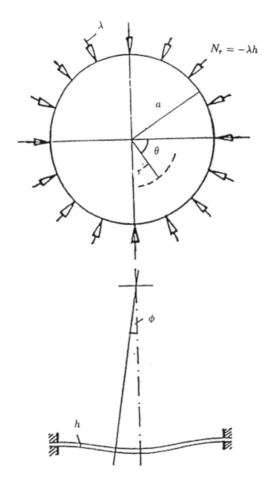

**FIGURE 3.1**
Circular plate in edge compression [32]. (From Bulson, P.S., *The Stability of Flat Plates*, American Elsevier, N.Y., 1969)

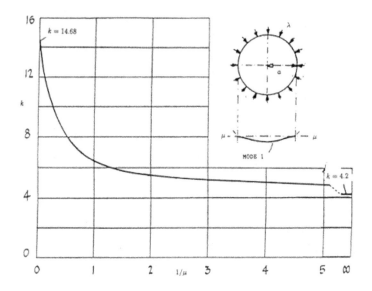

**FIGURE 3.2**
Buckling of circular plates with restrained edges. (Adopted, in modified form, from [146].)

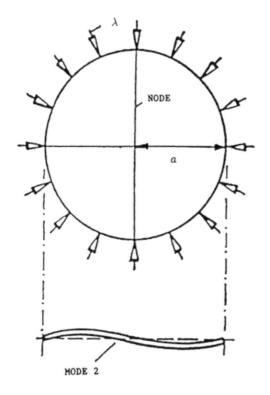

**FIGURE 3.3**
Second buckling mode of a circular plate [32]. (From Bulson,
P.S., *The Stability of Flat Plates*, American Elsevier, N.Y.,
1969)

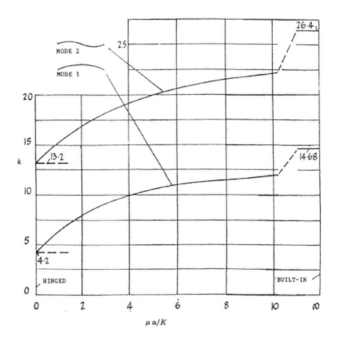

FIGURE 3.4
Buckling coefficient for circular plates [32]. (From Bulson, P.S., *The Stability of Flat Plates*, American Elsevier, N.Y., 1969)

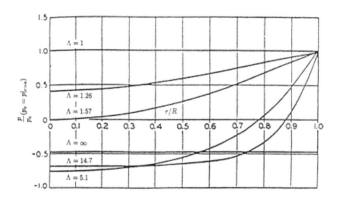

**FIGURE 3.5**
Radial membrane stresses. (Adopted, in modified form, from [143].)

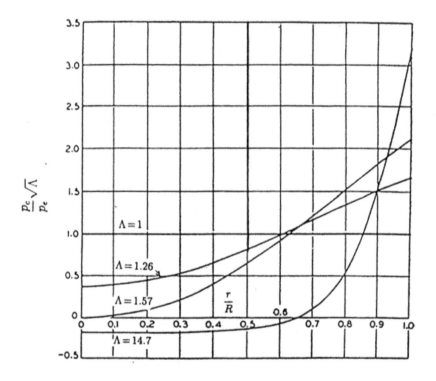

**FIGURE 3.6**
Circumferential membrane stress. (Adopted, in modified form, from [143].)

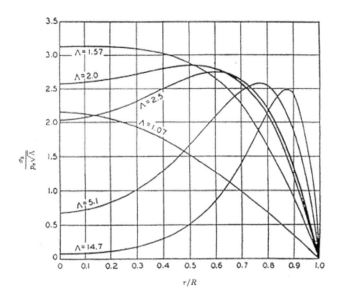

**FIGURE 3.7**
Radial bending stress.  (Adopted, in modified form, from [143].)

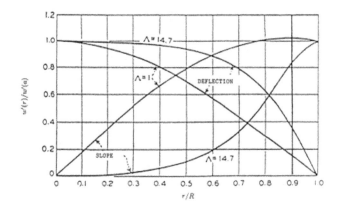

**FIGURE 3.8**
Slope and deflection. (Adopted, in modified form, from [143].)

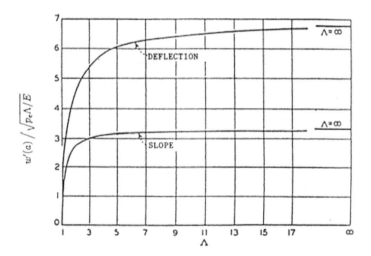

**FIGURE 3.9**
Slope and deflection. (Adopted, in modified form, from [143].)

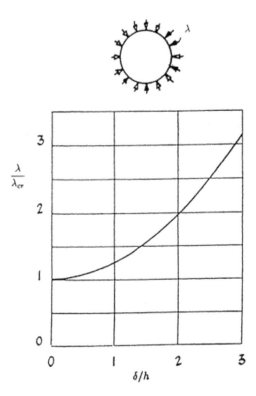

**FIGURE 3.10**
Large deflections of buckled circular plates. (Adopted, in modified form, from [147].)

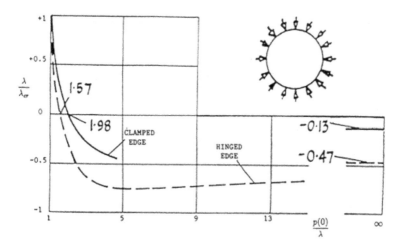

**FIGURE 3.11**
Radial membrane stresses for clamped and hinged circular
plates. (Adopted, in modified form, from [149].)

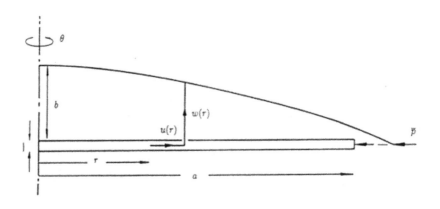

**FIGURE 3.12**
Notation for the circular plate [262] (Reprinted from *Int. J. Mech. Sci.*, **23**, Das, S., Note on Thermal Deflection of Regular Polygonal Viscoelastic Plates, 323-329, 1981, with permission from Elsevier Science.)

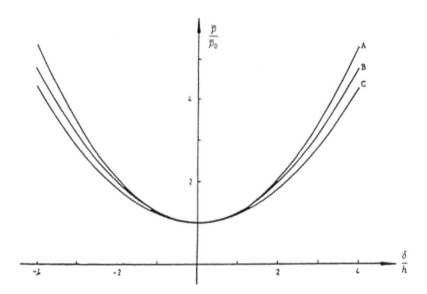

**FIGURE 3.13**
Postbuckling solutions: A, based on (3.55) for simple supports; B, Friedrichs and Stoker [143] with simple supports; C, based on (3.58) with clamped boundary conditions.

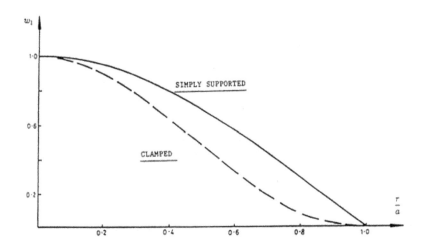

**FIGURE 3.14**
Form of the buckling mode $w_1(r)$. (Adopted, in modified form, from [143].)

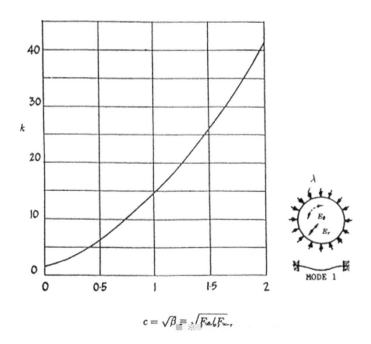

$$c = \sqrt{\beta} = \sqrt{F_{ab} F_{...r}}$$

**FIGURE 3.15**
Buckling of circular plate with cylindrical orthotropy. (Adopted, in modified form, from [151].)

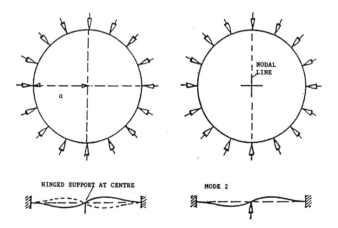

**FIGURE 3.16**
Clamped circular plates with anti-symmetric buckling modes
[32]. (From Bulson, P.S., *The Stability of Flat Plates*, American Elsevier, N.Y., 1969)

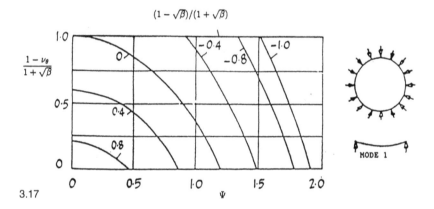

**FIGURE 3.17**
Simply supported circular plate buckling in first mode.
(Adopted, in modified form, from [151].)

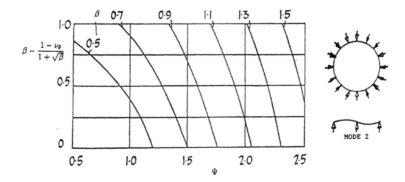

**FIGURE 3.18**
Simply supported circular plate buckling in anti-symmetric mode. (Adopted, in modified form, from [151].)

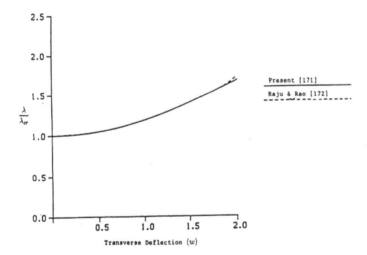

**FIGURE 3.19**
Comparative study: clamped plate $(\beta = 2)$ [155]. (From Sherbourne, A.N. and Panday, M.D., Postbuckling of Polar Orthotropic Circular Plates - Retrospective, *J. Eng. Mech. Div.*, ASCE, 118, 2087, 1992. Reproduced by permission of the publisher, ASCE.)

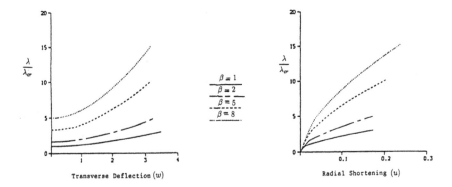

**FIGURE 3.20**
Postbuckling behavior of clamped plates: (a) load vs. transverse deflection (b) load vs. radial displacement [155]. (From Sherbourne, A.N. and Panday, M.D., Postbuckling of Polar Orthotropic Circular Plates - Retrospective, *J. Eng. Mech. Div.*, ASCE, 118, 2087, 1992. Reproduced by permission of the publisher, ASCE.)

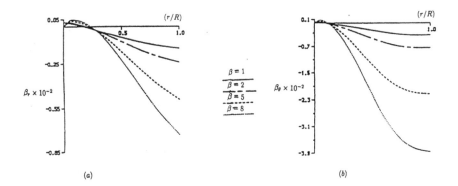

**FIGURE 3.21**
Membrane stresses in clamped plates ($\Lambda = 3$): (a) normalized radial stress and (b) normalized circumferential stress [155]. (From Sherbourne, A.N. and Panday, M.D., Postbuckling of Polar Orthotropic Circular Plates - Retrospective, *J. Eng. Mech. Div.*, ASCE, 118, 2087, 1992. Reproduced by permission of the publisher, ASCE.)

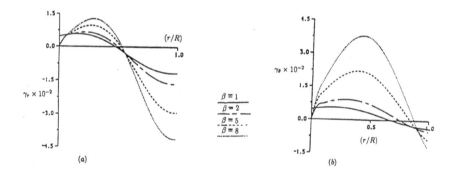

**FIGURE 3.22**
Bending moments in clamped plates ($\Lambda = 3$): (a) normalized radial moment and (b) normalized circumferential moment [155]. (From Sherbourne, A.N. and Panday, M.D., Postbuckling of Polar Orthotropic Circular Plates - Retrospective, *J. Eng. Mech. Div.*, ASCE, 118, 2087, 1992. Reproduced by permission of the publisher, ASCE.)

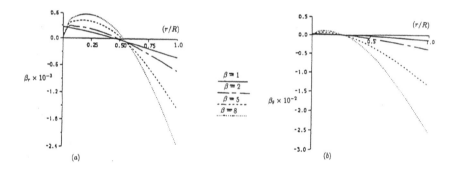

**FIGURE 3.23**
Membrane stresses in simply supported plates ($\Lambda = 3$): (a) normalized radial stress and (b) normalized circumferential stress [155]. (From Sherbourne, A.N. and Panday, M.D., Post-buckling of Polar Orthotropic Circular Plates - Retrospective, *J. Eng. Mech. Div.*, ASCE, 118, 2087, 1992. Reproduced by permission of the publisher, ASCE.)

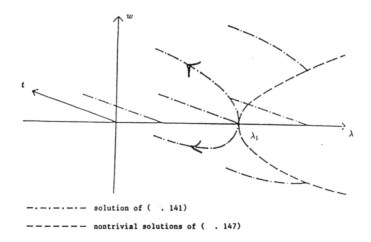

—·—·—·—·—  solution of ( . 141)

—————— nontrivial solutions of ( . 147)

**FIGURE 3.24**
Postbuckling behavior of a circular viscoelastic plate [140].
(Reprinted from *Mech. Res. Comm.*, 17, Brilla, J., Postbuckling Analysis of Circular Viscoelastic Plates, 263-269, 1990,
with permission from Elsevier Science.)

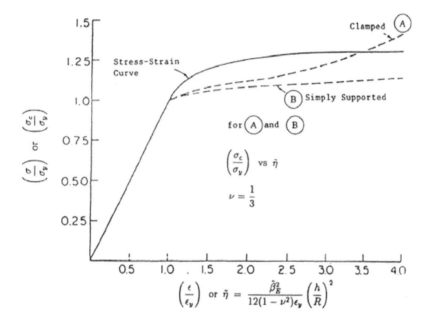

**FIGURE 3.25**
Uniaxial stress-strain curve and bifurcation stresses for clamped and simply supported circular plates ($n = 12$) [141]. (Reprinted from *Int. J. Mech. Sci.*, 17, Needleman, A., Postbifurcation Behavior and Imperfection Sensitivity of Elastic-Plastic Circular Plates, 1-13, 1975, with permission from Elsevier Science.)

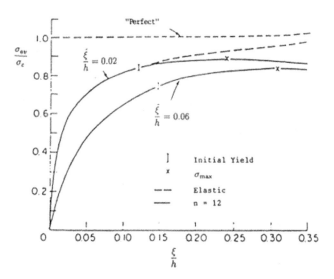

**FIGURE 3.26**
Curves of average applied stress vs. buckling deflection for
a simply supported circular plate $\left(\nu = \frac{1}{3}, \ \sigma_c/\sigma_y = 0.72\right)$ [141].
(Reprinted from *Int. J. Mech. Sci.*, 17, Needleman, A., Postb-
ifurcation Behavior and Imperfection Sensitivity of Elastic-
Plastic Circular Plates, 1-13, 1975, with permission from Elsevier Science.)

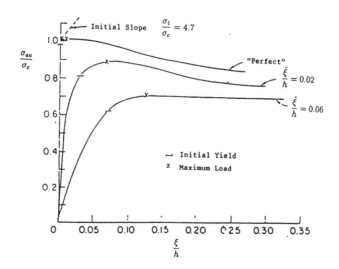

**FIGURE 3.27**
Curves of average applied stress vs. buckling deflection for a simply supported circular plate ($\nu = \frac{1}{3}, \bar{\eta} = 1.28$, $\sigma_c/\sigma_y = 1.02$) [141]. (Reprinted from *Int. J. Mech. Sci.*, 17, Needleman, A., Postbifurcation Behavior and Imperfection Sensitivity of Elastic-Plastic Circular Plates, 1-13, 1975, with permission from Elsevier Science.)

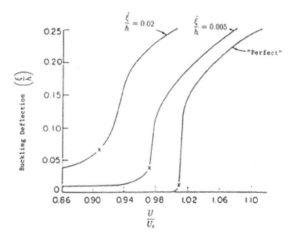

**FIGURE 3.28**
Curves of buckling deflection vs. applied displacement
for a simply supported circular plate ($\nu = \frac{1}{3}$, $n = 12$,
$\bar{\eta} = 1.28$, $\sigma_c/\sigma_y = 1.02$) [141]. (Reprinted from *Int. J. Mech.
Sci.*, 17, Needleman, A., Postbifurcation Behavior and Imper-
fection Sensitivity of Elastic-Plastic Circular Plates, 1-13, 1975,
with permission from Elsevier Science.)

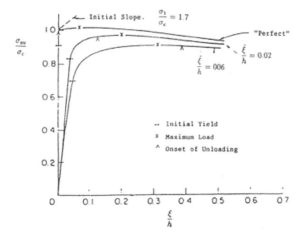

**FIGURE 3.29**
Curves of average applied stress vs. buckling deflection for a clamped circular plate ($\nu = \frac{1}{3}$, $n = 12$, $\bar{\eta} = 2.0, \sigma_c/\sigma_y = 1.12$) [141]. (Reprinted from *Int. J. Mech. Sci.*, 17, Needleman, A., Postbifurcation Behavior and Imperfection Sensitivity of Elastic-Plastic Circular Plates, 1-13, 1975, with permission from Elsevier Science.)

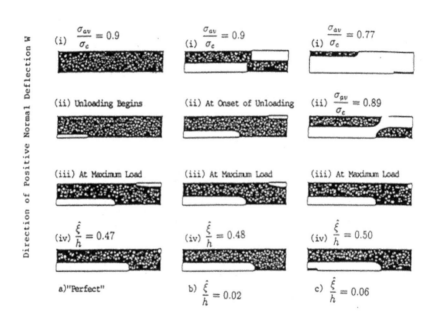

**FIGURE 3.30**
Development of plastic zone (shaded area) for a clamped circular plate ($\nu = \frac{1}{3}$, $n = 12$, $\tilde{\eta} = 2.0$, $\sigma_c/\sigma_y = 1.12$) [141]. (Reprinted from *Int. J. Mech. Sci.*, 17, Needleman, A., Postbifurcation Behavior and Imperfection Sensitivity of Elastic-Plastic Circular Plates, 1-13, 1975, with permission from Elsevier Science)

# CHAPTER 4

# Initial and Postbuckling Behavior of (Perfect) Thin Annular Plates

In this chapter, we present some of the basic results which pertain to the buckling and postbuckling behavior of annular plates. We begin by looking at annular plates which exhibit linear elastic behavior, treating first the problems of initial buckling for plates which possess either isotropic or cylindrically orthotropic symmetry. Next, we present an analysis of the postbuckling behavior of isotropic and cylindrically orthotropic linear elastic annular plates; this is followed by a discussion of the nature of the prebuckling stress distributions and a study of the initial buckling problem for a linearly elastic annular plate which exhibits rectilinearly orthotropic material symmetry. As was the case for buckling and postbuckling of circular plates possessing rectilinear orthotropic symmetry, the corresponding problem for annular plates is once again complicated by the mismatch between the geometry of the plate and its material symmetry behavior; indeed, for the linearly elastic annular plate, the problem of delineating the postbuckling behavior for the case where the plate exhibits rectilinear orthotropic symmetry appears to have not been treated at all. For plates which are annular and exhibit either nonlinear elastic, viscoelastic, or elastic-plastic behavior, very few results appear to exist in the literature; those results which do exist for annular plates which do not exhibit linear elastic response are presented in §4.2. In §4.3 we offer a comparison of the initial and postbuckling behavior of annular plates which results from varying the support conditions along the edges of the plate (inner and outer), the type of material symmetry, and the nature of the constitutive response.

## 4.1  Plates with Linear Elastic Behavior

Annular plates exhibiting linear elastic behavior may be classified according to whether the plate possesses isotropic symmetry, cylindrically orthotropic symmetry, or rectilinear orthotropic symmetry.

### 4.1.1  Isotropic Symmetry: Initial Buckling

The von Karman equations for a thin, linearly elastic, isotropic plate in polar coordinates are given by (1.78) and (1.79); we consider, in this chapter, an annular plate with inner radius $r = b$ and outer radius $r = a$. Initial buckling of the plate is governed by (1.78), which may be rewritten in the form

$$K\nabla^4 w = N_r w_{,rr} - 2N_{r\theta}\left(\frac{1}{r^2}w_{,\theta} - \frac{1}{r}w_{,r\theta}\right)$$

$$+N_\theta\left(\frac{1}{r}w_{,r} + \frac{1}{r^2}w_{,\theta\theta}\right)$$

with $N_r = \sigma_{rr}\cdot h$, $N_\theta = \sigma_{\theta\theta}\cdot h$, and $N_{r\theta} = \sigma_{r\theta}\cdot h$; this last equation is, of course, equivalent to (1.83) and $\nabla^4 w$ is given by

$$\nabla^4 w = \left(\frac{\partial^2}{\partial r^2} + \frac{1}{r}\frac{\partial}{\partial r} + \frac{1}{r^2}\frac{\partial^2}{\partial\theta^2}\right)\left(\frac{\partial^2}{\partial r^2} + \frac{1}{r}\frac{\partial}{\partial r} + \frac{1}{r^2}\frac{\partial^2}{\partial\theta^2}\right)w \quad (4.1)$$

In all of the problems we will consider in this subsection, the prebuckling stress distribution in the plate will satisfy $\sigma_{r\theta} = 0$ and, thus, in lieu of (1.83), we may write

$$K\nabla^4 w = h\sigma_{rr}w_{,rr} + h\sigma_{\theta\theta}\left(\frac{1}{r}w_{,r} + \frac{1}{r^2}w_{,\theta\theta}\right) \quad (4.2)$$

We begin with the case of an annular plate which is subject to uniform compression along both its inner and outer edges at $r = b$ and $r = a$, respectively. Denoting the edge stresses $\sigma_{rr}|_{r=a} = \sigma_{rr}|_{r=b}$ by $-\lambda, \lambda > 0$, we then have $\sigma_{rr} = \sigma_{\theta\theta} = -\lambda$ for the prebuckling stress distribution, in which case (4.2) reduces to

$$\nabla^4 w + \frac{\lambda h}{K}\nabla^2 w = 0 \quad (4.3)$$

where $\nabla^2 = \dfrac{\partial}{\partial r^2} + \dfrac{1}{r}\dfrac{\partial}{\partial r} + \dfrac{1}{r^2}\dfrac{\partial^2}{\partial\theta^2}$. In what follows, we will denote the aspect ratio of the annular plate by $\gamma = b/a$ and will employ the normalized radial coordinate $\rho = r/a$. We seek solutions of (4.3) in the form

$$w(\rho, \theta) = u_n(\rho) \cos n\theta, \qquad (4.4)$$

where $\gamma < \rho < 1$ and $0 \le \theta < 2\pi$. The index $n$ serves to delineate the number of diametric nodes in the buckled configuration of the annular plate. Substituting (4.4) into (4.3) we obtain, for $u_n(\rho)$, the equation

$$\nabla_n^4 u_n + \frac{\lambda h a^2}{K} \nabla_n^2 u_n = 0 \qquad (4.5)$$

where

$$\nabla_n^2 = \frac{d^2}{d\rho^2} + \frac{1}{\rho} \frac{d}{d\rho} - \frac{n^2}{\rho^2} \qquad (4.6)$$

The solutions to (4.5) may be expressed in terms of Bessel functions and then studied for a variety of support conditions along the edges at $r = b$ and $r = a$; the loading condition of uniform compression along both edges has been studied by a number of authors, but the most general treatment appears to belong to Yamaki [162], whose results we present below.

**(i) The edges at $r = a$ and $r = b$ are clamped.**

In this case, the boundary conditions are that $w = 0$ and $\dfrac{\partial w}{\partial \rho} = 0$, at both $r = a$ and $r = b$. The general solution of (4.5) has the form

$$u_0 = A_0 J_0 \left[ \left( \frac{\lambda h r^2}{K} \right)^{\frac{1}{2}} \right] + B_0 Y_0 \left[ \left( \frac{\lambda h r^2}{K} \right)^{\frac{1}{2}} \right] \qquad (4.7)$$

$$+ C_0 + D_0 \, ln\rho \quad ( \text{ for } n = 0), \ \rho = \frac{r}{a}$$

and

$$u_n = A_n J_n \left[ \left( \frac{\lambda h r^2}{K} \right)^{\frac{1}{2}} \right] + B_n Y_n \left[ \left( \frac{\lambda h r^2}{K} \right)^{\frac{1}{2}} \right] \qquad (4.8)$$

$$+ C_n \rho^n + D_n \rho^{-n} \quad ( \text{ for } n = 1, 2, \cdots), \ \rho = \frac{r}{a}$$

with the $A_0, B_0, A_n, B_n$, etc., arbitrary constants of integration and $J_n, Y_n$, respectively, the Bessel functions of the first and second kinds of order $n$.

By imposing the clamped edge boundary conditions on the general solution (4.7), (4.8) we obtain, for each fixed $n$, four homogeneous linear equations in the $A_n, B_n, C_n$, and $D_n$; the vanishing of the determinant of the matrix of coefficients of this system then provides the condition for determining the eigenvalues of the boundary value problem associated with (4.5) and, hence, the critical edge stress $\lambda_{cr}$. The solution of the initial buckling problem for this case may be displayed in graphical form as in Fig. 4.1, which is taken from Yamaki [161]; the situation in which both edges are clamped is represented by the curve $(i)$ where the buckling parameter $k$ is defined by $\lambda_{cr} = kK/a^2h$. In earlier work on the annular plate problem in which both edges were subject to uniform compression (i.e., Olsson [163] , Shubert [164]) only axisymmetric buckling was studied but, as we will see later in this subsection, the axisymmetric buckled state is not always necessarily associated with the lowest critical load. It should be noted from Fig. 4.1 that the number of diametric nodes increases as the hole diameter increases. As has been noted in [162], it can be shown that as $\dfrac{b}{a} \to 1$, the values of the buckling parameter $k$ approach those associated with a clamped infinite strip subject to uniform compression.

**(ii) The edge at $r = a$ is clamped, while the edge at $r = b$ is simply supported.**
In this case, as we have demonstrated in Chapter 1, the boundary conditions are $w = 0$, $\dfrac{\partial w}{\partial \rho} = 0$, at $r = a$, while $w = 0$ and

$$\frac{\partial^2 w}{\partial \rho^2} + \frac{\nu}{\rho}\left(\frac{\partial w}{\partial \rho} + \frac{1}{\rho}\frac{\partial^2 w}{\partial \theta^2}\right) = 0$$

at $r = b$. The initial buckling behavior is shown as curve (ii) in Fig. 4.1, which indicates that the annular plate will first buckle with one diametric note but that, as the hole size increases, this buckling mode changes to the one $(n = 0)$ which reflects axisymmetric buckling of the plate.

**(iii) The edge at $r = a$ is clamped, while the edge at $r = b$ is free.**
Once again, we have $w = 0$, $\dfrac{\partial w}{\partial \rho} = 0$, at $r = a$ while, as shown in Chapter 1, at the free edge at $r = b$ we require that

$$\frac{\partial^2 w}{\partial \rho^2} + \frac{\nu}{\rho}\left(\frac{\partial w}{\partial \rho} + \frac{1}{\rho}\frac{\partial^2 w}{\partial \theta^2}\right) = 0$$

and

$$\frac{\partial}{\partial\rho}\left(\frac{\partial^2 w}{\partial\rho^2} + \frac{1}{\rho}\frac{\partial w}{\partial\rho} + \frac{1}{\rho^2}\frac{\partial^2 w}{\partial\theta^2}\right)$$

$$+\frac{(1-\nu)}{\rho}\frac{\partial}{\partial\theta}\left(\frac{\partial}{\partial\rho}\left(\frac{1}{\rho}\frac{\partial w}{\partial\theta}\right)\right) + K\frac{\partial w}{\partial\rho} = 0$$

the second condition indicating that the edge shearing force at $r = b$ is equal to the normal component of the applied load at $r = b$. The initial buckling behavior is shown as curve (iii) in Fig. 4.1 and the annular isotropic plate, in this case, always buckles into the axisymmetric form $(n = 0)$ regardless of the hole size.

**(iv)  The edge at $r = a$ is simply supported, while the edge at $r = b$ is clamped.**
This case is the same as case (ii) but the roles of the edges at $r = a$, $r = b$ have been exchanged with respect to the boundary support conditions. The initial buckling behavior is described by curve (iv) in Fig. 4.2 and the annular plate first buckles into a mode with one diametric node but then changes to the axisymmetric mode as the hole size increases. If we compare the graphs for cases (ii) and (iv) we see that they are very similar but that the values of the buckling parameter $k$ for a fixed value of $b/a$ are slightly lower in case (iv).

**(v)  The edges at $r = a$ and $r = b$ are both simply supported.**
In this situation, the boundary conditions are just

$$\begin{cases} w = 0 \\ \dfrac{\partial^2 w}{\partial\rho^2} + \dfrac{\nu}{\rho}\left(\dfrac{\partial w}{\partial\rho} + \dfrac{1}{\rho}\dfrac{\partial^2 w}{\partial\theta^2}\right) = 0 \end{cases} \quad \text{at } r = a, \ r = b$$

The graph depicting the initial buckling behavior for this case is shown as curve (v) in Fig. 4.2; as $\dfrac{b}{a} \to 0$, $\sqrt{k} \to 3.625$.

**(vi)  The edge at $r = a$ is simply supported, while the edge at $r = b$ is free.**
The conditions at $r = a$ are delineated in case (v) above, while the conditions that the edge at $r = b$ be free are delineated in case (iii) above. The initial buckling behavior in this case is shown as curve (vi) in Fig. 4.2; the plate always buckles into the axisymmetric form and the buckling stress decreases as the hole size increases. This decrease of buckling stress with an increasing hole size does not happen for any

other case considered here except case (viii), which deals with the same boundary support conditions but with the roles of the edges at $r = a$ and $r = b$ interchanged.

**(vii)    The edge at $r = a$ is free and the edge at $r = b$ is clamped.**

The boundary support conditions here are the same as for case (iii) but with the roles of the edges at $r = a$ and $r = b$ interchanged. The initial buckling curve is shown as curve (vii) in Fig. 4.1 and the results indicate a behavior which differentiates this case from the others we have considered: as the hole size is increased, first one diametric node appears in the buckling mode, then two nodes, and again, one. Finally, at yet larger hole sizes, the annular isotropic plate buckles into the axisymmetric mode with $n = 0$.

**(viii)    The edge at $r = a$ is free, while the edge at $r = b$ is simply supported.**

At $r = b$, the boundary support conditions are those delineated in case (v), while at $r = a$ they are the conditions delineated in case (iii). The initial buckling curve is depicted as curve (viii) in Fig. 4.1 so that the initial buckling behavior in this situation is similar to that for case (vi).

Cases (i)–(viii), considered above, correspond to the situation in which the isotropic annular plate is subject to uniform compression along both its inner and outer edges; the basic reference, as noted earlier, is the work of Yamaki [161]. The next basic situation concerns the initial buckling behavior of an isotropic annular (linear elastic) plate which is subjected to compression or tension along its inner edge at $r = b$; the basic reference for this situation is the paper of Mansfield [165]. Because the solution in [165] is drawn from the results for an infinite plate supported along two concentric circles and subjected to radial stresses along the inner circle, it is applicable only if the outer edge is supported by a member possessing the required tensile stiffness. Mansfield [162] has noted that, when a uniform radial compressive stress is applied to the circumference of a circular hole in an infinite plate, the resulting radial stresses in the plate decay inversely as the square of the distance from the center of the circle; there are also tensile hoop stresses of the same magnitude and varying in the same way. An identical stress distribution exists in a similarly loaded annular plate of thickness $h$, say, which is bounded by circles of radii $b$ and $a$, if the outer circle is supported by a member of sectional area $ah/(1 + \nu)$. Mansfield, [165], also notes that such an annular plate will buckle as soon as the applied radial stress achieves a critical value which depends on the boundary support conditions; if the radial stress is compressive, the plate will buckle into an axially symmetric mode, while, if the radial stress is tensile, the plate will buckle into a number of circumferential waves because the hoop stresses are now compressive. The situation considered in [165] is depicted in Fig. 4.3. If we denote the hoop stress by $\sigma_{\theta\theta}$, then the equation of equilibrium for the infinite plate follows from (4.2), where it is now convenient to replace $\sigma_{rr} \to -\sigma_{rr}$ and $\sigma_{\theta\theta} \to -\sigma_{\theta\theta}$ so that

$$K\nabla^4 w + h\sigma_{rr} w_{,rr} + h\sigma_{\theta\theta}\left(\frac{1}{r}w_{,r} + \frac{1}{r^2}w_{,\theta\theta}\right) = 0 \qquad (4.9)$$

If we denote the radial stress (at $r = b$), which causes buckling, by $\lambda_{cr}$, then the radial and hoop stresses in the plate are of the same magnitude and are given by

$$\sigma_{rr} = -\sigma_{\theta\theta} = \left(\frac{b}{r}\right)^2 \lambda_{cr} \qquad (4.10)$$

Thus, as already noted above, these stresses decay inversely as the square of the distance $r$ from the center of the circle of radius $b$ and the same stress condition obtains if an outer circle of radius $a > b$ in the infinite plate is bounded by a member of sectional area equal to $ah/(1+\nu)$.

Using the stress distribution (4.10), we rewrite (4.9) in the form

$$\left(\frac{\partial^2}{\partial r^2} + \frac{1}{r}\frac{\partial}{\partial r} + \frac{1}{r^2}\frac{\partial^2}{\partial\theta^2}\right)\left(\frac{\partial^2 w}{\partial r^2} + \frac{1}{r}\frac{\partial w}{\partial r} + \frac{1}{r^2}\frac{\partial^2 w}{\partial\theta^2}\right) \qquad (4.11)$$

$$+\frac{k}{r^2}\left(\frac{\partial^2 w}{\partial r^2} - \frac{1}{r}\frac{\partial w}{\partial r} - \frac{1}{r^2}\frac{\partial^2 w}{\partial\theta^2}\right) = 0,$$

with $k = \dfrac{\lambda_{cr}b^2 h}{K}$, and again seek solutions of the form

$$w(r,\theta) = f(r)\cos n\theta \qquad (4.12)$$

with $f(r)$ the transverse deflection function and $n$ the number of diametric nodes; this problem has been solved by Mansfield, [165], for a number of boundary support conditions. When $\lambda_{cr}$ is compressive, the buckling mode, as we have already noted, is axially symmetric; the results obtained in [165] for this subcase are depicted in Fig. 4.4 with $k(1-\gamma)^2/\gamma$ graphed against $\dfrac{1}{\gamma}$ for various combinations of boundary support conditions at $r = a$ and $r = b$. Solutions of the initial buckling problem are also presented in [165] for the case where buckling occurs as a result of tensile stresses along the edge at $r = b$; such solutions are derived for those situations in which the edges at $r = a$ and $r = b$ are either both clamped or both simply supported. When buckling occurs because of tensile stresses applied along the edge at $r = b$, a number of diametric nodes are formed which decrease in number as the hole size is decreased; this behavior is depicted in Fig. 4.5. It is shown in [165] that, as $a \to \infty$, and $\gamma \to 0$, the compressive buckling stress for axially symmetric buckling converges to $K/b^2h$, i.e., $k \to 1$, provided we have zero radial slope at the inner edge at $r = b$; furthermore, the tensile buckling stress $\lambda_{cr} \to -3K/b^2h$. Therefore, the tensile buckling stress associated with radial pressure applied along the edge of a hole in an infinite plate is three times the compressive buckling stress.

**Remarks:** Using (4.11), (4.12), it may be shown that the equation governing the transverse deflection $f(r)$ has the form

$$\frac{d^4 f}{dr^4} + \frac{2}{r}\frac{d^3 f}{dr^3} - \frac{\left(1 + 2n^2 - \gamma\right)}{r^2}\frac{d^2 f}{dr^2}$$

$$+ \frac{\left(1 + 2n^2 - \gamma\right)}{r^3}\frac{df}{dr} - \frac{n^2 \left(4 - n^2 - \gamma\right)}{r^4} f = 0$$

(4.13)

and the particular case of buckling with axial symmetry is then obtained by setting $n = 0$ in (4.13). Solutions of (4.13), which is homogeneous in $r$, may be expressed in the form

$$f(r) = \sum_{i=1}^{r} A_i \rho^{1+\beta_i}$$

(4.14)

where $\rho = r/a$ and

$$\beta_i^2 = 1 + n^2 - \frac{1}{2}\gamma \pm \left(4n^2 - 2\gamma n^2 + \frac{1}{4}\gamma^2\right)^{\frac{1}{2}}$$

(4.15)

When $\lambda_{cr}$ is compressive, so that buckling with axial symmetry occurs, $f(r)$ assumes the form

$$f(r) = \alpha_1 + \alpha_2 \rho^2 + \alpha_3 \rho \sin\left(\beta_i ln\rho\right) + \alpha_4 \rho \cos\left(\beta_i ln\rho\right)$$

(4.16)

with $\beta_i$ given by (4.15) with $n = 0$. Solutions for particular cases are then determined by imposing the boundary support conditions at $r = a$, $r = b$, upon (4.16).

The last major case associated with the buckling of a linearly elastic isotropic circular annular plate concerns the situation in which a uniform radial compressive force is applied only at the outside edge at $r = a$; this problem has been studied, first, by Majumdar [166] and, later, by Troger and Steindl [66] and Machinek and Troger [167] with the outer edge of the plate clamped and the inner edge free. The relevant equilibrium equation is (1.78), with $N_{r\theta} = 0$, or

$$K\nabla^4 w - N_r w_{,rr} - N_\theta \left( \frac{1}{r} w_{,r} + \frac{1}{r^2} w_{,\theta\theta} \right) = 0 \qquad (4.17)$$

$N_r$ and $N_\theta$ being, as usual, the prebuckling (averaged) membrane stresses; the boundary conditions are just (1.139) with $i = 1$, and $R_1 = a$, and (1.140) and (1.141), with $i = 2$ and $R_2 = b$, i.e.

$$w = w_{,r} = 0, \quad \text{at } R = a \qquad (1.139')$$

and, at $r = b$,

$$\frac{\partial^2 w}{\partial r^2} + \nu \left( \frac{1}{4} \frac{\partial w}{\partial r} + \frac{1}{r^2} \frac{\partial^2 w}{\partial \theta^2} \right) = 0 \qquad (1.140')$$

$$\frac{\partial}{\partial r} \left( \frac{\partial^2 w}{\partial r^2} + \frac{1}{r} \frac{\partial w}{\partial r} + \frac{1}{r^2} \frac{\partial^2 w}{\partial \theta^2} \right)$$
$$+ \frac{1-\nu}{r} \frac{\partial}{\partial \theta} \left( \frac{1}{r} \frac{\partial^2 w}{\partial r \partial \theta} - \frac{1}{r^2} \frac{\partial w}{\partial \theta} \right) = 0 \qquad (1.141')$$

If we set

$$N_r|_{r=a} = \sigma_{rr}|_{r=a} h \equiv -N \, (= -\lambda h) \qquad (4.18)$$

then the prebuckling stress distribution in the annular circular plate is given by

$$N_r^0 = -N \frac{a^2}{a^2 - b^2} \left( 1 - \frac{b^2}{r^2} \right) \qquad (4.19a)$$

$$N_\theta^0 = -N \frac{a^2}{a^2 - b^2} \left( 1 + \frac{b^2}{r^2} \right) \qquad (4.19b)$$

and, of course, $N_{r\theta} = 0$. By substituting from (4.19 a,b) into (4.17) we obtain the equation which governs the initial buckling of the plate, namely,

$$\nabla^4 w + \mu \left[ \frac{1}{b^2} + \frac{1}{r^2} \right] \nabla^2 w - \frac{2\mu}{r^2} w_{,rr} = 0 \qquad (4.20)$$

where

$$\mu = \frac{Na^2b^2}{K\,(a^2 - b^2)} \tag{4.21}$$

and (4.20) is subject to the boundary conditions delineated above, i.e., (1.139′), (1.140′), (1.141′).

The critical load which is obtained by solving the eigenvalue problem (4.20), (4.21), (1.139′), (1.140′), and (1.141′) will be written in the form

$$N_{cr} = k\,(K/a^2) \tag{4.22}$$

with the buckling parameter $k$ related to the eigenvalue $\mu$ through the relation

$$k = \mu \left[ \frac{a^2}{b^2} - 1 \right] \tag{4.23}$$

Solutions of (4.20), (4.21) are now sought in the form

$$w(r, \theta) = A_n(r) \cos n\theta, \ n = 0, \ 1, \ 2 \cdots \tag{4.24}$$

in which case, substitution from (4.24) into (4.20), (4.21), and use of the transformation $\rho = \mu^{\frac{1}{2}} \left( \frac{r}{b} \right)$, yields the ordinary differential equations

$$\frac{d^4 A_n}{d\rho^4} + \frac{2}{\rho} \frac{d^3 A_n}{d\rho^3} - \left[ \frac{1 + 2n^2 + \mu}{\rho^2} - 1 \right] \frac{d^2 A_n}{d\rho^2}$$
$$+ \left[ \frac{1 + 2n^2 + \mu}{\rho^3} + \frac{1}{\rho} \right] \frac{dA_n}{d\rho} \tag{4.25}$$
$$+ \left[ \frac{n^2(n^2 - \mu - 4)}{\rho^4} - \frac{n^2}{\rho^2} \right] A_n = 0, \ n = 0, \ 1, \ 2, \cdots$$

For $n = 0$ the problem corresponds, of course, to an axially symmetric deformation which is then governed by the solutions of

$$\frac{d^3 \Psi}{d\rho^2} + \frac{2}{\rho} \frac{d^2 \Psi}{d\rho^2} + \left[ 1 - \frac{1 + \mu}{\rho^2} \right] \frac{d\Psi}{d\rho} + \left[ \frac{1 + \mu}{\rho^3} + \frac{1}{\rho} \right] \Psi = 0 \tag{4.26}$$

where $\Psi = \dfrac{dA_0}{d\rho}$. In terms of the variables $\Psi$ and $\rho$, equation (4.26) is subject to the boundary conditions

$$
\begin{cases}
\Psi = 0, \quad \text{at } \rho = \mu^{\frac{1}{2}}\left(\dfrac{a}{b}\right) \\[2mm]
\dfrac{d\Psi}{d\rho} + \left(\dfrac{\nu}{\rho}\right)\Psi = 0, \quad \text{at } \rho = \mu^{\frac{1}{2}} \\[2mm]
\dfrac{d}{d\rho}\left(\dfrac{d\Psi}{d\rho} + \dfrac{1}{\rho}\Psi\right) = 0, \quad \text{at } \rho = \mu^{\frac{1}{2}}
\end{cases}
\tag{4.27}
$$

A solution of (4.26), satisfying the last of the boundary conditions in (4.27), may be written in the form

$$
\Psi(\rho) = A J_p(\rho) + B J_{-p}(\rho) \tag{4.28}
$$

where $p = (\mu+1)^{\frac{1}{2}}$, $A$ and $B$ are arbitrary constants, and $J_p(\rho), J_{-p}(\rho)$ are Bessel functions of the first kind. The imposition of the first two boundary conditions in (4.27) then yields a determinental equation for $\mu$ and the corresponding buckling parameters $k$, when graphed against the aspect ratio $b/a$, provide the initial buckling curve depicted in Fig. 4.6. For the case in which $n = 1$, so that there is one wave around the circumference of the annular plate, equation (4.25) yields

$$
\frac{d^4 A_1}{d\rho^4} + \frac{2}{\rho}\frac{d^3 A_1}{d\rho^3} + \left[1 - \frac{3+\mu}{\rho^2}\right]\frac{d^2 A_1}{d\rho^2}
$$
$$
+ \left[\frac{1}{\rho} + \frac{3+\mu}{\rho^3}\right]\frac{dA_1}{d\rho} - \left[\frac{1}{\rho^2} + \frac{3+\mu}{\rho^4}\right]A_1 = 0
\tag{4.29}
$$

By using the transformations $A_1 = \rho\phi$ and $\Psi = \dfrac{d\phi}{d\rho}$, (4.29) may be reduced to

$$
\frac{d^2\Psi}{d\rho^2} + \frac{3}{\rho}\frac{d\Psi}{C\rho} + \left[1 + \frac{1-p^2}{\rho^2}\right]\Psi = \frac{C}{\rho^3} \tag{4.30}
$$

where $p^2 = \mu + 4$, and $C$ is an arbitrary constant. For the general solution of (4.30) we set

$$
S_{-1,p}(\rho) = \frac{\pi}{2\sin(p\pi)}\left[J_p(\rho)\int\frac{J_{-p}(\rho)}{\rho}d\rho - J_{-p}(\rho)\int\frac{J_p(\rho)}{\rho}d\rho\right] \tag{4.31}
$$

Then,

$$
\Psi(\rho) = \frac{1}{\rho}\left(A J_p(\rho) + B J_{-p}(\rho) + C S_{-1,p}(\rho)\right) \tag{4.32}
$$

with $A$ and $B$ arbitrary constants. If we rewrite $\Psi(\rho)$ in terms of $A_1(\rho)$, in (4.32), we find that

$$\rho \frac{dA_1}{d\rho} - A_1 = \rho\left[AJ_p(\rho) + BJ_{-p}(\rho) + CS_{-1,p}(\rho)\right] \tag{4.33}$$

and, in terms of the new variables, the boundary data becomes

$$
\begin{cases}
A_1 = \dfrac{dA_1}{d\rho} = 0, \quad \text{at } \rho = \mu^{\frac{1}{2}}\left(\dfrac{a}{b}\right) \\[2ex]
\dfrac{d^2 A_1}{d\rho^2} + \dfrac{\nu}{\rho^2}\left[\rho \dfrac{dA_1}{d\rho} - A_1\right] = 0, \quad \text{at } \rho = \mu^{\frac{1}{2}} \\[2ex]
\dfrac{d^3 A_1}{d\rho^3} + \dfrac{1}{\rho^2}\dfrac{d^2 A_1}{d\rho^2} - \dfrac{(3-\nu)}{\rho^2}\dfrac{dA_1}{d\rho} + \dfrac{(3-\nu)}{\rho^3}A_1 = 0, \quad \text{at } \rho = \mu^{\frac{1}{2}}
\end{cases}
\tag{4.34}
$$

If we impose the boundary conditions (4.34) on the general solution (4.33), we obtain the system of equations

$$
\begin{cases}
AJ_p\left(\mu^{\frac{1}{2}}\left(\dfrac{a}{b}\right)\right) + BJ_{-p}\left(\mu^{\frac{1}{2}}\left(\dfrac{a}{b}\right)\right) = 0 \\[2ex]
A\left[(1+\nu-p)J_{-p}\left(\mu^{\frac{1}{2}}\right) + \mu^{\frac{1}{2}}J_{p-1}\left(\mu^{\frac{1}{2}}\right)\right] \\[2ex]
+B\left[(1+\nu+p)J_{-p}\left(\mu^{\frac{1}{2}}\right) + \mu^{\frac{1}{2}}J_{-p-1}\left(\mu^{\frac{1}{2}}\right)\right] = 0
\end{cases}
\tag{4.35}
$$

and $C = 0$. For nontrivial solutions $A$ and $B$ of (4.35) we must have

$$J_p\left(\mu^{\frac{1}{2}}\left(\frac{a}{b}\right)\right) + (1+\nu-p)J_p\left(\mu^{\frac{1}{2}}\right) + \mu^{\frac{1}{2}}J_{p-1}\left(\mu^{\frac{1}{2}}\right) = 0 \tag{4.36}$$

and this leads to the initial buckling curve for $n = 1$, which is shown in Fig. 4.6.

Approximate solutions of the general problem in Majumdar [166] are generated by minimizing the potential energy using a Rayleigh-Ritz procedure where

$$V = \frac{1}{2} \int_b^a \int_0^{2\pi} \left\{ N_r \left( \frac{\partial w}{\partial r} \right)^2 + \frac{1}{r^2} N_\theta \left( \frac{\partial w}{\partial \theta} \right)^2 \right.$$

$$K \left[ \left( \frac{\partial^2 w}{\partial r^2} + \frac{1}{r} \frac{\partial w}{\partial r} + \frac{1}{r^2} \frac{\partial^2 w}{\partial \theta^2} \right)^2 \right.$$

$$-2 (1 - \nu) \frac{\partial^2 w}{\partial r^2} \left( \frac{1}{r} \frac{\partial w}{\partial r} + \frac{1}{r^2} \frac{\partial^2 w}{\partial \theta^2} \right)$$

$$\left. \left. +2 (1 - \nu) \left( \frac{1}{r} \frac{\partial^2 w}{\partial r \partial \theta} - \frac{1}{r^2} \frac{\partial w}{\partial \theta} \right)^2 \right] \right\} r \, d\theta \, dr$$

(4.37)

and $N_r, N_\theta$ are given by (4.19a,b); these approximate solutions, for $n >$
1, are shown in Fig. 4.7 and the experimental results of Majumdar [166]
are in good agreement with the calculated values.

A somewhat deeper treatment of the problem originally treated by
Majumdar [166] may be found in [167], as well as in [66], where for
the case of an annular circular plate, compressed along its outer edge
and free along the inner edge, group-theoretic concepts introduced by
Golubitsky and Shaeffer [13] are employed to study the postbuckling
problem and, in particular, to study the postbuckling of elastic annular
plates at multiple eigenvalues. In both [66] and [167] the outer radius of
the plate is normalized, i.e., $a = 1$, so that $1 \le r \le b$, and the full set of
von Karman plate equations have the form

$$\begin{cases} \Delta^2 w + N \frac{b^2}{1 - b^2} \left\{ \left( \frac{1}{b^2} + \frac{1}{r^2} \right) \Delta w - \frac{2}{r^2} w_{,rr} \right\} - [w, \Phi] = 0 \\ \Delta^2 \Phi + \frac{1}{2} [w, w] = 0 \end{cases}$$

(4.38)

if we nondimensionalize the deflection $w$ and the (extra) Airy function
$\Phi$. The brackets $[w, \Phi]$ and $[w, w]$ in (4.38) are given, respectively, in
polar coordinates by the expressions on the right-hand sides of (1.78) and
(1.79). As a discussion of the postbuckling behavior of elastic circular
annular plates will be postponed until later in this chapter, we will
content ourselves here with a brief discussion of some of those results in
[66] and [167] on the initial buckling behavior of annular plates which
amplify the original work of Majumdar [166].

For the first of the von Karman equations in (4.38), with the (extra)
Airy function $\Phi \equiv 0$, and the boundary conditions (1.139'), (1.140'), and

(1.141′), Troger and Steindl [66], [167] proceed as follows: first of all, the linearized equation is written in the form

$$Lw = N \cdot Mw \qquad (4.39)$$

with the operators $L$ and $M$ given by

$$Lw = \Delta^2 w, \quad Mw = -\frac{b^2}{1 - b^2}\left\{\left(\frac{1}{b^2} + \frac{1}{r^2}\right)\Delta w - \frac{2}{r^2}w_{,rr}\right\} \qquad (4.40)$$

Equation (4.39) is self-adjoint and positive definite; thus, the eigenfunctions are orthogonal and the eigenvalues $N$ are real and positive. Solutions of (4.39) may be sought in the form

$$w_{mn}(r, \theta) = g_{nm}(r)\left\{\begin{matrix}\cos n\theta \\ \sin n\theta\end{matrix}\right\}, \quad n = 0, 1, 2, \cdots, \quad m = 1, 2, \cdots \qquad (4.41)$$

In the separation of variables procedure there is no difference as to which of $\cos n\theta$ and $\sin n\theta$ is chosen so as to obtain a standard eigenvalue problem

$$L_n g_{nm} = N M_n g_{nm} \qquad (4.42)$$

for $g_{nm}(r)$, where the operators $L_n$ and $M_n$ are given by

$$L_n g = g_{,rrrr} + \frac{2}{r}g_{,rrr} - \frac{1 + 2n^2}{r^2}\left(g_{,rr} - \frac{1}{r}g_{,r}\right)$$
$$+ \frac{n^2(n^2 - 4)}{r^4}g \qquad (4.43a)$$

and

$$M_n g = -\frac{b^2}{1 - b^2}\left\{\left(\frac{1}{b^2} - \frac{1}{r^2}\right)g_{,rr}\right.$$
$$\left. + \left(\frac{1}{b^2} + \frac{1}{r^2}\right)\left(\frac{1}{r}g_{,r} - \frac{n^2}{r^2}g\right)\right\} \qquad (4.43b)$$

and the associated boundary conditions are

$$g_{nm} = 0, \quad g_{nm,r} = 0, \quad \text{at } r = 1 \qquad (4.44a)$$

and, at $r = b$,

$$
\begin{cases}
g_{nm,rr} + \nu \left( \dfrac{1}{r} g_{nm,r} + \dfrac{n^2}{r^2} g_{nm} \right) = 0 \\[2ex]
g_{nm,rrr} + \dfrac{1}{r} g_{nm,rr} - \dfrac{1 + n^2(2 - \nu)}{r^2} g_{nm,r} \\[2ex]
\quad + \dfrac{n^2(3 - \nu)}{r^3} g_{nm} = 0,
\end{cases}
\tag{4.44b}
$$

For $n = 0$ and $n = 1$ analytical solutions of (4.42), with the boundary data (4.44a,b), are given, as we have shown previously, in the form of Bessel functions, i.e., Majumdar [166]. On the other hand, for $n \geq 2$ analytical solutions of the eigenvalue problem (4.42), (4.43a,b), (4.44a,b) are not yet known; a numerical procedure has been used to calculate the eigenfunctions for $n \geq 2$ (Machinek and Steindl [168] ); these results are shown in Figs. 4.8–4.10 for $\nu = 0.3$. The index $m$ indicates that there is an infinite number of solutions of the eigenvalue problem for each $n$; however, from Fig. 4.8 it follows that only the case in which $m = 1$ must be studied, because this value corresponds to the lowest in-plane buckling load. In Fig. 4.9, the buckling mode shapes which correspond to the three buckling modes $m = 1, 2, 3$ are shown for $n = 0$ and $b = 0.3$; the physically relevant mode shape is $g_{01}(r)$. In Fig. 4.10, which contains only those curves for $m = 1$, we show the influence of the hole size $b$ on the critical load $N$ with the corresponding circumferential mode number $n$. For $b < 0.51$ the plate buckles in an axisymmetrical mode which corresponds to $n = 0$; this buckled configuration is depicted in Fig. 4.11a. At (approximately) $b = 0.51$ we see an intersection of the initial buckling curves corresponding to $n = 0$ and $n = 1$. For $n = 1$, two solutions, one with $\cos \theta$ and the other with $\sin \theta$, exist; the buckling mode is shown in Fig. 4.11b and the point of intersection of the curves corresponding to $n = 0$ and $n = 1$ is, therefore, associated with an eigenvalue of multiplicity three. As we increase the hole size $b$ even further, a point is reached where the initial buckling curves corresponding to $n = 1$ and $n = 2$ intersect; this point, and all subsequent points of intersection of the buckling curves, correspond to eigenvalues of the linearized buckling problem for the elastic, circular, annular plate of multiplicity four. Fig. 4.11c shows the essential structure of buckling mode when $m = 1$ and $n = 2$. The envelope of the buckling curves, which we show in Fig. 4.10, together with the points of intersection, form the stability boundary in the $N - b$ parameter space. Generically, the envelope corresponds to

double eigenvalues for $b$ & 0.51, while for $0 \leq b$ . 0.51 the axisymmetric buckling mode corresponds to a simple eigenvalue; the nongeneric
points of intersection correspond to three-and four-fold eigenvalues. As
noted in [66], the large number of multiple eigenvalues is due to the circular symmetry of the problem. We will return to a further discussion
of the results in [66] and [167] in the course of our examination of the
postbuckling behavior of annular plates.

## 4.1.2 Cylindrically Orthotropic Symmetry: Initial Buckling

In this subsection, we will delineate some of the basic results relative
to the initial buckling behavior of linearly elastic circular annular plates
which exhibit cylindrically orthotropic material symmetry; the full set
of von Karman equations in this case is given by (1.98), (1.99), in the
domain $0 \leq \theta < 2\pi$, $b < r < a$, where $D_r, D_\theta$, and $D_{r\theta}$ are given by
(1.96), (1.97), while $E_r, E_\theta, \nu_r, \nu_\theta$, and $G_{r\theta}$ are, respectively, the Young's
moduli, Poisson's ratios, and shear modulus. For the special case of a
cylindrically orthotropic annular plate undergoing an axisymmetric deformation, equations (1.98), (1.99) reduce to (1.101), (1.102) where $\Delta_r^2, \beta$
(the orthotropy ratio), and $F(w, \Phi)$ are given by (1.103). Because of the
complexity of the general system (1.98), (1.99), and the need to satisfy
boundary conditions along both the edges at $r = b$ and $r = a$, most of the
literature which deals with initial buckling and postbuckling of circular
annular plates is limited to a treatment of axisymmetric deformations.

For axisymmetric (initial) buckling of a cylindrically orthotropic annular plate, the governing equation is (1.101), where $\Phi = \Phi^0$ is the
(prebuckling) Airy stress function which is determined as the solution
of (1.102) with $w = 0$, subject to the loading conditions along the edges
at $r = a$ and $r = b$; this linearized stability equation may be written
in terms of the prebuckling stresses $\sigma_{rr}$ and $\sigma_{\theta\theta}$ associated with the
in-plane compressive loading along the edges, as

$$\Delta_r w = \frac{h}{D_r} \left( \sigma_{rr} \frac{d^2 w}{dr^2} + \sigma_{\theta\theta} \frac{1}{r} \frac{dw}{dr} \right) \tag{4.45}$$

where, by (1.103),

$$\Delta_r = \frac{d^2}{dr^4} + \frac{2}{r} \frac{d^3}{dr^3} - \frac{\beta}{r^2} \left( \frac{d^2}{dr^2} - \frac{1}{r} \frac{d}{dr} \right)$$

and $\beta = \dfrac{E_\theta}{E_r}$. The initial buckling problem for a cylindrically orthotropic elastic annular plate undergoing axially symmetric deformations has been studied by many authors, most notably Uthgenannt and Brand [70], [169], who determine the general form of the prebuckling stresses corresponding to in-plane compressive loading along the edges at $r = a$ and $r = b$, and then reduce (4.45) to the nondimensional form

$$\Delta_\rho w + \mu \left[ \left( C_1 \rho^{\sqrt{\beta}-1} + C_2 \rho^{-(\sqrt{\beta}+1)} \right) \frac{d^2 w}{d\rho^2} \right.$$

$$\left. + \frac{\sqrt{\beta}}{\rho} \left( C_1 \rho^{\sqrt{\beta}-1} - C_2 \rho^{-(\sqrt{\beta}+1)} \right) \frac{dw}{d\rho} \right] = 0 \qquad (4.46)$$

where $\rho = \dfrac{r}{a}$, $\Delta_\rho$ is just the operator $\Delta_r$, with $\rho$ in place of $r$, and

$$\mu = -\lambda h a^2 / D_r \qquad (4.47)$$

with $\lambda = \sigma_{rr}|_{r=a} = \sigma_{rr}|_{r=b}$. Nontrivial solutions of (4.46), subject to the boundary support conditions along $r = a$ and $r = b$, exist for particular values of $\mu$ (the eigenvalues), which then serve to determine the critical buckling (edge) stresses $\lambda_{cr}$. For axially symmetric deformations of a cylindrically orthotropic circular annular plate, the boundary conditions corresponding to the cases of clamped, simply supported, or free edges may be obtained by specializing the results in section 1.3, i.e.,

$$
\begin{cases}
w = \dfrac{dw}{dr} = 0, & \text{at a clamped edge} \\[2mm]
w = \dfrac{d^2 w}{dr^2} + \dfrac{\nu_\theta}{r}\dfrac{dw}{dr} = 0, & \text{at a simply supported edge} \\[2mm]
\dfrac{d^2 w}{dr^2} + \dfrac{\nu_\theta}{r}\dfrac{dw}{dr} = 0, & \\[2mm]
\dfrac{d^2 w}{dr^3} + \dfrac{1}{r}\dfrac{d^2 w}{dr^2} - \dfrac{\beta}{r^2}\dfrac{dw}{dr} = 0, & \text{at a free edge}
\end{cases}
\qquad (4.48)
$$

The solutions of (4.46), subject to various combinations of the conditions (4.48) along the edges at $r = a$ and $r = b$, have been studied, numerically, in [169] and the results are displayed in Fig. 4.12–4.17. For the cases corresponding to Figs. 4.14 and 4.15 only the outer edge is subject to an in-plane compressive loading, while the inner edge is free of any applied load. The boundary and loading conditions are indicated on each of the Figs. 4.12 through 4.17, which show the nondimensional critical buckling load $\mu$ graphed against the aspect ratio $b/a$ for various values of the orthotropy ratio $\beta = \dfrac{E_\theta}{E_r}$, with $\beta = 1$ corresponding to an isotropic elastic (circular) annular plate. In fact, $\beta$ ranges from $\beta = 1$ (isotropic) to $\beta = 30$, while $b/a$ ranges from zero (a circular plate) to $b/a = 0.8$, which approaches a ring. For all the calculations displayed, Poisson's ratios $\nu_r = \nu_\theta = 0.3$ were used. As noted by the authors in [70], each figure referenced above shows the pronounced effect of the orthotropic nature of the material on the critical buckling loads, and shows comparative results obtained by Shubert [164], Mossakowski [151], Yamaki [162], Woinowsky-Krieger [170], Roza [171], Meissner [172], and Pandalai and Patel [153] for those special subcases in which the cylindrically orthotropic annular plate is either a circular plate (i.e., $b = 0$) or an isotropic plate (or both).

### 4.1.3 Isotropic and Cylindrically Orthotropic Symmetry: Postbuckling Behavior

The postbuckling behavior of elastic (circular) annular plates possessing cylindrically orthotropic material symmetry, has been considered by a number of authors, with the case of isotropic material symmetry appearing as a special case for which the orthotropy ratio $\beta = \dfrac{E_\theta}{E_r} \equiv 1$. Among the main contributions have been those of Uthgenannt and Brand [70], Dumir, Nath and Gandhi [173], Huang [174], [175], Alwar and Reddy [176], and Machinek and Troger [167]; we will now review some of the more fundamental results contained in these references, beginning with the work of Uthgenannt and Brand [70].

In [70], the postbuckling behavior of orthotropic annular plates due to in-plane compressive loads applied along the outer edge at $r = a$, is considered under the assumption that deformations are restricted to be axially symmetric. The governing set of von Karman equations is solved, in nondimensional form, by a numerical technique originally employed by Keller and Reiss [177], and postbuckling results are obtained for the membrane and bending stresses, the deflections of the plate, and the slopes associated with those deflections. For axially symmetric deformations, the complete set of von Karman is, once again, given by (1.101), (1.102), and (1.103) equations (1.101), (1.102) may be integrated once so as to yield

$$\begin{cases} M(w) = \dfrac{1}{D_r}\dfrac{d\Phi}{dr}\dfrac{dw}{dr} + C_1 \\[4mm] M(\phi) = -\dfrac{E_\theta h}{2}\left(\dfrac{dw}{dr}\right)^2 + C_2 \end{cases} \tag{4.49}$$

with

$$M = r\left(\frac{d^3}{dr^3} + \frac{1}{r}\frac{d^2}{dr^2} - \frac{\beta}{r^2}\frac{d}{dr}\right) \tag{4.50}$$

For a circular plate, the constants of integration, $C_1$ and $C_2$, are zero because at $r = 0$ both $M(w)$ and $M(\phi)$ must vanish, as does $\dfrac{dw}{dr}$. For an annular plate, $C_1$ depends on the boundary support conditions although, as noted in [70], it can be shown that $C_2$ is zero by eliminating the $\theta$ dependence from the strain-displacement equations and rederiving the second equation in (4.49). The governing equations are further simplified and nondimensionalized in [70] by introducing the quantities $\rho = \dfrac{r}{a}$, $\dfrac{b}{a} \le \rho \le 1$, for $0 \le b < a$, and

$$\begin{cases} \rho V = -\dfrac{\alpha}{h}\dfrac{dw}{d\rho}, \quad S = \dfrac{1}{\rho D_r}\dfrac{d\Phi}{d\rho} \\[4mm] \alpha^2 = \dfrac{6\beta}{h^2}\left(1 - \nu_r\nu_\theta\right) \end{cases} \tag{4.51}$$

with $\rho$ the dimensionless radial coordinate, $V$ the dimensionless slope, and $S$ the dimensionless radial membrane stress; using these definitions in (4.49), (4.50), we obtain the equations

$$\begin{cases} L(V) = VS - \alpha\left(\dfrac{a}{\rho}\right)^2 C_1 \\[4mm] L(S) = -V^2 \end{cases} \tag{4.52}$$

with

$$L = \frac{d^2}{d\rho^2} + \frac{3}{\rho} + \frac{1 - \beta}{\rho^2} \tag{4.53}$$

which, together with the boundary support conditions, must be solved in order to determine the axisymmetric behavior of annular plates subjected to loads greater than the critical buckling load. In dimensionless form, the boundary conditions are

$$\text{at } \rho = \frac{b}{a}: \begin{cases} S = 0 \\[2mm] \rho\dfrac{dv}{d\rho} + (1 + \nu_r)V = 0 \end{cases}, \tag{4.54}$$

i.e., the inner edge is free and,

$$\text{at } \rho = 1 : \begin{cases} S = -P \\ V = 0 \end{cases}, \qquad (4.55)$$

i.e., the outer edge is clamped and loaded, with $P$ the magnitude of the external compressive load. A finite-difference scheme, similar to that employed by Keller and Reiss [177], is used to solve the boundary value problem consisting of (4.52)–(4.55) and to obtain results for the variation, with $\rho$, of the following quantities in nondimensional form: the deflection $w$, the membrane tangential stress $S_\theta^M$,

$$S_\theta^M = \rho \frac{dS}{d\rho} + S, \qquad (4.56a)$$

the bending radial stress $S_R^B$,

$$S_R^B = \rho \frac{dV}{d\rho} + (1 + \nu_r)V, \qquad (4.56b)$$

and the bending tangential stress $S_\theta^B$,

$$S_\theta^B = \nu_\theta \rho \frac{dV}{d\rho} + (1 + \nu_r)V \qquad (4.56c)$$

These results are depicted in Figs. 4.18–4.22 for an orthotropy ratio $\beta = 5.0$, an aspect ratio $b/a = 0.4$, and a number of dimensionless postbuckling loads $\Lambda$, $\Lambda = \dfrac{P}{P_{cr}}$; we recall that $S$ (Fig. 4.19) is the non-dimensional membrane radial stress. In Figs. 4.23–4.25, we depict those postbuckling results in [70] which show the variation of $w_{\max}$, the maximum plate deflection, and $S_\theta^M$, and $S_R^B$, with the aspect ratio $b/a$ for various values of both the orthotropy ratio $\beta$ and the nondimensional postbuckling load $\Lambda$; the value $\beta = 1$ corresponds, of course, to an isotropic linearly elastic annular plate. For all the calculations cited, Poisson's ratios of $\nu_r = \nu_\theta = 0.3$ were used. As the authors in [70] have noted, the most notable result obtained is that the membrane stresses for large values of $\Lambda$ become tensile stresses in the interior of the plate, and change abruptly to compressive stresses in a relatively narrow band at the edge of the plate, a boundary-layer effect that was previously noted by Friedrichs and Stoker [143] for a simply supported isotropic circular plate, and by Keller and Reiss [177] as well. As in the case of isotropic plates, the dominant stresses are the tangential membrane stresses, which reach a maximum at the outer edge of the plate. As Figs. 4.23 through 4.25 indicate, for a given $\Lambda$ and $\beta$, the various dependent variables are relatively constant for $\dfrac{b}{a} \leq 0.4$, but as the inner radius increases $\left( \dfrac{b}{a} > 0.4 \right)$ both the deflections and stresses increase markedly, particularly for $\Lambda$ large. Also, as the orthotropy ratio $\beta$ increases, the deflections and stresses both increase. Uthgenannt and Brand [70] point out that it can be shown that the magnitude of the in-plane compressive loads which are applied to the edges of elastic, cylindrically orthotropic, annular plates can be increased to several multiples of the critical buckling load without diminishing the ability of the plate to carry these loads; also, for such large values of the applied load, the maximum deflection amplitudes will be of the same order of magnitude as the plate thickness. In other words, as is the case with isotropic circular plates, although an initially flat elastic annular plate becomes unstable when the edge load reaches the critical buckling value, the plate continues to have a substantial load-carrying capacity well beyond the load associated with initial buckling.

While the results obtained by Uthgenannt and Brand [70] point out various interesting aspects of the postbuckling behavior of elastic cylindrically orthotropic annular plates, they must be viewed with some caution. As Huang [174], [175] has noted, the authors of [70] use the Poisson's ratios $\nu_r = \nu_\theta = 0.3$ in their finite-difference calculations, which violates the fact (see Chapter 1, section 2) that

$$\frac{\nu_r}{\nu_\theta} = \frac{E_\theta}{E_r} = \beta \qquad (4.57)$$

except for $\beta \equiv 1$ (an isotropic plate); Huang [174], therefore, reconsiders the problem of axially symmetric buckling of a cylindrically orthotropic, elastic, annular plate of inner radius $b$ and outer radius $a$ using a direct computational method devised by Keller, Keller, and Reiss [144] to prove the existence of buckled states of circular plates. The basic form of the (axisymmetric) von Karman equations is, of course, the same as in [70]. The boundary conditions considered are the third set of conditions in (4.48) at $r = b$, i.e., the inner edge is a free edge, and along the edge at $r = a$, either the first (clamped edge) or second (simply supported edge) set of conditions in (4.48) together with a constant compressive load of magnitude $P$ per unit length. Huang [174] introduces the dimensionless variables

$$\begin{cases} \rho = \dfrac{r}{a}, \; W = \dfrac{w}{a}, \; \phi = \Phi/haE_\theta \\ \qquad\qquad p = P/haE_\theta \end{cases} \tag{4.58}$$

and rewrites the governing system of von Karman equations in the form

$$\frac{d^4W}{d\rho^4} + \frac{2}{\rho}\frac{d^3W}{d\rho^3} - \frac{\beta}{\rho^2}\frac{d^2W}{d\rho^2} + \frac{\beta}{\rho^3}\frac{dW}{d\rho} \tag{4.59}$$

$$= 12\left(\frac{a}{h}\right)^2(\beta - \nu_\theta^2)\frac{1}{\rho}\frac{d}{d\rho}\left[\frac{dW}{d\rho}\phi\right]$$

$$\frac{d^2\phi}{d\rho^2} + \frac{1}{\rho}\frac{d\phi}{d\rho} - \frac{\beta}{\rho}\phi = -\frac{1}{2\rho}\left(\frac{dW}{d\rho}\right)^2 \tag{4.60}$$

with boundary support conditions

$$\begin{cases} \dfrac{d^2W}{d\rho^2} + \dfrac{\nu_\theta}{\rho}\dfrac{dW}{d\rho} = 0 \\[2mm] \left(\dfrac{d^3W}{d\rho^3} + \dfrac{1}{\rho}\dfrac{d^2W}{d\rho^2} - \dfrac{\beta}{\rho^2}\dfrac{dW}{d\rho}\right) \\[3mm] \qquad -12(\beta - \nu_\theta^2)\left(\dfrac{a}{h}\right)^2\dfrac{1}{\rho}\left(\dfrac{dW}{d\rho}\phi\right) = 0 \end{cases} \tag{4.61a}$$

at $\rho = b/a$ and either

$$W = \frac{dW}{d\rho} = 0, \quad \text{at } \rho = 1, \quad \text{for a clamped edge} \tag{4.61b}$$

or

$$W = \frac{d^2W}{d\rho^2} + \nu_\theta\frac{dW}{d\rho} = 0, \quad \text{at } \rho = 1, \tag{4.61c}$$

$$\text{for a simply supported edge}$$

Along the edge at $\rho = 1$ we also have the uniform, radial, compressive load of magnitude P. In the calculations presented in Huang [174], the aspect ratio $\gamma = b/a$ is taken to be 0.4, while the orthotropy ratio varies over the range $\beta = 1.0, 3.0, 5.0,$ and 10.0, for the case where the outer edge at $r = a$ is simply supported, while $\beta = 5.0$ for the case in which the outer edge at $r = a$ is clamped. With the buckling parameter $\lambda$ given by

$$\lambda = 12(\beta - \nu_\theta^2)\left(\frac{a}{h}\right)^2 p \tag{4.62}$$

the postbuckling (and initial buckling) behavior of the cylindrically orthotropic annular plate is indicated in Table 4.1 and displayed graphically in Figs. 4.26 and 4.27 where $\nu = \nu_\theta$; the corresponding distributions of (normalized) radial bending stresses, circumferential bending stresses, radial membrane stresses, and circumferential membrane
stresses are graphed in Figs. 4.28, 4.29, 4.30, and 4.31, respectively,
where $\delta = w(\gamma)/h$, while the rigidity $D_0$ is given by

$$D_0 = E_\theta h^3 /12 \left(1 - \nu_\theta^2\right) \tag{4.63}$$

Finally, based on the analysis in [174], we show in Fig. 4.32 the (normalized) transverse displacement shapes for the annular plate as functions
of the nondimensionalized plate radius $\rho$, thus indicating the influence
of postbuckling load magnitudes on the plate deflection shapes.

Two other noteworthy analyses of the axisymmetric buckling behavior of cylindrically orthotropic elastic annular plates are the papers of
Alwar and Reddy [176] and Dumir, Nath, and Gandhi [173]; the earlier
paper [176] presents a large deflection analysis of orthotropic annular
plates subjected to uniformly distributed loads using Chebyshev series
expansions, with a substantial part of the work geared toward dynamic
analysis. The work of Dumir, Nath and Gandhi [173] subsumes, for the
case of a static analysis, much of the work in [176]; in [173] both $w$ and $\Phi$
are expanded in finite power series and an orthogonal point collocation
method is used to obtain discretized algebraic equations from the governing set of von Karman differential equations. Results are presented
in [173] for annular plates with and without a rigid plug (in the hole of
radius $b$) and some of these results are displayed in Figs. 4.33 and 4.34.
In Fig. 4.33 we show the deflection versus load graphs for both the cases
of a clamped and a simply supported edge at $r = a$, and for orthotropy
ratios of $1, \frac{1}{3}$, and $\frac{1}{10}$, as well as aspect ratios of .5 and .25; for comparison purposes, the results obtained in these situations by Alwar and
Reddy [176] are shown on the same graphs. Figure 4.34 displays the
same kind of results as Fig. 4.33, but this time, for the case in which
a rigid plug occupies the region $0 \le r \le b$; for all the cases referenced
here the Poisson ratio $\nu_r = .25$. It can be easily observed from Fig. 4.33
that the effect of varying $\beta$ on the deflection response is very small in
the simply supported case as compared with the clamped case; this is
not the situation, of course, when the hole is occupied by a rigid plug,
as is made clear in Fig. 4.34.

**Remarks**: As we have noted earlier in this subsection, a somewhat
deeper analysis of the postbuckling behavior of elastic annular plates

subjected to a uniform compressive load along the edge at $r = a$, with the edge at $r = a$ clamped and the edge at $r = b$ free, has been carried out in Machinck and Troger [167] as well as in Troger and Steindl [66]; both analyses assume the plate to be isotropic and focus on the nongeneric situation in which buckling occurs at eigenvalues of multiplicity greater than one. In both [66] and [167] the operator $\Delta^2$ in (4.38) is inverted so as to yield

$$\Phi = -\frac{1}{2}\Delta^{-2}[w, w] \qquad (4.64)$$

and (4.64) is then inserted in (4.38) so as to provide the functional equation

$$F(w, N) = \Delta^2 w + N\frac{b^2}{1 - b^2}\left\{ \left( \frac{1}{b^2} + \frac{1}{r^2} \right)\Delta w - \frac{2}{r^2}w_{,rr} \right\}$$
$$+ \frac{1}{2}\left[ w, \Delta^{-2}\left[ w, w \right] \right] = 0 \qquad (4.65)$$

The buckling problem is then dealt with as a problem in bifurcation theory, after the linearized problem which governs initial buckling is analyzed: the functional equation is reduced locally in the neighborhood of a bifurcation point $N = N_{cr}$ to a system of algebraic equations

$$A_i\left( x_1, \cdots, x_k, \lambda \right) = 0, \quad i = 1, \cdots, k, \qquad (4.66)$$

with $\lambda = N - N_{cr}$, by employing the Liapunov–Schmidt reduction discussed in Chapter 1. In (4.66), the $x_i$ are the amplitudes of the buckling modes, while $k$ is equal to the multiplicity of the critical eigenvalue $N_{cr}$. The simplest set of bifurcation equations (4.66) is obtained by carrying out the derivation at a value of the inner edge radius $b$, which corresponds to a simple ($n = 0$) or double ($n = 1, 2, \cdots$) eigenvalue, i.e., (4.41) and Fig. 4.10. However, in [66] and [167], the postbuckling analysis is focused on deriving the bifurcation equations at the triple eigenvalue because small parameter variations in a neighborhood of such a multiple eigenvalue yield interesting and physically relevant phenomena such as mode jumping; also, all the basic properties of the simpler cases are included in the more complicated one. For the Liapunov–Schmidt reduction in [66], [167] the ansatz

$$w(r, \theta) = x_0\Psi_0 + x_{1c}\Psi_{1c} + x_{1s}\Psi_{1s} + W\left( x_0, x_{1c}, x_{1s} \right) \qquad (4.67)$$

is made where $x_0, x_{1c}, x_{1s}$ are the amplitudes of the buckling modes (or critical variables) and $\Psi_0 = g_{01}(r)$, $\Psi_{1c} = g_{11}(r)\cos\theta$, $\Psi_{1s} = g_{11}(r)\sin\theta$

are the buckling modes (eigenfunctions) corresponding to the triple eigenvalue. The function $W$ includes non-critical variables which correspond to modes that are orthogonal to the buckling modes and can be expressed in terms of the critical variables by means of the Liapunov-Schmidt reduction. It may be shown [66], [167] that

$$W(x_0, x_{1c}, x_{1s}) = \mathcal{O}\left(|x_0|^2 + |x_{1c}|^2 + |x_{1s}|^2\right) \tag{4.68}$$

By introducing (4.67) into (4.65), with $N = N_{cr}$, and projecting the resulting expression onto the three buckling modes, three equations are obtained; the process is simplified somewhat by using the work of Golubitsky and Schaeffer [13] to determine, in advance, based on the symmetry properties of the problem, those variables which may appear in the resulting bifurcation equations. These equations turn out to have the form

$$\begin{cases} Ax_0^3 + Cx_0\left(x_{1c}^2 + x_{1s}^2\right) - \alpha_0 \lambda x_0 = \gamma_0 \\ B\left(x_{1c}^2 + x_{1s}^2\right)x_{1c} + Cx_0^2 x_{1c} - \alpha_1 \lambda x_{1c} = \gamma_{1s} \\ B\left(x_{1c}^2 + x_{1s}^2\right)x_{1s} + Cx_0^2 x_{1s} - \alpha_1 \lambda x_{1s} = \gamma_{1s} \end{cases} \tag{4.69}$$

In (4.69) the terms $\gamma_0, \gamma_{1c}, \gamma_{1s}$ are derived from a transverse loading on the annular plate (and thus, vanish if such a loading (imperfection) is not present). With the inner product defined by

$$(\Psi_i, \Psi_j) \equiv \int_0^{2\pi} \int_b^1 \Psi_i \Psi_j r \, dr \, d\theta \tag{4.70}$$

it follows ([66], [167]) that

$$\begin{cases} A = \dfrac{1}{2}\left(\left[\Psi_0, \Delta^{-2}\left[\Psi_0, \Psi_0\right]\right], \Psi_0\right) = 0.00361 \\[2mm] B = \dfrac{1}{2}\left(\left[\Psi_{1c}, \Delta^{-2}\left[\Psi_{1c}, \Psi_{1c}\right]\right], \Psi_{1c}\right) = 0.00584 \\[2mm] C = \dfrac{1}{2}\left(\left[\Psi_0, \Delta^{-2}\left[\Psi_{1c}, \Psi_{1c}\right]\right], \Psi_0\right) \\[2mm] \qquad + \left(\left[\Psi_{1c}, \Delta^{-2}\left[\Psi_{1c}, \Psi_0\right]\right], \Psi_0\right) = 0.00919 \end{cases} \tag{4.71}$$

and

$$
\begin{cases}
\alpha_0 = -\dfrac{b^2}{1-b^2}\left(\left(\dfrac{1}{b^2}+\dfrac{1}{r^2}\right)\Delta\Psi_0 - \dfrac{2}{r^2}\Psi_{0,rr},\Psi_0\right) = 1 \\[2mm]
\alpha_1 = -\dfrac{b^2}{1-b^2}\left(\left(\dfrac{1}{b^2}+\dfrac{1}{r^2}\right)\Delta\Psi_{1c} - \dfrac{2}{r^2}\Psi_{1c,rr},\Psi_{1c}\right) = 1 \\[2mm]
\gamma_0 = (q,\Psi_0),\ \ \gamma_1 = \left((q,\Psi_{1c})^2 + (q,\Psi_{1s})^2\right)^{\frac{1}{2}}
\end{cases}
\tag{4.72}
$$

In (4.72), $q$ is the magnitude of the transverse loading (if any) on the elastic annular plate. With $M$ as given by (4.40), the values of $\alpha_0 = \alpha_1 = 1$ follow from the normalization of the eigenfunctions $\Psi_i$ given by

$$
\int_0^{2\pi}\int_1^b \Psi_i\,(M\Psi_i)\,r\,dr\,d\theta = 1
\tag{4.73}
$$

To compute the expressions $\Delta^{-2}[\Psi_i,\Psi_j]$ in (4.71), the eigenvalue problem $\Delta^2\Phi = \mu\Phi$ must be solved with the appropriate boundary conditions and then $[\Psi_i,\Psi_j]$ must be expanded in an infinite series in the eigenfunctions.

If we introduce polar coordinates $x_{1c} = x_1\cos\alpha$, $x_{1s} = x_1\sin\alpha$, $\gamma_{1c} = \gamma_1\cos\alpha$, and $\gamma_{1s} = \gamma_1\sin\alpha$, in (4.69), we obtain the new system

$$
\begin{cases}
Ax_0^3 + Cx_0 x_1^2 - \alpha_0\lambda x_0 = \gamma_0 \\[2mm]
Bx_1^3 + Cx_0^2 x_1 - \alpha_1\lambda x_1 = \gamma_1
\end{cases}
\tag{4.74}
$$

which are the so-called generic bifurcation equations of restricted generic type (Chow and Hale [12] ). If we sum the two buckling modes in (4.41) corresponding to $n = 1$, $m = 1$, we obtain

$$
x_{1c}g_{11}\cos\theta + x_{1s}g_{11}\sin\theta = x_1 g_{11}\cos(\theta - \alpha)
\tag{4.75}
$$

so that $x_1 = \left(x_{1c}^2 + x_{1s}^2\right)^{\frac{1}{2}}$ is the amplitude of the resulting mode, while $\alpha = \tan^{-1}(\gamma_{1s}/\gamma_{1c})$ describes its location on the circumference of the plate. Figure 4.35 is typical of the kind of postbuckling (bifurcation) diagrams obtained from the analysis in [66], [167]; diagram (a) in this figure is obtained for a perfect plate with no transversal loads ($\gamma_0 = \gamma_1 = 0$) and a hole radius $b < b_c$, and corresponds to the critical triple eigenvalue. The remarkable fact which is exhibited in diagram (a), Fig. 4.35, is that for $\lambda > \lambda_2$, besides the stable solution $x_0$, the $x_1$ solution is also stable; thus, in the domain $\lambda > \lambda_2$ one can obtain a jump in the model shape of the elastic annular plate by applying transversal loads and this has been noted by Suchy, Troger, and Weiss [178]. Diagrams (b)–(e) in Fig. 4.35 indicate what happens when an increasing imperfection $\gamma_1$ creates an initial preference for the mode $x_1$. As has been noted in [66], as long as $\gamma_1$ is still sufficiently small, a smooth transition into the axisymmetric mode $x_0$ is still possible; however, if $\gamma_1$ is sufficiently large, a transition from the initially preferred second mode $x_1$ to the first mode $x_0$, takes place only as a consequence of mode-jumping. We will elaborate further on the consequences (displayed in Fig. 4.35 ) of increasing $\gamma_1$ when we discuss the imperfection sensitivity of thin plates in the next chapter.

### 4.1.4 Rectilinear Orthotropic Symmetry: Initial Buckling and Postbuckling Behavior

In the previous chapter, we noted that there appears to be no work in the literature on the initial or postbuckling behavior of circular plates which possess rectilinearly orthotropic material symmetry; the same situation does not hold in the case of elastic annular plates having rectilinearly orthotropic symmetry, as this problem has been treated in [71] (at least with respect to a particular combination of load and support conditions). The analysis in [71] is limited to a study of prebuckling stress distributions and initial buckling modes as well as to a delineation of how those stress distributions, buckling modes, and associated buckling loads vary as one varies specific material properties of the elastic annular plate.

We assume, as in [71], that the plate occupies the domain $b \leq r \leq \infty$, $0 \leq \theta < 2\pi$, and that it is subjected to applied tensile loads along its inner edge at $r = b$; thus, the approach taken consists of modifying the work of Mansfield [165] for the isotropic annular plate, which we have discussed in section 1 of this chapter, i.e., use is made of the in-plane stress distribution which corresponds to the one present in an infinite annular plate ($b \leq r < \infty$, $0 \leq \theta < 2\pi$) subjected to a radial traction along the edge of the hole at $r = b$.

To make a reasonable analogy with the situation which occurs in the buckling of a finite annular plate, the analysis in [71] makes use of the same reasoning as that employed in [165] (see Fig. 4.3 or, equivalently, Fig. 4.36): the outer radius $a$ of the annular plate is not infinite but, rather, is much greater than the inner radius $b$ and, in addition, that part of the plate for which $r \geq a$, $0 \leq \theta < 2\pi$ is constrained so as to prevent lateral movement but not in-plane movement. Thus, buckling may occur in the region $b \leq r \leq a$, $0 \leq \theta < 2\pi$, when the applied load along the inner edge at $r = b$ reaches a critical value; in [71] the applied tractions are not restricted to a uniform radial loading, as in Mansfield [165], but are assumed to be self-equilibrating and symmetric about the principal material axes.

### 4.1.4.1 The Prebuckling Stress Distributions

The stress distribution in an unbuckled elastic annular plate possessing rectilinear orthotropic symmetry is determined from the second of the von Karman equations which pertain to the problem at hand, with $w \equiv 0$, and the associated load conditions along the inner edge at $r = b$. The second von Karman equation for a rectilinearly orthotropic annular plate has the same form, of course, as in the case of a circular plate; it is obtained by first inserting the functions $a_{ij}(\theta)$, as given by (2.122a)– (2.122i), into the constitutive relations (2.123 a,b,c) and then substituting the resulting expressions for $e_{rr}, e_{\theta\theta}$, and $\gamma_{r\theta}$ into (2.124). Because we are only interested, at this point, in computing the prebuckling in-plane stress distribution in the plate, $W$ may be set equal to zero in (2.124). Alternatively, as in [71], one may deal (in numerical computations) directly with the constitutive relations (2.111) in Cartesian coordinate form, as well as with (1.35), with $w = 0$, (1.49), and the definition (1.46) of $\Phi$, coupled with the transformations (1.105) and (1.118) of the strain and stress tensor components from Cartesian to polar coordinates. Along the inner edge of the annulus, at $r = b$, we have the applied traction boundary conditions analogous to (1.125), i.e.,

$$
\begin{cases}
p_r(\theta) = \left\{ N_x \cos^2 \theta + N_y \sin^2 \theta + 2N_{xy} \cos \theta \sin \theta \right\}\big|_{r=b} \\
p_\theta(\theta) = \left\{ (N_y - N_x) \cos \theta \sin \theta - N_{xy} \left( \cos^2 \theta - \sin^2 \theta \right) \right\}\big|_{r=b}
\end{cases}
\tag{4.76}
$$

with $p_r(\theta)$ the radial component, and $p_\theta(\theta)$ the tangential component, of the applied traction at $r = b$. All stresses are required to be bounded as $r \to \infty$ and to assume the same values at $\theta = 0$ and $\theta = 2\pi$. In [71] the boundary conditions (4.76) are rewritten in integral form and the components of the applied tractions in the negative $x$ and $y$ directions are then expanded as function of $\theta$ in Fourier series; the results are subsequently substituted back into the integral form of the boundary conditions (4.76) so as to yield equivalent conditions with respect to the Fourier coefficients.

The boundary value problem for the prebuckling stress distribution which we have delineated above was first solved by Lekhnitskii [51]. We will content ourselves here with presenting some of the basic results which address both uniform as well as nonuniform radial loading along the edge at $r = b$. The basic material parameters which enter into the analysis in [71] are the ratios $E_1/E_2, E_1/G_{12}$, and the Poisson ration $\nu_{21}$; furthermore, in all the calculations reported in [71], it is assumed that $E_2/E_1 \geq 1$ so that the $y$-direction is the direction of maximum stiffness.

The first problem considered in [71] vis à vis prebuckling stress distributions involves the effect of varying the ratio $E_2/E_1$ of Young's moduli; in that problem, the shear modulus ratio $E_1/G_{12}$ is fixed at 2.6, while $\nu_{21} = 0.3$. The applied loading at $r = b$ is taken to be a uniform radial loading with $p_r(\theta) = p_i$, $p_\theta(\theta) = 0$. For $E_2/E_1 = 1$, it can be shown that the prebuckling stress distribution is axisymmetric, and the magnitudes of both radial and tangential stress components are proportional to $r^{-2}$, while the polar stress component is identically zero. For $E_2/E_1 \neq 1$, the prebuckling stress distribution is not axisymmetric and the polar shear stress is non-zero; sample results are displayed in Figs. 4.37, 4.38, and 4.39 for a ratio $E_2/E_1 = 10$, with the stress contour lines indicated for the first quadrant, and $b \leq r \leq 26$. The most significant effect of varying $E_2/E_1$ (from the isotropic case) is on the distribution of the tangential normal stresses that now exhibit a large compressive stress concentration at $r = b$, $\theta = 0$ given by

$$N_\theta(b,0) = p_i \left\{ \sqrt{\frac{E_2}{E_1}} - \sqrt{\frac{E_2}{E_1}\left(\frac{E_1}{G_{12}} - 2\nu_{12} + 2\sqrt{\frac{E_1}{E_2}}\right)} \right\} \qquad (4.77)$$

In Fig. 4.40 we show the effect of varying the ratio of Young's moduli on the tangential normal stress distribution at $\theta = 0$; note that at large radial distances from the edge, at $r = b$, the (prebuckling) tangential stress distributions in the isotropic plate and the inextensible plate $(E_2/E_1 = \infty)$ are nearly the same. Figure 4.41 depicts a comparison of the tangential normal stresses along the line $\theta = \dfrac{\pi}{2}$ for the cases $E_2/E_1 = 1$ and $E_2/E_1 = \infty$, and shows that these compressive stresses are influenced by changes in $E_2/E_1$, only at small radial distances from the edge at $r = b$. In Fig. 4.42, we illustrate the effect of changing $E_2/E_1$ on the tangential stress distribution along the edge at $r = b$. As $E_2/E_1$ increases, stress concentrations develop at $\theta = 0$ and $\theta = \dfrac{\pi}{2}$. Figure 4.43 shows the effect of varying the ratio $E_2/E_1$ of Young's moduli on $N_y$ (along the line $x = b$). The anisotropy ratios vary in the range $1 \leq E_2/E_1 \leq 10^5$ and, as this ratio increases, the stresses become compressive for all $y$, with increasing magnitudes all along $x = b$, leading to a compressive stress concentration along the line $x = b$. As noted in [71], the line $x = b$ is quite significant in a highly anisotropic annular plate, of the type under consideration here; it tends to separate the plate into two regions: $|x| < b$, in which $N_y$ is tensile, and $|x| \geq b$ in which $N_y$ is compressive, with an abrupt change taking place at $x = b$. The above discussion, when coupled with Figs. 4.37 through 4.43, highlights the considerable influence on the prebuckling stress distribution in an elastic (infinite) annular plate, which is exerted by variations in the ratio $E_2/E_1$ of Young's moduli.

The second problem treated in [71], in relation to the nature of the prebuckling stress distributions, involves the effect of varying the shear modulus ratio $E_1/G_{12}$. Taking the load along the edge at $r = b$ to again be a uniform radial load, we show in Figs. 4.44, 4.45, and 4.46, respectively, the effects of varying the shear modulus ratio on the radial normal, tangential normal, and polar shear stress distributions. Figure 4.44 shows that, along the line $\theta = \dfrac{\pi}{4}$, the radial stress decreases rapidly, even becoming slightly compressive, while Fig. 4.45 indicates the formation of two compressive stress concentrations at the edge $r = b$, for $\theta = 0$ and $\theta = \dfrac{\pi}{2}$; for $E_2/E_1 = 1$ these stress concentrations are of magnitude

$$
N_\theta\,(b,0) = N_\theta\left(b, \frac{\pi}{4}\right) = p_i\left[1 - \sqrt{\frac{E_1}{G_{12}} + 2\,(1 - \nu_{12})}\right] \tag{4.78}
$$

In Fig. 4.47 we depict the radial stress distribution along the line $\theta = \dfrac{\pi}{4}$ for various values of $E_1/G_{12}$ which range from 2.6 to $10^8$; the graph indicates that a compressive stress concentration begins to form near $r = \sqrt{2}b$ as the ratio $E_1/G_{12}$ increases. Figure 4.48 depicts, for $b \leq r \leq 2b$, the relative displaced shape of the elastic annular plate for a low shear modulus ($E_1/G_{12} = 100$) when it is subjected to a uniform radial load along the edge at $r = b$; the plate, as in [71], is in a state of bi-axial compression for $r > 2b$.

Also considered in [71] are problems which correspond to nonuniform distributions of applied tractions along the edge at $r = b$, i.e., applied loads which are given in (Cartesian) vector form by $(p_1 \cos\theta, p_2 \sin\theta)$ where $p_1$ and $p_2$ both act in the negative $x$ and $y$ directions, respectively, in the first quadrant of the annular a plate. In polar coordinates, the applied load thus has the form

$$
\begin{cases}
p_r(\theta) = p_1 \cos^2 \theta + p_2 \sin^2 \theta \\
p_\theta(\theta) = [p_2 - p_1] \sin \theta \cos \theta
\end{cases}
\tag{4.79}
$$

If $p_2 = p_1$ then (4.79) reduces to a uniform radial load (along the edge at $r = b$). When

$$
\frac{p_1}{p_2} = \sqrt{\frac{E_1}{E_2}} \left\{ \frac{1 + \sqrt{\dfrac{E_1}{G_{12}} - 2\nu_{12} + 2\sqrt{\dfrac{E_1}{E_2}} - \nu_{12}\sqrt{\dfrac{E_1}{E_2}}}}{\sqrt{\dfrac{E_1}{E_2}} + \sqrt{\dfrac{E_1}{G_{12}} - 2\nu_{12} + 2\sqrt{\dfrac{E_1}{E_2}} - \nu_{12}\sqrt{\dfrac{E_1}{E_2}}}} \right\}
\tag{4.80}
$$

a uniform radial displacement is obtained; the relation (4.79) is graphed in Fig. 4.49 as a function of $E_1/E_2$ for various values of the ratio $E_1/G_{12}$. Figures 4.50, 4.51, and 4.52 show, respectively, the radial normal, tangential normal, and polar shear stress distributions, in the region $b \le r \le 2b$ for the case in which the applied load acts only in the $y$ direction, so that $p_2 = p_i$ and $p_1 = 0$. At $\theta = 0$ we have a compressive stress distribution, while at $\theta = \dfrac{\pi}{2}$ the tangential stresses are tensile. Finally, Fig. 4.53 depicts the variations in the tangential stress distributions along the inner edge of the plate at $r = b$ as the ratio $p_1/p_2$ varies between zero and one; we note that, as $p_1$ increases, the magnitude of the compressive stress concentration at $\theta = 0$ decreases and the tangential stress at $\theta = \dfrac{\pi}{2}$ changes over to a compressive stress.

### 4.1.4.2  Initial Buckling and Postbuckling Behavior

In the last subsection we delineated some of the results, presented in [71], concerning the nature of the in-plane prebuckling stress distributions in an infinite elastic annular plate which exhibits rectilinearly orthotropic material symmetry and which, for $r \ge a > b$ ($b$ the hole radius), $0 \le \theta < 2\pi$, is constrained so as to prevent lateral movement. Because the prebuckling stress distributions enter the partial differential equation governing the initial buckling of the annular plate, variations in the material anisotropy ratios, which greatly affect the prebuckling stress distributions, will also influence the plate's buckling behavior.

The partial differential equation which governs the initial buckling of the elastic, rectilinearly orthotropic, annular plate is (1.109) with $\Phi = \Phi^0(r, \theta)$ being the Airy function associated with the in-plane prebuckling stress distribution. In (1.109), $[w, \Phi]$ is given by (1.74) and (1.75b) while the $\tilde{D}_j, j = 1, 2, 3, 6$, and the coefficients $\tilde{D}_{12}, \tilde{D}_{16}$, and $\tilde{D}_{26}$, are given by (1.108) and (1.110) with $D_1 = D_{11}, D_2 = D_{22}$, and $D_3 = D_2\nu_{12} + 2D_{66}$, the principal rigidities, defined as in (1.63), (1.64). In Cartesian coordinates (1.109), with $\Phi = \Phi^0$, assumes the simpler form (1.68), with $\Phi = \Phi^0(x, y)$, where $b \leq \sqrt{x^2 + y^2} \leq a$.

**Remarks:**

(i) If $V$ denotes the potential energy function associated with the rectilinearly orthotropic annular plate, then for positive material coefficients $E_1, E_2, \nu_{21}$, and $G_{12}, \delta^4 V > 0$, which implies that at the smallest eigenvalue associated with the initial buckling equation the associated equilibrium state is stable with respect to small perturbations.

(ii) If $w(x, y)$ is decomposed as

$$w(x, y) = w_e(x, y) + w_0(x, y) \tag{4.81}$$

$w_e(x, y) = w_e(x, -y), w_0(x, y) = -w_0(x, -y)$, then $w_e(x, y)$ and $w_0(w, y)$ separately satisfy (1.68), with $\Phi = \Phi^0(x, y)$, if the applied load is even about the $x$-axis (the averaged stress components $N_x^0$ and $N_y^0$ being even functions, while $N_{xy}^0$ is an odd function about the $x$-axis). In addition, the $2\pi$-periodicity of $w$, as a function of $r$ and $\theta$, implies that $w_0$ and its even-order derivatives with respect to $y$ vanish at $y = 0$, while odd-order derivatives of $w_e$ with respect to $y$ also vanish at $y = 0$; this implies that the symmetric and antisymmetric components of the buckling modes are uncoupled from each other so that the buckled mode shape is either symmetric or antisymmetric about the $x$-axis. In a similar fashion, for applied loads which are symmetric about the $y$-axis, the buckled modes are either symmetric or antisymmetric with respect to the $y$-axis. When the loads are not symmetric with respect to the principal material directions in the rectilinearly orthotropic annular plate, the even and odd components of the buckling modes will be coupled together; such general conditions were not considered in [71].

(iii) Boundary support conditions must be associated with (1.109), with $\Phi = \Phi^0$, at $r = b$ and $r = a$ in order to have a well-defined eigenvalue problem. The analysis presented in [71] is valid only for the case of clamped supports at the inner and outer edges of the unsupported region of the plate; therefore, it is noted in [71] that the buckling behavior which we delineate, below, will change if different boundary support conditions are used or if we employ the prebuckling stress distributions that are associated with a finite annulus.

(iv) For the analysis reported in [71], a finite-difference scheme was deemed to be the most appropriate method of solution to solve the eigenvalue problem and determine the buckling mode shapes; because of the symmetry assumptions with respect to the loads, considerations could be limited to the first quadrant of the annular region $b \leq r \leq a$. As noted in [71], using a more sophisticated scheme such as a finite-element approach could, conceivably, allow for consideration of different loading and boundary support conditions along the edges of the plate, but any such work would be complicated by the expected appearance of high stress concentrations and singular behavior if the annular plate is highly anisotropic.

(v) The critical buckling load is taken to be the smallest positive eigenvalue of the reduced system obtained by applying a finite-difference approximation to (1.109), with $\Phi = \Phi^0$, and the associated clamped support conditions at $r = b$ and $r = a$. To verify the convergence of the finite-difference method, a study of the convergence of the scheme was conducted in [71] for an isotropic elastic annular plate with an aspect ratio of $a/b = 2.0$ and clamped edges at $r = b$ and $r = a$; for this particular case, the critical buckling load is determined from the relation $p_i b^2/K = 224.14$ (and a wave number of $n = 10$ is obtained), while the finite-difference scheme predicts values for the critical buckling load (see Table 4.2) which are lower than the exact value but appear to converge to it as the mesh size increases. In Table 4.2, $m$ denotes the number of divisions in the finite-difference scheme in the radial direction, while $n$ denotes the number of divisions in the tangential direction. In Fig. 4.54, we show the predicted mode shapes obtained for the isotropic annular plate as compared with the exact solution; excellent results are also obtained for the predicted mode shapes in the tangential direction.

For the isotropic annular plate, the convergence of the critical buckling load, with increasing mesh size, is reported in Table 4.3 for an aspect ratio of $a/b = 30.0$; once again, we have convergence to the exact value, which is given by $p_i b^2/K = 11.26$ (with an associated wave number of $n = 2$), but the convergence is slower than in the case where $a/b = 2.0$. The general buckling mode shape is already clear with the use of a $10 \times 10$ finite-difference mesh.

We now turn to the results obtained by applying the finite-difference scheme to approximate the initial buckling of the rectilinearly orthotropic elastic annular plate which is assumed to be subjected to tensile loads on its inner edge at $r = b$; we are interested, in particular, in the changes in the buckled mode shapes which are due to variations in the ratio $E_2/E_1$

of Young's moduli and variations in the shear modulus, as well as in the effect of using nonuniform applied loadings.

In Figs. 4.55 and 4.56, we displace contour plots of the lateral displacement associated with the buckled mode shapes for Young's moduli ratios of $E_2/E_1 = 1$ and $E_2/E_1 = 100$, respectively, an aspect ratio of $a/b = 2.0$ (assumed) symmetry about the $x$ and $y$ axes; the mesh size used was $10 \times 45$. Of course, Fig. 4.55 displays the results for an isotropic annular plate and the wave number obtained, i.e., $n = 10$, corresponds to what is expected from the exact solution; wrinkling, in the tangential direction, occurs at 18-degree intervals and we note that these wrinkles become wider as $n$ increases. For $\dfrac{E_2}{E_1} = 1$, the isotropic annular plate, 20 wrinkles occur over the entire plate; for $E_2/E_1 = 10$ this drops to 10 wrinkles in the buckled mode shape over the entire plate, and the wrinkles are no longer oriented in the radial direction as in the isotropic case. As the plate tends towards anisotropy, the wrinkles occurring in the buckled mode shapes become more prominent as $E_2$ increases in relation to $E_1$. For $E_2/E_1 = 100$ (Figure 4.56) the number of fully developed wrinkles in the buckled mode shape drops down to six. With $E_2/E_1 = 1, 10$, and $100$ the buckled mode shapes were again generated, but this time for an aspect ratio of $a/b = 10.0$ in place of $a/b = 2.0$; the results are displayed in Figs. 4.57, 4.58, and 4.59, respectively. For $E_2/E_1 = 1$ (isotropic annular plate) the wave number, in this situation, is $n = 4$. As the ratio of Young's moduli $E_2/E_1$ is increased, first to 10 and then to 100, the wrinkle which was oriented in the isotropic plate along the line $\theta = \dfrac{\pi}{4}$ becomes narrower and its orientation moves closer to the $y$ axis. On the other hand, the wrinkle centered on the $x$-axis in the buckled mode shape, for the isotropic plate, broadens out as the ratio $E_2/E_1$ is increased. The cases illustrated in Figs. 4.55 through 4.59 illustrate buckling modes which are symmetric with respect to the $x$-axis. In Fig. 4.60 we depict a mode shape which is anti-symmetric with respect to the $y$-axis; in this case, the calculations produce wrinkles which are somewhat narrower and more closely oriented toward the $y$-axis. One may summarize the results reported above by noting that, for a rectilinearly orthotropic elastic annular plate, as the ratio $E_2/E_1$ increases, wrinkles in the buckled mode shapes located near the $y$-axis tend to narrow, while those located near the $x$-axis tend to broaden and the orientation of the wrinkles shifts from the radial direction to the direction of the ray $\theta = \dfrac{\pi}{2}$.

The second problem considered in [71], with respect to the initial buckling behavior of an elastic rectilinearly orthotropic plate, concerns the effect on the buckling mode shapes of varying the shear modulus; some of the results obtained in [71] are depicted in Figs. 4.61, 4.62, and 4.63; in all these figures, the shear modulus ratio is 100, while the aspect ratio $a/b$ varies over 2, 5, and 10. For each of those cases in which the shear modulus $G_{12}$ is much lower than Young's moduli, the plate buckles along the rays $\theta = \pm\frac{\pi}{4}$ because, along these directions, the stiffness of the plate is a minimum and the compressive tangential and radial stresses combine so as to buckle the annular plate.

The effect of nonuniform applied tractions on the buckled mode shape of the rectilinearly orthotropic annular plate is depicted in Fig. 4.64 for the case in which $a/b = 2.0$, $E_2/E_1 = 10.0$, $E_1/G_{12} = 2.6$, and $\nu_{21} = 0.3$. The applied loads in this case assume the form (4.79) and the ratio $p_1/p_2$ goes from 0.0 to 0.8; it is easy to see from Fig. 4.64 that changes in the ratio $p_1/p_2$ significantly influence the buckled mode shapes, with increases in $p_1/p_2$ leading to wrinkles in the buckled mode shape which shift from being centered around the $x$-axis, and oriented in the $y$-direction, to being centered around the $y$-axis and oriented in the $y$-direction. Note that at the ratio of $p_1/p_2 = 0.712$ the wrinkles are distributed around the entire circumference of the annular plate. Because of the anisotropy of the plate, those wrinkles which are close to the $y$-axis are relatively narrow, while those which are close to the $x$-axis are relatively broad.

Figures 4.65 through 4.67 depict the results of calculations carried out in [71] to determine the buckled mode shapes for a rectilinearly orthotropic elastic annular plate with orthotropic in-plane properties and isotropic bending properties; in all three cases shown, $a/b = 2.0$, $\nu_{21} = 0.3$, and the assumption of isotropic bending means that $D_2/D_1 = 1.0$. In Fig. 4.65, the orthotropy is due to the large ratio $E_2/E_1 = 10^3$ of Young's moduli, and the plate is subjected to a uniform radial traction, while the same conditions apply in the case depicted in Fig. 4.66 except that the plate is subjected to a uniform radial displacement. In the case shown in Fig. 4.67 we again employ a uniform radial traction but now the orthotropic in-plane behavior is a consequence of the large shear modulus ratio, i.e., $E_1/G_{12} = 100.0$. As can easily be seen in Fig. 4.65 the largest buckle, in this case, occurs along the ray $\theta = \frac{\pi}{4}$; this may be compared with the buckling that can be expected to occur around the $y$-axis if the ratio of the bending properties were the same as the ratio of the in-plane material properties. When a uniform radial displacement is applied at the inner edge of the plate, at $r = b$, (Fig. 4.66) the largest buckle again appears near the ray $\theta = \dfrac{\pi}{4}$, with the wrinkle oriented uniformly in the $x$-and $y$-directions because of the equality of the two bending stiffnesses. Finally, in Fig. 4.67, we see that when the in-plane orthotropy results from the large ratio of $E_1/G_{12} = 100$, and the plate is subjected to a uniform radial traction at $r = b$, the most pronounced buckle still develops along the ray at $\theta = \dfrac{\pi}{4}$ because the combination of compressive radial and tangential stresses easily buckles the annular plate in the vicinity of this direction.

Finally, the issue of the way in which the critical buckling loads vary for the various cases studied has also been addressed in [71]. In Table 4.4, we show the changes which occur in the predicted values of the buckling loads as the ratio of Young's moduli varies; the results clearly indicate that the critical buckling load increases with increasing stiffness of the annular plate in the $y$-direction, i.e., with $E_1$ (and, thus, $D_1$ fixed ), that $p_i$ increases as $E_2$ increases, and that the critical buckling load also increases with increasing stiffness of the annular plate in the $x$-direction, i.e., with $E_2$ (and, thus, $D_2$ fixed), $p_i$ increases as $E_1$ increases. In Table 4.5, we record the effect on the critical buckling load of the annular plate as the shear stiffness $G_{12}$ decreases: when the overall stiffness (as measured by $E_2/G_{12}$) increases, the critical buckling loads increase very rapidly, while for $E_2$ (and hence, $D_2$) are fixed, with the shear stiffness $G_{12}$ decreasing, the critical buckling load exhibits a slow decrease.

## 4.2 Plates which Exhibit Nonlinear Elastic, Viscoelastic, or Elastic-Plastic Behavior

For neither finite, nor infinite annular plates does there appear to have been any serious attempt in the literature to deal with either initial or postbuckling behavior when the plate is constituted of a nonlinear elastic or a viscoelastic material; this state of affairs is not surprising when one considers the analogous situations for buckling of nonlinear elastic or viscoelastic circular plates. Certainly, the approach taken by Brilla [106], [140] in modeling the buckling behavior of viscoelastic circular plates, could be easily extended so as to consider both the initial and postbuckling behavior of viscoelastic annular plates, while the Johnson-Urbanik theory, that has been described in Chapter 1, should be susceptible to being extended to cover the initial and postbuckling behavior of nonlinear elastic annular plates (once it has been extended from rectangular to circular geometries).

The situation with regard to buckling of elastic-plastic annular plates is somewhat better than that for viscoelastic or nonlinear elastic annular plates, at least as far as initial buckling behavior goes; an important work in this area is that of Yu and Johnson [179]. In [179] the authors consider the problem of buckling of a finite elastic-plastic annular plate of inner radius $r = b$ and outer radius $r = a$ when the plate is subjected to a uniform radial tensile load along its inner edge.

To compute the elastic buckling load for this problem, it is assumed in [179] that the deflected shape of the annular plate has the form

$$w(r, \theta) = c(r - b)(1 + \cos n\theta) \tag{4.82}$$

Using (4.82), one computes for the bending energy of the plate

$$
\begin{aligned}
V_B^e &= \int_0^{2\pi} \int_b^a \left\{ \frac{1}{2} K \left( \frac{\partial^2 w}{\partial r^2} + \frac{1}{r} \frac{\partial w}{\partial r} + \frac{1}{r^2} \frac{\partial^2 w}{\partial \theta^2} \right)^2 \right. \\
&\quad - K(1 - \nu) \frac{\partial^2 w}{\partial r^2} \left( \frac{1}{r} \frac{\partial w}{\partial r} + \frac{1}{r^2} \frac{\partial^2 w}{\partial \theta^2} \right) \\
&\quad \left. + K(1 - \nu) \left( \frac{1}{r} \frac{\partial^2 w}{\partial r \partial \theta} - \frac{1}{r^2} \frac{\partial w}{\partial \theta} \right)^2 \right\} r \, dr \, d\theta \\
&= \frac{\pi}{2} K c^2 \left\{ \left[ 2 + (n^2 - 1)^2 \ln \frac{1}{\gamma} \right. \right. \\
&\quad \left. \left. + n^2 (1 - \gamma) \left[ (1 + \gamma) \left( \frac{n^2}{2} + 1 - \nu \right) - 2(n^2 - 1) \right] \right\} \right. \\
&\equiv \frac{\pi}{2} K c^2 F(n, \gamma)
\end{aligned}
\tag{4.83}
$$

where $\gamma = b/a$ and $F(n, \gamma)$ has the obvious definition which is implied by (4.83). If the annular plate is subjected to a radial tensile stress $\sigma_{rr}|_{r=b} = \lambda$ along its inner edge, then the prebuckling elastic stress distribution in the plate is given by

$$
\begin{cases}
\sigma_{rr}^0 = \dfrac{\lambda b^2}{a^2 - b^2} \left( \dfrac{a^2}{r^2} - 1 \right) \\[4mm]
\sigma_{\theta\theta}^0 = -\dfrac{\lambda b^2}{a^2 - b^2} \left( \dfrac{a^2}{r^2} + 1 \right)
\end{cases}
\tag{4.84}
$$

which is analogous to (but not identical with) the prebuckling stress distribution in an isotropic elastic annular plate subjected to a uniform radial compressive force applied at the outer edge at $r = a$, i.e., (4.19 a,b). By virtue of (4.82) and (4.84) the work done by the membrane forces in the plate is given by

$$
\begin{aligned}
\mathcal{W}^e &= -\frac{1}{2} \int_0^{2\pi} \int_b^a \left\{ \sigma_{rr}^0 h \left( \frac{\partial w}{\partial r} \right)^2 + \sigma_{\theta\theta}^0 h \left( \frac{1}{r} \frac{\partial w}{\partial \theta} \right)^2 \right\} r\,dr\,d\theta \\
&= \frac{\pi}{2} hc^2 \frac{\lambda b^2}{a^2 - b^2} \left\{ 3 \left[ \frac{1}{2} \left( a^2 - b^2 \right) - a^2 ln \frac{1}{\gamma} \right] \right. \\
&\qquad \left. + n^2 \left[ \left( b^2 - a^2 \right) + \left( b^2 + a^2 \right) \ln \frac{1}{\gamma} \right] \right\} \\
&= \frac{\pi}{2} hc^2 \frac{\lambda a^2 \gamma^2}{1 - \gamma^2} \left\{ 3 \left[ \frac{1}{2} \left( 1 - \gamma^2 \right) - \ln \frac{1}{\gamma} \right] \right. \\
&\qquad \left. + n^2 \left[ (\gamma^2 - 1) + (\gamma^2 + 1) \right] \right\} \\
&\equiv \frac{\pi}{2} hc^2 \lambda a^2 G(n, \gamma)
\end{aligned}
\tag{4.85}
$$

where $G(n, \gamma)$ has the obvious definition which is implied by (4.85). By equating $\mathcal{V}_B^e$, as given by (4.83), with $\mathcal{W}^e$, as given by (4.85), we obtain the critical condition for buckling in the elastic domain of the annular plate in the form

$$
\lambda_{cr} \frac{a^2 h}{K} = \frac{F(n, \gamma)}{G(n, \gamma)}
\tag{4.86}
$$

or, as $K - Eh^3/12(1 - \nu^2)$,

$$
\lambda_{cr} = \frac{1}{12(1 - \nu^2)} \left( \frac{h}{a} \right)^2 \cdot \frac{F(n, \gamma)}{G(n, \gamma)}
\tag{4.87}
$$

For $\nu = 0.3$, some numerical results for the problem delineated above are depicted in Fig. 4.68, as well as in Fig. 4.69, where $\xi = 1 - \frac{b}{a} \equiv 1 - \gamma$.

When the annular plate has a large outer radius $a$, but is relatively thin, i.e. $\frac{a - b}{a}$ is small, Yu and Johnson [179] note that initial buckling will usually be governed by the elastic model described above but, for $\frac{a - b}{a} \approx 1$, the stress distribution in the plate which is associated with $\lambda_{cr}$, as given by (4.87), will usually cause yielding before buckling; in this case

$$
(\sigma_{rr} - \sigma_{\theta\theta})|_{r=b} = \lambda_{cr} \frac{2a^2}{a^2 - b^2} = \lambda_{cr} \frac{2}{1 - \gamma^2} = Y
\tag{4.88}
$$

or

$$\lambda_{cr} = Y\frac{1-\gamma^2}{2} \equiv \frac{1}{2}\xi(2-\xi)Y \qquad (4.89)$$

where $Y$ is the yield-stress associated with the elastic-plastic material constituting the plate; plastic yield will first occur at the inner edge of the plate at $r = b$. Equation (4.89) imposes a limitation with respect to the applicability of (4.87) for determining the critical buckling load of the elastic-plastic annular plate; using (4.87), in conjunction with (4.89), this limitation can be expressed in the form

$$\frac{a}{h} \geq \sqrt{\frac{F(n,\gamma)}{G(n,\gamma)} \cdot \frac{E}{Y} \frac{1}{G(1-\nu^2)(1-\gamma^2)}} \qquad (4.90)$$

As an example of the utility of (4.90), Yu and Johnson [179] offer the following example: with $E/Y = 500$, $\nu = 0.3$, and $\gamma = 0.6$ computations yield the result $F(n,\gamma)/G(n,\gamma) = 59.1$ so that (4.90) is equivalent to $a/h \geq 92.0$; as a consequence, for $\xi a/h \geq 36.8$ elastic buckling of the annular plate occurs, while for $\xi a/h < 36.8$ the tensile force given by (4.87) will cause yield before elastic buckling occurs.

If initial elastic buckling does not occur, and the annular plate is in a fully plastic stress state, as loading proceeds, the plastic stress distribution in the plate will be given by

$$\sigma_{rr} - \sigma_{\theta\theta} = Y \qquad (4.91)$$

with

$$\begin{cases} \sigma_{rr} = Y\ln\left(\frac{a}{r}\right) > 0 \\ \sigma_{\theta\theta} = Y\left[\ln\left(\frac{a}{r}\right) - 1\right] < 0 \end{cases} \qquad (4.92)$$

Assuming, once again, that the deflection has the form (4.82) one may compute the work done by the membrane stresses to be

$$W^p = \frac{\pi}{8}hYc^2a^2H(n,\gamma) \qquad (4.93)$$

where

$$H(n,\gamma) = (n^2-3)(1-\gamma^2-2\gamma^2\ln\frac{1}{\gamma}) - 2n^2\gamma^2\left(\ln\frac{1}{\gamma}\right)^2 \qquad (4.94)$$

For plastic buckling, the strain energy of bending takes the same form as $V_B$ in (4.83) but the bending rigidity $K$ needs to be changed now to

$$K^p = E_0h^3/12(1-0.5^2) \equiv E_0h^3/9 \qquad (4.95)$$

where the buckling modulus $E_0$ is given by

$$E_0 = \frac{4EE_t}{(\sqrt{E} + \sqrt{E_t})^2} \tag{4.96}$$

with $E_t$ the tangent modulus of the elastic-plastic material constituting the annular plate. Thus, for the strain energy of bending, we now have

$$\mathcal{V}_B^p = \frac{\pi}{2} K^p c^2 F^p(n, \gamma) \tag{4.97}$$

with

$$F^p(n, \gamma) = \left[2 + (n^2 - 1)^2\right] \ln(1/\gamma)$$
$$+ n^2 (1 - \gamma) \left[\frac{1}{2}(1 + \gamma)(n^2 + 1) - 2(n^2 - 1)\right] \tag{4.98}$$

The critical condition for plastic buckling is now obtained by setting $\mathcal{V}_B^p = \mathcal{W}^p$ which yields

$$\frac{E_0}{Y} \left(\frac{h}{a}\right)^2 = \frac{9}{4} \frac{H(n, \gamma)}{F^p(n, \gamma)} \tag{4.99}$$

and implies that plastic buckling will occur when

$$\zeta = \sqrt{\frac{E_0}{Y}} \left(\frac{h}{a}\right) < \frac{3}{2} \sqrt{\frac{H(n, \gamma)}{F^p(n, \gamma)}} \tag{4.100}$$

Computational results of Yu and Johnson [179] for various values of $n$ are shown in Figs. 4.70 and 4.71 and a comparison with the analytical and experimental results obtained earlier by Geckeler [180] and Senior [181] is illustrated in Fig. 4.72.

---

## 4.3 Comparisons of Initial and Postbuckling Behavior of Annular Plates

In this section, we examine the effects that varying the boundary support conditions, the material symmetry, and the constitutive response have on the initial buckling and postbuckling behavior of thin annular plates; because of the relative lack of information regarding the non-linear elastic, viscoelastic, and plastic behavior of annular plates, our attention is focused on variations in boundary support and material symmetry assumptions for linearly elastic plates.

We begin our discussion by looking at the problem of initial buckling for isotropic linearly elastic annular plates and, in particular, at the case of a finite annular plate of inner radius $b$ and outer radius $a > b$, which is subjected to uniform compression along both edges; the effect of varying the boundary support conditions, in this situation, is depicted on the graphs in Figs. 4.1 and 4.2 where $\sqrt{k}$ is graphed against $\gamma = b/a$, $k = \lambda_{cr} \dfrac{a^2 h}{K}$, for the following cases:

(i)  The edges at $r = a$, $r = b$ are both clamped.

(ii)  The edge at $r = a$ is clamped; the edge at $r = b$ is simply supported.

(iii)  The edge at $r = a$ is clamped; the edge at $r = b$ is free.

(iv)  The edge at $r = a$ is simply supported; the edge at $r = b$ is clamped.

(v)  The edges at $r = a, r = b$ are both simply supported.

(vi)  The edge at $r = a$ is simply supported; the edge at $r = b$ is free.

(vii)  The edge at $r = a$ is free; the edge at $r = b$ is clamped.

(viii)  The edge at $r = a$ is free; the edge at $r = b$ is simply supported.

For most of the cases delineated above, it is clear from Figs. 4.1 and 4.2 that the critical edge buckling stress $\lambda_{cr}$ increases as $\gamma = b/a$ increases, i.e., as the hole size increases relative to the radius of the outer edge of the plate; only in cases (vi) and (viii) is the opposite behavior observed, i.e., when one edge is free, while the other edge is simply supported. In most of the cases listed, it is noted that the plate buckles, initially, into the axially symmetric mode; noted exceptions may be found in case (i), both edges clamped, where the number of diametric nodes in the initial buckling mode increases as $\gamma = b/a$ increases, case (ii) where the plate first buckles with one diametric node but then buckles (initially) into the axisymmetric mode with increasing relative hole size, cases (iv) and (v) where a similar behavior is observed, and case (vii), in which the outer edge is free, while the inner edge is clamped, and where the plate first buckles into a mode with one diametric node, then two, as $\gamma$ is increased, then back to one, and finally into the axially symmetric mode for sufficiently large $\gamma$. Figures 4.1 and 4.2 also indicate that, for a fixed $\gamma$, $0 \le \gamma < 1$, the critical buckling load increases as the edges are more severely restricted with respect to lateral movement, i.e., at $\gamma = b/a = 0.4$ we have $\sqrt{k} = 10$ when both edges are clamped, $\sqrt{k} = 8$

when the edge at $r = a$ is clamped but the edge at $r = b$ is simply supported, $\sqrt{k} \simeq 7.2$ when the edge at $r = a$ is simply supported while the edge at $r = b$ is clamped, $\sqrt{k} \simeq 5.4$ when the edges at both $r = a$ and $r = b$ are simply supported, $\sqrt{k} \simeq 3.4$ when the edge at $r = a$ is clamped while the edge at $r = b$ is free, and so forth.

The second basic situation associated with the initial buckling of an isotropic annular plate is the one considered first by Mansfield [165], i.e., the buckling of an infinite annular plate of hole size $b$, which is supported by a member of sectional area $ah/(1 + \nu)$ along the circle of radius $r = a$, and which is subjected to either tension or compression along the edge at $r = b$. If we refer to Figs. 4.4 and 4.5, we see that, with compressive loading along the edge at $r = b$, the plate will always buckle into the axially symmetric mode, while with tensile loading along the edge at $r = b$ the plate buckles initially into modes with several diametric nodes. From Fig. 4.4, it is clear that (for all the combinations of boundary support conditions indicated) increases in $\gamma = b/a$ (i.e., decreases in $1/\gamma = a/b$) lead to decreases in the normalized buckling parameter $k(1-\gamma)^2/\gamma$; however, as $\gamma = b/a$ increases, $f(\gamma) = (1-\gamma)^2/\gamma$ decreases; and, thus, it is difficult to gauge from Fig. 4.4 the effect on the buckling parameter $k = \lambda_{cr}\dfrac{a^2h}{K}$ of increasing the ratio $b/a$. What is clear from Fig. 4.4 is that the critical buckling edge stress $\lambda_{cr}$ increases, for any fixed $\gamma$, $0 \leq \gamma < 1$, as one further restricts the joint lateral movement of the plate along the edge at $r = b$ and along the circle of radius edge at $r = a$. In Fig. 4.5 we show some results associated with applying an (inward) tensile stress along the edge at $r = b$ ; for both the case in which the plate is clamped at $r = a$ and $r = b$, and the case in which the plate is simply supported at $r = a$ and $r = b$, the parameter $k\dfrac{(1-\gamma)^2}{\gamma}$ again decreases as $\gamma = b/a$ increases. Now, however, the annular plate buckles into modes possessing several diametrical nodes. At any fixed value of $\gamma = a/b$ more diametrical nodes appear in the initial buckling modes when the plate is clamped at both $r = a$ and $r = b$ as opposed to the case where the plate is simply supported at $r = a$ and $r = b$ ; in both cases, the number of diametrical nodes in an initial buckling mode increases as the aspect ratio $\gamma = a/b$ of the plate increases. By comparing Figs. 4.4 and 4.5 it is easy to see that, for any given aspect ratio $\gamma$, the critical buckling loads for the case when the annular plate is subjected to a tensile load along the edge at $r = b$ are much higher than the critical buckling loads which are indicated for the case in which the plate is subjected to a compressive loading along the edge at $r = b$; this is true both for the clamped boundary conditions at $r = a$, $r = b$ and for the simply supported boundary conditions at $r = a$ and $r = b$. For both types of loading along the edge at $r = b$, and any fixed $\gamma$, the clamped boundary conditions at $r = a$ and $r = b$ lead to higher values of $\lambda_{cr}$ than the simply supported boundary conditions at $r = a$ and $r = b$.

The last major subcase associated with the initial buckling of an isotropic linearly elastic annular plate is the one in which the plate is subjected to a uniform radial compressive loading only along the edge at $r = a$, with the outer edge clamped and the edge of the hole at $r = b$ free; this is the case first considered by Majumdar [166] and, later, by Troger and Steindl [66] and Machinek and Troger [167]. The initial buckling curves are depicted in Fig. 4.7, which is (roughly) equivalent to Fig. 4.10. Because results appear to exist in the literature for only one set of boundary support conditions, we note only that as the aspect ratio $\gamma = b/a$ is increased, the critical buckling edge stress $\lambda_{cr}$ increases and the number of diametrical nodes in the initial buckling mode also increases. Up to an aspect ratio of approximate 0.5, the buckling mode is axially symmetric.

Figures 4.12 through 4.17 provide the basis for making a comparative study of the initial buckling behavior of cylindrically orthotropic annular elastic plates as one varies the material symmetry assumptions (i.e., the degree of orthotropy) as well as the boundary support conditions. If we recall that $\beta = E_\theta/E_r$, then the case of an isotropic elastic annular plate corresponds to $\beta = 1.0$ so that the magnitude of $\beta$ gauges the deviation from isotropy. In Figs. 4.12 through 4.17 the buckling parameter

$$\mu = -\lambda_{cr}ha^2/D_r,$$

$$D_r = E_r h^3/12(1 - \nu_r\nu_\theta)$$

is graphed against the plate aspect ratio $b/a$ for various combinations of boundary support conditions along the edges of the plate at $r = a$ and $r = b$ ; the plate is assumed to be subjected to an in-plane compressive loading along one or both edges. For all the combinations of boundary support conditions considered, it is easy to see that for any fixed value of $\gamma = b/a$, the critical buckling edge stress $\lambda_{cr}$ increases as the degree, $\beta$, of orthotropy increases. For the first three cases considered (both edges clamped and loaded, both edges simply supported and loaded, outer edge fixed and loaded and inner edge free), $\lambda_{cr}$ increases monotonically with $\gamma = b/a$ for any degree of orthotropy $\beta$. For the fourth and sixth case (outer edge simply supported and loaded, inner edge free and outer edge simply supported, inner edge free, with both edges loaded) $\lambda_{cr}$ decreases monotonically with increasing $\gamma = b/a$ for any degree of orthotropy. Finally, in the fifth case depicted, where the outer edge is fixed, the inner edge is free, and both edges are loaded, it is only in the isotropic case that $\lambda_{cr}$ increases monotonically with $\gamma = b/a$; in all the cases $(\beta > 1.0)$, $\lambda_{cr}$ first decreases and then increases as $\gamma = b/a$ increases with the initial rate of decrease of $\lambda_{cr}$ higher the larger the degree, $\beta$, of orthotropy. If one fixes any particular value of $\beta$ in Figs. 4.12 through 4.17 and looks at the values of $\mu$ (i.e. $\lambda_{cr}$) for a given value of the aspect ratio, then it is easy to isolate the effect that varying the boundary support conditions along the edges at $r = a$, $r = b$ has on the initial buckling of the elastic annular plate; for example, with $\beta = 5.0$ and $\gamma = 0.6$ the (approximate) values of $\mu$ are

1. $\mu \simeq 200$ (both edges clamped and loaded)

2. $\mu \simeq 75$ (both edges simply supported and loaded)

3. $\mu \simeq 50$ (outer edge clamped and loaded, inner edge free)

4. $\mu \simeq 15$ (outer edge simply supported and loaded, inner edge free)

5. $\mu \simeq 25$ (outer edge clamped, inner edge free, both edges loaded)

6. $\mu \simeq 8$ (outer edge simply supported, inner edge free, both edges loaded)

so that we once again see the trend toward higher (initial) buckling loads as the (joint) degree of lateral movement allowed at the edges of the plate is increasingly restricted.

Because of the noted mismatch between the circular geometry of annular elastic plates and the other deviation from isotropic material symmetry we have discussed, namely, rectilinear orthotropy, it is not possible to directly mesh this latter variation from isotropic response into the comparative study delineated above; this is due, analytically, to our inability to compare, in a simple way, the measure $\bar{\beta} = E_2/E_1$ of deviation from isotropy in the rectilinearly orthotropic case with the measure $\beta = E_r/E_\theta$ of deviation from isotropy in the cylindrically orthotropic case. Also, for an elastic (circular) annular plate exhibiting rectilinear orthotropic material symmetry, the available data base, even for the problem of initial buckling, appears to be limited to the results in [71] ; these results, in turn, are limited to the type of loading and support conditions first studied for isotropic material response by Mansfield [165] and are, in essence, delineated in Tables 4.4 and 4.5 (critical loads versus the ratios of Young's moduli and shear moduli) as well as in Figures 4.55 through 4.67 (variations in buckled mode shapes).

We now want to look at the effect of varying the nature of the material symmetry assumption (i.e. degree of orthotropy $\beta$) and the type of boundary support conditions imposed along the edges, on the postbuckling behavior of elastic annular plates. Because there appear to be no results available for (circular) elastic annular plates with rectilinear orthotropic material symmetry, our discussion will be confined to those postbuckling results which are available for cylindrically orthotropic plates; all such results appear to be restricted to axially symmetric deformations of the plate. In [70] the outer edge at $r = a$ is clamped and subjected to a uniform, in-plane, compressive loading, while the inner edge at $r = b$ is free; from Fig. 4.23 it is easily seen that with $\Lambda = \lambda/\lambda_{cr}$, (the normalized postbuckling load), the maximum deflection not only increases with $\Lambda$ for each fixed $\beta = E_r/E_\theta$ and fixed value of $\gamma = b/a$, but for each fixed value of the aspect ratio $\gamma = b/a$ and each fixed postbuckling load $\lambda > \lambda_{cr}$, the maximum deflection $W_{\max}$ increases as the variance from isotropy increases. The very best comparative results which are available in connection with the postbuckling behavior of elastic annular plates may be those which follow from the work of Huang [175]; in this case, the inner edge at $r = b$ is free, the outer edge at $r = a$ is subjected to a uniform radial compressive loading, and that edge may be either clamped or simply supported. Thus, in this case, it is possible to gauge the effect on the postbuckling behavior of the plate of varying both the boundary support conditions and the degree of deviation from isotropic material response. Taking the measure of plate deflection to be $|w(\gamma)/h|$, $\gamma = b/a \equiv 0.4$, Table 4.1 shows the effect on the postbuckling deflection of varying the postbuckling load $\lambda > \lambda_{cr}$ for the case of a simply supported edge at $r = a$, for $\beta = 1$, 3, 5, and 10, and for the case of a clamped edge at $r = a$ for $\beta = 5$. For $\beta = E_r/E_\theta = 5$, we see that achieving particular values of $|w(\gamma)/h|$ in the case where the edge at $r = a$ is clamped requires a postbuckling load about three times larger than that which is needed to produce the same deflection when the edge at $r = a$ is simply supported. Also, in addition to the obvious conclusion that for both types of boundary support conditions at $r = a$, and any fixed $\beta$, most buckling deflections increase monotonically with increasing $\lambda > \lambda_{cr}$, we also note (at least for the case in which the outer edge of the plate at $r = a$ is simply supported) that achieving any particular value of $|w(\gamma)/h|$ requires an increasing postbuckling edge stress

$\lambda > \lambda_{cr}$ as the degree, $\beta$, of orthotropy increases. For $\beta = 5.0$, Fig. 4.26 depicts the effect of varying the boundary support conditions at $r = a$ on the postbuckling behavior; a considerably higher postbuckling load is needed to obtain a given postbuckling deflection for the case of a clamped edge as opposed to a simply supported edge. In Fig. 4.27, the normalized postbuckling loads $\lambda/\lambda_{cr}$ are graphed against $|w(\gamma)/h|$ for both types of boundary support condition at $r = a$ and the same variations in $\beta$ as depicted in Table 4.1; here we note that for a fixed value of $|w(\gamma)/h|$, in the simply supported case, higher values of $\beta$ correspond to lower values of the normalized load $\lambda/\lambda_{cr}$. Finally, for $\beta = 5.0$, Fig. 4.32 displays the postbuckled (axially symmetric) displacement shapes for the two types of boundary support conditions along the outer edge at $r = a$ and variations in the postbuckling loads $\lambda/\lambda_{cr}$ (which are indicated on these graphs as variations in the postbuckling deflections $w(\gamma)$); other comparative results concerning the postbuckling behavior of cylindrically orthotropic (circular) annular plates may be read directly off the graphs in Figs. 4.28 through 4.31, which depict, for $\beta = 5.0$ and $\gamma = 0.4$, in both the clamped and simply supported cases at $r = a$, the distributions of radial bending stresses, circumferential bending stresses, radial membrane stresses, and circumferential membrane stresses and their respective variations with the postbuckling edge stress $\lambda > \lambda_{cr}$ (as gauged by variations in $\delta = w(\gamma)/h$).

As we have already noted, very little (if any) serious work appears to have been done on the initial or postbuckling behavior of nonlinear elastic, viscoelastic, or elastic-plastic annular plates; a good treatment within the domain of elastic-plastic response is that of Yu and Johnson [179], but this work treats only initial buckling and does not allow for a comparative study based on either variations in loading conditions/boundary support conditions at $r = a$ and $r = b$, or on variations in the degree of deviation from isotropic material symmetry response within the elastic domain of the plate. Variations in both initial and postbuckling response for (circular) annular plates, as a consequence of variations in constitutive response, may constitute a good area for future research efforts. For additional work on the buckling and postbuckling behavior of (circular) annular plates, one may consult [182]–[190]; none of these references concern buckling in the presence of imperfections, which will be dealt with in the next chapter.

## 4.4 Initial Buckling/Postbuckling Figures, Graphs, and Tables: Annular Plates

Table 4.1  Postbuckling Loads of an Orthotropic Annular Plate [174] (Reprinted from the *Int. J. Nonlinear Mech.*, 10, Huang, C.L., On postbuckling of orthotropic annular plates, 63-74, 1974, with permission from Elsevier Science)

| $\dfrac{|w(\gamma)|}{h}$ | $\lambda$ (Free-hinged) | | | | $\lambda$ (Free-Clamped) |
|---|---|---|---|---|---|
| | $\beta = 1^\dagger$ | $\beta = 3$ | $\beta = 5$ | $\beta = 10$ | $\beta = 5$ |
| 0.0 | 2.740 | 8.847 | 14.725 | 28.509 | 47.900 |
| 0.3 | 2.845 | 9.129 | 15.144 | 29.239 | 48.988 |
| 0.6 | 3.158 | 9.967 | 16.392 | 31.434 | 52.198 |
| 0.9 | 3.674 | 11.334 | 18.441 | 35.056 | 57.392 |
| 1.2 | 4.383 | 13.199 | 21.259 | 40.133 | 64.391 |
| 1.5 | 5.274 | 15.531 | 24.821 | 46.666 | 73.022 |
| 1.8 | 6.334 | 18.306 | 29.115 | 54.683 | 83.154 |
| 2.1 | 7.553 | 21.509 | 34.142 | 64.230 | 94.705 |
| 2.4 | 8.918 | 25.135 | 39.917 | 75.365 | 107.640 |
| 2.7 | 10.422 | 29.188 | 46.461 | 88.153 | 121.956 |

$^\dagger$ When $\beta = 1$, the material of the annulus is isotropic.

**Table 4.2   Convergence of the Eigenvalue for Increasing Mesh Size. (Based on [71])**

| Isotropic Plate: $a/b = 2.0 : (p_i b^2/K)_{exact} = 224.136$ | | |
|---|---|---|
| $m \times n$ | $p_i b^2/K$ | percent error |
| $5 \times 5$ | 249.01 | 11.1 |
| $10 \times 10$ | 207.80 | $-7.3$ |
| $20 \times 20$ | 218.41 | $-5.7$ |
| $30 \times 30$ | 221.386 | $-1.2$ |

**Table 4.3  Convergence of the Eigenvalue for Increasing Mesh Size. (Based on [71])**

| $m \times n$ | Isotropic plate: $a/b = 30. : (p_i b^2 / K)_{exact} = 11.26$ | |
|---|---|---|
| | $p_i b^2 / K$ | percent error |
| $5 \times 5$ | 8.92 | 20.8 |
| $10 \times 10$ | 8.50 | 24.5 |
| $20 \times 20$ | 9.35 | 16.9 |
| $30 \times 30$ | 9.94 | 11.7 |
| $50 \times 10$ | 10.54 | 6.4 |
| $60 \times 10$ | 10.70 | 5.0 |

**Table 4.4** Comparison of Critical Loads Versus Ratio of Young's Moduli. (Based on [71])

| $E_2/G_{12} = 2.6, \nu_{12} = 0.3, m = n = 20$ | | |
|---|---|---|
| $E_2/E_1$ | $p_i b^2/D_1$ | $p_i b^2/D_2$ |
| 1.0 | 218.4 | 218.4 |
| 2.0 | 273.8 | 136.9 |
| 5.0 | 363.3 | 72.7 |
| 10.0 | 455.6 | 45.6 |
| 100.0 | 1043.8 | 10.4 |

**Table 4.5** Comparison of Critical Loads Versus Shear Moduli. (Based on [71])

| $E_2/E_1 = 1.0, \nu_{12} = 0.3, m = n = 20$ | | |
|---|---|---|
| $E_2/G_{12}$ | $p_i b^2/D_G$ | $p_i b^2/D_2$ |
| 2.6 | 624.0 | 218.4 |
| 5 | 1070.3 | 194.8 |
| 10.0 | 1829.7 | 166.5 |
| 100.0 | 13692. | 124.6 |

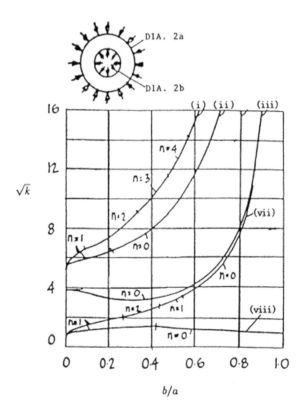

**FIGURE 4.1**
Buckling of annular plates. (Adopted, in modified form, from [162].)

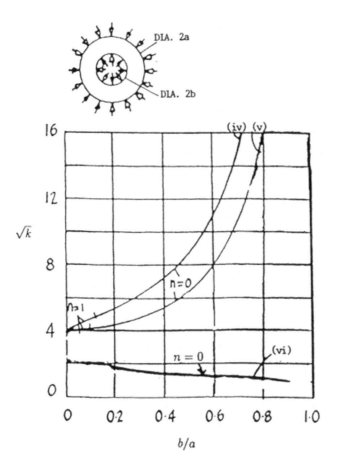

**FIGURE 4.2**
Buckling of annular plates. (Adopted, in modified form, from [162].)

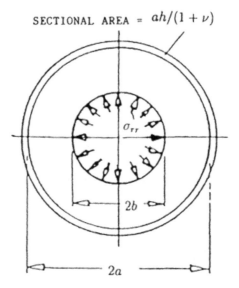

SECTIONAL AREA $= ah/(1 + \nu)$

$\sigma_{rr}$

$2b$

$2a$

FIGURE 4.3
Annular plate stiffened along outer edge [165]. (From Mansfield, E.H., "On the Buckling of an Annular Plate," *Quart. J.
Mech. Appl. Math. XIII*, 16, 1960. Reprinted by permission
of Oxford University Press.)

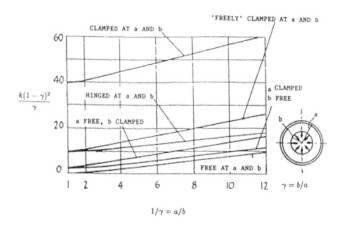

**FIGURE 4.4**
Buckling of annular plate with stiffener along outer edge [165].
(From Mansfield, E.H., "On the Buckling of an Annular Plate,"
*Quart. J. Mech. Appl. Math. XIII*, 16, 1960. Reprinted by
permission of Oxford University Press.)

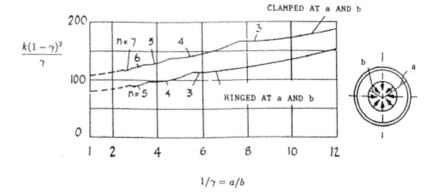

**FIGURE 4.5**
Buckling of annular plates under inward tensile stress [165].
(From Mansfield, E.H., "On the Buckling of an Annular Plate",
*Quart. J. Mech. Appl. Math. XIII*, 16, 1960. Reprinted by
permission of Oxford University Press.)

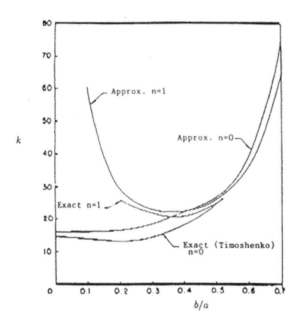

**FIGURE 4.6**
Comparison between exact and approximate buckling loads.
(Adopted, in modified form, from [166].)

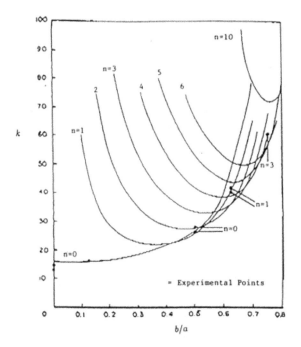

**FIGURE 4.7**
Buckling load for various wave numbers. (Adopted, in modified form, from [166].)

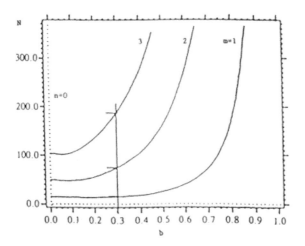

**FIGURE 4.8**
Eigenvalue curves showing for $n = 0$ the influence of $m$ on the critical
load $N$. The physically relevant value is $m = 1$. (Machinek, A. and
Troger, H., "Postbuckling of Elastic Annular Plates," *Dynamics and
Stability of Systems*, 3, 1988, 79–88. With permission of the Taylor
& Francis Group plc.)

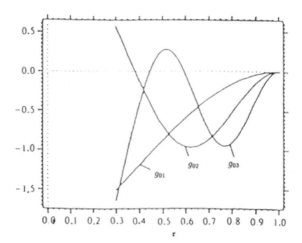

**FIGURE 4.9**
Eigenfunctions for $b = 0.3$; the physically relevant eigenfunction is $g_{01}$.
(Machinek, A. and Troger, H., "Postbuckling of Elastic Annular Plates,"
*Dynamics and Stability of Systems*, 3, 1988, 79–88. With permission
of the Taylor & Francis Group plc.)

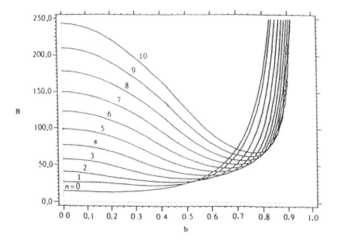

**FIGURE 4.10**
Stability boundary in parameter space $(N, b)$ for $m = 1$ and $n = 0, 1, \ldots, 10$. (Machinek, A. and Troger, H., "Postbuckling of Elastic Annular Plates," *Dynamics and Stability of Systems*, 3, 1988, 79–88. With permission of the Taylor & Francis Group plc.)

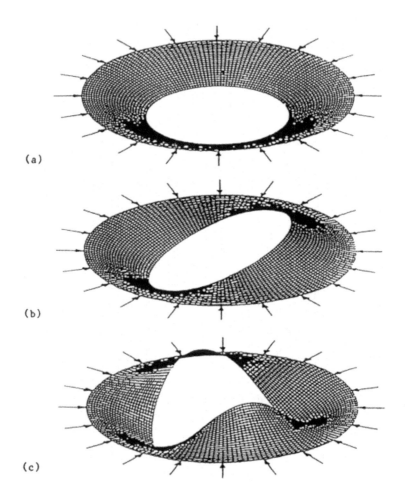

FIGURE 4.11
Axonometrical representation of the deflection surface of the annular
plate for $m = 1$ and $n = 0, 1, 2$. (Machinek, A. and Troger, H., "Postbuckling of Elastic Annular Plates," *Dynamics and Stability of Systems*, 3, 1988, 79–88. With permission of the Taylor & Francis Group
plc.)

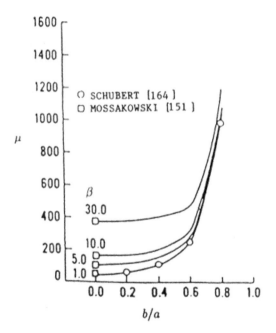

**FIGURE 4.12**
Critical buckling loads: buckling parameter $\mu = -\lambda_{cr}ha^2/D_r$ versus $b/a$, both edges fixed and loaded. (Adopted, in modified form, from [169].)

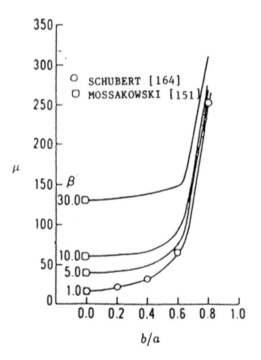

**FIGURE 4.13**
Critical buckling loads: buckling parameter $\mu = -\lambda h a^2 / D_r$ versus $b/a$ with edges simply supported and loaded. (Adopted, in modified form, from [169].)

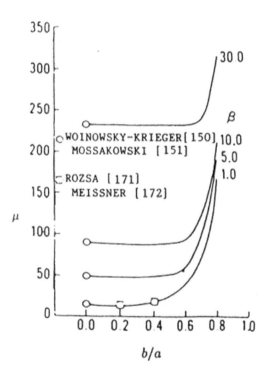

**FIGURE 4.14**
Critical buckling loads: buckling parameter $\mu = -\lambda h a^2 / D_r$ versus $b/a$, outer edge fixed and loaded, inner edge free. (Adopted, in modified form, from [169].)

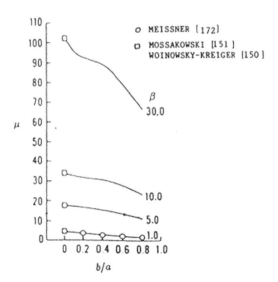

**FIGURE 4.15**
Critical buckling loads: buckling parameter $\mu = -\lambda h a^2 / D_r$ versus $b/a$, outer edge simply supported and loaded, inner edge free. (Adopted, in modified form, from [169].)

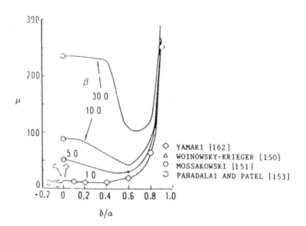

**FIGURE 4.16**
Critical buckling loads: buckling parameter $\mu = -\lambda h a^2/D_r$
versus $b/a$, outer edge fixed, inner edge free; both edges loaded.
(Adopted, in modified form, from [169].)

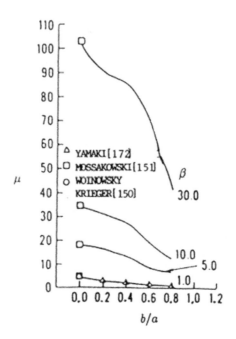

**FIGURE 4.17**
Critical buckling loads: buckling parameter $\mu = -\lambda h a^2 / D_r$ versus $b/a$, outer edge simply supported, inner edge free, both edges loaded. (Adopted, in modified form, from [169].)

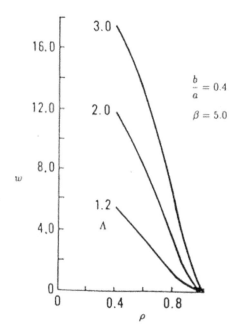

**FIGURE 4.18**
Nondimensional deflection ($w$) versus nondimensional radius
($\rho$) [70]. (From Uthgenannt, E. and Brand, R., *J. Appl. Mech.*,
40, 559, 1973. With permission.)

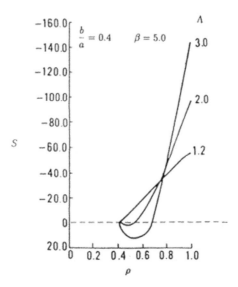

**FIGURE 4.19**
Nondimensional membrane radial stress ($S$) versus nondimensional radius ($\rho$) [70]. (From Uthgenannt, E. and Brand, R., *J. Appl. Mech.*, 40, 559, 1973. With permission.)

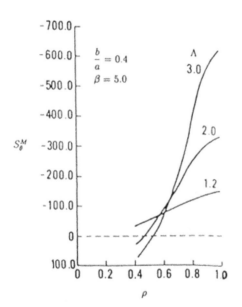

**FIGURE 4.20**
Nondimensional membrane tangential stress $(S_{\theta}^{M})$ versus nondimensional radius $(\rho)$ [70]. (From Uthgenannt, E. and Brand, R., *J. Appl. Mech.*, 40, 559, 1973. With permission.)

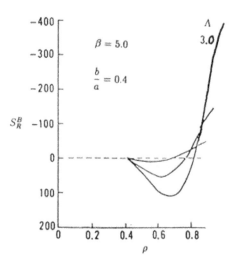

**FIGURE 4.21**
Nondimensional bending radial stress ($S_R^B$) versus nondimensional radius ($\rho$)[70]. (From Uthgenannt, E. and Brand, R., *J. Appl. Mech.*, 40, 559, 1973. With permission.)

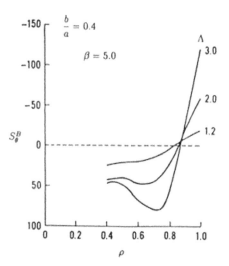

**FIGURE 4.22**
Nondimensional bending tangential stress ($S_\theta^B$) versus nondimensional radius ($\rho$) [70]. (From Uthgenannt, E. and Brand, R., *J. Appl. Mech.*, 40, 559, 1973. With permission.)

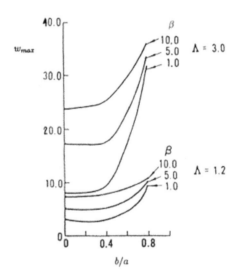

**FIGURE 4.23**
Nondimensional maximum deflection ($w$) versus Annular plate
parameter ($b/a$) [70]. (From Uthgenannt, E. and Brand, R.,
*J. Appl. Mech.*, 40, 559, 1973. With permission.)

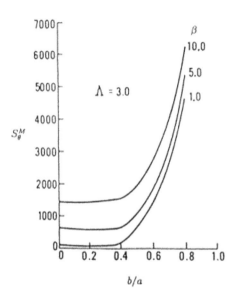

**FIGURE 4.24**
Nondimensional maximum membrane tangential stress $(S_\theta^M)$ versus annular plate parameter $(b/a)$ [70]. (From Uthgenannt, E. and Brand, R., *J. Appl. Mech.*, 40, 559, 1973. With permission.)

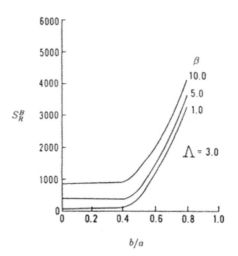

**FIGURE 4.25**
Nondimensional maximum bending radial stress $(S_R^B)$ versus
annular plate parameter $(b/a)$ [70]. (From Uthgenannt, E. and
Brand, R., *J. Appl. Mech.*, 40, 559, 1973. With permission.)

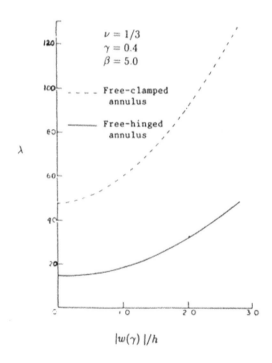

**FIGURE 4.26**
Postbuckling loads and transverse displacement parameter for
an orthotropic annulus [174]. (Reprinted from the *Int. J.
Nonlinear Mech.*, 10, Huang, C.L., On Postbuckling of Or-
thotropic Annular Plates, 63-74, 1974, with permission from
Elsevier Science)

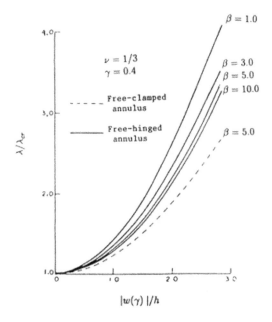

**FIGURE 4.27**
Ratio of postbuckling load to linear buckling load and transverse displacement parameter [174]. (Reprinted from the *Int. J. Nonlinear Mech.*, 10, Huang, C.L., On Postbuckling of Orthotropic Annular Plates, 63-74, 1974, with permission from Elsevier Science)

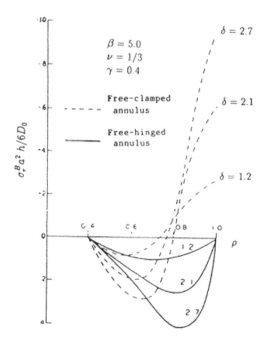

**FIGURE 4.28**
The distribution of radial bending stresses [174]. (Reprinted from the *Int. J. Nonlinear Mech.*, 10, Huang, C.L., On Post-buckling of Orthotropic Annular Plates, 63-74, 1974, with permission from Elsevier Science)

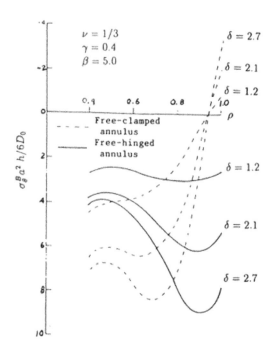

**FIGURE 4.29**
The distribution of circumferential bending stresses [174].
(Reprinted from the *Int. J. Nonlinear Mech.*, 10, Huang, C.L.,
On Postbuckling of Orthotropic Annular Plates, 63-74, 1974,
with permission from Elsevier Science)

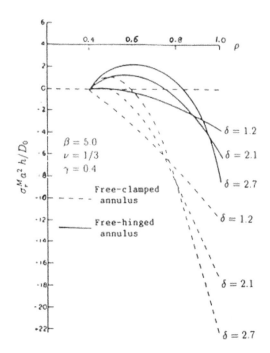

**FIGURE 4.30**
The distribution of radial membrane stresses [174]. (Reprinted from the *Int. J. Nonlinear Mech.*, 10, Huang, C.L., On Post-buckling of Orthotropic Annular Plates, 63-74, 1974, with permission from Elsevier Science)

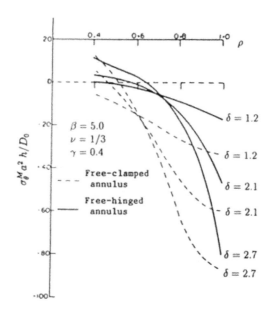

**FIGURE 4.31**
The distribution of circumferential membrane stresses [174].
(Reprinted from the *Int. J. Nonlinear Mech.*, 10, Huang, C.L.,
On Postbuckling of Orthotropic Annular Plates, 63-74, 1974,
with permission from Elsevier Science)

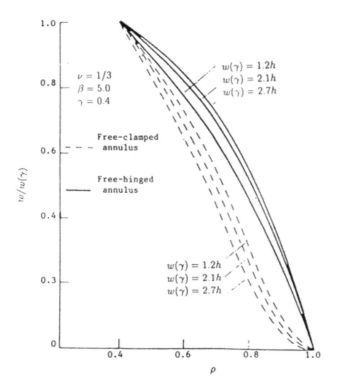

**FIGURE 4.32**
Transverse displacement shape of the buckled annular plate
[174]. (Reprinted from the *Int. J. Nonlinear Mech.*, 10,
Huang, C.L., On Postbuckling of Orthotropic Annular Plates,
63-74, 1974, with permission from Elsevier Science)

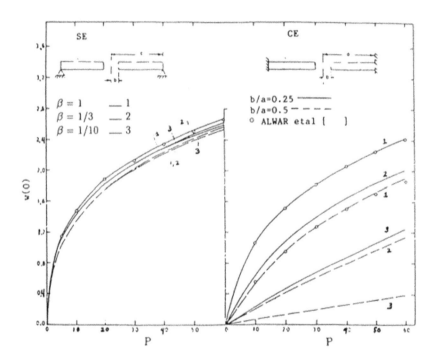

**FIGURE 4.33**
Deflection response for a uniformly distributed load [173].
(Reprinted from the *Int. J. Nonlinear Mech.*, 19, Dumir, P.C.,
Nath, Y., and M. Gandhi, Non-Linear Axisymmetric Static
Analysis of Orthotropic Thin Annular Plates, 255-272, 1984,
with permission from Elsevier Science)

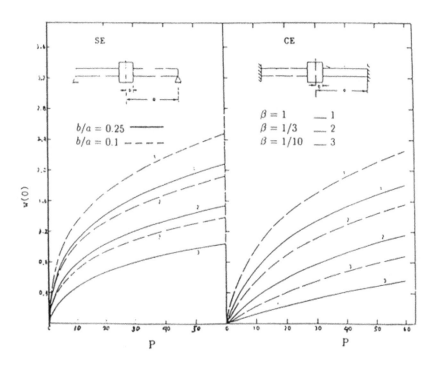

**FIGURE 4.34**
Deflection response for a plugged hole under a uniformly distributed load [173]. (Reprinted from the *Int. J. Nonlinear Mech.*, 19, Dumir, P.C., Nath, Y., and M. Gandhi, Non-Linear Axisymmetric Static Analysis of Orthotropic Thin Annular Plates, 255-272, 1984, with permission from Elsevier Science)

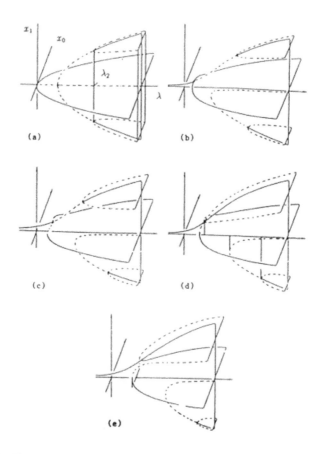

**FIGURE 4.35**
Branches of the bifurcated solutions for an annular plate $b < b_c$
(a) no imperfections present ($\gamma_0 = \gamma_1 = 0$); (b) $0 < \gamma_1 <$
0.36, $\gamma_0 = 0$, smooth transition from $x_1$ mode to $x_0$ mode; (c)
$0.36 < \gamma_1 < 0.51$, $\gamma_0 = 0$, jump transition from $x_1$ Mode to $x_0$
mode: (d) transition point $\gamma_1 = 0.51$ and (e) $\gamma_1 > 0.51$, $\gamma_0 = 0$,
plate stays in $x_1$ mode [66].

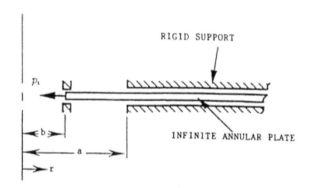

**FIGURE 4.36**
Sectional view of one quadrant of a partially restrained infinite annular plate. (Original figure from the second author's Ph.d. Thesis [71].)

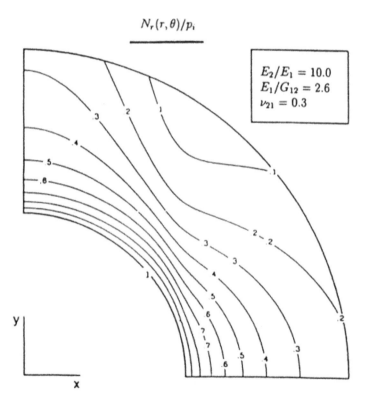

**FIGURE 4.37**
Radial normal stress distribution in an annular plate with orthotropy due to variation in Young's Moduli. (Original figure from the second author's Ph.d. Thesis [71].)

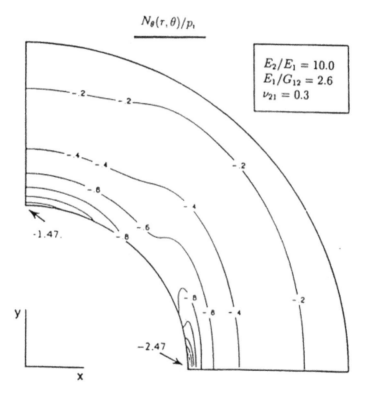

**FIGURE 4.38**
Tangential normal stress distribution in an annular plate with orthotropy due to variation in Young's moduli. (Original figure from the second author's Ph.d. Thesis [71].)

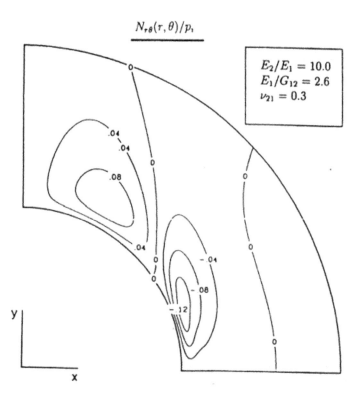

$$N_{r\theta}(r,\theta)/p_1$$

$$E_2/E_1 = 10.0$$
$$E_1/G_{12} = 2.6$$
$$\nu_{21} = 0.3$$

**FIGURE 4.39**
Polar shear stress distribution in an annular plate with orthotropy due to variation in Young's moduli. (Original figure from the second author's Ph.d. Thesis [71].)

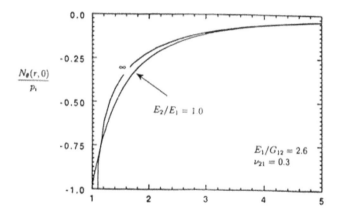

**FIGURE 4.40**
Effect of the ratio of Young's moduli on the tangential normal
stress distribution at $\theta = 0$. (Original figure from the second
author's Ph.d. Thesis [71].)

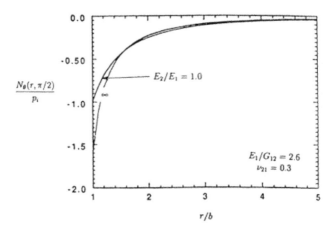

**FIGURE 4.41**
Effect of the ratio of Young's moduli on the tangential normal stress distribution at $\theta = \pi/2$. (Original figure from the second author's Ph.d. Thesis [71].)

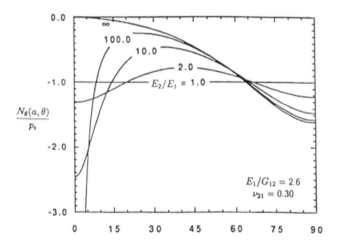

**FIGURE 4.42**
Effect of the ratio of Young's moduli on the tangential stress distribution along the inner edge of the annular plate. (Original figure from the second author's Ph.d. Thesis [71].)

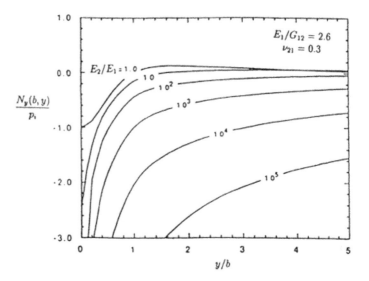

**FIGURE 4.43**
Effect of the ratio of Young's moduli on the $y$ component of the stress distribution at $x = b$. (Original figure from the second author's Ph.d. Thesis [71].)

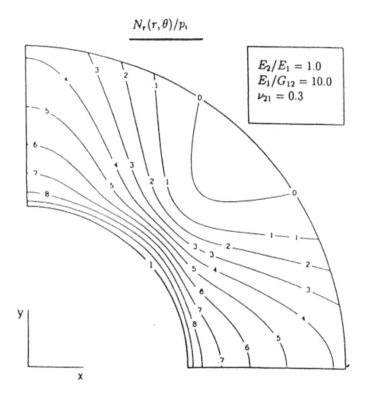

FIGURE 4.44
Radial normal stress distribution in an annular plate with orthotropy due to variation in the shear modulus ratio. (Original
figure from the second author's Ph.d. Thesis [71].)

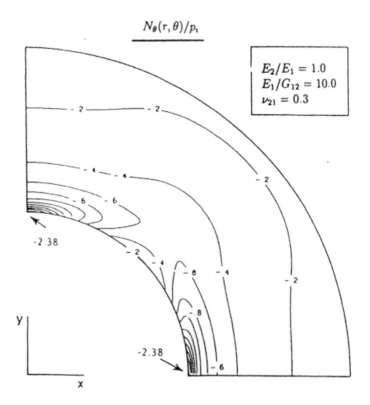

**FIGURE 4.45**
Tangential normal stress distribution in an annular plate with orthotropy due to variation in the shear modulus ratio. (Original figure from the second author's Ph.d. Thesis [71].)

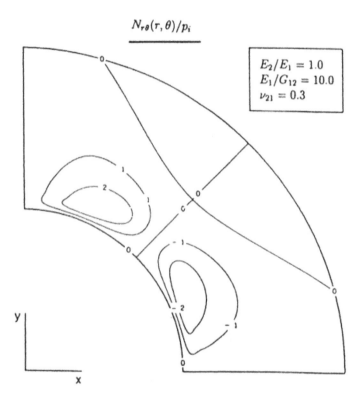

$$N_{r\theta}(r,\theta)/p_i$$

$$E_2/E_1 = 1.0$$
$$E_1/G_{12} = 10.0$$
$$\nu_{21} = 0.3$$

**FIGURE 4.46**
Polar shear stress ddstribution in an annular plate with orthotropy due to variation in the shear modulus ratio. (Original
figure from the second author's Ph.d. Thesis [71].)

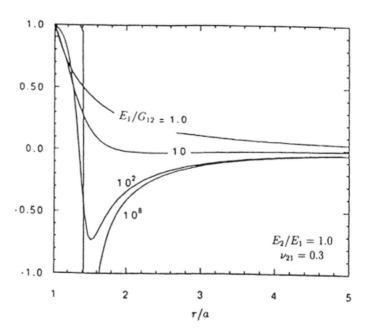

**FIGURE 4.47**
Effect of shear modulus ratio on the radial normal stress distribution at $\theta = \pi/4$. (Original figure from the second author's Ph.d. Thesis [71].)

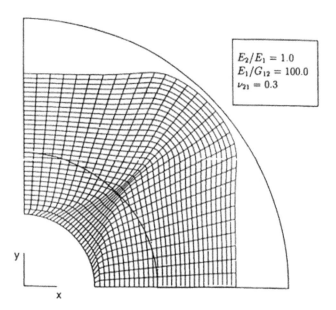

$E_2/E_1 = 1.0$
$E_1/G_{12} = 100.0$
$\nu_{21} = 0.3$

**FIGURE 4.48**
Relative displacement shape of an annular plate with low shear modulus subjected to uniform load $(b \leq r \leq 2b)$. (Original figure from the second author's Ph.d. Thesis [71].)

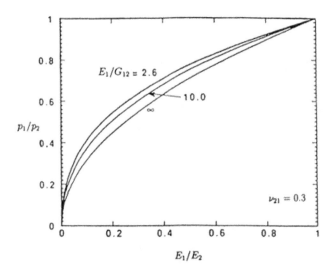

**FIGURE 4.49**
Ratio of Young's moduli versus ratio of traction parameters
yielding uniform radial displacement. (Original figure from
the second author's Ph.d. Thesis [71].)

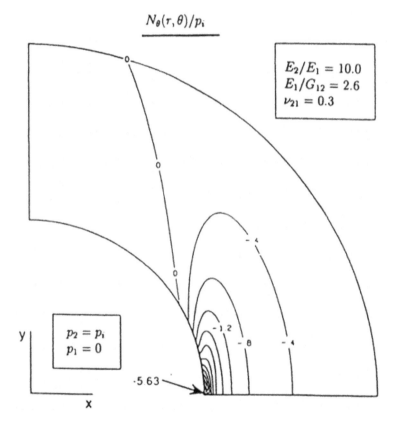

**FIGURE 4.50**
Tangential normal Stress distribution in an orthotropic annular plate with the applied traction acting in direction of maximum stiffness. (Original figure from the second author's Ph.d. Thesis [71].)

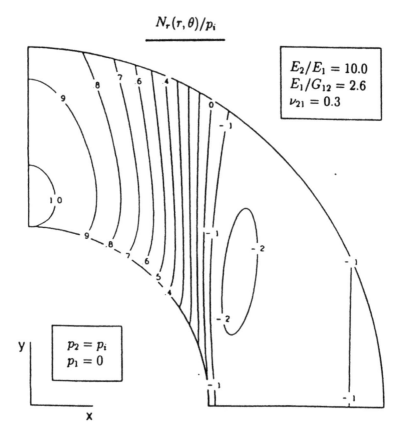

FIGURE 4.51
Radial normal stress distribution in an orthotropic annular plate with the applied traction acting in the direction of maximum stiffness. (Original figure from the second author's Ph.d. Thesis [71].)

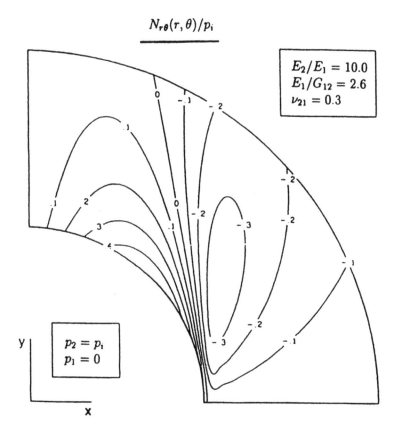

**FIGURE 4.52**
Polar shear stress distribution in an orthotropic annular plate
with the applied traction acting in the direction of maximum
stiffness. (Original figure from the second author's Ph.d. Thesis [71].)

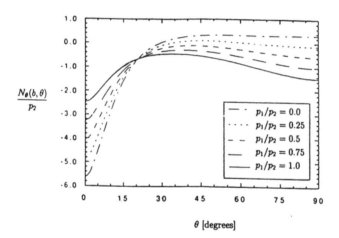

**FIGURE 4.53**
Effect of variation in the ratio of traction parameters on the
tangential normal stress distribution along the inner edge of
the annular plate. (Original figure from the second author's
Ph.d. Thesis [71].)

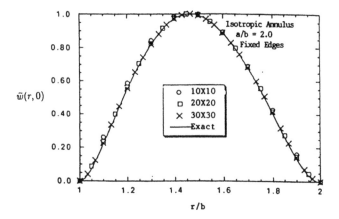

**FIGURE 4.54**
Convergence of the finite-difference solution of buckled mode shape in the radial direction for a plate aspect ratio of 2. (Original figure from the second author's Ph.d. Thesis [71].)

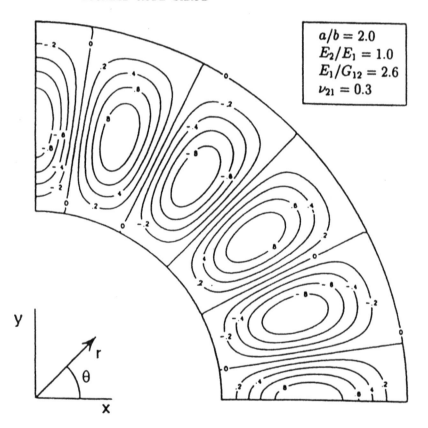

**FIGURE 4.55**
Buckled mode shape for an isotropic plate with plate aspect
ratio of 2. (Original figure from the second author's Ph.d.
Thesis [71].)

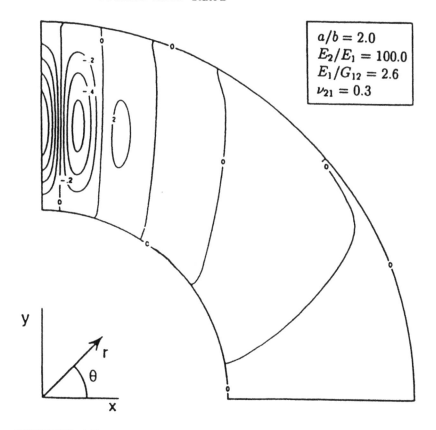

BUCKLED MODE SHAPE

$a/b = 2.0$
$E_2/E_1 = 100.0$
$E_1/G_{12} = 2.6$
$\nu_{21} = 0.3$

FIGURE 4.56
Buckled mode shape for an orthotropic plate with ratio of
Young's moduli equal to 100 and a plate aspect ratio of 2.
(Original figure from the second author's Ph.d. Thesis [71].)

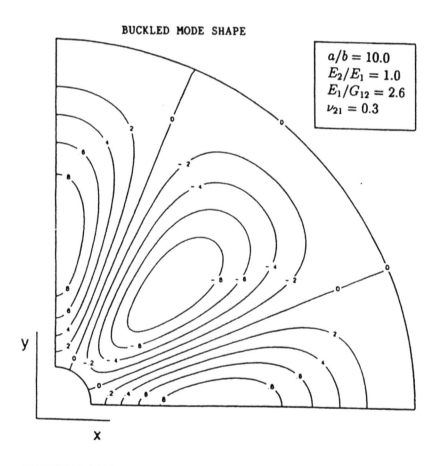

BUCKLED MODE SHAPE

$a/b = 10.0$
$E_2/E_1 = 1.0$
$E_1/G_{12} = 2.6$
$\nu_{21} = 0.3$

**FIGURE 4.57**
Buckled mode shape for an isotropic plate with a plate aspect
ratio of 10. (Original figure from the second author's Ph.d.
Thesis [71].)

BUCKLED MODE SHAPE

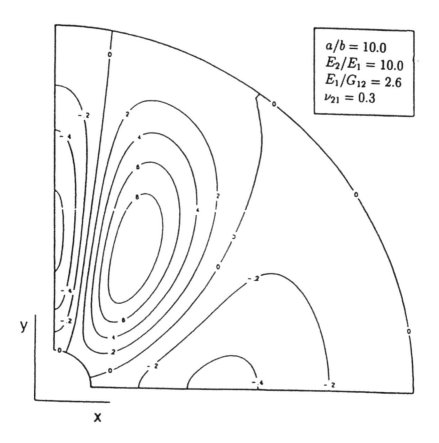

FIGURE 4.58
Buckled mode shape for an orthotropic plate with ratio of
Young's moduli equal to 10 and a plate aspect ratio of 10.
(Original figure from the second author's Ph.d. Thesis [71].)

BUCKLED MODE SHAPE

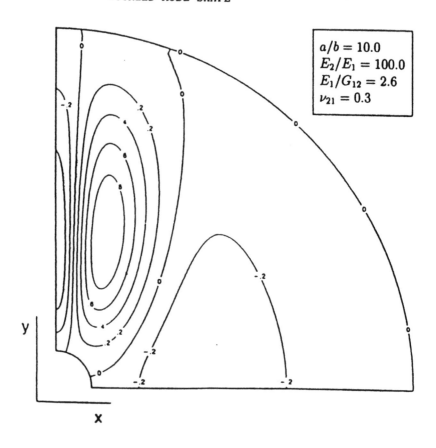

FIGURE 4.59
Buckled mode shape for an orthotropic plate with ratio of
Young's moduli equal to 100 and a plate aspect ratio of 101.
(Original figure from the second author's Ph.d. Thesis [71].)

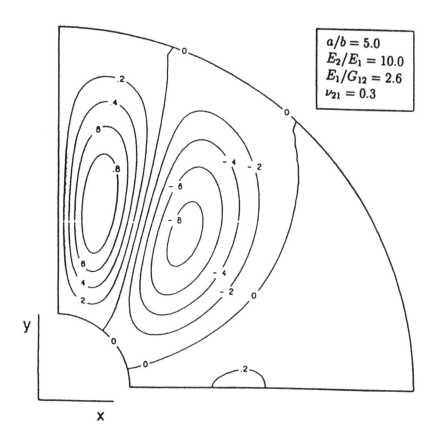

**FIGURE 4.60**
Buckled mode shape for an orthotropic plate with ratio of Young's moduli equal to 10 and a plate sspect ratio of 5 (symmetric about $x$ axis and antisymmetric about $y$ axis ). (Original figure from the second author's Ph.d. Thesis [71].)

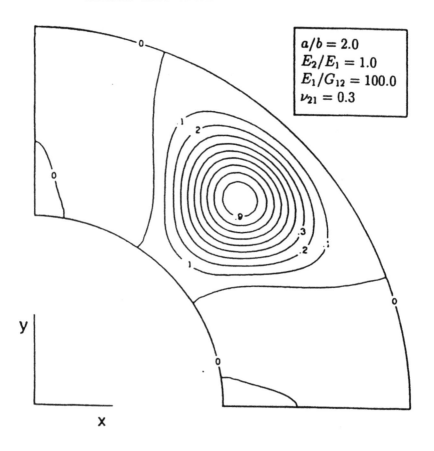

FIGURE 4.61
Buckled mode shape for an orthotropic plate with a shear modulus ratio of 100 and a plate aspect ratio of 2. (Original figure
from the second author's Ph.d. Thesis [71].)

BUCKLED MODE SHAPE

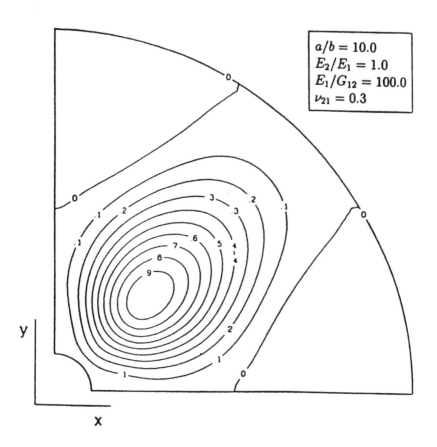

FIGURE 4.62
Buckled mode shape for an orthotropic plate with a shear modulus ratio of 100 and a plate aspect ratio of 10. (Original figure from the second author's Ph.d. Thesis [71].)

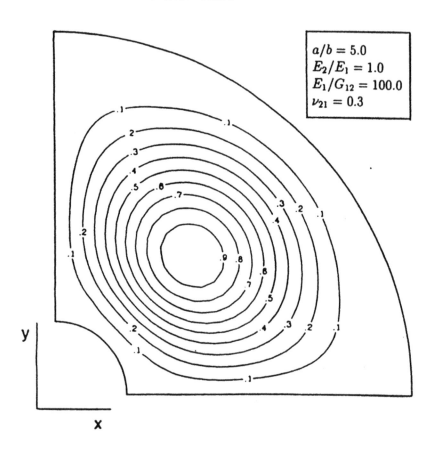

BUCKLED MODE SHAPE

$a/b = 5.0$
$E_2/E_1 = 1.0$
$E_1/G_{12} = 100.0$
$\nu_{21} = 0.3$

**FIGURE 4.63**
Buckled mode shape for an orthotropic plate with a shear modulus ratio of 100 and a plate aspect ratio of five (antisymmetric about the $x$ and $y$ axes). (Original figure from the second author's Ph.d. Thesis [71].)

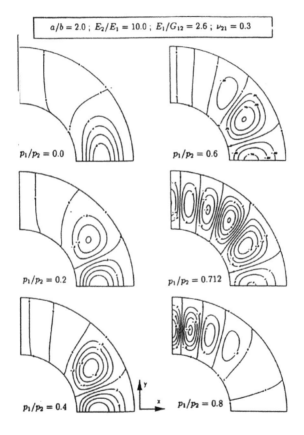

**FIGURE 4.64**
Effect of nonuniform applied tractions on the buckled mode
shape. (Original figure from the second author's Ph.d. Thesis
[71].)

BUCKLED MODE SHAPE

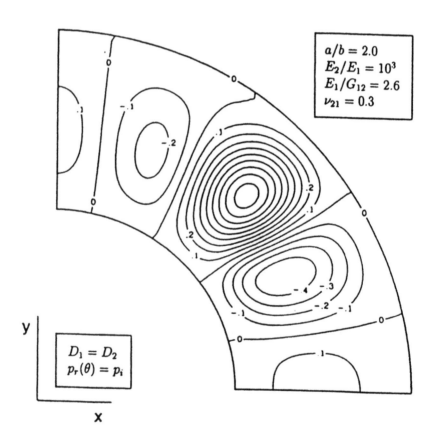

FIGURE 4.65
Buckled mode shape for a plate with orthotropic in-plane properties and isotropic bending properties and subjected to a uniform radial traction. Orthotropy due to ratio of Young's moduli. (Original figure from the second author's Ph.d. Thesis [71].')

BUCKLED MODE SHAPE

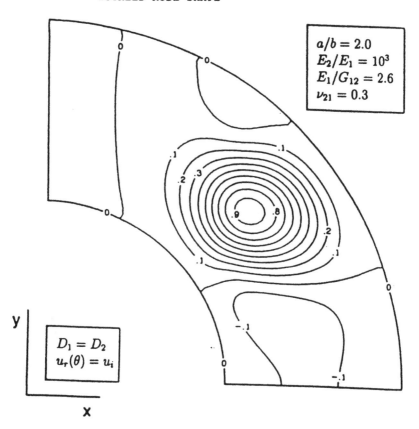

**FIGURE 4.66**
Buckled mode shape for a plate with orthotropic in-plane properties and isotropic bending properties and subected to a uniform radial displacement. Orthotropy due to ratio of Young's moduli. (Original figure from the second author's Ph.d. Thesis [71].)

BUCKLED MODE SHAPE

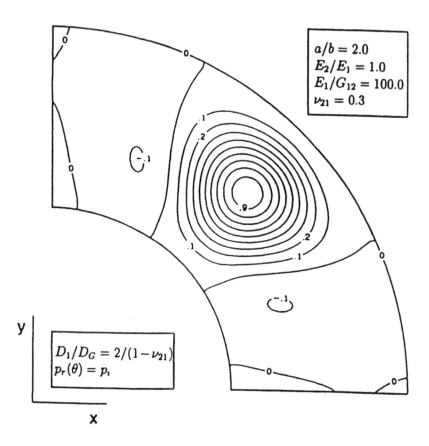

FIGURE 4.67
Buckled mode shape for a plate with orthotropic in-plane properties and isotropic bending properties and subjected to a uniform radial traction. Orthotropy due to the shear modulus ratio. (Original figure from the second author's Ph.d. Thesis [71].)

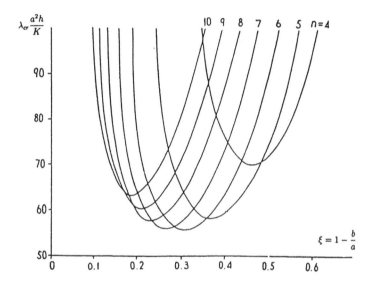

**FIGURE 4.68**
The Critical load for the elastic buckling of a flange [179].
(Reprinted from *Int. J. Mech. Sci.*, 24, Yu, T.X. and John-
son, W., The Buckling of Annular Plates in Relation to the
Deep-Drawing Process, 175-188, 1972, with permission from
Elsevier Science)

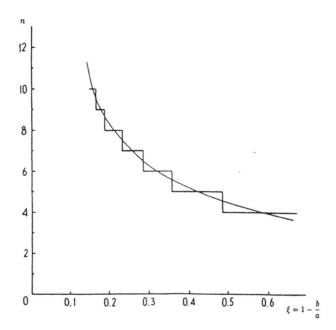

**FIGURE 4.69**
The number of waves for the elastic buckling of a flange [179].
(Reprinted from *Int. J. Mech. Sci.*, 24, Yu, T.X. and Johnson,
W., The Buckling of Annular Plates in Relation to the Deep-
Drawing Process, 175-188, 1972, with permission from Elsevier
Science)

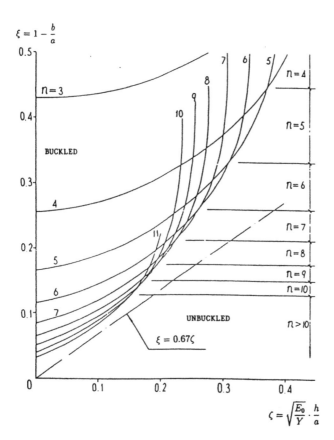

FIGURE 4.70
The critical condition for the plastic buckling of a flange when $n \leq 11$ [179]. (Reprinted from *Int. J. Mech. Sci.*, 24, Yu, T.X. and Johnson, W., The Buckling of Annular Plates in Relation to the Deep-Drawing Process, 175-188, 1972, with permission from Elsevier Science)

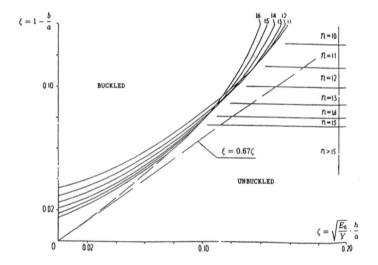

**FIGURE 4.71**
The critical condition for the plastic buckling of a flange when
$n \geq 10$ [179]. (Reprinted from *Int. J. Mech. Sci.*, 24, Yu, T.X.
and Johnson, W., The Buckling of Annular Plates in Relation
to the Deep-Drawing Process, 175-188, 1972, with permission
from Elsevier Science)

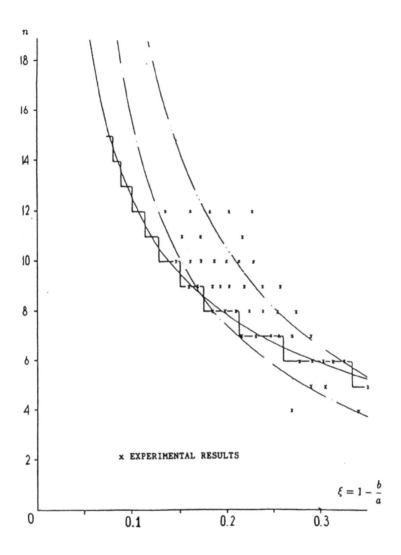

FIGURE 4.72
The number of waves for the plastic buckling of a flange [179].
(Reprinted from *Int. J. Mech. Sci.*, 24, Yu, T.X. and Johnson,
W., The Buckling of Annular Plates in Relation to the Deep-
Drawing Process, 175-188, 1972, with permission from Elsevier
Science)

# CHAPTER 5

# Postbuckling Behavior of Imperfect Plates and Secondary Plate Buckling

In this chapter we examine, briefly, the issue of imperfection sensitivity for rectangular, circular, and annular plates exhibiting both elastic and elastic-plastic response; although our discussion of behavior within the elastic regime is limited to linear elastic constitutive response we do consider both isotropic and orthotropic material symmetry. The chapter also includes a limited discussion of some of the work in the research literature on the phenomena of secondary bifurcation for thin plates; in this case, our considerations are confined to an examination of branching off of the primary postbuckling curve for isotropic, linearly elastic, rectangular and circular plates.

## 5.1   Imperfection Sensitivity

Imperfection buckling of thin plates occurs when the plate is either subjected to a small transverse loading in addition to compressive, tensile, or shear forces along its edge(s), or possesses an initial deviation from the flat, unbuckled state. Indeed, when nonlinear bifurcation results are at variance with observations it is frequently due to small deviations from the ideal configuration assumed in the nonlinear theory, i.e., the unloaded plate may not be exactly plane or the assumption of no lateral load normal to the face of the plate may be invalid because, e.g., of the gravitational force on a horizontal plate. The mathematical branch of buckling theory which includes such initial imperfections is usually called perturbed bifurcation theory and many general expositions may be found in the literature (i.e., Keener and Keller [191]). The first general rigorous nonlinear theory of postbuckling behavior which is directly applicable to the behavior of structures, and which takes into account

the case of small finite initial deflections from the fundamental (or unloaded) state appears to be the work of Koiter [3]; here it is made clear
that the primary effect of initial imperfections is that the fundamental
state of the idealized perfect structure does not represent a configuration
of equilibrium of the actual imperfect structure. As a direct consequence
of the presence of imperfections, the experimental values for the critical buckling load of a structure (e.g., a thin plate) may be considerably
smaller than the theoretical values for the idealized model. Prior to
looking at some of the specific results which are available for imperfection buckling of rectangular, circular, and annular plates exhibiting both
elastic and elastic-plastic response, we will illustrate the phenomena of
imperfection buckling by looking at a simple model problem which has
been employed by both Koiter [3] and Budiansky [16].

### 5.1.1   An Example of Imperfection Bifurcation

We consider the simple model depicted in Fig. 5.1, in which a rigid
(vertical) rod of length $L$ is fixed with respect to translation at its base
and is elastically constrained against rotation through an angle $\xi$ by a
spring which supplies a restoring moment $f(\xi)$; if we assume the presence
of the vertical load $\lambda$ as indicated in Fig. 5.1, then the condition implying
static equilibrium is

$$\lambda L \sin \xi = f(\xi) \tag{5.1}$$

If we take $f(\xi)$ in the form

$$f(\xi) = K_1 \xi + K_2 \xi^2 + K_3 \xi^3 + \cdots (K_1 > 0) \tag{5.2}$$

then, for all $\lambda$, we always have the fundamental solution $\xi = 0$; however,
we have, in addition, "buckled" states which correspond to those values
of $\xi \neq 0$ which satisfy (5.1), i.e.,

$$\lambda = f(\xi)/L \sin \xi \tag{5.3}$$

If we assume small angular deviations $\xi$, so that $\sin \xi \approx \xi$, then (5.2)
and (5.3) combine so as to yield

$$\lambda \simeq \frac{1}{L} \left( K_1 + K_2 \xi + K_3 \xi^2 + \cdots \right) \tag{5.4}$$

$$= \lambda_c + \lambda_1 \xi + \lambda_2 \xi^2 + \cdots$$

with

$$\lambda_c = \frac{K_1}{L}, \lambda_1 = \frac{K_2}{L}, \ \lambda_2 = \frac{K_3}{L} \tag{5.5}$$

so that

$$\lambda_1/\lambda_c = K_2/K_1 \tag{5.6}$$

The solutions of (5.3) lie on an equilibrium path in the $\lambda - \xi$ plane (see Fig. 5.2 ) which is connected to the fundamental path $\xi = 0$ at the critical load $\lambda_c$; for sufficiently small values of $\xi$ (in fact, for $|\xi| < \pi$) these buckled states lie on the path given by (5.4). We note that, for $\lambda_1 \neq 0$, the bifurcation which occurs is asymmetric, while symmetric bifurcations occur for $\lambda_1 = 0$, $\lambda_2 \neq 0$.

Now, suppose that in its unloaded state the rod suffers some initial deviation $\bar{\xi}$ from the perfectly vertical position and let $\xi$ be the additional rotation produced by the load $\lambda$ (see Fig. 5.3); then equilibrium states are now defined by the solutions of

$$\lambda = f(\xi)/L \sin \left( \xi + \bar{\xi} \right) \tag{5.7}$$

and these solution paths in the $\lambda - \xi$ plane are depicted in Figs. 5.4a and 5.4b for both positive and negative values of the initial deviation $\bar{\xi}$: the intersecting equilibrium paths for the perfect structure $(\bar{\xi} = 0)$, which are shown as dotted curves, have been deformed into disjointed branches by the presence of the initial imperfection $\bar{\xi} \neq 0$ and, with respect to a loading that increases from $\lambda = 0$, those branches that do not pass through the origin have become irrelevant. For various values of $\bar{\xi}$, the significant branches which do contain the origin make up the two families of curves depicted in Figs. 5.5a and 5.5b; in these graphs, as $|\bar{\xi}|$ converges to zero, the members of each family approach different pairs of branches of the equilibrium paths which are associated with the perfect structure. We note that, irrespective of the sign of $\bar{\xi}$, paths from the origin in the $\lambda - \xi$ plane never approach the fundamental path above $\lambda = \lambda_c$.

With a sufficiently small positive value of $\bar{\xi}$ the corresponding equilibrium path (Fig. 5.5a) possesses a local maximum with respect to the load at the point $(\xi_s, \lambda_s)$, with $\lambda_s < \lambda_c$, and with the rotation $\xi_s$ depending on $\bar{\xi}$. Now, suppose that the actual loading is increased beyond the load $\lambda_s$; for $\lambda > \lambda_s$ static equilibrium becomes impossible at values of $\xi$ in a neighborhood of $\xi_s$: a dynamic process must ensue, and the final static configuration (if one exists) cannot be arbitrarily close to $\xi_s$ no matter how slightly $\lambda_s$ is exceeded. In accord with this observation, $\lambda_s$ is designated as the buckling load of the imperfect structure; this kind of buckling, which is associated with a local maximum on a static equilibrium path in the $\lambda - \xi$ plane, and an expected jump to a non-neighboring equilibrium configuration when the local maximum is exceeded, is known as snap-buckling (or, sometimes, limit-point buckling). On the other hand, the graph depicted in Fig. 5.5b for the case $\bar{\xi} < 0$, suggests a somewhat less drastic behavior: for loads in a neighborhood of $\lambda_c$, significant increases occur in the rate of the growth of rotation with load; however, $\lambda_c$ still warrants being designated the buckling load of the structure and the buckling cannot, in any case, be compared with the shape and possible catastrophic snap buckling which occurs at $\lambda = \lambda_s$, when $\bar{\xi} > 0$.

While Figs. 5.5a and 5.5b depict the situation at hand when $\lambda_1/\lambda_c \equiv K_2/K_1 < 0$, other combinations of the parameters in (5.2) are also of interest. For $\lambda_1 > 0$, small values of $\bar{\xi} < 0$ may lead to snap-buckling (i.e., Fig. 5.6a), while for $\lambda_1 = 0$, where we have a symmetric bifurcation for the perfect structure, snap-buckling may be induced by initial rotations with either $\bar{\xi} > 0$ or $\bar{\xi} < 0$ if $\lambda_2 < 0$ (i.e. Fig. 5.6b); the case $\lambda_2 > 0$ yields only mild buckling (i.e. Fig. 5.6c). It is, therefore, the postbuckling behavior of the perfect structure, with $\bar{\xi} = 0$, and, in particular, the issue of whether the load increases or decreases after initial buckling, which determines what kind of buckling behavior one may expect for the imperfect structure with $\bar{\xi} \neq 0$. We now turn to a discussion of imperfection sensitivity for the simple structure pictured in Fig. 5.3.

When snap-buckling occurs, the magnitude of $\lambda_s$ (see Fig. 5.5a) may be well below that of $\lambda_c$. If we maximize $\lambda(\xi)$ in (5.7), with respect to $\xi$, then we find that $(\lambda'(\xi_s) = 0$ and $\lambda(\xi_s) = \lambda_s)$

$$\lambda_s L \cos(\xi_s + \bar{\xi}) = f'(\xi_s) \tag{5.8}$$

which, in conjunction with (5.2), and the definitions of $\lambda_c, \lambda_1$, yields the asymptotic result

$$\lambda_s/\lambda_c \approx 1 - 2 \left[ -(\lambda_1/\lambda_c) \bar{\xi} \right]^{\frac{1}{2}} \tag{5.9}$$

for small $\bar{\xi}$, with $\lambda_1 \bar{\xi} < 0$. In an analogous fashion, for $\lambda_1 = 0$ and $\lambda_2 < 0$ we have, for small $\bar{\xi}$,

$$\lambda_s/\lambda_c \simeq 1 - 3 \left( -\lambda_2/\lambda_c \right)^{\frac{1}{3}} \left( \frac{1}{2} \bar{\xi} \right)^{\frac{2}{3}} \tag{5.10}$$

In both of the cases considered above, therefore, relatively small imperfections (small initial rotations $\bar{\xi}$) can produce a snap-buckling load $\lambda_s$ which is much lower than the critical load which is predicted for the perfect structure with $\bar{\xi} = 0$; structures in which this phenomena can occur are said to be imperfection sensitive. For the model depicted in Fig. 5.3, imperfection sensitivity is implied by the conditions $\lambda_1 \neq 0$ or $\lambda_1 = 0$, with $\lambda_2 < 0$, and the extent of the imperfection sensitivity in these cases then depends on the magnitudes of $\lambda_1/\lambda_c$ or $\lambda_2/\lambda_c$.

**Remarks:** In lieu of prescribing the load $\lambda$ which acts on the imperfect structure in Fig. 5.3, one could impose a vertical displacement (or shortening) $\Delta$ at the point of the application of the load (see Fig. 5.1) of the form

$$\Delta/L = \cos(\xi + \bar{\xi}) - \cos\bar{\xi} \tag{5.11}$$

For the limiting case of the perfect model, i.e. $\bar{\xi} = 0, \Delta/L \approx \dfrac{1}{2}\xi^2$ and then, by virtue of (5.4),

$$\begin{cases} \lambda/\lambda_c \approx 1 + (\lambda_1/\lambda_c)\,(2\Delta/L)^{\frac{1}{2}}, (\xi > 0) \\[2ex] \lambda/\lambda_c \approx 1 - (\lambda_1/\lambda_c)\,(2\Delta/L)^{\frac{1}{2}}, (\xi < 0) \\[2ex] \lambda/\lambda_c \approx 1 + 2\,(\lambda_2/\lambda_c)\,(\Delta/L), \begin{pmatrix} \lambda_1 = 0 \\ \lambda_2 \neq 0 \end{pmatrix} \end{cases} \tag{5.12}$$

for $\Delta$ small. Some typical load-shortening relations are depicted in Fig. 5.7 along with the associated $\lambda - \xi$ graphs; from these graphs one notes that asymmetric branching is associated with an initial vertical slope on the $\lambda - \Delta$ curves, while $d\lambda/d\Delta$ is finite for the case of symmetrical branching. For the imperfect structure, equations (5.7) and (5.11) give the $\lambda - \Delta$ relationship parametrically in terms of $\xi$ and these graphs are depicted as the dotted curves in Fig. 5.7.

### 5.1.2   Rectangular Plates

#### (a)   Plates Exhibiting Elastic Response

The full system of von Karman equations for an isotropic linearly elastic plate which is subjected to a distributed load $t(x, y)$ normal to the undeflected plane of the plate, and which possesses an initial deflection $w_0(x, y)$, is given in (1.58), where $\tilde{w}(x, y)$ represents the net deflection of the plate. Suppose that the small imperfection has the form $\alpha w_0(x, y)$, with $\alpha$ a small parameter and $w_0$ a given function and that $\bar{\Phi}_0(x, y) = \lambda\Phi_0(x, y)$ is the Airy function corresponding to the state of generalized plane stress in the plate (when $w \equiv 0$), which is generated by the loading conditions in effect along the edges of the plate; here the parameter $\lambda$ gauges the magnitude of the loading and $\tilde{\Phi}(x, y)$ is the extra Airy stress function. Let $t(x, y) = \epsilon \tau(x, y)$, with $\epsilon$ a small parameter and $\tau(x, y)$ a given function; then in non-dimensional form we have

$$\begin{cases} \Delta^2 \tilde{\Phi} = -\dfrac{1}{2} \, [w, w] - \alpha \, [w, w_0] \\[2mm] \Delta^2 \tilde{w} = \left[ \tilde{w} + \alpha w_0, \tilde{\Phi} + \lambda \Phi_0 \right] + \epsilon \tau \end{cases} \tag{5.13}$$

where $w(x, y) = \tilde{w}(x, y) + \alpha w_0(x, y)$ yields the final shape of the plate; in rectilinear coordinates, the $[ \, , \, ]$ is given by (1.51a) and the form of (5.13) results from assuming that the thin plate is both linearly elastic and isotropic.

The influence of small deviations $w_0(x, y)$, from initial flatness, on the out-of-plane deflections of simply supported isotropic elastic square plates which are subjected to unidirectional edge compression, has been investigated by Hu, Lundquist, and Batdorf [109], while other types of boundary support conditions along the edges have been studied by Yamaki [94] and Coan [192]; only in a neighborhood of the critical elastic buckling stress do the resulting lateral deflections differ substantially from those of an ideal flat plate during the loading process. Thus, very large postbuckling deflections appear not to be greatly affected by small initial irregularities. For the square plate which is simply supported along all four edges, the postbuckling curves of Hu, Lundquist, and Batdorf [109] are depicted in Fig. 5.8, where $\delta_0$ is the initial deflection at the center of the plate while $\tilde{\delta}$ is the additional deflection at the center of the plate; the type of imperfection buckling displayed in this figure is similar to that pictured in Fig. 5.7d for the case $\bar{\xi} > 0$. In [94] Yamaki computed the values of $\tilde{\delta}/h$ for other types of boundary support conditions along the edges of a square, linearly elastic isotropic thin plate with an initial deviation (imperfection) given by $\delta_0/h = 0.1$; these results are depicted in Figs. 5.9 and 5.10 along with the corresponding dotted curves for the perfect plate in which there is no initial deflection. It is also worthwhile to comment, briefly, on the type of influence on the buckling of a rectangular, isotropic linearly elastic plate, which is exerted by the presence of a distributed load $t(x, y)$ normal to the (initially) flat configuration for the plate. As noted in [32], "the critical stress of a rectangular plate increases when normal loads of sufficient magnitude are applied ... this increase is due to the effect of the additional tensile membrane stresses caused by large deflections of the plate under normal loads and is related to the magnitude of deflection that can be allowed."

An analysis of the initial buckling of isotropic rectangular elastic plates subjected not only to edge compression, but also to a distributed normal load (with associated pressure function $p$), and having an aspect ratio $a/b = 4$, was carried out by Levy, Goldenburg, and Zibritosky [193]; the plate was assumed to be simply supported and the lateral deflection $w$ was approximated by

$$w(x,y) = A_{11} \sin \frac{\pi x}{a} \sin \frac{\pi y}{b} + A_{31} \sin \frac{3\pi x}{a} \sin \frac{\pi y}{b}$$
$$+ A_{51} \sin \frac{5\pi x}{a} \sin \frac{\pi y}{b} + A_{71} \sin \frac{7\pi x}{a} \sin \frac{\pi y}{b} \qquad (5.14)$$

The analysis in [193] leads to four simultaneous equations relating $\frac{pb^4}{Eh^2}$, $h^2$, and the coefficients $A_{11}, A_{31}, A_{51}$, and $A_{71}$, in (5.14), which are shown to have more than one real solution; thus, the plate may be in equilibrium for a given combination of lateral and end loading in more than one buckle pattern, and this phenomenon is depicted in Fig. 5.11, which shows the relationship between the parameter $\lambda b^2 / Eh^2$, (which is based on the edge stress $\lambda = \sigma_{xx}|_{x=0,a}$) and the parameter $\epsilon_{xx} b^2 / h^2$, which is based on the axial strain for the constant normal pressure given by $p = 24.03 \frac{Eh^4}{b^4}$. We note that the shift in the buckle pattern is associated with a drop in the axial load and that the plate can be in equilibrium with either one, three, or seven buckles in the initial buckling mode. In Fig. 5.12 we show the development of longitudinal buckles in the initial buckling mode, thus indicating that, for low values of $\lambda$, the resulting deflection is a long shallow bulge. When high values of $\lambda$ prevail, one obtains the regular initial buckling pattern again and any initial deflection of the plate has all but disappeared. For all the conditions investigated by Levy, Goldenburg, and Zibritosky [193], it was noted that the presence of a distributed normal load leads to an increase in the critical buckling load of the plate. When the edges of the rectangular plate are clamped instead of simply supported, the existence of a distributed normal load leads to results similar to those obtained in [193]; this follows from the analysis in Woolley, Corrick, and Levy [120] in which the same gradual transition from a long bulge to a regular initial buckling pattern was observed as in the case of a simply supported plate. One notable difference between the cases of a simply supported and a clamped rectangular plate subject to distributed normal loading as well as edge loads, is that the increase in the critical buckling load is much less in the latter case; this result is illustrated in Fig. 5.13, where the critical buckling strain parameter $\epsilon_{cr}b^2/h^2$ is compared with the parameter $pb^4/Eh^4$ for the two different types of boundary support conditions.

We now turn to a description of the postbuckling behavior of general linear elastic anisotropic rectangular plates possessing imperfections in the form of an initial prebuckling displacement $w_0(x,y)$. The analysis, which is based on the work of Romeo and Frulla [114], includes the case of rectilinear orthotropic symmetry as a particular subcase of the general anisotropic response considered, and applies to several types of boundary support conditions for plates subjected to combined biaxial compression and shear loads. As indicated in [114], plates possessing initial imperfections $w_0$ have been studied on the basis of the Marguerre approximate nonlinear theory by Chia [33]; the resultant strain-displacement relations are, essentially, (1.34) with $W$ replaced by $w_0$, i.e.,

$$
\begin{cases}
\epsilon_{xx} = u_{,x} + \dfrac{1}{2}w_{,x}^2 + w_{0,x}w_{,x} - zw_{,xx} \\[2mm]
\epsilon_{yy} = v_{,y} + \dfrac{1}{2}w_{,y}^2 + w_{0,y}w_{,y} - zw_{,yy} \\[2mm]
\gamma_{xy} = u_{,y} + v_{,x} + w_{,x}w_{,y} + w_{0,y}w_{,x} \\[2mm]
\qquad\quad +w_{0,x}w_{,y} - 2zw_{,xy}
\end{cases}
\tag{5.15}
$$

while the equilibrium equations consist of (1.42) and the analog of (1.45) with $W$ replaced by $w_0$, i.e.

$$
M_{x,xx} + M_{y,yy} + 2M_{xy,xy} + N_x w_{,xx}
$$
$$
+2N_{xy}w_{,xy} + N_x w_{0,xx} + N_y w_{0,yy}
\tag{5.16}
$$
$$
+2N_{xy}w_{0,xy} = 0
$$

Introducing the Airy function as in (1.46), the first two equilibrium equations (1.42) are, once again, satisfied identically, while (5.16) becomes

$$
M_{x,xx} + M_{y,yy} + 2M_{xy,xy} + \Phi_{,yy}w_{,xx}
$$
$$
+\Phi_{,xx}w_{,yy} - 2\Phi_{,xy}w_{,xy} + \Phi_{,yy}w_{0,xx}
\tag{5.17}
$$
$$
+\Phi_{,xx}w_{0,yy} - 2\Phi_{,xy}w_{0,xy} = 0
$$

If $\epsilon_{xx}^0$, $\epsilon_{yy}^0$, and $\epsilon_{xy}^0$ denote the strain components of the middle surface of the plate (i.e., (1.35) with $w \equiv 0$), the relevant compatibility relation, i.e., (1.49), when combined with (5.15), and averaged over the thickness of the plate, yields

$$\epsilon^0_{xx,yy} + \epsilon^0_{yy,xx} - \epsilon^0_{xy,xy} = w^2_{,xy} - w_{,xx}w_{,yy}$$

$$+2w_{0,xy}w_{,xy} - w_{0,yy}w_{,xx} - w_{0,xx}w_{,yy}$$

$$(5.18)$$

The stress-strain relations in [114] are taken in the form

$$\left\{\epsilon^0_{xx}, \epsilon^0_{yy}, \epsilon^0_{xy}\right\}^t = [\boldsymbol{A}]^{-1} \left\{\Phi_{,yy}, \Phi_{,xx}, -\Phi_{,xy}\right\}^t \qquad (5.19)$$

which leads to the relations

$$\left\{M_x, M_y, M_{xy}\right\}^t = [\boldsymbol{D}] \left\{-w_{,xx}, -w_{,yy}, -2w_{,xy}\right\}^t \qquad (5.20)$$

in which $[\boldsymbol{A}]$ and $[\boldsymbol{D}]$ are, respectively, the extensional and bending stiffness matrices. Employing (5.19) and (5.20) in (5.17) and (5.18) and introducing the normalizations

$$\xi = \frac{x}{a}, \eta = \frac{y}{b}, \zeta = \frac{z}{h}, \phi = a/b \qquad (5.21\text{a})$$

$$F = \Phi/A_{22}h^2, W_0 = w_0/h, W = w/h \qquad (5.21\text{b})$$

$$[\boldsymbol{A}^*] = [\boldsymbol{A}]^{-1}, \left[\bar{\boldsymbol{A}}^*\right] = A_{22}[\boldsymbol{A}^*] \qquad (5.21\text{c})$$

$$[\boldsymbol{D}^*] = [\boldsymbol{D}], \left[\bar{\boldsymbol{D}}^*\right] = \frac{1}{h^2 A_{22}}[\boldsymbol{D}^*] \qquad (5.21\text{d})$$

the following generalized von Karman system is obtained:

$$\bar{D}^*_{11}W_{,\xi\xi\xi\xi} + 4\bar{D}^*_{16}\phi W_{,\xi\xi\xi\eta} + 2\left(\bar{D}^*_{12} + 2\bar{D}^*_{66}\right)\phi^2 W_{,\xi\xi\eta\eta}$$

$$+4\bar{D}^*_{26}\phi^3 W_{,\xi\eta\eta\eta} + \bar{D}^*_{22}\phi^4 W_{,\eta\eta\eta\eta}$$

$$-\phi^2 \left(F_{,\eta\eta}W_{,\xi\xi} + F_{,\xi\xi}W_{,\eta\eta} - 2F_{,\xi\eta}W_{,\xi\eta}\right)$$

$$+F_{,\eta\eta}W_{0,\xi\xi} + F_{,\xi\xi}W_{0,\eta\eta} - 2F_{,\xi\eta}W_{0,\xi\eta}\right) = 0$$

$$(5.22)$$

and

$$\bar{A}_{11}^* \phi^4 F_{,\eta\eta\eta\eta} - 2\bar{A}_{16}^* \phi^3 F_{,\xi\eta\eta\eta} + \left(2\bar{A}_{12}^* + \bar{A}_{66}^*\right) \phi^2 F_{,\xi\xi\eta\eta}$$

$$-2\bar{A}_{26}^* \phi F_{,\xi\xi\xi\eta} + \bar{A}_{22}^* F_{1\xi\xi\xi\xi}$$

$$= \phi^2 \left( W_{,\xi\eta}^2 - W_{,\xi\xi} W_{,\eta\eta} + 2W_{0,\xi\eta} W_{,\xi\eta} \right.$$

$$\left. - W_{0,\eta\eta} W_{,\xi\xi} - W_{0,\xi\xi} W_{,\eta\eta} \right)$$

(5.23)

If we let $\lambda_\xi, \lambda_\eta$, and $\lambda_{\xi\eta}$ denote the nondimensional applied external loads along the edges of the plate, i.e.

$$\lambda_\xi = \frac{N_x b^2}{h^2 A_{22}}, \lambda_\eta = \frac{N_y b^2}{h^2 A_{22}}, \lambda_{\xi\eta} = \frac{N_{xy} b^2}{h^2 A_{22}}$$

(5.24)

then the four different kinds of boundary conditions considered in [114] are

(i) ends simply supported–sides simply supported (BC-1):

$$\begin{cases} \xi = 0, 1 : \Phi_{,\eta\eta} = \lambda_\xi, \Phi_{,\xi\eta} = -\phi\lambda_{\xi\eta}, W = 0, M_\xi = 0 \\ \eta = 0, 1 : \Phi_{,\xi\xi} = \phi^2 \lambda_\eta, \Phi_{,\xi\eta} = -\phi\lambda_{\xi\eta}, W = 0, M_\eta = 0 \end{cases}$$

(5.25)

(ii) sides simply supported–ends clamped (BC -2):

$$\begin{cases} \xi = 0, 1 : \Phi_{,\eta\eta} = \lambda_\xi, \Phi_{,\xi\eta} = -\phi\lambda_{\xi\eta}, W = 0, W_{,\xi} = 0 \\ \eta = 0, 1 : \Phi_{,\xi\xi} = \phi^2 \lambda_\eta, \Phi_{,\xi\eta} = -\phi\lambda_{\xi\eta}, W = 0, M_\eta = 0 \end{cases}$$

(5.26)

(iii) ends clamped–sides clamped (BC-3):

$$\begin{cases} \xi = 0, 1 : \Phi_{,\eta\eta} = \lambda_\xi, \Phi_{,\xi\eta} = -\phi\lambda_{\xi\eta}, W = 0, W_{,\xi} = 0 \\ \eta = 0, 1 : \Phi_{,\xi\xi} = \phi^2 \lambda_\eta, \Phi_{,\xi\eta} = -\phi\lambda_{\xi\eta}, W = 0, W_{,\eta} = 0 \end{cases}$$

(5.27)

(iv) sides clamped–ends simply supported (BC - 4):

$$\begin{cases} \xi = 0, 1 : \Phi_{,\eta\eta} = \lambda_\xi, \Phi_{,\xi\eta} = -\phi\lambda_{\xi\eta}, W = 0, M_\xi = 0 \\ \eta = 0, 1 : \Phi_{,\xi\xi} = \phi^2 \lambda_\eta, \Phi_{,\xi\eta} = -\phi\lambda_{\xi\eta}, W = 0, W_{,\eta} = 0 \end{cases}$$

(5.28)

To satisfy the boundary conditions, the displacement and Airy functions are chosen in the respective forms

$$W = \sum_{p=1}^{i}\sum_{q=1}^{j} C_{pq}\omega(\xi,\eta)$$

(5.29)

and

$$\Phi = \frac{\eta^2}{2}\lambda_\xi + \frac{\xi^2\phi^2}{2}\lambda_\eta - \phi\xi\eta\lambda_{\xi\eta}$$
$$+ \sum_{h=1}^{m}\sum_{k=1}^{n} F_{nk}X_h(\xi)Y_k(\eta)$$

(5.30)

where $X$ and $Y$ are the characteristic functions

$$X_h(\xi) = \cosh(\mu_h\xi) - \cos(\mu_h\xi)$$
$$-\alpha_h\left(\sinh(\mu_h\xi) - \sin(\mu_h\xi)\right)$$

(5.31a)

and

$$Y_k(\eta) = \cosh(\mu_k\eta) - \cos(\mu_k\eta)$$
$$-\alpha_k\left(\sinh(\mu_k\eta) - \sin(\mu_k\eta)\right)$$

(5.31b)

with the constants $\mu_h, \mu_k, \alpha_h$, and $\alpha_k$ chosen so as to satisfy the boundry conditions

$$\begin{cases} X_h(0) = X_h(1) = X_h'(0) = X_h'(1) = 0 \\ Y_k(0) = Y_k(1) = Y_k'(0) = Y_k'(1) = 0 \end{cases}$$

(5.32)

Also, in (5.29), the function $\omega(\xi,\eta)$ is chosen according to the boundary conditions, i.e.,

$$(1)BC - 1 : \omega(\xi,\eta) = \sin(h\pi\xi)\sin(k\pi\eta)$$

(5.33a)

$$(2)BC - 2 : \omega(\xi,\eta) = X_h(\xi)\sin(k\pi\eta)$$

(5.33b)

$$(3)BC - 3 : \omega(\xi,\eta) = X_h(\xi)Y_k(\eta)$$

(5.33c)

$$(4)BC - 4 : \omega(\xi,\eta) = \sin(h\pi\xi)Y_k(\eta)$$

(5.33d)

In [114] a set of nonlinear algebraic equations is obtained, involving the coefficients $C_{pq}$ and $F_{hk}$, by applying the Galerkin procedure to (5.22) and (5.23); the initial imperfections are taken in the form

$$W_0(\xi, \eta) = T \sin \pi \xi \sin \pi \eta \qquad (5.34)$$

In Fig. 5.14, the results obtained in [114] are displayed and compared with the results obtained by Yamaki [94] for an isotropic square plate under uniaxial compression, both with an initial imperfection of the form (5.34) with $T = 0.1$, and without the imperfection under various constraint conditions: I is a plate with all edges simply supported, II is the plate with the sides clamped and the ends simply supported, for III the sides are simply supported and the ends are clamped, while IV represents a plate clamped along all edges. The results obtained in [114] are very close to those of Yamaki [94]. In Fig. 5.15, results are shown for the clamped/simply supported isotropic plate under axial compression, for plate aspect ratios of $\phi = 1$ and $\phi = 3$, and these are compared with those of Sheinman, etal. [194] for $T = 0.1$ with $m = n = i = j = 4$ or 6 in (5.29), (5.30); it may be noted that, as the plate aspect ratio increases, the initial imperfection causes a postbuckling behavior which is very different from the theoretical behavior without the imperfections. In Fig. 5.16, results are shown for an anisotropic square plate under uniaxial compression which has its edges clamped and its sides simply supported; these results are compared with those of Minguet, et. al. [195]. Finally, in Fig. 5.17, we show a comparison of analytical and experimental results (based on the work in [114]) for a rectangular anisotropic plate under uniaxial compression. The plate has clamped edges, simply supported sides, and an initial imperfection of the form (5.34) with $T = 0.05$; the results are compared with those of Engelstad, et. al. [196].

### (b)  Plates Exhibiting Elastic-Plastic Response

In §3 of Chapter 2 we discussed the initial and postbuckling behavior of square elastic-plastic plates subjected to axial compression; that discussion was based, to a large extent, on the work presented by Needleman and Tvergaard [104] and concerned a plate material which is strain-hardening and characterized by a flow theory of plasticity with a smooth yield surface. The constitutive theory employed in [104] is given by (2.221)–(2.225). Two types of in-plane boundary conditions were considered in 2.3: one which requires all four edges of the plate to remain straight throughout the loading history and one which leaves the edges unconstrained; in the elastic range, these in-plane boundary conditions play a major role in determining the initial postbuckling behavior of the plate but the work in [104] also investigates the effects of the in-plane boundary conditions on the postbuckling behavior and imperfection sensitivity in the plastic range. For buckling of the perfect plate in the plastic range, the authors [104] employ Hutchinson's postbuckling theory [17] to obtain an asymptotically exact description of the initial postbuckling behavior, i.e., (2.229); they also note that for a plate with a small initial curvature compressed into the plastic range, there is no available general theory which relates the behavior of the imperfect plate to the postbuckling behavior of the corresponding perfect plate as in, e.g., Koiter's theory of elastic stability [5]. Thus, in [104], an approximate analysis based on the approach used by Hutchinson and Budiansky [197], which neglects elastic unloading, is employed in order to assess the imperfection sensitivity of simply supported square plates when buckling of the perfect plate occurs in the plastic range.

Following the discussion in 2.3, we confine our attention to imperfections in the shape of the initial buckling mode (2.227) with the amplitude of the imperfection denoted by $\bar{\xi}$. For $\lambda < \lambda_c$ we have noted, in Chapter 2, that the lowest order effect of the imperfection on the magnitude of the lateral deflection $\xi$ is given by (2.230) where $\Psi \geq 1$ and $p(\lambda)$ is finite at $\lambda = \lambda_c$. For $\lambda$ close to $\lambda_c$, a singular perturbation analysis produces (2.231) and matching (2.230) and (2.231) produces

$$ c = -\lambda_c \left(p(\lambda_c)\right)^{\frac{1}{\Psi}}, \gamma = 2\Psi + 1 \qquad (5.35) $$

When $\lambda_2 < 0$, (2.231) produces a maximum support stress $\lambda_{\max}$ which satisfies (2.232) with $\mu$ given by (2.233). For $\lambda_c < \sigma_y, \lambda_2$ reduces, as noted in [104], to the appropriate value for a linear elastic plate.

Figures 2.51, 2.52 and 2.53, which we have already referenced in Chapter 2, for the buckling behavior of the perfect (i.e., $\bar{\xi} = 0$) elastic-plastic square plate, are based on numerical results obtained in [104]. Figures 2.51 and 2.52 display results which correspond to buckling of the square plate occurring in the plastic range; in both cases, $h/a = 0.035$ and $\sigma_y/E = 0.0037$. In Fig. 2.51, the hardening parameter $n$ in (2.225) is 10 while $\lambda_c/\sigma_y = 1.196$. In Fig. 2.52, we have $n = 3$ and $\lambda_c/\sigma_y = 1.259$. Although initial buckling occurs well into the plastic range of the material, it turns out that the predictions of $J_2$-flow theory and those of the simplest deformation theory, i.e. $J_2$-deformation theory, do not differ by a large amount; for example, in Fig. 2.51, with $n = 10$, the discrepancy noted in [104] is 5.8%, while in Fig. 2.52, with $n = 3$, the discrepancy is just 2.3%. In Fig. 2.53 we have $h/a = 0.031, \nu = 0.3$, and $\sigma_y/E = 0.00337$ and initial buckling of the perfect plate occurs just after the onset of plastic yield; as noted in [104], in this case there exists virtually no discrepancy between the bifurcation predictions of $J_2$-flow theory and those of $J_2$-deformation theory. In Fig. 2.53, $n = 10$ and $\lambda_c/\sigma_y = 1.015$. As we have already noted in Chapter 2, for the perfect plates the behavior in the immediate neighborhood of initial buckling is described by (2.229) in that buckling takes place under increasing load and the curvature of the load deflection curve is large and negative; also, in the immediate neighborhood of the initial buckling point, the initial postbuckling behavior is seen to be independent of the boundary conditions. We now turn to the imperfection buckling behavior described in figures 2.51, 2.52, and 2.53.

In Fig. 2.51, where the hardening parameter $n = 10$, the load falls off rapidly after the maximum load point is reached. The postbuckling behavior is nearly identical for both sets of in-plane boundary conditions considered, as is the behavior of a plate possessing a small initial imperfection given by $\bar{\xi} = 0.01$. However, as the amplitude of the initial imperfection is increased, the effect of the in-plane boundary conditions is more pronounced. For the largest imperfection considered in Fig. 2.51, i.e. $\bar{\xi} = 1.0$, the plate subjected to the constrained boundary conditions has a maximum support load which is about 25% higher than the one with the unconstrained in-plane boundary condition.

In Fig. 2.52, the postbuckling behavior of the perfect plate depends very strongly on the in-plane boundary conditions. Under the unconstrained in-plane boundary conditions, the load reaches a maximum and then decreases monotonically, whereas with the constrained boundary condition, the load continues to increase; in this case the mode amplitude $\xi_{\max}$, which is obtained from (2.229), is much smaller than the numerical value obtained for the plate subjected to the unconstrained boundary condition. The load-deflection behavior for a plate with an initial imperfection also depends, in this case, on the in-plane boundary conditions. An imperfect plate subjected to the constrained in-plane boundary condition is capable of supporting loads in excess of the buckling load, while plates with the unconstrained boundary condition clearly exhibit a mild imperfection sensitivity.

In Fig. 2.53, in which $n = 10$, the load on the perfect plate is seen to reach a maximum shortly after buckling and then decrease monotonically. When compared with the example considered in Fig. 2.51, in which initial buckling of the perfect plate occurs further into the plastic range, the effect of the in-plane boundary conditions shows up much earlier in Fig. 2.53. Even with an initial imperfection amplitude of $\bar{\xi} = 0.01$, the effect of the in-plane boundary conditions on the load-deflection behavior of the plate is apparent. We also note, that in Fig. 2.53, for each initial imperfection amplitude $\bar{\xi}$ considered, the imperfection sensitivity is less than that for the corresponding example in Fig. 2.51 and that the effect of the in-plane boundary conditions on the load-deflection behavior of the plate increases with the amplitude of the initial imperfection. Overall, the results described above for square elastic-plastic plates exhibit a far smaller degree of imperfection sensitivity than the comparable results for circular elastic-plastic plates, which are presented in the next section.

### 5.1.3   Circular and Annular Plates

#### (a)  Plates Exhibiting Elastic Response

While the buckling behavior of initially flat, linearly elastic circular and annular plates has been widely studied, especially for isotropic plates, the response of such plates in the presence of imperfections and in-plane compression has been less thoroughly researched; the first attempt to deal with the imperfection buckling of circular and annular plates appears to have been made by Massonnet [198] who employed a power series method to produce solutions for the postbuckling response of clamped isotropic plates with a single half-wave imperfection. Tani and Yamaki [199] have analyzed the postbuckling response of isotropic annular plates with axisymmetric initial deflections. In this subsection, we will present the results of Turvey and Drinali [157] on the elastic post-buckling of orthotropic circular and annular plates with imperfections; their results are valid for clamped and simply supported plates subjected to uniform compression at the outer edge and are presented for three orthotropy ratios $\beta \, (\equiv E_\theta / E_r) = \dfrac{1}{3}, 1$, and 3, a wide range of aspect ratios $\phi = b/a$, where $b$ is, once again, the hole radius and $a$ is the radius of the outer boundary of the plate, and several different imperfection amplitudes. The case $\beta = 1$ yields, as usual, the specialization to the case of an isotropic plate; the aspect ratio $\phi = 0$ covers the specialization to a circular plate. The imperfection amplitude will again be denoted by $\bar{\xi}$, which is taken to represent the amplitude of the imperfection at the origin of the plate coordinate system; of course, only in the case of a complete circular plate does this imperfection amplitude correspond to the actual maximum initial plate deflection (see Fig. 5.18.) Two types of single half-wave initial deflection profiles are employed in

[157]: for the case of a plate simply supported along its outer edge, the initial deflection profile was taken to be

$$w_0(r) = \bar{\xi}(1 - 0.1\bar{r} - 0.9\bar{r}^2), \quad \bar{r} = \frac{r}{a} \tag{5.36}$$

while for the case of a clamped plate

$$w_0(r) = \bar{\xi}\cos\left(\frac{1}{2}\pi\bar{r}\right) \tag{5.37}$$

In the analysis presented in [157] the initial deflection parameter $\bar{\xi}$ was varied over the values $0, 0.1, 0.5$, and $1.0$. For the analyses of the imperfection buckling of an annular plate, the inner edge of the plate was assumed to be free; at the center of a circular plate one imposes, at $r = 0$, the usual conditions, i.e., $u(0) = w'(0) = 0$ where $u$ is the plate displacement in the radial direction. The axially symmetric large deformation states for cylindrically orthotropic circular and annular plates possessing an initial deflection $w_0(r)$ are governed by that straightforward modification of (1.101), (1.102), (1.103) which is analogous, e.g., in the case of the first von Karman equation, in rectilinear Cartesian coordinates, to retaining $W = w_0$ in (1.45); the relevant constitutive relations are given by the obvious modifications of (1.104 a,b) where, e.g., $\sigma_{rr}, e_{rr}$, etc., must be replaced by $\sigma_{rr} - \sigma_{rr}^0, e_{rr} - e_{rr}^0$, with $\sigma_{rr}, e_{rr}$ the stress and strain components in the present state of the plate, while $\sigma_{rr}^0, e_{rr}^0$ correspond to the state given by the initial deflection $w_0$. The plates in the examples considered in [157] are buckled by controlling the radial shortening applied uniformly at the outer circumference at $r = a$; the flexural conditions assumed to exist along this boundary are those which correspond to either the simply supported or clamped edge conditions.

In the work presented in [157] the authors employ a finite difference discretization of the governing equations and first consider the computation of the initial buckling loads and subsequent postbuckling behavior of flat isotropic circular and annular plates; the results are summarized in Table 5.1 and are compared with earlier results of Nadai [186] and Meissner [172], where the first column in the table refers to the number of mesh divisions in the finite difference scheme. A comparison of the accuracy of two different mesh divisions is shown in Fig. 5.19, the results being those in the postbuckling regime (i.e. load-radial shortening response) for initially flat, clamped, and simply supported circular plates. To check that the initial deflections had been correctly incorporated into the program, the computations were checked against Massonnet's [198] results for imperfect clamped isotropic circular plates; in this case, the edge compression versus additional plate center deflection comparison is shown in Fig. 5.20 and there appears to be good agreement between the two sets of results over the range of initial deflection amplitudes considered. To check that the cylindrical orthotropy of the plates had been appropriately incorporated into the program, yet another set of computations were performed for the postbuckling response of flat orthotropic annular plates (with aspect ratio $\phi = 0.4$) having both simply supported and clamped outer edges; the benchmark in this case was the work of Huang [174] and the edge compression versus free edge deflection comparison is shown in Fig. 5.21. In Figs. 5.22 and 5.23 we reproduce the results obtained by Turvey and Drinali [157] relative to the postbuckling response of imperfect polar orthotropic annular plates having an aspect ratio of $\phi = 0.4$. The postbuckling response was computed for three orthotropy ratios and four imperfection amplitudes, i.e., $\bar{\xi} = 0, 0.1, 0.5$, and $1.0$ and $\beta = 3, 1$, and $\frac{1}{3}$ and the edge compression versus additional deflection (at $r = 0.4a$) responses are shown in Figs. 5.22 and 5.23 for clamped and simply supported edges, respectively. The general postbuckled response for the clamped plates, as depicted in Figs. 5.22a through 5.22c does not appear to depend on the orthotropy parameter and, as the edge compression becomes large, the magnitude of the additional deflection depends less and less on the initial deflection amplitude, i.e., the postbuckling response of the imperfect plate tends to that of the flat plate with the trend being more pronounced as the orthotropy parameter increases. However, the edge compression versus edge deflection response of the simply supported plates, as noted on Figs. 5.23a through 5.23c is quite different from those of clamped plates: in the simply supported case the additional imperfect plate deflection does not converge to that of the flat plate deflection; indeed, the additional imperfect plate deflection becomes less than the flat plate value once the edge compression exceeds a value between $\frac{5}{4}$ and $\frac{3}{2}$ times the critical buckling load value for the flat plate. Additionally, if one compares the clamped and simply supported postbuckling plate responses shown in Figs. 5.22 and 5.23, it is observed that (for edge compression ratios greater than unity) the additional deflection is generally smaller for the simply supported plate than it is for the clamped plate. It may, therefore, be deduced that the presence of the initial deflection has a greater transverse stiffening effect, in the postbuckling regime, on simply supported plates as opposed to clamped plates. Finally, it has been noted in [157] that none of the computed flat or imperfect plate responses exhibited any buckling mode change tendencies as either the imperfection amplitude and/or the edge compression increased, i.e., the initial axisymmetric single half-wave buckling mode was maintained throughout.

## (b) Plates Exhibiting Elastic-Plastic Response

In Chapter 3 we considered the initial buckling and postbuckling behavior of (perfect) thin elastic-plastic circular plates, basing our analysis on the work of Needleman [141] which is, *a priori* , restricted to axisymmetric deformations. As noted in [141], however, the postbuckling behavior of elastic-plastic circular plates is characterized by a high degree of imperfection sensitivity, which is the issue that we will be concerned with in this subsection. We begin by reviewing, briefly, the mathematical formulation of the buckling problem for an elastic-plastic circular plate as presented in §2 of Chapter 3.

As in Chapter 3, the plate is assumed to have radius $a$ and thickness $h$ and is subjected to an in-plane radial compressive loading. The strain rates $\dot{e}_{rr}, \dot{e}_{\theta\theta}$ are given by (3.140) with $u(r)$ and $w(r)$, respectively, the radial in-plane and lateral displacements. Stress rates $\dot{\sigma}$ are related to the strain rates $\dot{e}$ by $\dot{\sigma} = \boldsymbol{L}\dot{e}$ where $\boldsymbol{L}$, the elastic part of the tensor of moduli $\boldsymbol{L}$, is given by (3.141). A $J_2$ flow theory with isotropic hardening is employed so that the uniaxial stress-strain behavior is given by (3.145) with $n$ the strain-hardening exponent and $\sigma_y$ the yield stress. The full tensor $\boldsymbol{L}$ is then represented by (3.142) with $\sigma_e, s_{ij}$, and $c$ given, respectively, by (3.143) and (3.144). The tangent modulus is a function of $\sigma_e$ and represents the slope of the uniaxial stress-strain curve as depicted in Fig. 3.25. As noted in Chapter 3, each surface in the plate which is parallel to the middle surface is in an (approximate) state of plane stress, implying that only the in-plane stresses need to be admitted into the constitutive theory; this observation leads, in turn, to the introduction of the tensor $\boldsymbol{K}$ of in-plane moduli which is defined as in (3.146). For a simply supported plate, the relevant conditions at $r = a$ are given by (3.147) while for a clamped plate (3.148) holds. The equilibrium equations are given by the variational principle $\delta\mathcal{I} = 0$, where the functional $\mathcal{I}$ is given by (3.149) and all those incremental fields $\dot{u}(r)$ and $\dot{w}(r)$ which satisfy the kinematic boundary condition at $r = a$ are admissible.

The critical buckling stress in the plastic range is given by (3.150) where $E/E_f$ and $\nu_f$ are defined, respectively, by (3.151), (3.152); for a clamped plate $\tilde{\beta}$ is the smallest nonzero root of $J_1(\tilde{\beta}) = 0$ while, for a simply supported plate, it is the smallest non-zero root of $\tilde{\beta}J_0(\tilde{\beta}) - (1 - \nu_f)_c J_1(\tilde{\beta}) = 0$, the $c$ subscript denoting evaluation of the effective

Poisson ratio $\nu_f$ at buckling. For both types of boundary support conditions, the associated buckling mode $w_c(r)$ has the form (3.153). The critical buckling stress $\sigma_c$ for both the clamped and simply supported cases is depicted in Fig. 3.25 as a function of the parameter $\tilde{\eta}$, which is defined by (3.154), where $\tilde{\beta}_E$ is the value of $\tilde{\beta}$ when buckling occurs in the elastic range.

Employing the amplitude $\xi$ of the relevant eigenmode as the expansion variable in a postbuckling perturbation analysis, the average radial stress applied at the edge of the plate at buckling is given by (3.155), where $\sigma_c$ is the critical buckling stress as given by (3.150) and $\sigma_{av}$ is defined by (3.156). The initial slope $\sigma_1$ in the expansion (3.155) is uniquely determined by the condition that plastic loading takes place throughout the plate except at one point where neutral loading occurs; this leads to the relation (3.157) between $\sigma_1$ and $\sigma_c$. As a consequence of (3.157), $\sigma_1 > 0$. Carrying the expansion for the average applied compressive stress one term further leads to (3.158), where $\sigma_2$ is given by (3.159), (3.160) and (3.161a,b). The first three terms that are depicted in the expansion (3.158) for $\sigma_{av}$ may be used to estimate the maximum applied stress $\sigma_{max}$ that can be supported by the plate as well as the amplitude $\xi$ of the buckling mode at $\sigma_{max}$; these results are given by (3.162a,b). As $\xi$ increases from zero, the neutral loading surface expands out from the point $r = 0, z = \dfrac{1}{2}h$ and has the form (3.163).

In Chapter 3 we have referenced Figs. 3.26 thorough 3.29 and presented a fairly thorough discussion of the postbuckling behavior for the perfect elastic-plastic circular plate for each of the respective cases considered in these graphs; the graphs in question depict the (postbuckling) curves of average applied stress vs. buckling deflection for two different cases: the (postbuckling) curves governing buckling deflection vs. applied displacement for a simply supported plate, and the (postbuckling) curves depicting average applied stress vs. buckling deflection for a clamped plate. The results displayed in Figs. 3.26 through 3.29 are based on a finite-element analysis of the variational equation $\delta \mathcal{I} = 0$ where $\mathcal{I}$ is given by (3.149); in this analysis the author [141] allows for an initial imperfection $w_0(r)$ which is proportional to the (initial) buckling mode (3.153) and the amplitude is then specified by $\hat{\xi} = w_0(0)$. The results for perfect plates, which have been discussed in detail in Chapter 3, actually result from a calculation in which the small imperfection amplitude $\hat{\xi} = 10^{-4}h$ has been employed.

For $\sigma_c \ll \sigma_y$ the elastic postbuckling analysis shows that the initial postbuckling behavior is stable in the sense that applied stresses higher than the buckling stress $\sigma_c$ can be supported by the plate; in this range, the plate is imperfection insensitive and these observations have already been made in Chapter 3. However, for $\sigma_c > \sigma_y$ the maximum support load of the circular plate is not much larger than $\sigma_c$ and the analysis by Hutchinson [45], [46] of imperfection sensitivity for buckling in the plastic range indicates that, in general, one may expect imperfection sensitivity to be prevalent when $\sigma_c > \sigma_y$.

In Fig. 5.24 we show the transition from imperfection insensitive to imperfection sensitive behavior for the simply supported circular plate with $\nu = \dfrac{1}{3}$ and $n = 12$. The maximum support stress, $\sigma_{\max}$, of an imperfect plate is graphed against the critical buckling load $\sigma_c$ for three different values of the imperfection amplitude $\hat{\xi}$, i.e., $\hat{\xi} = .02h, \hat{\xi} = .06h$, and $\hat{\xi} = .15h$. The graph in Fig. 5.24 is limited to the range $\sigma_c \leq \sigma_y$. As the buckling stress $\sigma_c$ approaches the yield stress $\sigma_y$ the simply supported plate becomes imperfection sensitive, e.g., for $\sigma_c/\sigma_y = .99, \nu = \dfrac{1}{3}, n = 12$, and $\hat{\xi} = .15h$ we find that $\sigma_{\max}$ is only $.6\sigma_c$. Clamped plates were determined in [141] to be less imperfection sensitive than simply supported plates, e.g., for $\sigma_c/\sigma_y = .99, \nu = \dfrac{1}{3}, n = 12$, and $\hat{\xi} = .15h, \sigma_{\max}$ for the clamped plate was found to be $.77\sigma_c$.

The role of plastic deformation in reducing the load-carrying capacity of circular plates which buckle in the elastic range is shown in some of the Figs. 3.26 through 3.29. In Fig. 3.26 the curves of $\sigma_{av}$ vs. the buckling deflection are shown for the simply supported plate with $\sigma_c/\sigma_y = .72, \nu = \dfrac{1}{3}$, and $n = 12$; here, and in the discussion which follows, the buckling deflection is measured by the parameter $\xi = w(0) - \hat{\xi}$. For purposes of comparison the postbuckling curves for a similar elastic plate with $\sigma_c/\sigma_y = .72$ and $\nu = \dfrac{1}{3}$ are also shown in Fig. 3.26 and similar numerical methods were employed for both the elastic and the elastic-plastic calculations. With $\hat{\xi} = .02h$, initial yield occurs at $.84\sigma_c$ and the curve for the elastic plate serves to illustrate the imperfection insensitivity which is characteristic of elastic response. However, the elastic-plastic plate has a maximum support load $\sigma_{\max}$ which is only $.89\sigma_c$, i.e., $\sigma_{\max}$ is only a little higher than the value of the applied stress at which initial plastic yield occurs. In Fig. 3.27 we show a case in which buckling of the perfect plate occurs outside the elastic range of the plate; this plate is simply supported with $\sigma_c/\sigma_y = 1.02$ and $\tilde{\eta}$, as given by (3.154) has the value 1.28 which, $n = 12$, and $\nu = \frac{1}{3}$. The maximum load is reached at a deflection of $\xi = .008h$ and has the value $\sigma_{\max} = 1.006\sigma_c$. After the maximum load point is attained, the load drops off very sharply and then levels off at larger buckling deflections. At an imperfection amplitude of $\hat{\xi} = .02h$ the maximum load occurs at a larger value of the deflection $\xi$ than for the perfect plate and the difference between the maximum load and the load at which the postbuckling curve levels off is less than that for a perfect plate. At $\hat{\xi} = .06h$, $\sigma_{\max}$ occurs at yet a larger value of $\xi$ and, in this case, $\xi$ increases at almost a constant load. Also shown on Fig. 3.27 is the value of $\sigma_y$ and, as in the previous example, the maximum is only slightly above the point at which initial yield occurs.

In Fig. 3.28 we depict the buckling deflection vs. applied edge displacement $U$ for the same values of the geometric and material parameters as those employed in Fig. 3.27; the applied displacement has been normalized by its value $U_c$ at buckling. For the perfect plate, of course, $\xi = 0$ prior to achieving $U = U_c$; at $U = U_c$ the buckling deflection begins to grow, albeit very slowly. At the maximum load point, $\xi$ is growing with $U$ very nearly constant and this is the range in which the load, as shown in Fig. 3.27, is decreasing rapidly. As the imperfection amplitude $\hat{\xi}$ is increased, the jump in the postbuckling curves shown in Fig. 3.28 is smoothed out. As noted in [141], the jump referenced above was found to occur for small values of the imperfection amplitude $\hat{\xi}$ if buckling occurred near $\sigma_y$; this is true for both the clamped and simply supported edge conditions.

In Fig. 3.29 we depict the postbuckling curves for a case in which buckling occurs further into the plastic range of the circular plate; in this case, the plate is clamped, $\sigma_c/\sigma_y = 1.12$, $\tilde{\eta} = 2.0$, $\nu = \frac{1}{3}$, and $n = 12$. The maximum load, given by $\sigma_{\max} = 1.018\sigma_e$, occurs at a larger value of the deflection $\xi$(i.e., at $\xi = .065h$) than in the previous example. The postbuckling curve in Fig. 3.29, which corresponds to the perfect plate, indicates that the load decreases more gradually after $\sigma_{\max}$ is reached than when $\sigma_c \approx \sigma_y$. Also, it may be noted that the buckling deflection-end shortening curves, in this case, do not exhibit the jumps characteristic of those cases in which buckling occurs near $\sigma_y$. At $\hat{\xi} = .06h$ most of the buckling deflection is associated with a nearly constant load. The example illustrated by Fig. 3.29 also shows that, when buckling of the perfect plate occurs well into the plastic range, the value of $\sigma_{\max}$ for an imperfect plate can be well above the value of

$\sigma_y$. We also remark that both Fig. 3.27 and Fig. 3.29 indicate that for a sufficiently small imperfection, the maximum value of $\sigma_{av}$ occurs shortly after the onset of elastic unloading. However, if $\hat{\xi}$ is sufficiently large, then the tensile bending stresses which are developed in the plate are large enough to inhibit plastic loading in compression and then $\sigma_{max}$ occurs before unloading (with the amplitude $\hat{\xi}$ required for this to occur depending on how far into the plastic range of the plate initial buckling occurs).

In Fig. 3.30, which we have already (partially) discussed in Chapter 3, the development of the plastic zone and unloading region for the example depicted in Fig. 3.29 are shown. For the perfect plate unloading begins at $r = 0$ and $z = \frac{1}{2}h$ and, as this region of unloading grows, a region near the clamped edge unloads. The unloading region near the clamped edge does not have a large effect on the overall deformation of the plate inasmuch as this region is highly constrained by the boundary condition; once the maximum load point is reached, this region near the clamped edge begins to reload. For plates possessing an initial imperfection, plastic loading proceeds as follows: at the maximum load, $\sigma_{max}$, all of the plate has undergone plastic yielding except for a small region near the clamped edge and a region which is almost identical to the primary unloading region of the perfect plate. At the last stage of the deformation, as shown in Fig. 3.30, for the case in which $\hat{\xi} = .06h$, there exists a small region near $r = 0$ which has yielded in tension.

Following [141], the results concerning imperfection buckling of elastic-plastic circular plates may be summarized as follows:

(i) For initial buckling occurring in the elastic range of the perfect plate, imperfections can lead to a substantial reduction in the load-carrying capacity of the plate if $\sigma_c \simeq \sigma_y$.

(ii) For initial buckling occurring in the plastic range of the perfect plate, $\sigma_{max}$ is only slightly higher than $\sigma_c$; furthermore, an asymptotic analysis of the postbuckling behavior of the perfect plate in the plastic range, which is based on the work of Hutchinson [45], [46], yields a good estimate of $\sigma_{max}$ as well as of the development of the unloading region after buckling.

(iii) The imperfection sensitivity of circular plates in the plastic range is significant and is present even when the difference between the buckling stresses predicted by the simplest flow theory and those predicted by the simplest deformation theory is very small.

## 5.2 Secondary Buckling for Thin Plates

When a uniform radial compressive load $\lambda > 0$ is applied to the edge, say, of a clamped circular plate, the uniformly compressed plate is in an equilibrium state, the unbuckled state, for all values of $\lambda$; for $\lambda < \lambda_{cr} = \lambda_1, \lambda_1$ being the smallest eigenvalue of the corresponding linearized problem, this unbuckled state is the only equilibrium state of the plate and $\lambda_1$ is also known as the primary (or critical) buckling load. For $\lambda > \lambda_1$, we have seen that the unbuckled state is unstable and the circular plate will buckle into its first or primary initial buckling mode which, in this case, is axisymmetric. Using the von Karman theory, Friedrichs and Stoker [72] proved that the primary (buckled) state exists for all $\lambda > \lambda_1$ for the simply supported (isotropic, elastic) circular plate while Wolkowsky [145], employing a different technique, established the same result for the clamped plate as well. In [201] Cheo and Reiss, using numerical arguments, pointed at the existence of critical values of $\lambda$, called secondary buckling loads, at which equilibrium states (which are axially unsymmetric) may branch from the primary buckled state of the circular plate; these unsymmetric states are now referred to as secondary buckled states and the bifurcation process itself is now commonly referred to as secondary buckling. In this section, we will present a brief survey of some of the results which follow from the work of Cheo and Reiss [201], [202] on the secondary buckling of circular plates, as well as some related work of Stroebel and Warner [203] on stability and secondary bifurcation for rectangular plates. Several authors have examined the relationship between the existence of multiple eigenvalues for the linearized branching problem and the phenomena of secondary bifurcation; most notable among these papers is the review article of Bauer, Keller, and Reiss [204] who point out that while, for many problems, the primary branching points are easily determined because the associated linearized problem may be explicitly solved, secondary branching points and the associated secondary states cannot usually be determined explicitly and must be found by numerical and perturbation methods. The phenomena of mode jumping which is observed, experimentally, in the compressive and shear buckling of rectangular plates is also thought to be an example of secondary buckling; as the authors of [204] point out, such secondary buckling was observed in the numerical study by Bauer and Reiss [205] of the compressive buckling of rectangular plates and was subsequently analyzed in the work of Stroebel and Warner [203], that will be reviewed a little further on in this section. In other pivotal work on this problem, both Knightly and Sather [128], as well as Matkowsky and Putnick [206] have studied bifurcation near multiple bifurcation points and have shown that, in some cases, the multiplicity of these points as primary branching points exceeds their multiplicity as eigenvalues of the corresponding linearized problem, thus suggesting that secondary branching may occur. Bauer, Keller, and Reiss [204] have pointed out that the earliest discovery of secondary bifurcation appears to have been made by Poincaré in his well-known work on the ellipsoidal figures of equilibrium of a rotating inviscid fluid; a survey of this work may be found in the monograph [207] by Chandrasekhar.

### 5.2.1   Secondary Buckling for Circular Plates

As Cheo and Reiss [201] have indicated, the physical mechanism that initiates secondary buckling in the case of an isotropic elastic circular plate is directly related to the following property of the primary axisymmetric buckled state as $\lambda$ increases beyond the value $\lambda_1$ : a circumferential strip possessing large circumferential compressive membrane stress develops in the vicinity of the edge as $\lambda$ increases; the width of this strip (adjacent to the edge of the plate) is decreasing, and the compressive stress intensity increasing, as $\lambda$ continues to increase. Then for $\lambda$ sufficiently large, the strip may buckle, like a ring, into circumferential waves and, as a consequence, the plate buckles unsymmetrically about the axisymmetric buckled state by wrinkling near its edge.

In Cheo and Reiss [202], the secondary buckling loads for the isotropic elastic circular plate were determined, numerically, by linearizing the solutions of von Karman equations about the primary axisymmetric state; thus, it was found in [202] that the minimum secondary buckling load is (approximately) 7.5 times the primary buckling load. Subsequently, in [201], a formal perturbation method was used to study the secondary states for $\lambda$ near the secondary buckling loads and the resulting linear problems were solved numerically; this work, which will be reviewed below, points to the fact that secondary buckled states can exist for $\lambda$ slightly less than the secondary buckling loads. We note that Friedrichs and Stoker, in [143], had already conjectured that secondary buckling might occur for the isotropic elastic circular plate, while Morozov [208] had shown, for the simply supported plate that, for sufficiently large $\lambda$, there exist unsymmetric equilibrium states which have less potential energy than all axisymmetric states for the same value of $\lambda$. As Cheo and

Reiss [201] note, however, Morozov [207] did not prove that secondary buckling actually occurs. Yanowitch [209], on the other hand, did show that buckling of a circular elastic plate can take place from an initially deformed state of the plate by proving that the radially symmetric primary buckled state is unstable for $\lambda$ sufficiently large.

In [201] the following (dimensionless) form of the von Karman theory for clamped circular plates is used:

$$\begin{cases} \Delta^2 w = [\Phi, w] \\ \\ \Delta^2 \Phi = -\dfrac{1}{2}[w, w] \end{cases} \quad (0 \leq r < 1, \ 0 \leq \theta < 2\pi) \qquad (5.38)$$

$$\begin{cases} w(1, \theta) = w_{,r}(1, \theta) = 0 \\ \\ \Phi(1, \theta) = 0, \Phi_{,r}(1, \theta) = -\lambda \end{cases} \quad (0 \leq \theta < 2\pi) \qquad (5.39)$$

where $r$ is the dimensionless radial coordinate, $\Phi(r, \theta)$ the dimensionless Airy function, $w(r, \theta)$ the dimensionless lateral deflection of the plate, $\Delta$ the Laplacian in polar coordinates, and the nonlinear (bracket) operator $[\Phi, w]$ is given by (1.163). A dimensionless edge thrust $\lambda'$ may be defined by

$$\lambda' = c^2(\lambda/E)(R/h)^2, \ c^2 = \{12(1 - \nu^2)\}^{\frac{1}{2}} \qquad (5.40)$$

with $R$ and $h$, respectively, the radius and thickness of the plate, and $E$ and $\nu$ Young's modulus and Poisson ratio. Dropping the $\prime$ on $\lambda$, the unbuckled state is given by

$$w \equiv 0, \ \Phi \equiv \bar{\Phi} = \frac{1}{2}\lambda(1 - r^2) \qquad (5.41)$$

and is a solution of(5.38), (5.39), for all values of $\lambda$, which corresponds to a state of uniform compressive stress. Substituting

$$\Phi = \bar{\Phi} + \tilde{\Phi}, w = \tilde{w} \qquad (5.42)$$

into (5.38), (5.39), and subsequently dropping the $\tilde{}$ on $\Phi$ and $w$, we obtain the reduced forms, i.e.,

$$\begin{cases} \Delta^2 w + \lambda \Delta w = [\Phi, w] \\ \\ \Delta^2 \Phi = -\dfrac{1}{2}[w, w] \end{cases} \qquad (5.43)$$

and (5.39), where $\Phi$ now denotes the extra Airy stress function. The classical linearized buckling problem is, of course, obtained from (5.43), (5.39) by deleting the nonlinear terms in (5.43). We have already noted, in our work in Chapter 3, that the smallest eigenvalue of the linearized problem for the clamped isotropic elastic circular plate, which corresponds to the primary buckling load, is given by $\lambda_1 \approx 14.7$ and that $\lambda_1$ is the square of the smallest root of the Bessel function $J_1(z)$; also, the eigenfunction corresponding to $\lambda_1$ is radially symmetric, i.e., the primary buckling mode is axisymmetric. In fact, two axisymmetric solutions of (5.43), (5.39), which differ only in the sign of the deflection $w$, branch at $\lambda = \lambda_1$, and either of them may be referred to as the primary buckled state. The primary buckled state will be denoted by

$$\{\Phi_0(r; \lambda), w_0(r; \lambda)\}$$

and is known to exist for all $\lambda > \lambda_1$.

In order to study the secondary buckling of an isotropic elastic circular plate near a secondary buckling load $\lambda^{(0)}$, the authors in [201] seek unsymmetric solutions of (5.43), (5.39) of the form

$$w(r, \theta; \lambda) = w_0(r; \lambda) + \epsilon \sum_{m=0}^{\infty} w^{(m)}(r, \theta)\epsilon^m \qquad (5.44)$$

$$\Phi(r, \theta; \lambda) = \Phi_0(r; \lambda) + \epsilon \sum_{m=0}^{\infty} \varphi^{(m)}(r, \theta)\epsilon^m \qquad (5.45)$$

$$\lambda(\epsilon) = \lambda^{(0)} + \epsilon \sum_{m=0}^{\infty} \lambda^{(m)}\epsilon^m \qquad (5.46)$$

where $\epsilon$ is a small parameter which is defined by

$$\epsilon^2 = \iint_{\Omega} \left[ (w - w_0)^2 + (\Phi - \Phi_0)^2 \right] r\, dr\, d\theta \qquad (5.47)$$

and $\Omega$ denotes the unit circle. The choice of the perturbation parameter is not unique, of course. To determine the coefficients $w^{(m)}, \varphi^{(m)}$, and $\lambda^{(m)}$ in the expansions (5.44)–(5.46), the primary state is first expanded about $\lambda = \lambda^{(0)}$, i.e., about $\epsilon = 0$. Thus, it is assumed in [201] that

$$\begin{cases} w_0\,(r; \lambda(\epsilon)) = \displaystyle\sum_{m=0}^{\infty} w_{0m}(r)\epsilon^m \\ \Phi_0\,(r; \lambda(\epsilon)) = \displaystyle\sum_{m=0}^{\infty} \Phi_{0m}(r)\epsilon^m \end{cases} \qquad (5.48)$$

The coefficients $w_{om}$ and $\Phi_{om}, m = 0, 1 \cdots$ are determined using a numerical evaluation of the primary state with $\{w_{oo}, \Phi_{oo}\}$ the primary state evaluated at $\lambda = \lambda^{(0)}$. The expansion coefficients $w^{(m)}, \varphi^{(m)}$, and $\lambda^{(m)}$ are computed by substituting (5.44)–(5.46), and (5.48), into (5.43), (5.39), and (5.47); this yields a sequence of linear boundary value problems with associated normalizing conditions, e.g., $w^{(0)}, \varphi^{(0)}$, and $\lambda^{(0)}$ must satisfy the eigenvalue problem:

$$
\begin{cases}
\mathcal{L}_1\left(w^{(0)}, \varphi^{(0)}\right) \equiv \Delta^2 w^{(0)} + \lambda^{(0)} \Delta w^{(0)} \\[4pt]
\qquad + \left[w_{00}, \varphi^{(0)}\right] + \left[\Phi_{00}, w^{(0)}\right] = 0 \\[6pt]
\mathcal{L}_2\left(w^{(0)}, \varphi^{(0)}\right) \equiv \Delta^2 \varphi^{(0)} + \left[w_{00}, w^{(0)}\right] = 0 \\[6pt]
w^{(0)} = w_{,r}^{(0)} = \varphi^{(0)} = \varphi_{,r}^{(0)} = 0, \text{ at } r = 1 \\[6pt]
\displaystyle\iint_\Omega \left[w^{(0)2} + \varphi^{(0)2}\right] r\,dr\,d\theta = 1
\end{cases}
\tag{5.49}
$$

Because $\{w_{00}, \Phi_{00}\}$ is a primary buckled state, the eigenvalue parameter $\lambda^{(0)}$ appears nonlinearly in (5.49); also, the coefficients in (5.49) are only functions of $r$. Thus, the eigenfunctions of (5.49) are given, for $n = 1, 2, \cdots$, by

$$
w^{(0)} = a_n(r) \sin n\theta, \quad \varphi^{(0)} = b_n \sin n\theta \tag{5.50}
$$

Inserting (5.50) into (5.49) one obtains, for each $n$, an ordinary differential equation (eigenvalue problem) for the functions $a_n(r), b_n(r)$ and the corresponding eigenvalues are the secondary buckling loads which were determined, numerically, by Cheo and Reiss in [202]. The integer $n$ in (5.50) is the circumferential wave number of the eigenfunctions; the amplitudes of the numerically determined eigenfunctions were observed, in [201], to be small near the center of the plate but increased very quickly near the edge of the plate. Therefore, when the plate is in a secondary buckled state, and $\lambda$ is close to a secondary buckling load, the interior of the plate is almost axisymmetric and the edge is wrinkled.

For the coefficients $w^{(m)}$ and $\varphi^{(m)}, m > 1$, the boundary value problems have the form

$$
\begin{cases}
\mathcal{L}_1\left(w^{(m)}, \varphi^{(m)}\right) = \alpha^{(m)} \\[4pt]
\mathcal{L}_2\left(w^{(m)}, \varphi^{(m)}\right) = \beta^{(m)} \\[4pt]
w^{(m)} = w_{,r}^{(m)} = \varphi^{(m)} = \varphi_{,r}^{(m)} = 0, \text{ at } r = 1
\end{cases}
\tag{5.51}
$$

and $w^{(m)}, \varphi^{(m)}$ must satisfy the normalization conditions which are implied by (5.44)–(5.47). In (5.51), the inhomogeneous terms $\alpha^{(m)}, \beta^{(m)}$ depend on $\lambda^{(m)}$ as well as on the expansion coefficients with indices $< m$. Because the homogeneous problems which correspond to (5.51) have nontrivial solutions, e.g., (5.49), the terms $\alpha^{(m)}, \beta^{(m)}$ must satisfy solvability conditions which give the $\lambda^{(m)}$ as integrals of $w^{(k)}, \varphi^{(k)}$ for $k < m$. In this manner, Cheo and Reiss [201] have determined that $\lambda^{(1)} = 0$ while $\lambda^{(2)}$ is given by integrals involving the $w^{(k)}$ and $\varphi^{(k)}$, for $k = 0, 1$, and the primary solution (buckled state). In order to solve the problems (5.51), the $\theta$-dependence is first separated out and then the resulting ordinary differential equations and associated boundary conditions are solved by shooting and parallel shooting methods (as described in Bauer, Reiss, and Keller [204] ).

A sketch of the postbuckling behavior of the circular plate as $\lambda$ is increased from zero is shown in Fig. 5.25. For $\lambda < \lambda_1$ the compressed, unbuckled state (5.41) is the unique solution and the linear theory of dynamic stability may be employed to show that this prebuckled equilibrium state is stable. For $\lambda > \lambda_1$ the compressed prebuckled state is, of course, unstable, (as we have already seen in Chapter 3) and the circular plate buckles into its primary mode as $\lambda$ increases through $\lambda_1$. The secondary buckling loads are now denoted by $\lambda_1^{(0)} < \lambda_2^{(0)} < \cdots$ and as $\lambda \to \lambda_1^{(0)}$, from below, there are secondary buckled states arbitrarily near the primary state because $\lambda^{(2)} < 0$. When $\lambda$ is slightly larger than $\lambda_1^{(0)}$ a secondary buckled state arbitrarily close to the primary buckled state does not exist. Therefore, as $\lambda$ increases through $\lambda_1^{(0)}$, or any other $\lambda_i^{(0)}$ that was computed by Cheo and Reiss [202], a smooth branching from the primary buckled state to a secondary buckled state is not possible; such a smooth secondary branching may occur only if $\lambda^{(2)}$ were positive for some secondary buckling load. If, on the other hand, one applies external forces to the plate as $\lambda \to \lambda_1^{(0)}$, then the plate could jump from the primary buckled state to either the secondary state which branches from $\lambda_1^{(0)}$ or, perhaps, to another, nonadjacent, unsymmetric state; such a possibility is contingent on the dynamic stability behavior of the primary and secondary buckled states as well as on the amplitude of the external disturbances. It was conjectured in [201] that when the potential energy of the secondary buckled state which branches from $\lambda_1^{(0)}$ is less than the potential energy of the primary buckled state, for the same value of $\lambda$ near $\lambda_1^{(0)}$, jumping due to a finite amplitude external disturbance might be more easily achieved.

## 5.2.2 Secondary Buckling for Rectangular Plates.

In [203] the Movchan-Liapunov theorem, which we describe in the analysis to follow, is used to establish a sufficiency condition for the stability of equilibrium configurations of a rectangular plate under edge loads and displacements; this sufficiency criterion may, as we will see, also be used to characterize the possible occurrence of secondary bifurcation from the first buckled state of the perfect plate. A postbuckling analysis of the secondary solution can then be used to formulate an explanation of the mode-jumping phenomena which is experimentally observed for rectangular plates. The Movchan-Liapunov criterion is dynamical in nature and the relevant dynamical equations are those for small motion near equilibrium; they are derived from the full set of equations by an appropriate perturbation method.

Experimental studies indicate not only the presence, for this problem, of the primary buckling solutions but also that other secondary solutions may exist as well. As the authors [203] point out, work by Stein [210], et al., with rectangular plates under longitudinal compression, have shown that, as the load is slowly increased, in the postbuckled domain, the plate appears to suddenly snap from the equilibrium configuration which corresponds to the original lowest mode to one which corresponds to a higher mode; this apparent mode-jumping has led several investigators to speculate that transition paths, which are, perhaps, unstable, connect the primary buckled solutions. The phenomena is illustrated in Fig. 5.26 in which equilibrium configurations corresponding to combinations of a load parameter $\lambda$ and some characteristic measure $w^*$ of the deflection are depicted. The $\lambda$ axis represents, of course, the zero deflection solution which is possible under all loads for the perfect system; the primary buckled states $C_i(\lambda)$ of the perfect system branch from the trivial solution as the load levels exceed $\lambda_i$, which are the eigenvalues of the classical linear buckling problem for the plate. The states $C_i(\lambda)$ exist for all loads $\lambda \geq \lambda_i$ but may (as we saw for the circular plate, i.e., Fig. 5.25) change their local stability characteristics by branching again, thus yielding the secondary buckled states $C_{i,j}(\lambda)$; or snapping may occur, i.e., local extrema may appear, such as at point $A$, in Fig. 5.26, without branching. The local behavior, at branching, for these buckled states, can also be related to the imperfection configurations of the plate which are represented in Fig. 5.26 by curves such as $C^*(\lambda)$. Several numerical and analytical studies have supported the hypothesis that secondary instability on the first primary branch corresponds to branching; in particular, the work of Bauer and Reiss

[205], in which the postbuckling behavior of rectangular plates was numerically analyzed, yields an asymmetric solution for the problem under consideration, which appears to branch from a primary solution.

To formulate the dynamical theory to which the Movchan-Liapunov stability theorem may be applied, we let $w(x_\alpha)$ and $\Phi(x_\alpha), \alpha = 1, 2$, denote, as usual, the transverse displacement (deflection) and Airy function corresponding to an equilibrium configuration of the plate; for a "motion" configuration, close to this equilibrium configuration, the displacement and stress functions will be given by

$$\begin{cases} \tilde{w}(x_\alpha, t) = w(x_\alpha) + \epsilon \hat{w}(x_\alpha, t) \\ \tilde{\Phi}(x_\alpha, t) = \Phi(x_\alpha) + \epsilon \hat{\Phi}(x_\alpha, t) \end{cases} \tag{5.52}$$

where $\epsilon$ is some appropriate small parameter associated with the plate. The von Karman equations for the (assumed) isotropic rectangular plate are (1.53a,b) or, in non-dimensionalized form, with $t \equiv 0$,

$$\begin{cases} \Delta^2 w - [\Phi, w] = 0 \\ \Delta^2 \Phi + \dfrac{1}{2}[w, w] = 0 \end{cases} \tag{5.53}$$

The linear equations which govern small motions near solutions of (5.53) are

$$\begin{cases} \Delta^2 \hat{w} - [\Phi, \hat{w}] - [\hat{\Phi}, w] + \dfrac{\partial^2 \hat{w}}{\partial t^2} = 0 \\ \Delta^2 \hat{\Phi} + [w, \hat{w}] = 0 \end{cases} \tag{5.54}$$

The boundary conditions relative to $w$ (or $\hat{w}$) are the ones associated with either clamped edges, i.e., (1.130a,b) for a plate with aspect ratio $\phi = a/b$, or simply supported edges, i.e., (1.130a), (1.131), or some combination of clamped and simply supported edges. The boundary conditions for the in-plane equilibrium problem, which are connected with the edge loading of the (rectangular) plate, are those associated with either prescribed edge tractions or prescribed edge displacements. We now denote by $N_{\alpha\beta}, E_{\alpha\beta}$ the stress resultants and strain components associated with the equilibrium configurations determined by $(w, \Phi)$ and by $n_{\alpha\beta}, e_{\alpha\beta}$ the same quantities associated with the perturbed configurations determined by $(\hat{w}, \hat{\Phi})$. Also, we let $U_\alpha$ denote the displacement components associated with an equilibrium configuration while the $u_\alpha$ are the corresponding quantities for a perturbed configuration. Then

$$\begin{cases} E_{\alpha\beta} = \dfrac{1}{2}\left(U_{\alpha,\beta} + U_{\beta,\alpha} + w_{,\alpha}w_{,\beta}\right) \\[3mm] e_{\alpha\beta} = \dfrac{1}{2}\left(u_{\alpha,\beta} + u_{\beta,\alpha} + \hat{w}_{,\alpha}w_{,\beta} + \hat{w}_{,\beta}w_{,\alpha}\right) \end{cases} \tag{5.55}$$

and the constitutive relations for the linear isotropic plate may be expressed in the form

$$\begin{cases} N_{\alpha\beta} = (1-\sigma^2)^{-1}\left[(1-\sigma)E_{\alpha\beta} + \sigma E_{\gamma\gamma}\delta_{\alpha\beta}\right] \\[2mm] n_{\alpha\beta} = (1-\sigma^2)^{-1}\left[(1-\sigma)e_{\alpha\beta} + \sigma e_{\gamma\gamma}\delta_{\alpha\beta}\right] \end{cases} \tag{5.56}$$

To derive a quasi-static principle which is sufficient for establishing the stability of solutions to the equilibrium problem (5.53) with given loads, we study the boundedness of solutions to the dynamical equation (5.54) and introduce the two energy functionals $\mathcal{E}$ and $E$:

$$\mathcal{E} = \frac{1}{2}\int\int_R \left[\{(\Delta w)^2 - (1-\sigma)[w,w]\} \right.$$
$$+ (1-\sigma^2)^{-1}\left\{(1-\sigma)E_{\alpha\beta}E_{\alpha\beta} + \sigma E_{\alpha\alpha}E_{\beta\beta}\right\}\Big]\, dxdy \tag{5.57}$$
$$- \oint \bar{T}_\alpha U_\alpha ds$$

with $R$ denoting the domain occupied by the rectangular plate, and the $\bar{T}_\alpha$ being the components of the applied traction on $\partial R$. Also,

$$E = \frac{1}{2}\int\int_R \left\{ \left(\frac{\partial \hat{w}}{\partial t}\right)^2 + [(\Delta \hat{w})^2 - (1-\sigma)[\hat{w},\hat{w}]] \right.$$
$$+ (1-\sigma^2)^{-1}\left[(1-\sigma)e_{\alpha\beta}e_{\alpha\beta} + \sigma e_{\alpha\alpha}e_{\beta\beta}\right]$$
$$\left. + N_{\alpha\beta}\hat{w}_{,\alpha}\hat{w}_{,\beta} \right\} dxdy \tag{5.58}$$

The functional $\mathcal{E}$ represents the total potential energy associated with an equilibrium configuration of the plate while $E$ is an energy associated with the perturbation $(\hat{w}, \hat{\Phi})$ and is, essentially, equal to the difference (up to terms of second order) between the total energy of a perturbed configuration and that of the (base) equilibrium configuration. If $(w, \Phi)$ denotes a solution pair for (5.53), then it may be shown that $E$ is constant on solutions $(\hat{w}, \hat{\Phi})$ of (5.54).

As an appropriate Liapunov functional for the stability theorem, the authors in [202] take the functional $E(\hat{w})$; this functional may also be chosen as the metric $\rho_0$ which is used to measure the size of an initial disturbance. A second metric $\rho$, which is used to measure the "distance" between the perturbed state and the base state, is given by

$$\rho = (\dot{\hat{w}}, \dot{\hat{w}}) + (\hat{w}, \Delta^2 \hat{w}) + (\hat{w}, \hat{w}) \qquad (5.59)$$

where

$$(f, g) = \int \int_R f(x_\alpha) g(x_\alpha) dx dy \qquad (5.60)$$

To ensure that $E \equiv \rho_o \geq k^2 \rho$, it suffices that $V(\theta) \geq 0$ for all functions $\theta(x_\alpha)$ which are twice continuously differentiable on $R$ and satisfy the relevant boundary conditions with respect to the deflection on $\partial R$; here $V(\theta)$ is the functional corresponding to the "static" part of $E(\hat{w})$, i.e.,

$$\begin{aligned} V(\theta) = \frac{1}{2} \bigg( & \int \int_R \left\{ (\Delta^2 \theta) - (1 - \sigma)\,[\theta, \theta] \right\} dx dy \\ + & \int \int_R (1 - \sigma^2)^{-1} \left\{ (1 - \sigma) e_{\alpha\beta} e_{\alpha\beta} + \sigma e_{\alpha\alpha} e_{\beta\beta} \right\} dx dy \\ + & \int \int_R N_{\alpha\beta} \theta_{,\alpha} \theta_{,\beta} dx dy \bigg) \end{aligned} \qquad (5.61)$$

The three distinct terms in $V(\theta)$ correspond to variations in bending energy, membrane energy, and the additional virtual work of the equilibrium membrane forces; also, in (5.61), $e_{\alpha\beta} = \frac{1}{2}(u_{\alpha,\beta} + u_{\beta,\alpha} + \theta_{,\alpha}\hat{w}_{,\beta} + \theta_{,\beta}\hat{w}_{,\alpha})$. Introducing the perturbation stress function $\hat{\zeta}(\theta)$, which corresponds to the equilibrium displacement $w$ (and each admissible $\theta$), through the homogeneous boundary value problem

$$\begin{cases} \Delta^2 \hat{\zeta}(\theta) + [w, \theta] = 0, & \text{in } R \\ \nu_\alpha n_{\alpha\beta}(\hat{\zeta}) = 0, & \text{on } \partial R_T \\ u_\alpha(\hat{\zeta}, \theta, w) = 0, & \text{on } \partial R_u \end{cases} \qquad (5.62)$$

where $\nu$ is the exterior unit normal on $\partial R$, and $\partial R_T, \partial R_u$ denote the subregions of $\partial R$ on which traction and displacement conditions are applied, $V(\theta)$ may also be written in terms of the inner product representation,

$$V(\theta) = \frac{1}{2} \left\{ (\theta, \Delta^2 \theta) - (\theta, [\Phi, \theta]) - (\theta, [\hat{\zeta}(\theta), w]) \right\} \qquad (5.63)$$

As the stability criteria has been formulated without restricting the form of the in-plane boundary conditions in an equilibrium configuration of the plate, it may be used to study either the stability of the trivial solution $w = 0$ or of any buckled state. To pursue the study of secondary bifurcation in [203], the authors now restrict the form of the boundary conditions to those that give the classical initial buckling loads. Thus, we set

$$\bar{T}_\alpha = \lambda T_\alpha^0, \ \bar{U}_\alpha = \lambda U_\alpha^0 \tag{5.64}$$

where the $\bar{U}_\alpha$ are the components of the applied displacements on $\partial R_u$ and $\lambda$ is the proportional loading parameter. In (5.64), $T_\alpha^0, U_\alpha^0$ are fixed boundary tractions and displacements corresponding to the prebuckled in-plane equilibrium state of the plate. If we set $\bar{\Phi} = \lambda \overset{\circ}{\Phi}$, where $\overset{\circ}{\Phi}$ is the Airy function associated with the prebuckled in-plane state, the linearized equilibrium eigenvalue problem which follows from (5.53) is governed by

$$\Delta^2 w - \lambda [\overset{\circ}{\Phi}, w] = 0 \tag{5.65}$$

which, (together with the associated clamped/simply supported edge conditions) has only the trivial solution $w \equiv 0$, except for the eigenvalues $\lambda = \lambda_i$ at which $w = c\psi_i$ the $\psi(x_\alpha)$ being the associated eigenfunctions; the same result may be obtained from the energy criterion $V(\theta) \geq 0$, for with $w = 0$ one has $(\theta[\hat{\zeta}(\theta), w]) = 0$ and $(\theta, [\Phi, \theta]) = -\lambda \int \int_R \overset{\circ}{N}_{\alpha\beta} \theta_{,\alpha} \theta_{,\beta} dx dy$. Thus, the functional $V(\theta)$ assumes the form of the classical potential energy functional for the plate problem; we note that $V(\theta)$ ceases to be positive definite at $\lambda = \lambda_1, \theta = c\psi_1$.

To investigate secondary branching from a primary buckled state, one must compute the critical value $\lambda_c$ of the load parameter $\lambda$, the form of the primary buckling state, and the branching behavior of the solution for loads close to the critical load. Thus, for $\lambda \geq \lambda_1$ we let $(w, \Phi)$ represent the first buckled solution, set

$$\Phi = \lambda_1 \overset{\circ}{\Phi} + \Phi^* \tag{5.66}$$

and introduce the linear operator

$$\mathcal{L}w = \Delta^2 w - \lambda_1 [\overset{\circ}{\Phi}, w] \tag{5.67}$$

Using (5.66), (5.67), (5.53) may be written in the form

$$\begin{cases} \mathcal{L}w = [\Phi^*, w] \\ \Delta^2 \Phi^* + \dfrac{1}{2}[w, w] = 0, \end{cases} \tag{5.68}$$

in $\mathcal{R}$, to which we append the boundary conditions

$$\nu_\alpha N_{\alpha\beta}(\Phi^*) = (\lambda - \lambda_1)\overset{\circ}{T_\beta}, \text{ on } \partial R_T$$
$$U_\alpha^*(\Phi^*, w) = (\lambda - \lambda_1)\overset{\circ}{U_\alpha}, \text{ on } \partial R_u \tag{5.69}$$

As the perfect plate may deflect either up or down, $w$ must be odd in the load parameter $\lambda - \lambda_1$. Thus, we let

$$\mu^2(\lambda) = (\lambda - \lambda_1)/\lambda_1 \tag{5.70}$$

and set

$$\begin{cases} w(x, y; \lambda) = \mu W(x, y; \mu^2) \\ \Phi(x, y; \lambda) = \mu^2 \phi(w, y; \mu^2) \end{cases} \tag{5.71}$$

with $W(x, y; 0) \neq 0, \phi(x, y; 0) \neq 0$. The functions $W$ and $\phi$ satisfy

$$\begin{cases} \mathcal{L}W = \mu^2[\phi, W] \\ \Delta^2\phi = -\dfrac{1}{2}[W, W] \end{cases} \tag{5.72}$$

and

$$\begin{cases} \nu_\alpha N_{\alpha\beta}(\phi) = \lambda_1 \overset{\circ}{T_\beta}, \text{ on } \partial R_T \\ U_\alpha(\phi, W) = \lambda_1 \overset{\circ}{U_\alpha}, \text{ on } \partial R_u \end{cases} \tag{5.73}$$

The in-plane boundary conditions involve fixed traction and displacement vectors which are the same as those that hold for $\Phi = \lambda_1 \overset{\circ}{\Phi}$ at the first branching. Problem (5.72), (5.73) for $(W, \phi)$ clearly admits a zero eigenvalue and the energy functional takes the form

$$V(\theta, \mu^2) = \frac{1}{2}(\theta, \mathcal{L}\theta) - \frac{1}{2}\mu^2\{(\theta, [\phi, \theta]) + (\theta, [\zeta, W])\} \tag{5.74}$$

where $\zeta = \mu^{-1}\hat{\zeta}$ satisfies

$$\Delta^2\zeta + [W, \theta] = 0, \text{ in } R \tag{5.75}$$

A criterion is now sought that determines the first positive value $\mu_c^2$ of $\mu^2$ such that $V(\theta, \mu_c^2) = 0$. Using the work of Rogers [211] it may be established that the real eigenvalues of the Euler-Lagrange equations for $V$ have a variational characterization; to establish that characterization, an orthogonality condition is imposed to eliminate the effects of the zero eigenvalue, i.e., $\theta(x_\alpha)$ is split as

$$\theta = \alpha_1\psi_1 + \alpha_2\bar{\theta} \tag{5.76}$$

where

$$(\psi_1, [\phi, \bar{\theta}] + [\bar{\zeta}, W]) = 0 \tag{5.77}$$

and the auxiliary function $\bar{\zeta}$ is defined below. In fact (5.77) can also be written in the form

$$(\bar{\theta}, [\phi, \psi_1] + [F_1, W]) = 0 \tag{5.78}$$

where

$$\begin{cases} \Delta^2 F_1 + [\psi_1, W] = 0 \\ \Delta^2 \bar{\zeta} + [\bar{\theta}, W] = 0 \end{cases} \tag{5.79}$$

Using the splitting (5.76) we compute that

$$\begin{aligned} 2V(\theta; \mu^2) = &-\alpha_1^2 \mu^2 \left(\psi_1, [\phi, \psi_1] + [F_1, W]\right) \\ &+\alpha_2^2 \left\{ (\bar{\theta}, \mathcal{L}\bar{\theta}) - \mu^2 \left(\bar{\theta}, [\phi, \bar{\theta}] + [\bar{\zeta}, W]\right) \right\} \end{aligned} \tag{5.80}$$

but $(\psi_1, [F_1, W]) \leq 0$, while $(\psi_1, [\phi, \psi_1]) < 0$ for $\mu^2$ sufficiently small. Therefore, the first term in $V$ is positive, as the first bifurcating branch of solutions is traversed; this means that the first critical value of $\mu^2$ comes from the second term in $V$, i.e., at the first positive $\mu^2 = \mu_c^2$, $V$ becomes zero for $\theta = \alpha_2 \bar{\theta}$ with $\bar{\theta}$ the eigenfunction corresponding to the eigenvalue $\mu_c^2$. The variational equation for the restricted function

$$\bar{V}(\bar{\theta}) = (\bar{\theta}, \mathcal{L}\bar{\theta}) - \mu^2(\bar{\theta}, [\phi, \bar{\theta}]) + [\bar{\zeta}, W]) \tag{5.81}$$

is

$$\mathcal{L}\bar{\theta} - \mu^2[\phi, \bar{\theta}] - \mu^2[\bar{\zeta}, W] = 0, \text{ in } R \tag{5.82}$$

where $\bar{\zeta}$ satisfies the second equation in (5.79). Equations (5.82), (5.79) are the linearized variational equations (of the von Karman equations), at fixed load, for small static displacements $\bar{\theta}$ superposed on the base displacement $W$; therefore, a solution to these equations may be given an interpretation as a neighboring equilibrium solution.

To examine the postbuckling behavior of the plate we let $(W, \Phi)$, $\Phi = \lambda_1 \overset{\circ}{\Phi} + \Phi^*$, be the primary buckling solution for (5.68), (5.69) and we denote by $(\bar{w}, \bar{\Phi})$ the bifurcating solution at loads $\lambda$ near $\lambda_c$, i.e. at parameter values of $\mu^2 = (\lambda - \lambda_1)/\lambda_1$ and $\mu_c^2 = (\lambda_c - \lambda_1)/\lambda_1$. We now introduce the new load parameter

$$\eta = \frac{\lambda - \lambda_c}{\lambda_c} = \frac{\lambda_1(\mu^2 - \mu_c^2)}{\lambda_c}$$

and denote the first eigensolution for (5.82), (5.79) by $(\bar{\theta}, \bar{\zeta})$. Following [203] we set

$$\bar{w}(x, y; \gamma, \xi) = w(x, y; \eta) + \xi\theta(x, y; \eta, \xi) \qquad (5.83)$$

so that $\bar{w}(x, y; \eta, 0) = w(x, y; \eta)$, write $\theta$ as a perturbation power series in the new scalar bifurcation parameter $\xi$, i.e.,

$$\theta(x, y; \xi) = \bar{\theta}(x, y) + \xi\Theta(x, y; \xi)$$

$$= \bar{\theta}(x, y) + \xi\left(\sum_{n=0}^{\infty} \Theta^{(n)}(x, y)\xi^n\right) \qquad (5.84)$$

and impose the orthogonality condition

$$(\bar{\theta}, \mathcal{L}(\theta - \bar{\theta})) = 0 \qquad (5.85)$$

on $\theta$ so as to ensure that $\eta \to 0$ as $\xi \to 0$ so that two solutions branch from $\eta = 0$. Imposing the condition that both $w$ and $\bar{w}$ are solutions of the von Karman equations produces a series of perturbation equations for the terms of successive order in (5.84). The relation implied by (5.85) between $\eta$ and $\xi$ is then written down as a power series, i.e.,

$$\eta(\xi) = \bar{a}\xi + \cdots, \quad \bar{a} = \eta'(0) \qquad (5.86)$$

and the perturbation equations are then used so as to infer that $\bar{a} = 0$ if $(\bar{\theta}, [\hat{\zeta}_c, \bar{\theta}]) = 0$; in this case, therefore, $\eta = \bar{b}\xi^2$, in the first approximation, and by using the equations governing the term $\Theta^{(1)}$ in (5.84) a formula may be generated for the bifurcation parameter $\bar{b}$. If $\bar{a} \neq 0$, then the bifurcating solution has a portion which exists under decreasing load and a bifurcation diagram of some norm of the solution against the loading parameter would show the primary and secondary branches crossing at an angle; thus, a secondary branch exists for both decreasing and increasing load levels, which indicates, as the authors in [203] point out, a destabilization of the primary solution with an imperfection-sensitive type of behavior on the part of the secondary branch associated with a decreasing load parameter. For $\bar{a} = 0$, the sign of $\bar{b}$ is crucial, i.e., $\bar{b} > 0$ indicates a relatively stable secondary branch, and an imperfection-insensitive instability on the primary branch, while $\bar{b} < 0$ indicates imperfection-sensitive behavior on the primary branch, while the secondary branch exists only for decreasing loads.

With $\bar{a} = 0, \bar{b} < 0$, a dynamic jump may be experimentally observed under loading: the bifurcating equilibrium branch, in this case, can be maintained only under decreasing load levels and is, therefore, unlikely to be observed. In [205], Bauer and Reiss proposed an energy mechanism based on total potential energy for the phenomenon of mode-jumping; in their scenario, there exist energy-crossing parameter values $\lambda_c$ at which one mode shape is preferred over another, i.e., small disturbances in a mode shape at such values may cause the given state to jump to a state of lower energy through an intermediate state. The authors [205] speculated that there may exist secondary branches which connect the bifurcation points on two primary branches with the bifurcation point on the second branch at a lower load level than that on the first.

To illustrate the method employed in [203], the example of a rectangular simply supported plate in longitudinal compression (with edges constrained so that the normal, in-plane edge displacement is uniform, and the edge shear force is zero) is treated; the plate occupies the region $0 \leq x_1 \leq \ell, 0 \leq y \leq 1$. The vanishing of the second normal derivative of the deflection on all four edges of the plate may be shown to be equivalent to the physical requirement that the edge moment be zero. The load parameter $\lambda$ is just the negative of the resultant load on the edges at $x = 0, x = \ell$, i.e.

$$\lambda = -\int_0^1 N_{11}(0, y)dy = -\int_0^1 N_{11}(\ell, y)dy \qquad (5.87)$$

while the mean load on the edges at $y = 0$ and $y = 1$ is zero. The requirements that the normal edge displacements be constant while the edge stresses vanish may be expressed by the conditions

$$\begin{cases} u_{1,2}(0, y) = u_{1,2}(\ell, y) = u_{2,1}(x, 0) = u_{2,1}(x, 1) = 0 \\ u_{2,1}(0, y) = u_{2,1}(\ell, y) = u_{1,2}(x, 0) = u_{1,2}(x, 1) = 0 \end{cases} \qquad (5.88)$$

The corresponding initial buckling problem has a well-known solution, i.e., the critical values at which solutions branch from the undeflected state are given by

$$\lambda_{m,n} = \left[ \left( \frac{m\pi}{\ell} \right)^2 + (n\pi)^2 \right] \Big/ \left( \frac{m\pi}{\ell} \right)^2 \qquad (5.89)$$

while the corresponding eigenmodes are of the form

$$\psi_{m,n} \approx \sin \frac{m\pi x}{\ell} \sin n\pi y \qquad (5.90)$$

The lowest critical load is obtained as the minimum of the $\lambda_{m,n}$ over all positive integers $m, n$ and always occurs for $n = 1$ and values of $m$

which, as we have seen in our earlier work in Chapter 2, is dependent on the aspect ratio $\ell$; the appropriate value of $m$ satisfies $\sqrt{m(m-1)} < \ell < \sqrt{m(m+1)}$. If $0 < \ell < \sqrt{2}$, then $\lambda_1 = \lambda_{1,1}$ while $\psi_1 \approx \sin(\pi x/\ell)\sin(\pi y)$.

The base von Karman solution is now represented, following our general discussion, above, in terms of an expansion in the parameter $\mu^2 = (\lambda - \lambda_1)/\lambda_1$. By comparing their expansion with the (approximate) Fourier series solution of Levy [212], the authors in [203] are able to show that the first two terms of the expansion yield center-line deflections $w(x, \frac{1}{2}; \lambda)$ that begin to deviate only slightly from those of [212] at loads which are several times larger than the critical buckling load. By using the representations of the base deflection, $w = \mu W(x, y; \mu^2)$, through terms of order $\mu^3$, and the difference stress function $\Phi^* = \mu^2\phi$, through terms of order $\mu^4$, and employing those representations in the secondary buckling problem, i.e., (5.82), (5.79), both first and second order approximations for the solution to this problem are computed in [203], which have the form

$$
\begin{cases}
w^{(1)} \approx \mu w_0(x, y) \\
w^{(2)} \approx \mu w_0(x, y) + \mu^3 w_1(x, y)
\end{cases}
\tag{5.91}
$$

Then, for each of the approximations in (5.91) a Galerkin approximation method is used to determine the eigenfunction $\bar{\theta}$; specifically, $\bar{\theta}$ is approximated by using a linear combination of eigenfunctions of the linearized problem, viz.,

$$
\bar{\theta} \approx \sum_i \sum_j a_{ij} \sin \frac{i\pi x}{\ell} \sin j\pi y
\tag{5.92}
$$

The critical eigenvalues and eigenfunctions are computed in [203] using up to nine terms in the Galerkin series; four groups of terms in $\bar{\theta}(x, y)$ appear to arise in a natural manner, which correspond to the parity of the indices $i$ and $j$, and the buckling loads may be determined for each class or grouping. It was determined in [203] that, for some aspect ratios approaching $\ell_1 = \sqrt{2}$, where the base buckling mode transitions to the second primary mode, there is sensitivity to the group or class in which $i$ is odd and $j$ is even. The computation of the parameters $\bar{a}$ and $\bar{b}$ can be carried out in the postbuckling regime and it turns out that for $\bar{\theta}$ in the same $i$ odd, $j$ odd group as the primary mode, $w_0(x, y) = \sin \frac{\pi x}{\ell} \sin \pi y$ and we have $\bar{a} \neq 0$; for all other groupings $\bar{a} = 0$ but $\bar{b} \neq 0$. In most cases, $\bar{b} < 0$ thus indicating imperfection-sensitivity to perturbations and pointing to the likelihood of snapping occurring at an unstable bifurcation. The authors [203] also compute the snapping loads for the imperfect systems and the energy-crossing loads for the first few primary branches. In Table 5.2 we have summarized, following the work in [203], the numerical values of the secondary buckling loads and the values of the parameters $\bar{a}, \bar{b}$, governing the (approximate) postbuckling analysis, for the four types of $\bar{\theta}$ series; three different aspect ratios $\ell$ are considered in this table.

## 5.3 Imperfection Buckling and Secondary Buckling Figures, Graphs, and Tables

**Table 5.1** Computed Initial Buckling Loads and Associated Radial Edge Shortenings for Isotropic Circular and Annular Plates $(\nu = 0.3)$ [157]

| Number of divisions | Buckling load $(\bar{N}_r, N_r^*)^3$ | Radial edge shortening $(\bar{u})$ | Outer edge condition | Plate type |
|---|---|---|---|---|
| 7.5 | 0.38 | 0.25 | Simply supported | Circular |
| 12.5 | 0.38 | 0.255 | Simply | Circular |
|  | 0.3846 | 0.269 | supported |  |
| 7.5 | 1.34 | 0.95 | Clamped | Circular |
| 12.5 | 1.34 | 0.95 | Clamped | Circular |
|  | 1.3445 | 0.941 | Clamped | Circular |
| 8 | 2.3 | 0.3 | Simply supported | Annular $(\phi = \frac{1}{2})$ |
|  | 2.527 | 0.317 | Simply supported | Annular $(\phi = \frac{1}{2})$ |
| 8 | 24.4 | 3.13 | Clamped | Annular $(\phi = \frac{1}{2})$ |
|  | 24.80 | $3.11^2$ | Clamped | Annular $(\phi = \frac{1}{2})$ |

**Table 5.2** Numerical Values of the Secondary Buckling Loads for a Rectangular Plate [203] (From Stroebel, G.J. and Warner, W.H., *Stability and Secondary Bifurcation for von Karman Plates*, J. Elast., 3, 185, 1973. Reprinted with kind permission from Kluwer Academic Publishers.)

| | $\ell = .75$ | $\ell = 1.00$ | $\ell = 1.25$ |
|---|---|---|---|
| $\lambda_1$ : | $\overline{4.37\pi^2}$ | $\overline{4.00\pi^2}$ | $\overline{4.19\pi^2}$ |
| $i, j$ odd | | | |
| $\lambda_c$ : | $29.60\pi^2$ | $25.62\pi^2$ | $21.46\pi^2$ |
| $\bar{a}$ : | $.876\pi^2$ | $.697\pi^2$ | $.487\pi^2$ |
| $i$ even, $j$ odd | | | |
| $\lambda_c$ : | $29.95\pi^2$ | $26.05\pi^2$ | $20.95\pi^2$ |
| $\bar{b}$ : | $-4.656\pi^2$ | $-18.178\pi^2$ | $-22.44\pi^2$ |
| $i$ odd, $j$ even | | | |
| $\lambda_c$ : | $31.71\pi^2$ | $27.42\pi^2$ | $18.80\pi^2$ |
| $\bar{b}$ : | $-381.47\pi^2$ | $-129.31\pi^2$ | $-.827\pi^2$ |
| $i$ even, $j$ even | | | |
| $\lambda_c$ : | $30.70\pi^2$ | $25.70\pi^2$ | $21.59\pi^2$ |
| $\bar{b}$ : | $-99.345\pi^2$ | $85.29\pi^2$ | $57.43\pi^2$ |

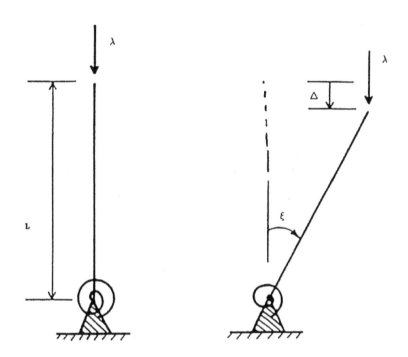

**FIGURE 5.1**
Simple model for illustrating imperfection buckling [16]. (From
"Theory of Buckling and Post-Buckling Behavior of Elastic
Structures" by B. Budiansky in ADVANCES IN APPLIED
MECHANICS, Volume 14, copyright ©1974 by Academic
Press, reproduced by permission of the publisher.)

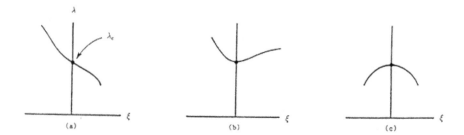

FIGURE 5.2

Equilibrium paths. (a) $\lambda_1 < 0$; (b) $\lambda_1 = 0$, $\lambda_2 > 0$; (c) $\lambda_1 = 0$, $\lambda_2 < 0$ [16]. (From "Theory of Buckling and Post-Buckling Behavior of Elastic Structures" by B. Budiansky in ADVANCES IN APPLIED MECHANICS, Volume 14, copyright ©1974 by Academic Press, reproduced by permission of the publisher.)

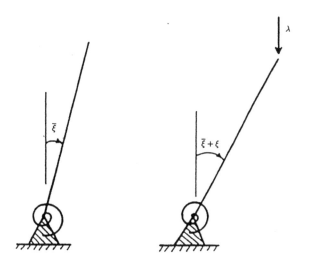

**FIGURE 5.3**
Initially imperfect simple model [16]. (From "Theory of Buckling and Post-Buckling Behavior of Elastic Structures" by B. Budiansky in ADVANCES IN APPLIED MECHANICS, Volume 14, copyright ©1974 by Academic Press, reproduced by permission of the publisher.)

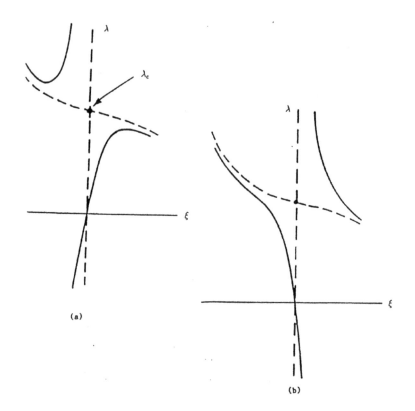

FIGURE 5.4
Equilibrium paths for imperfect model. (a) $\bar{\xi} > 0$; (b) $\bar{\xi} < 0$
[16]. (From "Theory of Buckling and Post·Buckling Behavior
of Elastic Structures" by B. Budiansky in ADVANCES IN AP-
PLIED MECHANICS, Volume 14, copyright ©1974 by Academic Press, reproduced by permission of the publisher.)

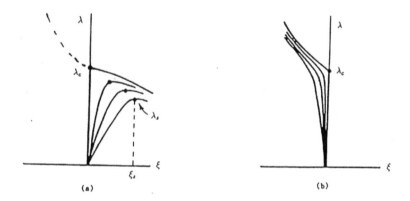

**FIGURE 5.5**
Equilibrium paths for various values of initial imperfection $\bar{\xi}(\lambda_1 < 0)$ (a) $\bar{\xi} > 0$; (b) $\bar{\xi} < 0$ [16]. (From "Theory of Buckling and Post-Buckling Behavior of Elastic Structures" by B. Budiansky in ADVANCES IN APPLIED MECHANICS, Volume 14, copyright ©1974 by Academic Press, reproduced by permission of the publisher.)

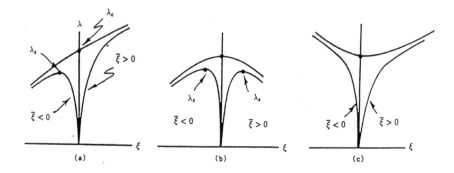

**FIGURE 5.6**
Equilibrium paths for perfect and imperfect models. (a) $\lambda_1 > 0$; (b) $\lambda_1 = 0$, $\lambda_2 < 0$; (c) $\lambda_1 = 0$, $\lambda_2 > 0$ [16]. (From "Theory of Buckling and Post·Buckling Behavior of Elastic Structures" by B. Budiansky in ADVANCES IN APPLIED MECHANICS, Volume 14, copyright ©1974 by Academic Press, reproduced by permission of the publisher.)

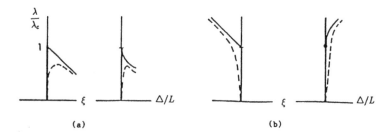

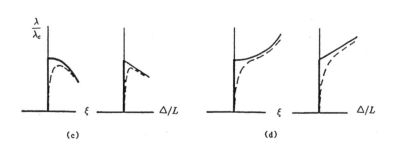

**FIGURE 5.7**
Load-shortening relations for the simple model (a) $\lambda_1 < 0$, $\bar{\xi} \geq 0$; (b) $\lambda_1 < 0$, $\bar{\xi} \leq 0$; (c) $\lambda_1 = 0$, $\lambda_2 < 0$, $\bar{\xi} \geq 0$; (d) $\lambda_1 = 0$, $\lambda_2 > 0$, $\bar{\xi} \geq 0$ [16]. (From "Theory of Buckling and Post·Buckling Behavior of Elastic Structures" by B. Budiansky in ADVANCES IN APPLIED MECHANICS, Volume 14, copyright ©1974 by Academic Press, reproduced by permission of the publisher.)

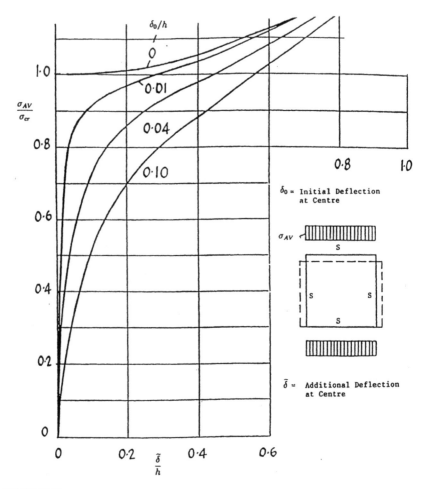

**FIGURE 5.8**
Deflection of plates with initial deviation from flatness.
(Adopted, in modified form, from [109].)

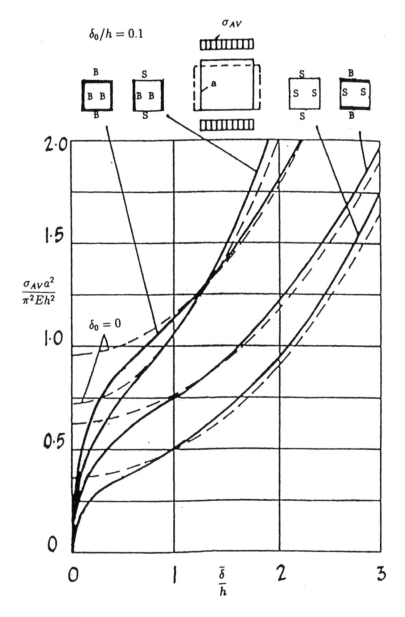

**FIGURE 5.9**
Deflections of plates with $\delta_0/t = 1$, for various support conditions. (Adopted, in modified form, from [94].)

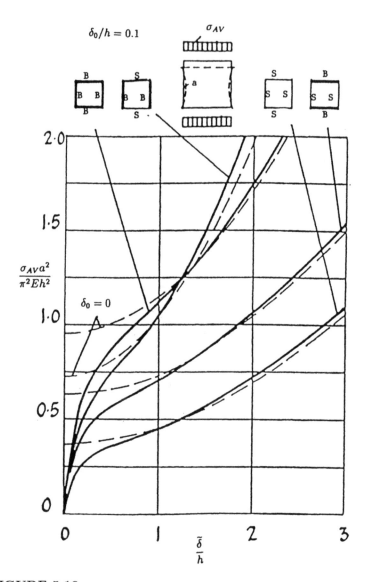

**FIGURE 5.10**
Deflections of plates with $\delta_0/t = 1$, and sides free to wave. (Adopted, in modified form, from [94].)

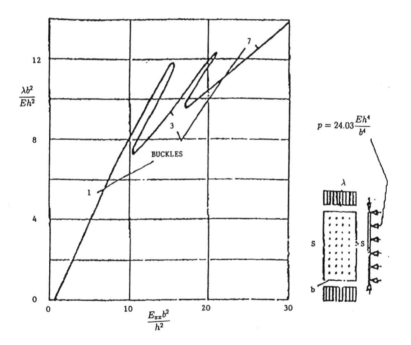

**FIGURE 5.11**
Shifting of the buckle pattern in plates under lateral and end loads. (Adopted, in modified form, from [193].)

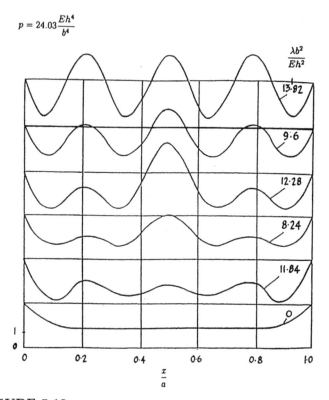

$$p = 24.03 \frac{Eh^4}{b^4}$$

$\frac{\lambda b^2}{Eh^2}$

13·82

9·6

12·28

8·24

11·84

O

0          0·2          0·4          0·6          0·8          1·0

$\frac{x}{a}$

**FIGURE 5.12**
Development of longitudinal buckles. (Adopted, in modified form, from [193].)

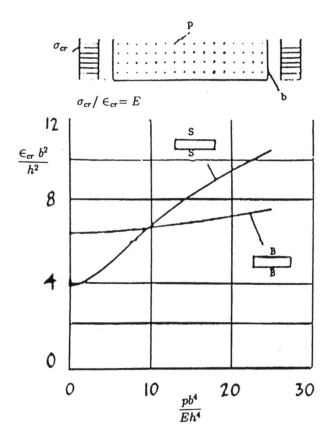

**FIGURE 5.13**
Critical buckling strain of plates under lateral and end loading.
(Adopted, in modified form, from [120].)

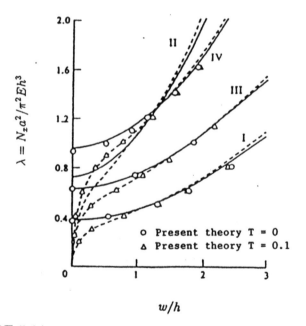

**FIGURE 5.14**
Load-deflection curves for various boundary conditions isotropic plate under uniaxial compression. (Adopted, in modified form, from [94].)

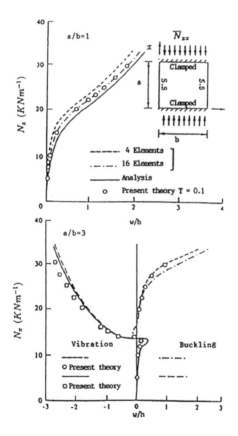

**FIGURE 5.15**
Load-deflection curves for isotropic plate with various aspect
ratios [114]. (Reprinted from the Int. J. Solids and Structures, 31, Romeo, G. and Frulla, G., Nonlinear Analysis
of Anisotropic Plates with Initial Imperfections and Various
Boundary Conditions Subjected to Combined Biaxial Compression and Shear Loads, 763-783, 1994, with permission from
Elsevier Science)

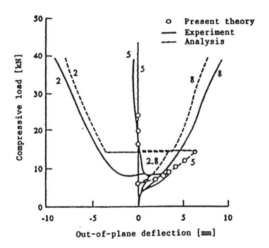

**FIGURE 5.16**
Load-deflection behavior for an anisotropic plate under uniaxial compression [114]. (Reprinted from the Int. J. Solids and Structures, 31, Romeo, G. and Frulla, G., Nonlinear Analysis of Anisotropic Plates with Initial Imperfections and Various Boundary Conditions Subjected to Combined Biaxial Compression and Shear Loads, 763-783, 1994, with permission from Elsevier Science)

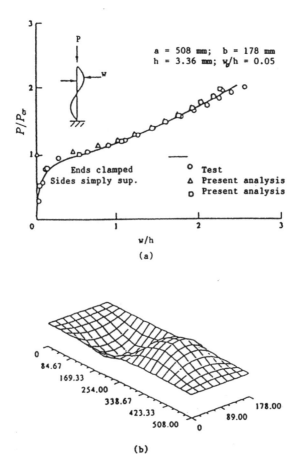

**FIGURE 5.17**
Load-deflection curves of an anisotropic plate under uniaxial compression. (a) Out-of-plane deflection at a quarter length. (b) Overall deflection at $P/P_{cr} = 2$ [114]. (Reprinted from the Int. J. Solids and Structures, 31, Romeo, G. and Frulla, G., Nonlinear Analysis of Anisotropic Plates with Initial Imperfections and Various Boundary Conditions Subjected to Combined Biaxial Compression and Shear Loads, 763-783, 1994, with permission from Elsevier Science)

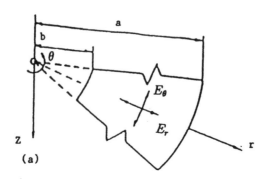

INITIAL DEFLECTION PROFILE

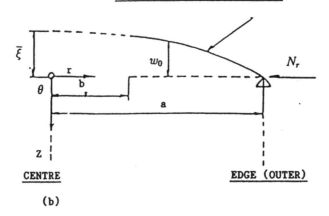

**FIGURE 5.18**
Plate geometry and coordinate system. (a) Annular sector showing positive polar coordinate and principal orthotropy directions. (b) Radial section through annular plate with an initial deflection. (Turvey, G.J. and Drinali, H., "Elastic Postbuckling of Circular and Annular Plates with Imperfections," *Proc. 3rd. Int. Conf. Composite Struct.* Appl. Sci. Pub. 1985, 315–335. With permission of Elsevier Science.)

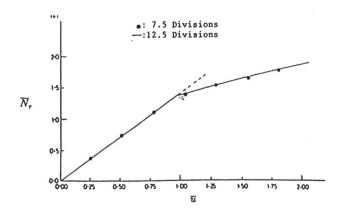

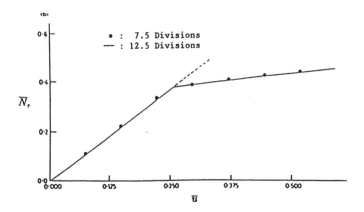

**FIGURE 5.19**
Edge compression-radial shortening response of an isotropic flat circular plate: $(\nu = 0.3)$ (a) clamped edge (b) simply supported edge. (Turvey, G.J. and Drinali, H., "Elastic Postbuckling of Circular and Annular Plates with Imperfections," *Proc. 3rd. Int. Conf. Composite Struct.* Appl. Sci. Pub. 1985, 315–335. With permission of Elsevier Science.)

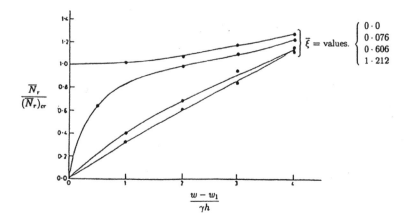

**FIGURE 5.20**
Edge compression-plate center additional deflection response of isotropic
clamped imperfect circular plates ($\nu = 0.3$). (Turvey, G.J. and Drinali,
H., "Elastic Postbuckling of Circular and Annular Plates with Imperfections," *Proc. 3rd. Int. Conf. Composite Struct.* Appl. Sci. Pub.
1985, 315–335. With permission of Elsevier Science.)

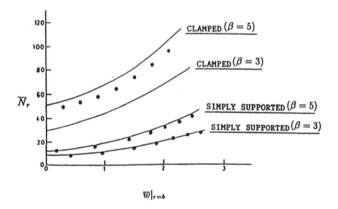

**FIGURE 5.21**
Edge compression-hole-edge deflection response of orthotropic flat annular plates ($\phi = 0.4$) and ($\nu_{r\theta} = 0.3333$). (Turvey, G.J. and Drinali, H., "Elastic Postbuckling of Circular and Annular Plates with Imperfections," *Proc. 3rd. Int. Conf. Composite Struct.* Appl. Sci. Pub. 1985, 315–335. With permission of Elsevier Science.)

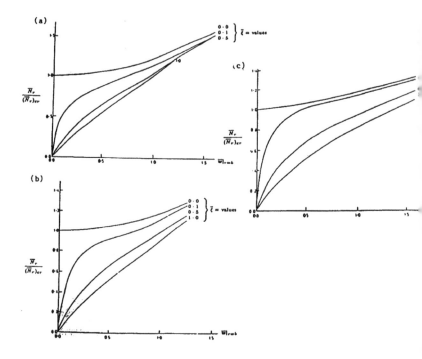

**FIGURE 5.22**
Edge compression-hole-edge additional deflection response for clamped orthotropic imperfect annular plates ($\phi = 0.4$) (a) $\beta = 1.0$, $\nu_{r\theta} = 0.1$ (b) $\beta = 1/3$, $\nu_{r\theta} = 0.3$. (Turvey, G.J. and Drinali, H., "Elastic Post-buckling of Circular and Annular Plates with Imperfections," *Proc. 3rd. Int. Conf. Composite Struct.* Appl. Sci. Pub. 1985, 315–335. With permission of Elsevier Science.)

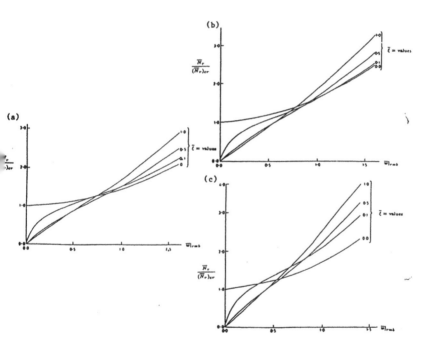

**FIGURE 5.23**
Edge compression-hole-edge additional deflection response for clamped orthotropic imperfect annular plates ($\phi = 0.4$) (a) $\beta = 3 \, \nu_{r\theta} = 0.1$ (b) $\beta = 1, \nu_{r\theta} = 0.3$ (c) $\beta = 1/3$, $\nu_{r\theta} = 0.3$. (Turvey, G.J. and Drinali, H., "Elastic Postbuckling of Circular and annular Plates with Imperfections," *Proc. 3rd. Int. Conf. Composite Struct. Appl. Sci. Lab.* 1985, 315–335. With permission of Elsevier Science.)

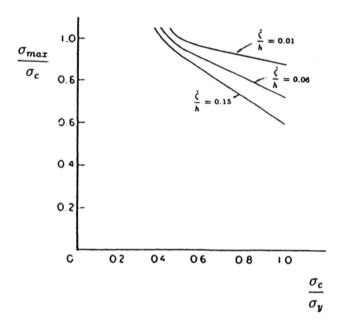

**FIGURE 5.24**
Maximum average applied stress of an elastic-plastic simply
supported circular plate when buckling takes place in the elastic range ($\nu = \frac{1}{3}$, $n = 12$) [141]. (Reprinted from Int. J. Mech.
Sci., 17, Needleman, A., Postbifurcation Behavior and Imperfection Sensitivity of Elastic-Plastic Circular Plates, 1-13, 1975,
with permission from Elsevier Science)

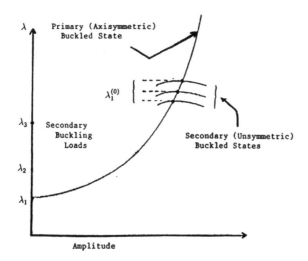

**FIGURE 5.25**
A sketch of the response of the radially compressed and clamped circular plate [201]. (From Cheo, L.S. and Reiss, E.L., Secondary Buckling of Circular Plates, SIAM J. Appl. Math., 490, 26, 1974. With permission)

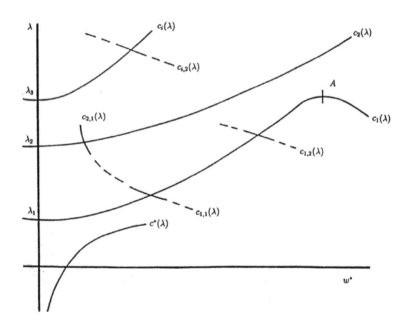

**FIGURE 5.26**
Schematic bifurcation diagram [203]. (From Stroebel, G.J. and Warner, W.H., "Stability and Secondary Bifurcation for von Karman Plates", J. Elast., 3, 185, 1973. Reprinted with kind permission from Kluwer Academic Publishers.)

# CHAPTER 6

# Generalized von Karman Equations for Elastic Plates Subject to Hygroexpansive or Thermal Stress Distributions

We begin our study of hygroexpansive and thermal buckling and bending of thin plates by deriving the generalized von Karman equations for elastic isotropic and orthotropic plates; the pertinent results will be presented in both rectilinear and polar coordinates for, respectively, rectangular plates and plates with a circular geometry. In Chapter 8 the equations governing the bending and buckling behavior of thin plates either exhibiting viscoelastic (creep) behavior or undergoing plastic deformations will be presented. The chapter also includes a discussion of boundary conditions for plates undergoing hygrothermal expansion or contraction.

## 6.1   Rectilinear Coordinates

We consider an isotropic thin plate of constant thickness $h$ which occupies the domain $\Omega$ in the $x, y$ plane (Fig. 1.11); as in Chapter 1, we employ the Kirchoff hypothesis, i.e., sections $x = $ const., $y = $ const. of the undeformed plate remain plane after deformation and also maintain their angle with respect to the deformed middle surface of the plate. In terms of the displacement components $u, v, w$ of the middle surface of the plate, we have the following generalization of equations (1.34) which applies when either hygroexpansive or thermal strains (or both) must be taken into account:

$$\begin{cases} \tilde{\epsilon}_{xx} = \epsilon_{xx} - \epsilon_{HT} \\ \tilde{\epsilon}_{yy} = \epsilon_{yy} - \epsilon_{HT} \\ \tilde{\gamma}_{xy} = 2\tilde{\epsilon}_{xy} = 2\epsilon_{xy} \end{cases} \tag{6.1}$$

where $\epsilon_{xx}, \epsilon_{xy}, \epsilon_{yy}$ are given by (1.34), i.e.,

$$
\begin{cases}
\epsilon_{xx} = u_{,x} + \dfrac{1}{2}w_{,x}^2 - \zeta w_{,xx} \\[2mm]
\epsilon_{xy} = \dfrac{1}{2}(u_{,y} + v_{,x} + w_{,x}w_{,y}) - \zeta w_{,xy} \\[2mm]
\epsilon_{yy} = v_{,y} + \dfrac{1}{2}w_{,y}^2 - \zeta w_{,yy}
\end{cases}
$$

with $-h \leq \zeta \leq h$ the (normal) distance from the middle surface of the plate, and

$$\epsilon_{HT} = \beta \delta H + \alpha \delta T \tag{6.2}$$

In (6.2), $\beta$ is the (assumed constant) coefficient of hygroscopic expansion, $\alpha$ the thermal expansion coefficient, $\delta H$ the change in moisture content and $\delta T$ the change in temperature. For a static problem we have, in general

$$
\begin{cases}
\delta H = H(x,y,z) - H_o \\
\delta T = T(x,y,z) - T_o
\end{cases} \tag{6.3}
$$

with $H_0, T_0$, respectively, reference moisture and temperature levels. In writing down (6.1), (6.2) we have already assumed isotropy, i.e. $(\epsilon_{HT})_{xx} = (\epsilon_{HT})_{yy}$; for the case of rectilinear orthotropy we will have to introduce coefficients $\alpha_i, \beta_i, i = 1, 2$. We also note that for a purely hygroscopic problem $\delta T = 0$, while for an entirely thermal problem $\delta H = 0$.

**Remarks:** To simplify the presentation, and because almost all of the literature, to date, has dealt solely with problems of thermal buckling and bending as opposed to hygroscopic buckling and bending (or a combination of both mechanisms) we will often write $\epsilon_{HT} \equiv \epsilon_T = \alpha \delta T$ or the natural generalization with respect to orthotropic response. In almost all the cases that will be discussed $\alpha \delta T$ (in the isotropic case, for example) will be interchangeable with $\beta \delta H$.

For rectilinear isotropic response, the constitutive relations are given by (see, e.g., [216])

$$
\begin{cases}
\tilde{\epsilon}_{xx} = \dfrac{1}{E}(\sigma_{xx} - \nu\sigma_{yy}) \\[2mm]
\tilde{\epsilon}_{yy} = \dfrac{1}{E}(\sigma_{yy} - \nu\sigma_{xx}) \\[2mm]
\tilde{\gamma}_{xy} = \dfrac{2(1+\nu)}{E}\sigma_{xy}
\end{cases}
$$

or, in view of (6.1)

$$\begin{cases} \epsilon_{xx} - \epsilon_{HT} = \dfrac{1}{E}(\sigma_{xx} - \nu\sigma_{yy}) \\[2mm] \epsilon_{yy} - \epsilon_{HT} = \dfrac{1}{E}(\sigma_{yy} - \nu\sigma_{xx}) \\[2mm] \epsilon_{xy} = (\dfrac{1+\nu}{E})\sigma_{xy} \end{cases} \qquad (6.4)$$

with $\nu$ the Poisson's ratio and $E$ the Young's modulus. Alternatively, by solving (6.4) for the stress components, we have

$$\begin{cases} \sigma_{xx} = \dfrac{E}{1-\nu^2}(\epsilon_{xx} + \nu\epsilon_{yy}) = \dfrac{E}{1-\nu}\epsilon_{HT} \\[2mm] \sigma_{yy} = \dfrac{E}{1-\nu^2}(\epsilon_{yy} + \nu\epsilon_{xx}) - \dfrac{E}{1-\nu}\epsilon_{HT} \\[2mm] \sigma_{xy} = \dfrac{E}{1+\nu}\epsilon_{xy} \equiv 2G\epsilon_{xy} \end{cases} \qquad (6.5)$$

where $G$ is the shear modulus of the plate. With, e.g., $\epsilon_{HT} = \alpha\delta T \equiv \epsilon_T$, (6.4) says that as a consequence of a change in the heat content of the plate strains $\epsilon_{xx}, \epsilon_{xy}$, and $\epsilon_{yy}$ are caused by thermal expansion of the material composing the plate as well as by stresses that might arise from applied loads or other sources. As in Chapter 1, the averaged stresses over the plate thickness $h$ (assumed to be small) and the bending moments are given by (1.37), (1.38) which we repeat here as

$$\begin{cases} N_x = \displaystyle\int_{-h/2}^{h/2} \sigma_{xx}dz \\[3mm] N_y = \displaystyle\int_{-h/2}^{h/2} \sigma_{yy}dz \\[3mm] N_{xy} = \displaystyle\int_{-h/2}^{h/2} \sigma_{xy}dz \end{cases} \qquad (6.6a)$$

and

$$\begin{cases} M_x = \displaystyle\int_{-h/2}^{h/2} \sigma_{xx}zdz \\[3mm] M_y = \displaystyle\int_{-h/2}^{h/2} \sigma_{yy}zdz \\[3mm] M_{xy} = \displaystyle\int_{-h/2}^{h/2} \sigma_{xy}zdz \end{cases} \qquad (6.6b)$$

In view of the definitions of $\epsilon_{xx}, \epsilon_{xy}$, and $\epsilon_{yy}$, if we define, in the usual way, the middle surface strains by (see (1.34), (1.35))

$$
\begin{cases}
\epsilon_{xx}^o = u_{,x} + \dfrac{1}{2} w_{,x}^2 \\[2mm]
\epsilon_{yy}^o = v_{,y} + \dfrac{1}{2} w_{,y}^2 \\[2mm]
\epsilon_{xy}^o = \dfrac{1}{2}(u_{,y} + v_{,x} + w_{,x} w_{,y})
\end{cases}
\tag{6.7}
$$

so that

$$
\begin{cases}
\epsilon_{xx} = \epsilon_{xx}^o - \zeta w_{,xx} \\[1mm]
\epsilon_{xy} = \epsilon_{xy}^o - \zeta w_{,xy} \\[1mm]
\epsilon_{yy} = \epsilon_{yy}^o - \zeta w_{,yy}
\end{cases}
\tag{6.8}
$$

then, by virtue of (6.5), (6.6a), and (6.8)

$$
\begin{cases}
N_x = \dfrac{Eh}{1-\nu^2}(\epsilon_{xx}^o + \nu \epsilon_{yy}^o) - N_{HT} \\[2mm]
N_y = \dfrac{Eh}{1-\nu^2}(\epsilon_{yy}^o + \nu \epsilon_{xx}^o) - N_{HT} \\[2mm]
N_{xy} = 2Gh\epsilon_{xy}^o
\end{cases}
\tag{6.9}
$$

where

$$
N_{HT} = \frac{E}{1-\nu} \cdot \int_{-h/2}^{h/2} \epsilon_{HT} dz
\tag{6.10}
$$

For the purely thermal situation in which $\epsilon_{TH} \equiv \epsilon_T = \alpha \triangle T(x, y, z)$

$$
N_{HT} = N^T = \frac{\alpha E}{1-\nu} \int_{-h/2}^{h/2} \delta T(x, y, z) dz
\tag{6.11a}
$$

while, in the entirely hygroscopic case with $\epsilon_{TH} = \epsilon_H$

$$
N_{HT} = N^H = \frac{\beta E}{1-\nu} \int_{-h/2}^{h/2} \delta H(x, y, z) dz
\tag{6.11b}
$$

Equations (6.9) may be found, e.g., in §9.4 of [217], with $N_{HT} = N^T$.

In an analogous fashion, we may compute that, by virtue of (6.5), (6.6a), and (6.8), the bending moments are given by

$$
\begin{cases}
M_x = -K(w_{,xx} + \nu w_{,yy}) - M_{HT} \\[1mm]
M_y = -K(w_{,yy} + \nu w_{,xx}) - M_{HT} \\[1mm]
M_{xy} = -(1-\nu)K w_{,xy}
\end{cases}
\tag{6.12}
$$

where $K = \dfrac{Eh^2}{12(1-\nu^2)}$ is the usual plate stiffness for the isotropic case while the hygrothermal moment $M_{HT}$ is given by

$$M_{HT} = \frac{E}{1-\nu} \int_{-h/2}^{h/2} \epsilon_{HT} z \, dz \qquad (6.13)$$

For $\epsilon_{HT} \equiv \epsilon_T$,

$$M_{HT} = M^T = \frac{\alpha E}{1-\nu} \cdot \int_{-h/2}^{h/2} \delta T(x,y,z) z \, dz \qquad (6.14)$$

while for $\epsilon_{HT} = \epsilon_H$,

$$M_{HT} = M^H = \frac{\beta E}{1-\nu} \cdot \int_{-h/2}^{h/2} \delta H(x,y,z) z \, dz \qquad (6.15)$$

The relations (6.12), with $M_{HT} = M^T$ may also be found, e.g., in §9.4 of [217].

In certain situations it may be the case that the coefficients $\alpha$ and/or $\beta$ are field-dependent, i.e., $\alpha = \alpha(\delta T), \beta = \beta(\delta H)$. In such case, one would have, e.g.,

$$\begin{cases} N^T = \dfrac{E}{1-\nu} \displaystyle\int_{-h/2}^{h/2} \alpha(\delta T) \cdot \delta T \, dz \\[4mm] M^T = \dfrac{E}{1-\nu} \displaystyle\int_{-h/2}^{h/2} \alpha(\delta T) \cdot \delta T z \, dz \end{cases} \qquad (6.16)$$

with analogous expressions for $N^H, M^H$.

The equilibrium equations which apply in the present situation (see Figs. 1.8, 1.9) are precisely the same relations which hold in the absence of hygrothermal strains, i.e., (1.42), (1.43), and (1.44); we write these (in the absence of an initial deflection and a distributed normal loading) in the form

$$\begin{cases} N_{x,x} + N_{xy,y} = 0 \\ N_{xy,x} + N_{y,y} = 0 \end{cases} \qquad (6.17a)$$

$$Q_{xz,x} + Q_{yz,y} + N_x w_{,xx} + N_y w_{,yy} + 2N_{xy} w_{,xy} = 0 \qquad (6.17b)$$

$$\begin{cases} M_{xy,x} + M_{y,y} - Q_{yz} = 0 \\ M_{yx,y} + M_{x,x} - Q_{xz} = 0 \end{cases} \quad (M_{xy} = M_{yx}) \qquad (6.17c)$$

Eliminating $Q_{yz}$ and $Q_{xz}$ from among the equations in (6.17b, c) we obtain

$$M_{x,xx} + 2M_{xy,xy} + M_{y,yy} + N_x w_{,xx} + 2N_{xy} w_{,xy} + N_y w_{,yy} = 0 \quad (6.18)$$

Substituting for the moments $M_x, M_y$, and $M_{xy}$ in (6.18) then yields

$$K\triangle^2 w = N_x w_{,xx} + 2N_{xy} w_{,xy} + N_y w_{,yy} - \triangle M_{HT} \quad (6.19)$$

where $\triangle = \dfrac{\partial^2}{\partial x^2} + \dfrac{\partial^2}{\partial y^2}$ is the two-dimensional Laplacian while $\triangle^2$ is the biharmonic operator. Modifications (which will be discussed later) must be made to (6.19) if imperfection buckling is considered, i.e., if the plate possesses an initial prebuckling deflection $w_0 = w_0(x, y)$ or is subject to a transverse normal loading. As in Chapter 1, we may introduce the Airy stress function $\Phi(x, y)$ by

$$N_x = \Phi_{,yy}, N_y = \Phi_{,xx}, N_{xy} = -\Phi_{,xy} \quad (6.20)$$

in which case equations (6.17a) are satisfied identically, while (6.19) becomes

$$K\triangle^2 w = \Phi_{,yy} w_{,xx} - 2\Phi_{,xy} w_{,xy} + \Phi_{,xx} w_{,yy} - \triangle M_{HT} \quad (6.21)$$

From the compatibility equation

$$\frac{\partial^2}{\partial y^2} \epsilon_{xx}^0 + \frac{\partial^2}{\partial x^2} \epsilon_{yy}^0 - 2\frac{\partial^2 \epsilon_{xy}^0}{\partial x \partial y} = (\frac{\partial^2 w}{\partial x \partial y})^2 - \frac{\partial^2 w}{\partial x^2} \frac{\partial^2 w}{\partial y^2}, \quad (6.22)$$

where $\epsilon_{xx}^0, \epsilon_{yy}^0, \epsilon_{xy}^0$ are the middle surface strains, as given by (6.7) and the constitutive relations (6.9), we easily obtain the second of the two generalized von Karman equations which apply in the case of hygrothermal buckling, i.e.,

$$\triangle^2 \Phi = Eh\{(\frac{\partial^2 w}{\partial x \partial y})^2 - \frac{\partial^2 w}{\partial x^2} \frac{\partial^2 w}{\partial y^2}\} - (1 - \nu)\triangle N_{HT} \quad (6.23)$$

If, as in Chapter 1, we introduce the nonlinear (bracket) differential operator by

$$[f, g] = f_{,yy} g_{,xx} - 2f_{,xy} g_{,xy} + f_{,xx} g_{,yy}$$

we may write (6.23) in the form

$$\triangle^2 \Phi = -\frac{1}{2} Eh[w, w] - (1 - \nu)\triangle N_{HT} \quad (6.24)$$

With $\Delta M_{HT} = \Delta N_{HT} = 0$, the system (6.21), (6.24) reduces to the standard system of von Karman equations which apply in the case of linear isotropic elastic response in rectilinear Cartesian coordinates in the absence of both an initial prebuckling deflection and an applied transverse normal loading.

For the case in which the thin plate exhibits linear elastic behavior but possesses rectilinear orthotropic symmetry, the generalized von Karman equations governing hygrothermal buckling may be derived as follows: for a constant thickness orthotropic thick plate, in which the $x$ and $y$ axes coincide with the principle directions, the constitutive equations have the form (compare with (1.60))

$$\left\{ \begin{array}{c} \sigma_{xx} \\ \sigma_{yy} \\ \sigma_{xy} \end{array} \right\} = \left[ \begin{array}{ccc} c_{11} & c_{12} & 0 \\ c_{21} & c_{22} & 0 \\ 0 & 0 & c_{66} \end{array} \right] \left\{ \begin{array}{c} \epsilon_{xx} - \epsilon_{HT}^{1} \\ \epsilon_{yy} - \epsilon_{HT}^{2} \\ \gamma_{xy} \end{array} \right\} \tag{6.25}$$

where the hygrothermal strains have the form

$$\left\{ \begin{array}{l} \epsilon_{HT}^{1} = \beta_{1}\delta H + \alpha_{1}\delta T \\ \epsilon_{HT}^{2} = \beta_{2}\delta H + \alpha_{2}\delta T \end{array} \right. \tag{6.26}$$

with $\alpha_1, \alpha_2$ the coefficients of linear thermal expansion along the $x$ and $y$ axes, respectively, and $\beta_1, \beta_2$ the coefficients of hygroscopic expansion. For now, we shall assume that the $\alpha_i$ and $\beta_i$ are constant, $i = 1, 2$. In (6.25) the elastic constants are given by (1.61) with $E_1, E_2, \nu_{12}, \nu_{21}$, and $G_{12}$, respectively, Young's moduli, Poisson's ratios, and shear modulus associated with the principal directions:

$$\left\{ \begin{array}{l} c_{11} = E_1/(1 - \nu_{12}\nu_{21}) \\ c_{12} = E_2\nu_{21}/(1 - \nu_{12}\nu_{21}) \\ c_{21} = E_1\nu_{12}/(1 - \nu_{12}\nu_{21}) \\ c_{22} = E_2/(1 - \nu_{12}\nu_{21}) \\ c_{66} = G_{12} \end{array} \right. \tag{1.61}$$

Also, $E_1\nu_{12} = E_2\nu_{21}$ so that $c_{12} = c_{21}$. As in Chapter 1, the constants $D_{ij} = c_{ij}h^3/12$ are the associated rigidities (i.e., stiffness ratios) of the orthotropic plate. Recalling (1.63), we have for the bending rigidities about the $x$ and $y$ axes, respectively,

$$D_{11} = \frac{E_1 h^3}{12(1 - \nu_{12}\nu_{21})} \text{ and } D_{22} = \frac{E_2 h^3}{12(1 - \nu_{12}\nu_{21})} \tag{1.63}$$

while, from (1.64), i.e.,

$$D_{66} = \frac{G_{12}h^3}{12} \tag{1.64}$$

is the twisting rigidity. Often, the ratios $D_{12}/D_{22}$ and $D_{26}/D_{11}$ are termed reduced Poisson's ratios.

We write (6.25) out in the form

$$
\left\{
\begin{aligned}
\sigma_{xx} &= c_{11}(\epsilon^0_{xx} - \zeta w_{,xx}) + c_{12}(\epsilon^0_{yy} - \zeta w_{,yy}) \\
&\quad - c_{11}\epsilon^1_{HT} - c_{12}\epsilon^2_{HT} \\
\sigma_{yy} &= c_{21}(\epsilon^0_{xx} - \zeta w_{,xx}) + c_{22}(\epsilon^0_{yy} - \zeta w_{,yy}) \\
&\quad - c_{21}\epsilon^1_{HT} - c_{22}\epsilon^2_{HT} \\
\sigma_{xy} &= 2c_{66}(\epsilon^0_{xy} - \zeta w_{,xy})
\end{aligned}
\right.
\tag{6.27}
$$

where the middle surface strains are given by (6.7).

The averaged stresses (over the plate thickness) are still given by (6.6a) so that, by virtue of (1.61), and (6.27)

$$
\left\{
\begin{aligned}
N_x &= \left\{\frac{E_1 h}{1 - \nu_{12}\nu_{21}}\right\}\epsilon^0_{xx} + \left\{\frac{E_2\nu_{21}h}{1 - \nu_{12}\nu_{21}}\right\}\epsilon^0_{yy} \\
&\quad - (N^{11}_{HT} + N^{12}_{HT}) \\
N_y &= \left\{\frac{E_1\nu_{12}h}{1 - \nu_{12}\nu_{21}}\right\}\epsilon^0_{xx} + \left\{\frac{E_2 h}{1 - \nu_{12}\nu_{21}}\right\}\epsilon^0_{yy} \\
&\quad - N^{21}_{HT} + N^{22}_{HT}) \\
N_{xy} &= (2G_{12}h)\epsilon^0_{xy}
\end{aligned}
\right.
\tag{6.28}
$$

where

$$
\left\{
\begin{aligned}
N^{11}_{HT} &= \left(\frac{E_1}{1 - \nu_{12}\nu_{21}}\right)\int_{-h/2}^{h/2}\epsilon^1_{HT}\,dz \\
N^{12}_{HT} &= \left(\frac{E_2\nu_{21}}{1 - \nu_{12}\nu_{21}}\right)\int_{-h/2}^{h/2}\epsilon^2_{HT}\,dz \\
N^{21}_{HT} &= \left(\frac{E_1\nu_{12}}{1 - \nu_{12}\nu_{21}}\right)\int_{-h/2}^{h/2}\epsilon^1_{HT}\,dz \\
N^{22}_{HT} &= \left(\frac{E_2}{1 - \nu_{12}\nu_{21}}\right)\int_{-h/2}^{h/2}\epsilon^2_{HT}\,dz
\end{aligned}
\right.
\tag{6.29}
$$

**Remarks:** In the special case of rectilinear isotropic elastic response (6.28), (6.29) reduce as follows: we set $E_1 = E_2 = E$, $\nu_{12} = \nu_{21} = \nu$, and $G_{12} = G$. Also

$$\epsilon^1_{HT} = \epsilon^2_{HT} \equiv \beta\delta H + \alpha\delta T \equiv \epsilon_{HT}$$

as $\alpha_1 = \alpha_2 = \alpha$, $\beta_1 = \beta_2 = \beta$. Then

$$
\begin{cases}
N^{11}_{HT} + N^{12}_{HT} = \dfrac{E}{1 - \nu} \displaystyle\int_{-h/2}^{h/2} \epsilon_{HT}\, dz \\[3mm]
N^{21}_{HT} + N^{22}_{HT} = \dfrac{E}{1 - \nu} \displaystyle\int_{-h/2}^{h/2} \epsilon_{HT}\, dz
\end{cases}
\tag{6.30}
$$

and it is clear that (6.28) reduces to (6.9) with $N_{HT} = N^{11}_{HT} + N^{12}_{HT} = N^{21}_{HT} + N^{22}_{HT}$ given by (6.10).

From (6.28) we have, immediately, that

$$\epsilon^0_{xy} = \left(\frac{1}{2G_{12}h}\right) N_{xy} \tag{6.31a}$$

while the linear algebraic system

$$
\begin{cases}
c_{11}\epsilon^0_{xx} + c_{12}\epsilon^0_{yy} = \dfrac{1}{h}(N_x + \tilde{N}^1_{HT}) \\[3mm]
c_{21}\epsilon^0_{xx} + c_{22}\epsilon^0_{yy} = \dfrac{1}{h}(N_y + \tilde{N}^2_{HT})
\end{cases}
$$

with

$$
\begin{cases}
\tilde{N}^1_{HT} = N^{11}_{HT} + N^{12}_{HT} \\[2mm]
\tilde{N}^2_{HT} = N^{21}_{HT} + N^{22}_{HT}
\end{cases}
$$

yields

$$
\epsilon^0_{xx} = \frac{1}{|c_{ij}|h} \left\{ \left(\frac{E_2}{1 - \nu_{12}\nu_{21}}\right)(N_x + \tilde{N}^1_{HT}) \right.
$$
$$
\left. - \left(\frac{E_2\nu_{21}}{1 - \nu_{12}\nu_{21}}\right)(N_y + \tilde{N}^2_{HT}) \right\}
\tag{6.31b}
$$

$$
\epsilon^0_{yy} = \frac{1}{|c_{ij}|h} \left\{ \left(\frac{E_1}{1 - \nu_{12}\nu_{21}}\right)(N_y + \tilde{N}^2_{HT}) \right.
$$
$$
\left. - \left(\frac{E_1\nu_{12}}{1 - \nu_{12}\nu_{21}}\right)(N_x + \tilde{N}^1_{HT}) \right\}
\tag{6.31c}
$$

with

$$|c_{ij}| = \begin{vmatrix} \dfrac{E_1}{1 - \nu_{12}\nu_{21}} & \dfrac{E_2\nu_{21}}{1 - \nu_{12}\nu_{21}} \\[3mm] \dfrac{E_1\nu_{12}}{1 - \nu_{12}\nu_{21}} & \dfrac{E_2}{1 - \nu_{12}\nu_{21}} \end{vmatrix},$$

i.e.,

$$|c_{ij}| = E_1 E_2/(1 - \nu_{12}\nu_{21}) \qquad (6.32)$$

Employing (6.32) in (6.31b,c) and recalling (6.31a) we easily find that the inverted constitutive relations assume the form

$$\begin{cases} \epsilon^0_{xx} = \dfrac{1}{hE_1}\left\{ (N_x + \tilde{N}^1_{HT}) - \nu_{21}(N_y + \tilde{N}^2_{HT}) \right\} \\[4mm] \epsilon^0_{yy} = \dfrac{1}{hE_2}\left\{ (N_y + \tilde{N}^2_{HT}) - \nu_{12}(N_x + \tilde{N}^1_{HT}) \right\} \\[4mm] \epsilon^0_{xy} = \dfrac{1}{2hG_{12}} N_{xy} \end{cases} \qquad (6.33)$$

To compute the bending moments $M_x$, $M_y$, and $M_{xy}$ we employ the constitutive relations (6.27) in (6.6b) and obtain

$$\begin{cases} M_x = -\dfrac{h^3}{12}(c_{11}w_{,xx} + c_{12}w_{,yy}) - (M^{11}_{HT} + M^{12}_{HT}) \\[4mm] M_y = -\dfrac{h^3}{12}(c_{21}w_{,xx} + c_{22}w_{,yy}) - (M^{21}_{HT} + M^{22}_{HT}) \\[4mm] M_{xy} = -\dfrac{h^3}{6}c_{66}w_{,xy} \end{cases}$$

where the hygrothermal moments $M^{ij}_{HT}$ are given by

$$\begin{cases} M^{11}_{HT} = c_{11} \displaystyle\int_{-h/2}^{h/2} \epsilon^1_{HT}(x,y,z)z\,dz \\[4mm] M^{12}_{HT} = c_{12} \displaystyle\int_{-h/2}^{h/2} \epsilon^2_{HT}(x,y,z)z\,dz \\[4mm] M^{21}_{HT} = c_{21} \displaystyle\int_{-h/2}^{h/2} \epsilon^1_{HT}(x,y,z)z\,dz \\[4mm] M^{22}_{HT} = c_{22} \displaystyle\int_{-h/2}^{h/2} \epsilon^2_{HT}(x,y,z)z\,dz \end{cases} \qquad (6.34)$$

Employing the rigidities $D_{ij} = c_{ij}h^3/12$, the bending moments may be written in the form

$$\begin{cases} M_x = -(D_{11}w_{,xx} + D_{12}w_{,yy}) - (M_{HT}^{11} + M_{HT}^{12}) \\ M_y = -(D_{21}w_{,xx} + D_{22}w_{,yy}) - (M_{HT}^{21} + M_{HT}^{22}) \\ M_{xy} = -2D_{66}w_{,xy} \end{cases} \quad (6.35)$$

Equations (6.34), (6.35) generalize the relations (1.67a, b, c) of Chapter 1 and reduce to the latter expressions for the bending moments when $\alpha_i = \beta_i = 0, i = 1, 2$.

For the orthotropic case, the equilibrium equations (6.17a), (6.18) still apply, with the averaged stresses given by (6.28), (6.29) and the bending moments by (6.34), (6.35). As in the (rectilinear) isotropic case, we introduce the Airy function $\Phi$, which is defined by (6.20) and the pair of equations (6.17a) is satisfied identically. Substituting into (6.18), from (6.20) and (6.35), we next find that (6.18) implies that

$$D_{11}w_{,xxxx} + \{D_{12} + 4D_{66} + D_{21}\}w_{,xxyy} + D_{22}w_{,yyyy} = [\Phi, w]$$
$$(6.36)$$
$$-\{(\tilde{M}_{HT}^1)_{,xx} + (\tilde{M}_{HT}^2)_{,yy}\}$$

where

$$\begin{cases} \tilde{M}_{HT}^1 = M_{HT}^{11} + M_{HT}^{12} \\ \tilde{M}_{HT}^2 = M_{HT}^{21} + M_{HT}^{22} \end{cases} \quad (6.37)$$

With $\alpha_i = \beta_i = 0, i = 1, 2$, (6.36) reduces to the first of the von Karman equations for the non-hygrothermal case, e.g., (1.68).

The second generalized von Karman equation for hygrothermal bending and bucking for the case of rectilinear orthotropic symmetry, follows as in the isotropic case from the compatibility equation (6.22), i.e.,

$$\frac{\partial^2}{\partial y^2}\epsilon_{xx}^o + \frac{\partial^2}{\partial x^2}\epsilon_{yy}^o - 2\frac{\partial^2\epsilon_{xy}^o}{\partial x\partial y} = -\frac{1}{2}[w, w]$$

In the present situation, the middle surface strains $\epsilon_{xx}^o, \epsilon_{xy}^o$, and $\epsilon_{yy}^o$ are given by (6.33); introducing the Airy function into (6.33), we now write this system in the form

$$\begin{cases} \epsilon_{xx}^o = \frac{1}{hE_1} \cdot (\Phi_{,yy} - \nu_{21}\Phi_{,xx}) + \frac{1}{hE_1}(\tilde{N}_{HT}^1 - \nu_{21}\tilde{N}_{HT}^2) \\ \epsilon_{yy}^o = \frac{1}{hE_2} \cdot (\Phi_{,xx} - \nu_{12}\Phi_{,yy}) + \frac{1}{hE_2}(\tilde{N}_{HT}^2 - \nu_{12}\tilde{N}_{HT}^1) \qquad (6.38) \\ \epsilon_{xy}^o = -\frac{1}{2hG_{12}}\Phi_{,xy} \end{cases}$$

and then substitute into (6.22) so as to obtain

$$\frac{1}{E_1 h}\Phi_{,yyyy} + \frac{1}{h}\Big(\frac{1}{G_{12}} - \frac{2\nu_{12}}{E_2}\Big)\Phi_{,xxyy}$$

$$+\frac{1}{E_2 h}\Phi_{,xxxx} = -\frac{1}{2}[w, w] \qquad (6.39)$$

$$-\frac{1}{E_1 h}(\tilde{N}_{HT}^1 - \nu_{21}\tilde{N}_{HT}^2)_{,yy} - \frac{1}{E_2 h}(\tilde{N}_{HT}^2 - \nu_{12}\tilde{N}_{HT}^1)_{,xx}$$

Equation (6.39) reduces to (1.69) for the non-hygrothermal case when $\alpha_i = \beta_i = 0$, $i = 1, 2$, in which case $\tilde{N}_{HT}^2 = \tilde{N}_{HT}^2 = 0$. Various special cases of (6.39) will appear in Chapter 7, when we consider specific (mostly, thermal buckling and bending) problems that have appeared in the literature. For a thin linear elastic plate, which exhibits rectilinear orthotropic symmetry, the complete system of generalized von Karman equations incorporating hygrothermal expansion and contraction, consists of (6.36) and (6.39); these equations hold in the absence of initial deflections and a transverse (normal) loading. Systems of equations which are similar to (but less general than) (6.36), (6.39) for the case of thermal buckling and bending of thin linearly elastic rectilinearly orthotropic plates have appeared in several places in the literature, e.g., [218], [219], and [220], as well as in §7.2 and §9.2 of [217].

---

## 6.2 Polar Coordinates

In this section, we will present versions of the generalized von Karman equations in polar coordinates for thin linearly elastic plates exhibiting isotropic response as well as cylindrically orthotropic response. Because of the rather complex structure of the buckling equations for a circular geometry coupled to an assumption of rectlinear orthotropic response, even in the absence of hygrothermal strains (i.e., see Chapters 1 and 3), no attempt will be made here to explicitly write down the buckling equations for this situation.

In a cylindrical coordinate system $(r, \theta, z)$, with the polar coordinates $(r, \theta)$ describing the middle surface of the plate, the expression in (6.2) for the hygrothermal strain remains, essentially, invariant except that

§H and §T are now functions of $(r, \theta, z)$, i.e.,

$$e_{HT}^* = \beta \delta H(r, \theta, z) + \alpha \delta T(r, \theta, z) \qquad (6.40)$$

In lieu of (6.1) we have

$$\begin{cases} \tilde{e}_{rr} = e_{rr} - e_{HT}^* \\ \tilde{e}_{\theta\theta} = e_{\theta\theta} - e_{HT}^* \\ \tilde{\gamma}_{r\theta} = \gamma_{r\theta} \end{cases} \qquad (6.41)$$

where $e_{rr}, e_{\theta\theta}$, and $\gamma_{r\theta}$ are given in terms of the displacement components $u_r, u_\theta$ in the middle surface of the plate and the out-of-plane displacement $w = w(r, \theta)$ by the relations in (1.71), i.e.,

$$\begin{cases} e_{rr} = \dfrac{\partial u_r}{\partial r} + \dfrac{1}{2}\left(\dfrac{\partial w}{\partial r}\right)^2 - \zeta \dfrac{\partial^2 w}{\partial r^2} \\[2mm] e_{\theta\theta} = \dfrac{u_r}{r} + \dfrac{1}{r}\dfrac{\partial u_\theta}{\partial \theta} + \dfrac{1}{r^2}\left(\dfrac{\partial w}{\partial \theta}\right)^2 - \zeta\left(\dfrac{1}{r}\dfrac{\partial w}{\partial r} + \dfrac{1}{r^2}\dfrac{\partial^2 w}{\partial \theta^2}\right) \\[2mm] \gamma_{r\theta} = \dfrac{\partial u_\theta}{\partial r} - \dfrac{u_\theta}{r} + \dfrac{1}{r}\left(\dfrac{\partial u_r}{\partial \theta}\right) + \dfrac{1}{r}\left(\dfrac{\partial w}{\partial r}\right)\left(\dfrac{\partial w}{\partial \theta}\right) - 2\zeta\left(\dfrac{1}{r}\dfrac{\partial^2 w}{\partial r \partial \theta} - \dfrac{1}{r^2}\dfrac{\partial w}{\partial \theta}\right) \end{cases}$$

Middle surface strains $e_{rr}^o, e_{\theta\theta}^o, \gamma_{r\theta}^o$ are obtained from these relations by simply setting $\zeta = 0$.

The stress components $\sigma_{rr}, \sigma_{\theta\theta}$, and $\sigma_{r\theta}$ must satisfy the equilibrium equations delineated, e.g., in (1.72a, b); they are related to the stress components in rectangular Cartesian coordinates by the equations in (1.73), i.e.,

$$\sigma_{rr} = \sigma_{xx} \cos^2 \theta + \sigma_{yy} \sin^2 \theta + 2\sigma_{xy} \sin \theta \cos \theta$$
$$\sigma_{\theta\theta} = \sigma_{xx} \sin^2 \theta + \sigma_{yy} \cos^2 \theta - 2\sigma_{xy} \sin \theta \cos \theta$$
$$\sigma_{r\theta} = (\sigma_{yy} - \sigma_{xx}) \sin \theta \cos \theta + \sigma_{xy}(\cos^2 \theta - \sin^2 \theta)$$

Assuming that the thin plate exhibits linearly elastic response, we have, in lieu of (6.4), the constitutive relations

$$\begin{cases} e_{rr} - e_{HT}^* = \dfrac{1}{E}(\sigma_{rr} - \nu\sigma_{\theta\theta}) \\[2mm] e_{\theta\theta} - e_{HT}^* = \dfrac{1}{E}(\sigma_{\theta\theta} - \nu\sigma_{rr}) \\[2mm] e_{r\theta} = \left(\dfrac{1+\nu}{E}\right)\sigma_{r\theta} \end{cases} \qquad (6.42)$$

whose inverted form is

$$
\begin{cases}
\sigma_{rr} = \dfrac{E}{1-\nu^2}(e_{rr} + \nu e_{\theta\theta}) - \dfrac{E}{1-\nu}e^*_{HT} \\[2mm]
\sigma_{\theta\theta} = \dfrac{E}{1-\nu^2}(e_{\theta\theta} + \nu e_{rr}) - \dfrac{E}{1-\nu}e^*_{HT} \\[2mm]
\sigma_{r\theta} = (\dfrac{E}{1+\nu})e_{r\theta} \equiv G\gamma_{r\theta}
\end{cases}
\tag{6.43}
$$

The averaged stresses and bending moments in polar coordinates are defined, in the obvious way, as the natural counterparts of (6.6a, b), i.e.

$$
\begin{cases}
N_r = \displaystyle\int_{-h/2}^{h/2} \sigma_{rr}dz \\[3mm]
M_r = \displaystyle\int_{-h/2}^{h/2} \sigma_{rr}zdz
\end{cases}
\quad (etc.)
$$

So that, by virtue of (6.43)

$$
\begin{cases}
N_r = \dfrac{Eh}{1-\nu^2}(e^0_{rr} + \nu e^0_{\theta\theta}) - N^*_{HT} \\[2mm]
N_\theta = \dfrac{Eh}{1-\nu^2}(e^0_{\theta\theta} + \nu e^0_{rr}) - N^*_{HT} \\[2mm]
N_{r\theta} = 2Ghe^0_{r\theta}
\end{cases}
\tag{6.44}
$$

where $N^*_{HT} = \dfrac{E}{1-\nu}\displaystyle\int_{-h/2}^{h/2} e^*_{HT}dz$.

It is easily shown that the polar coordinate equivalent of the equilibrium equations (6.17a), i.e., (compare with (1.72a))

$$
\begin{cases}
N_{r,r} + \dfrac{1}{r}N_{r\theta,\theta} + \dfrac{1}{r}(N_r - N_\theta) = 0 \\[2mm]
N_{r\theta,r} + \dfrac{1}{r}N_{\theta,\theta} + \dfrac{2}{r}N_{r\theta} = 0
\end{cases}
\tag{6.45}
$$

is satisfied by introducing the Airy (stress) function $\Phi = \Phi(r,\theta)$ defined by

$$
\begin{cases}
N_r = \dfrac{1}{r}\Phi_{,r} + \dfrac{1}{r^2}\Phi_{,\theta\theta} \\[2mm]
N_\theta = \Phi_{,rr} \\[2mm]
N_{r\theta} = \dfrac{1}{r^2}\Phi_{,\theta} - \dfrac{1}{r}\Phi_{,r\theta}
\end{cases}
\tag{6.46}
$$

which is, of course, identical with (1.74). For the isotropic situation under consideration, the generalized von Karman equations (in polar coordinates) can now be derived in a direct manner by noting that the differential operators present in the system (6.21), (6.24) are invariant with respect to linear transformations of the coordinate system; in particular we have, in lieu of (6.21), (6.24), the system

$$\begin{cases} K\Delta^2 w(r,\theta) = [\Phi(r,\theta), w(r,\theta)] - \Delta M^*_{HT} \\ \Delta^2 \Phi(r,\theta) \quad = -\dfrac{1}{2} Eh[w(r,\theta), w(r,\theta)] - (1-\nu)\Delta N^*_{HT} \end{cases} \tag{6.47}$$

where

$$\begin{cases} N^*_{HT} = \dfrac{E}{1-\nu} \displaystyle\int_{-h/2}^{h/2} e^*_{HT}\,dz \\ \\ M^*_{HT} = \dfrac{E}{1-\nu} \displaystyle\int_{-h/2}^{h/2} e^*_{HT} z\,dz \end{cases} \tag{6.48}$$

while $\Delta^2 w$ is given by (1.75a), i.e.,

$$\Delta^2 w = w_{,rrrr} + \frac{2}{r} w_{,rrr} - \frac{1}{r^2} w_{,rr}$$
$$+ \frac{2}{r^2} w_{,rr\theta\theta} + \frac{1}{r^3} w_{,r} - \frac{2}{r^3} w_{,r\theta\theta}$$
$$+ \frac{1}{r^4} w_{,\theta\theta\theta\theta} + \frac{4}{r^4} w_{,\theta\theta},$$

with an analogous expression for $\Delta^2 \Phi$, while

$$[\Phi, w] = w_{,rr}\left(\frac{1}{r}\Phi_{,r} + \frac{1}{r^2}\Phi_{,\theta\theta}\right)$$
$$+ \left(\frac{1}{r} w_{,r} + \frac{1}{r^2} w_{,\theta\theta}\right)\Phi_{,rr}$$
$$- 2\left(\frac{1}{r} w_{,r\theta} - \frac{1}{r^2} w_{,\theta}\right)\left(\frac{1}{r}\Phi_{,r\theta} - \frac{1}{r^2}\Phi_{,\theta}\right),$$
$$= N_r w_{,rr} - 2N_{r\theta}\left(\frac{1}{r^2} w_{,\theta} - \frac{1}{r} w_{,r\theta}\right)$$
$$+ N_\theta\left(\frac{1}{r} w_{,r} + \frac{1}{r^2} w_{,\theta\theta}\right)$$
$$[w, w] = 2\{w_{,rr}\left(\frac{1}{r} w_{,r} + \frac{1}{r^2} w_{,\theta\theta}\right)$$
$$- \left(\frac{1}{r} w_{,r\theta} - \frac{1}{r^2} w_{,\theta}\right)^2\}$$

and

$$\triangle M_{HT}^*(r,\theta) = (M_{HT}^*)_{,rr} + \frac{1}{r}(M_{HT}^*)_{,r}$$
$$+ \frac{1}{r^2}(M_{HT}^*)_{,\theta\theta}$$

(6.49)

with an analogous expression for $\triangle N_{HT}^*(r,\theta)$. If $\delta H_{,\theta} = \delta T_{,\theta} = 0$ so that

$$e_{HT}^*(r,z) = \beta\delta H(r,z) + \alpha\delta T(r,z),$$

(6.50)

and we consider only radial symmetric deformations of the plate, then the generalized von Karman system for the case of linearly elastic isotropic response reduces to the pair of equations

$$K[w'''' + \frac{2}{r}w''' - \frac{1}{r}w']$$
$$= N_r w'' + \frac{N_\theta}{r}w' - \{M_{HT}^{*''} + \frac{1}{r}M_{HT}^{*'}\},$$

(6.51a)

where $' = \dfrac{d}{dr}$, $N_r = \dfrac{1}{r}\Phi'$, $N_\theta = \Phi''$ $M_{HT}^*(r) = \dfrac{E}{1-\nu}\displaystyle\int_{-h/2}^{h/2} e_{HT}^*(r,z)z\,dz$,

and

$$\frac{1}{Eh}[\Phi'''' + \frac{2}{r}\Phi''' - \frac{1}{r^2}\Phi'' + \frac{1}{r^3}\Phi']$$
$$= -\frac{1}{r}w'w'' - \frac{(1-\nu)}{Eh}\{N_{HT}^{*''} + \frac{1}{r}N_{HT}^{*'}\}$$

(6.51b)

Both the systems (6.47) and (6.51a, b) neglect the effects of initial (pre-buckling) deflections and an applied transverse loading; special cases of these systems, which have appeared in the literature in connection with problems associated with the thermal bucking of isotropic linearly elastic circular plates, will be analyzed later in this report.

For a linearly elastic orthotropic body exhibiting cylindrical orthotropy there exist three planes of elastic symmetry; one of these is normal to the plane of anisotropy, the second passes through that axis, and the third is orthogonal to the first two. For a thin plate, the first plane of elastic symmetry, for a cylindrically orthotropic material, is chosen parallel to the middle plane of the plate; with this convention, the constitutive equations generalize those in (6.42) for an isotropic plate (in polar coordinates) and assume the form

$$
\begin{cases}
e_{rr} - e_{HT}^r = \dfrac{1}{E_r}\sigma_{rr} - \dfrac{\nu_\theta}{E_\theta}\sigma_{\theta\theta} \\[2ex]
e_{\theta\theta} - e_{HT}^\theta = -\dfrac{\nu_r}{E_r}\sigma_{rr} + \dfrac{1}{E_\theta}\sigma_{\theta\theta} \\[2ex]
\gamma_{r\theta} = \dfrac{1}{G_{r\theta}}\sigma_{r\theta}
\end{cases}
\tag{6.52}
$$

where the radial and angular hygrothermal strains $e_{HT}^r$ and $e_{HT}^\theta$ are defined, respectively, by

$$
\begin{cases}
e_{HT}^r = \beta_r \delta H(r,\theta,z) + \alpha_r \delta T(r,\theta,z) \\[1ex]
e_{HT}^\theta = \beta_\theta \delta H(r,\theta,z) + \alpha_\theta \delta T(r,\theta,z)
\end{cases}
\tag{6.53}
$$

with $\alpha_r, \alpha_\theta$ the coefficients of thermal expansion in the radial and angular directions, respectively, and $\beta_r, \beta_\theta$ the coefficients of hygroexpansion in the radial and angular directions, respectively. Also, in (6.52), $E_r$ and $E_\theta$ are the Young's moduli for tension (or compression) in the radial and tangential directions, respectively, while $\nu_r$ and $\nu_\theta$ are the corresponding principal Poisson's ratios and $G_{r\theta}$ is the shear modulus which characterizes the change of angle between the radial and angular directions. We note that for a cylindrically (or polar) orthotropic body we have $E_r \nu_\theta = E_\theta \nu_r$ so that the constitutive equations (6.52) may be rewritten in the form

$$
\begin{cases}
e_{rr} = \dfrac{1}{E_r}(\sigma_{rr} - \nu_r \sigma_{\theta\theta}) + e_{HT}^r \\[2ex]
e_{\theta\theta} = \dfrac{1}{E_\theta}(\sigma_{\theta\theta} - \nu_\theta \sigma_{rr}) + e_{HT}^\theta \\[2ex]
\gamma_{r\theta} = \dfrac{1}{G_{r\theta}}\sigma_{r\theta}
\end{cases}
\tag{6.54}
$$

For the (degenerate) case of isotropic symmetry, $E_r = E_\theta = E, \nu_r = \nu_\theta = \nu, \alpha_r = \alpha_\theta = \alpha$, and $\beta_r = \beta_\theta = \beta$ in which case $e_{HT}^r = e_{HT}^\theta = e_{HT}^*$, as given by (6.40), and the constitutive relations (6.54) redue to those in (6.42) as $G_{r\theta} = G$ for isotropy. The strains $e_{rr}, e_{\theta\theta}$, and $e_{r\theta} = \dfrac{1}{2}\gamma_{r\theta}$ are still given by the relations following (6.41) or, in terms of the middle surface strains, by

$$
\begin{cases}
e_{rr} = e^o_{rr} - \zeta w_{,rr} \\
e_{\theta\theta} = e^o_{\theta\theta} - \zeta(\dfrac{1}{r}w_{,r} + \dfrac{1}{r^2}w_{,\theta\theta}) \\
\gamma_{r\theta} = \gamma^o_{r\theta} - 2\zeta(\dfrac{1}{r}w_{,r\theta} - \dfrac{1}{r^2}w_{,\theta})
\end{cases}
\tag{6.55}
$$

where

$$
\begin{cases}
e^o_{rr} = u_{r,r} + \dfrac{1}{2}(w_{,r})^2 \\
e^o_{\theta\theta} = \dfrac{1}{r}u_r + \dfrac{1}{r}u_{\theta,\theta} + \dfrac{1}{2r^2}(w_{,\theta})^2 \\
\gamma^o_{r\theta} = u_{\theta,r} - \dfrac{1}{r}u_\theta + \dfrac{1}{r}u_{r,\theta} + \dfrac{1}{r}w_{,r}w_{,\theta}
\end{cases}
\tag{6.56}
$$

Inverting the relations (6.54) we obtain

$$
\sigma_{rr} = \frac{E_r}{1 - \nu_r \nu_\theta} e_{rr} + \frac{\nu_r E_\theta}{1 - \nu_r \nu_\theta} e_{\theta\theta}
$$
$$
- \frac{1}{1 - \nu_r \nu_\theta}\{E_r e^r_{HT} + \nu_r E_\theta e^\theta_{HT}\}
\tag{6.57a}
$$

$$
\sigma_{\theta\theta} = \frac{\nu_\theta E_r}{1 - \nu_r \nu_\theta} e_{rr} + \frac{E_\theta}{1 - \nu_r \nu_\theta} e_{\theta\theta}
$$
$$
- \frac{1}{1 - \nu_r \nu_\theta}\{\nu_\theta E_r e^r_{HT} + E_\theta e^\theta_{HT}\}
\tag{6.57b}
$$

$$
\sigma_{r\theta} = G_{r\theta}\gamma_{r\theta}
\tag{6.57c}
$$

For the cylindrically orthotropic case, the averaged stresses and bending moments are still defined by the relations following (6.43); thus, by (6.57a, b, c)

$$
\begin{cases}
N_r = \dfrac{E_r h}{1 - \nu_r \nu_\theta} e^o_{rr} + \dfrac{\nu_r E_\theta h}{1 - \nu_r \nu_\theta} e^o_{\theta\theta} - N^r_{HT} \\
N_\theta = \dfrac{\nu_\theta E_r h}{1 - \nu_r \nu_\theta} e^o_{rr} + \dfrac{E_\theta h}{1 - \nu_r \nu_\theta} e^o_{\theta\theta} - N^\theta_{HT} \\
N_{r\theta} = 2G_{r\theta} h e^o_{r\theta}
\end{cases}
\tag{6.58}
$$

where

$$
\begin{cases}
N_{HT}^r = \dfrac{E_r}{1-\nu_r\nu_\theta} \displaystyle\int_{-h/2}^{h/2} e_{HT}^r \, dz + \dfrac{\nu_r E_\theta}{1-\nu_r\nu_\theta} \displaystyle\int_{-h/2}^{h/2} e_{HT}^\theta \, dz \\[3mm]
N_{HT}^\theta = \dfrac{\nu_\theta E_r}{1-\nu_r\nu_\theta} \displaystyle\int_{-h/2}^{h/2} e_{HT}^r \, dz + \dfrac{E_\theta}{1-\nu_r\nu_\theta} \displaystyle\int_{-h/2}^{h/2} e_{HT}^\theta \, dz
\end{cases}
\tag{6.59}
$$

If we denote by $D_r$ and $D_\theta$, respectively, the bending stiffnesses around axes in the $r$ and $\theta$ directions passing through a given point in the plate, and by $\tilde{D}_{r\theta}$ the twisting rigidity, as given by (1.95), (1.96), i.e.,

$$
\begin{cases}
D_r = E_r h^3/12(1-\nu_r\nu_\theta) \\
D_\theta = E_\theta h^3/12(1-D_r\nu_\theta) \\
\tilde{D}_{r\theta} = G_{r\theta} h^3/12
\end{cases}
\tag{6.60}
$$

while $D_{r\theta}$ is given by (1.97), which we repeat here as

$$
D_{r\theta} = D_r\nu_\theta + 2\tilde{D}_{r\theta}
\tag{6.61}
$$

then

$$
\begin{cases}
M_r = -D_r[w_{,rr} + \nu_\theta(\dfrac{1}{r}w_{,r} + \dfrac{1}{r^2}w_{,\theta\theta}) - M_{HT}^r \\[3mm]
M_\theta = -D_\theta[\nu_r w_{,rr} + (\dfrac{1}{r}w_{,r} + \dfrac{1}{r^2}w_{,\theta\theta})] - M_{HT}^\theta \\[3mm]
M_{r\theta} = -2\tilde{D}_{r\theta}(\dfrac{w}{r})_{,r\theta}
\end{cases}
\tag{6.62}
$$

with

$$
\begin{cases}
M_{HT}^r = \dfrac{E_r}{1-\nu_r\nu_\theta} \displaystyle\int_{-h/2}^{h/2} e_{HT}^r z \, dz + \dfrac{\nu_r E_\theta}{1-\nu_r\nu_\theta} \displaystyle\int_{-h/2}^{h/2} e_{HT}^\theta z \, dz \\[3mm]
M_{HT}^\theta = \dfrac{\nu_\theta E_r}{1-\nu_r\nu_\theta} \displaystyle\int_{-h/2}^{h/2} e_{HT}^r z \, dz + \dfrac{E_\theta}{1-\nu_r\nu_\theta} \displaystyle\int_{-h/2}^{h/2} e_{HT}^\theta z \, dz
\end{cases}
\tag{6.63}
$$

The expressions in (6.58) for the averaged stresses and (6.62), for the bending moments generalize, for the case of cylindrical (polar) orthotropic behavior, the corresponding relations (1.93) and (1.94) for the non-hygrothermal case.

The first of the generalized von Karman equations for a plate possessing cylindrically orthotropic symmetry is obtained by substituting the expressions in (6.62), (6.63), for the bending moments, into the polar coordinate equivalent form of (6.18), namely,

$$\frac{1}{r}(rM_r)_{,rr} + \frac{1}{r^2}M_{\theta,\theta\theta} - \frac{1}{r}M_{\theta,r}$$

$$+\frac{1}{r}M_{r\theta,r\theta} + N_r w_{,rr}$$

$$+N_\theta(\frac{1}{r}w_{,r} + \frac{1}{r^2}w_{,\theta\theta}) + 2N_{r\theta}(\frac{w}{r})_{,r\theta} = 0 \qquad (6.64)$$

where the stress resultants $N_r$, $N_\theta$, and $N_{r\theta}$ are, again, given by (6.46) in terms of the Airy stress function $\Phi(r,\theta)$. Equation (6.64) is identical with (1.106) of Chapter 1. Carrying out the process indicated above, we obtain

$$D_r w_{,rrrr} + 2D_{r\theta}\frac{1}{r^2}w_{,rr\theta\theta} + D_\theta\frac{1}{r^4}w_{,\theta\theta\theta\theta}$$

$$+2D_r\frac{1}{r}w_{,rrr} - 2D_{r\theta}\frac{1}{r^3}w_{,r\theta\theta} - D_\theta\frac{1}{r^2}w_{,rr} \qquad (6.65)$$

$$+2(D_\theta + D_{r\theta})\frac{1}{r^4}w_{,\theta\theta} + D_\theta\frac{1}{r^3}w_{,r}$$

$$= (\frac{1}{r}\Phi_{,r} + \frac{1}{r^2}\Phi_{,\theta\theta})w_{,rr}$$

$$+\Phi_{,rr}(\frac{1}{r}w_{,r} + \frac{1}{r^2}w_{,\theta\theta})$$

$$+2(\frac{1}{r^2}\Phi_{,\theta} - \frac{1}{r}\Phi_{,r\theta})(\frac{1}{r}w_{,r\theta} - \frac{1}{r^2}w_{,\theta})$$

$$+\frac{1}{r}(rM_{HT}^r)_{,rr} + \frac{1}{r^2}(M_{HT}^\theta)_{,\theta\theta} - \frac{1}{r}(M_{HT}^\theta)_{,r}$$

To obtain the polar coordinate equivalent form of the strain compatibility relation without transforming this relation directly, we proceed as follows: with respect to the linearized strains

$$\begin{cases} e_{rr}^\ell = u_{r,r} \\ e_{\theta\theta}^\ell = \frac{1}{r}u_r + \frac{1}{r}u_{\theta,\theta} \\ \gamma_{r\theta}^\ell = u_{\theta,r} + \frac{1}{r}u_{r,\theta} - \frac{1}{r}u_\theta \end{cases} \qquad (6.66)$$

it is easy to check that

$$(r\gamma_{r\theta,\theta}^\ell)_{,r} - e_{rr,\theta\theta}^\ell - (r^2 e_{\theta\theta,r}^\ell)_{,r} + re_{rr,r}^\ell = 0 \qquad (6.67)$$

Moreover, in view of (6.56),

$$
\begin{cases}
e^o_{rr} = e^\ell_{rr} + \dfrac{1}{2}(w_{,r})^2 \\[2mm]
e^o_{\theta\theta} = e^\ell_{\theta\theta} + \dfrac{1}{2r^2}(w_{,\theta})^2 \\[2mm]
\gamma^o_{r\theta} = \gamma^\ell_{r\theta} + \dfrac{1}{r}w_{,r}w_{,\theta}
\end{cases}
\tag{6.68}
$$

Thus, strain compatibility, when written in terms of the middle surface strains, requires that

$$
(r\gamma^o_{r\theta,\theta})_{,r} - e^o_{rr,\theta\theta} - (r^2 e^o_{\theta\theta,r})_{,r} + r e^o_{rr,r}
$$

$$
= [r(\tfrac{1}{r}w_{,r}w_{,\theta})_{,\theta}]_{,r} - [\tfrac{1}{2}(w_{,r})^2]_{,r}
\tag{6.69}
$$

$$
- [r^2(\tfrac{1}{2r^2}(w_{,\theta})^2_{,r}]_{,r} + r(\tfrac{1}{2}(w_{,r})^2)_{,r}
$$

Expanding the right-hand side of (6.69), and simplifying, we obtain

$$
(r\gamma^o_{r\theta,\theta})_{,r} - e^o_{rr,\theta\theta} - (r^2 e^o_{\theta\theta,r})_{,r} + r e^o_{rr,r}
$$

$$
= w_{,rr}(rw_{,r} + w_{,\theta\theta}) - (w_{,r\theta} - \tfrac{1}{r}w_{,\theta})^2
\tag{6.70}
$$

If we invert the relations in (6.44) we find that

$$
\begin{cases}
e^o_{rr} = \dfrac{1}{E_r h}\{(N_r + N^r_{HT}) - \nu_r(N_\theta + N^\theta_{HT})\} \\[3mm]
e^o_{\theta\theta} = \dfrac{1}{E_\theta h}\{(N_\theta + N^\theta_{HT}) - \nu_\theta(N_r + N^r_{HT})\} \\[3mm]
\gamma^o_{r\theta} = \dfrac{1}{G_{r\theta} h}N_{r\theta}
\end{cases}
\tag{6.71}
$$

Computing, in succession, therefore, the expressions on the left-hand side of the compatibility relation (6.70), we now obtain, through the use of (6.71)

$$
(r\gamma^o_{r\theta,\theta})_{,r} = \dfrac{1}{G_{r\theta} h}\{N_{r\theta,\theta} + rN_{r\theta,r\theta}\}
\tag{6.72a}
$$

$$
e^o_{rr,\theta\theta} = \dfrac{1}{E_r h}\{(N_{r,\theta\theta} + N^r_{HT,\theta\theta}) - \nu_r(N_{\theta,\theta\theta} + N^\theta_{HT,\theta\theta})\}
\tag{6.72b}
$$

$$(r^2 e^o_{\theta\theta,r}),_r = \frac{r^2}{E_\theta h} \{N_\theta + N^\theta_{HT}),_{rr} - \nu_\theta(N_r + N^r_{HT}),_{rr}\}$$

$$+ \frac{2r}{E_\theta h} \{(N_\theta + N^\theta_{HT}),_r - \nu_\theta(n_r + N^r_{HT}),_r\}$$

(6.72c)

and

$$re^o_{rr,r} = \frac{r}{E_r h} \{(N_r + N^r_{HT}),_r - \nu_r(N_\theta + N^\theta_{HT}),_r\} \qquad (6.72d)$$

Finally, by substituting (6.72a, b, c, d) into (6.70), and simplifying, we obtain

$$\frac{1}{E_r h}[(N_r - \nu_r N_\theta),_{\theta\theta} + r(N_{r,r} - \nu_r N_{\theta,r})]$$

$$+ \frac{1}{2G_{r\theta}h}(N_{r\theta,\theta} - rN_{r\theta,r\theta})$$

$$+ \frac{1}{E_\theta h}[(r^2(N_{\theta,rr} - \nu_\theta N_{r,rr}) + 2r(N_{\theta,r} - \nu_\theta N_{r,r})]$$

$$= w,_{rr}(rw,_r + w,_{\theta\theta}) - (w,_{r\theta} - \frac{1}{r}w,_\theta)^2 \qquad (6.73)$$

$$- \frac{1}{E_r h}[(N^r_{HT} - \nu_r N^\theta_{HT}),_{\theta\theta} + r(N^r_{HT,,r} - \nu_r N^\theta_{HT,r})]$$

$$- \frac{1}{E_\theta h}[r^2(N^\theta_{HT,rr} - \nu_\theta N^r_{HT,rr})$$

$$+ 2r(N^\theta_{HT,r\theta} - \nu_\theta N^r_{HT,r})]$$

To rewrite (6.73) in terms of the Airy function $\Phi(r,\theta)$, we substitute for $N_r, N_\theta$, and $N_{r\theta}$ in (6.73) from (6.46), multiply the resulting equation through by $-h/r^2$, and simplify; there results the following equation:

$$\frac{1}{E_\theta}\Phi_{,rrrr} + (\frac{1}{G_{r\theta}} - \frac{2\nu_r}{E_r})\frac{1}{r^2}\Phi_{,rr\theta\theta}$$

$$+\frac{1}{E_r}\frac{1}{r^4}\Phi_{,\theta\theta\theta\theta} + \frac{2}{E_\theta}\frac{1}{r}\Phi_{,rrr}$$

$$-(\frac{1}{G_{r\theta}} - \frac{2\nu_r}{E_r})\frac{1}{r^3}\Phi_{,r\theta\theta} - \frac{1}{E_r}\frac{1}{r^2}\Phi_{,rr}$$

$$+(2\frac{1-\nu_r}{E_r} + \frac{1}{G_{r\theta}})\frac{1}{r^4}\Phi_{,\theta\theta} + \frac{1}{E_r}\frac{1}{r^3}\Phi_{,r}$$

$$= -h[w_{,rr}(\frac{1}{r}w_{,r} + \frac{1}{r^2}w_{,\theta\theta})$$

$$-(\frac{1}{r}w_{,r\theta} - \frac{1}{r^2}w_{,\theta})^2]$$

$$+\frac{1}{E_r}[\frac{1}{r^2}(N_{HT}^r - \nu_r N_{HT}^\theta)_{,\theta\theta} + \frac{1}{r}(N_{HT,r}^r - \nu_r N_{HT,r}^\theta)]$$

$$+\frac{1}{E_\theta}[(N_{HT,rr}^\theta - \nu_\theta N_{HT,rr}^r) + \frac{2}{r}(N_{HT,r\theta}^\theta - \nu_\theta N_{HT,r}^\theta)]$$

(6.74)

Therefore, the generalized von Karman equations governing the bending and buckling of (hygrothermal) cylindrically orthotropic, linearly elastic plates consist of (6.65) and (6.74); in the absence of hygrothermal strains these equations reduce to (1.98), (1.99). In (6.65), (6.74) $M_{HT}^r, M_{HT}^\theta$ are given by (6.63) and $N_{HT}^r, N_{HT}^\theta$ by (6.59) where $e_{HT}^r, e_{HT}^\theta$ are defined by (6.53). For a radially symmetric problem all derivatives with respect to $\theta$ in both (6.65) and (6.74) would be deleted.

---

## 6.3 Boundary Conditions

The discussion of boundary conditions for thin elastic plates undergoing hygrothermal expansion or contraction parallels that for edge-loaded plates in Chapter 1; the important difference is that the expressions for the (resultant) forces and moments along the edge of a plate now include the contributions induced by stresses which result directly from hygrothermal strains. As in Chapter 1, if the thin plate occupies a region $\Omega$ in the $x, y$ plane with a smooth (or piecewise smooth) boundary $\partial\Omega$, we let $\vec{n}$ denote the unit normal to the boundary at any arbitrary point on the boundary, while $\vec{t}$ denotes the unit tangent vector to the boundary at that point. The normal derivative of a function $f$ on $\partial\Omega$ is denoted by $\dfrac{\partial f}{\partial n} equiv f_{,n}$ while the tangential derivative is given by $\dfrac{\partial f}{\partial s}$, $s$ being a measure of arc length along the boundary. Thus, e.g., if $\Omega$ is a disk centered at $(0,0)$ of radius $R > 0$ and $f = f(r\theta)$ is defined on $\Omega$ and is of class $C^1(\Omega)$ with first derivative continuous up to $\partial\Omega$, then $\dfrac{\partial f}{\partial n} = f_{,r}$ while $\dfrac{\partial f}{\partial s} = \dfrac{1}{r} f_{,\theta}$. As noted in Chapter 1, the three most prevalent types of boundary conditions in the buckling literature are those which correspond to clamped edges, simply supported edges, and free edges; regardless of whether we are considering plates with isotropic or orthotropic symmetry (either cylindrical or rectilinear), the basic forms assumed by these various sets of boundary conditions are still the same as those delineated in (1.129a, b, c), i.e.,

(i) $\partial\Omega$ is clamped: $w = 0$ and $\dfrac{\partial w}{\partial n} = 0$, on $\partial\Omega$

(ii) $\partial\Omega$ is simply supported: $w = 0$ and $M_n = 0$, on $\partial\Omega$

(iii) $\partial\Omega$ is free: $M_n = 0$ and $Q_n + \dfrac{\partial M_{tn}}{\partial s} = 0$, on $\partial\Omega$

where $M_n$ is the bending moment on $\partial\Omega$ in the direction normal to $\partial\Omega$, $M_{tn}$ is the twisting moment on $\partial\Omega$, with respect to the tangential and normal directions on $\partial\Omega$, and $Q_n$ is the shearing force associated with the direction normal to $\partial\Omega$.

For the work to be considered in this report only rectangular and circular (or annular) domains $\Omega$ will be covered. If $\partial\Omega$ is clamped, therefore, or if, as in the case of a rectangular plate, one or more edges are clamped, the pertinent boundary conditions will be exactly the same as for the non-hygrothermal case, i.e., for a rectangle of width $a$ and length $b$, exhibiting isotropic response, the clamped boundary conditions are expressed by (1.130a, b). For a circular plate (or annular plate) exhibiting isotropic response the relevant conditions are those in (1.139). Conditions (1.130a, b) apply equally well to a rectangular plate exhibiting rectilinear orthotropic response when all the edges are clamped, while (1.139) still applies for circular (or annular) plates with clamped edge(s) when the plate exhibits cylindrically orthotropic behavior.

We now consider thin elastic plates subject to hygrothermal expansion or contraction which have one or more edges simply supported. For a rectangular plate of width $a$ and length $b$ the condition $M_n = 0$ on $\partial\Omega$

becomes

$$\begin{cases} M_x = 0, \ x = 0, x = a; \ 0 \le y \le b \\ M_y = 0, \ y = 0, y = b; \ 0 \le x \le a \end{cases} \tag{6.75}$$

Thus, by virtue of (6.12),

$$K(w_{,xx} + \nu w_{,yy}) + M_{HT} = 0,$$
$$x = 0, x = a; 0 \le y \le b \tag{6.76a}$$

and

$$K(w_{,yy} + \nu w_{,xx}) + M_{HT} = 0$$
$$y = 0, y = b; 0 \le x \le a, \tag{6.76b}$$

where

$$\begin{cases} M_{HT} = \dfrac{E}{1-\nu} \displaystyle\int_{-h/2}^{h/2} \epsilon_{HT} z \, dz \\ \epsilon_{HT} = \beta \delta H(x, y, z) + \alpha \delta T(x, y, z) \end{cases} \tag{6.77}$$

For the same rectangular elastic plate, now assumed to exhibit rectilinear orthotropic symmetry, the conditions in (6.75) take, as a direct
consequence of (6.35) and (6.37), the following form:

$$(D_{11} w_{,xx} + D_{12} w_{,yy}) + \tilde{M}_{HT}^1 = 0,$$
$$x = 0, x = a; 0 \le y \le b \tag{6.78a}$$

$$(D_{21} w_{,xx} + D_{22} w_{,yy}) + \tilde{M}_{HT}^2 = 0,$$
$$y = 0, y = b; 0 \le x \le a \tag{6.78b}$$

where $D_{ij} = c_{ij} h^3/12$, the constitutive constants $c_{ij}$ are given by (1.61)
and

$$\tilde{M}_{HT}^1 = c_{11} \int_{-h/2}^{h/2} \epsilon_{HT}^1(x, y, z) z \, dz$$
$$+ c_{12} \int_{-h/2}^{h/2} \epsilon_{HT}^2(x, y, z) z \, dz \tag{6.79a}$$

$$\tilde{M}_{HT}^2 = c_{21} \int_{-h/2}^{h/2} \epsilon_{HT}^1(x, y, z) z \, dz$$
$$+ c_{22} \int_{-h/2}^{h/2} \epsilon_{HT}^2(x, y, z) z \, dz \tag{6.79b}$$

with, as per (6.26),

$$\begin{cases} \epsilon_{HT}^1 = \beta_1 \delta H(x, y, z) + \alpha_1 \delta T(x, y, z) \\ \epsilon_{HT}^2 = \beta_2 \delta H(x, y, z) + \alpha_2 \delta T(x, y, z) \end{cases}$$

For an annular elastic plate with (circular) boundaries at $r = R_i, i = 1, 2, R_1 = a, R_2 = b > a$, exhibiting isotropic material symmetry, the condition $M_n = 0$ on $\partial\Omega$ translates into

$$M_r = K[w_{,rr} + \nu(\frac{1}{r^2}w_{,\theta\theta} + \frac{1}{r}w_{,r})]$$
$$+ M_{HT}^* = 0, r = R_i, i = 1, 2.$$
(6.80)

where

$$\begin{cases} M_{HT}^* = \dfrac{E}{1 - \nu} \displaystyle\int_{-h/2}^{h/2} e_{HT}^* z\, dz \\ e_{HT}^* = \beta \delta H(r, \theta, z) + \alpha \delta T(r, \theta, z) \end{cases}$$
(6.81)

On the other hand, for a circular (or annular) plate with edge(s) at $r = R_i, i = 1, 2$, which exhibits cylindrically orthotropic symmetry, we have as the expression of $M_n = 0$ on $\partial\Omega$ the condition

$$D_r[w_{,rr} + \nu_\theta(\frac{1}{r}w_{,r} + \frac{1}{r^2}w_{,\theta\theta})] + M_{HT}^r = 0,$$
$$r = R_i, i = 1, 2$$
(6.82)

where, by virtue of (6.63) and (6.53),

$$\begin{cases} M_{HT}^r = \dfrac{E_r}{1 - \nu_r \nu_\theta} \displaystyle\int_{-h/2}^{h/2} e_{HT}^r dz + \dfrac{\nu_r E_\theta}{1 - \nu_r \nu_\theta} \displaystyle\int_{-h/2}^{h/2} e_{HT}^\theta z\, dz \\ \begin{cases} e_{HT}^r = \beta_r \delta H(r, \theta, z) + \alpha_r \delta T(r, \theta, z) \\ e_{HT}^\theta = \beta_\theta \delta H(r, \theta, z) + \alpha_\theta \delta T(r, \theta, z) \end{cases} \end{cases}$$

and $D_r = E_r h^3 / 12(1 - \nu_r \nu_\theta)$.

Along any edge of a thin plate which is free, we must have $M_n = 0$ as well as $Q_n + \dfrac{\partial M_{tn}}{\partial s} = 0$. Conditions equivalent to $M_n = 0$ along a portion of $\partial\Omega$ (or all of $\partial\Omega$) for various cases of interest have been elucidated above. For rectangular plates of width $a$ and length $b$, it has been shown in Chapter 1 that the condition $Q_n + \dfrac{\partial M_{tn}}{\partial s} = 0$ on $\partial\Omega$ is equivalent to the following relations

$$M_{y,y} + 2M_{xy,x} = 0,$$
$$y = 0, y = b; 0 \leq x \leq a \tag{6.83a}$$

$$M_{x,x} + 2M_{yx,y} = 0,$$
$$x = 0, x = a; 0 \leq y \leq b \tag{6.83b}$$

If the rectangular plate exhibits isotropic response then the bending moments $M_x, M_y$, and $M_{xy}$ are given by (6.12), in which case, (6.83a,b) become

$$K[w_{,yyy} + (2-\nu)w_{,xxy}] + M_{HT,y} = 0$$
$$y = 0, y = b; 0 \leq x \leq a \tag{6.84a}$$

and

$$K[w_{,xxx} + (2-\nu)w_{,xyy}] + M_{HT,x} = 0$$
$$x = 0, x = a; 0 \leq y \leq b \tag{6.84b}$$

with $M_{HT} = \dfrac{E}{1-\nu} \displaystyle\int_{-h/2}^{h/2} \epsilon_{HT} z \, dz$ and $\epsilon_{HT} = \beta \delta H + \alpha \delta T$.

For the case in which the rectangular plate exhibits rectilinear orthotropic symmetry, (6.83a,b) still represent the conditions equivalent to $Q_n + \dfrac{\partial M_{en}}{\partial s} = 0$ along all four edges, but now, the bending moments $M_x, M_y$, and $M_{xy}$ are given by (6.35). An easy computation then shows that in lieu of (6.84a, b) for the isotropic case we have

$$D_{21}w_{,xxy} + D_{22}w_{,yyy} + 4D_{66}w_{,xxy} + \tilde{M}^2_{HT,y} = 0,$$
$$y = 0, b; 0 \leq x \leq a \tag{6.85a}$$

and

$$D_{11}w_{,xxx} + D_{12}w_{,xyy} + 4D_{66}w_{,xyy} + \tilde{M}^1_{HT,x} = 0,$$
$$x = 0, a; 0 \leq y \leq b \tag{6.85b}$$

where $\tilde{M}^1_{HT}, \tilde{M}^2_{HT}$ are defined by (6.34) and (6.37) with $\epsilon^1_{HT}, \epsilon^2_{HT}$ as given by (6.26).

To elucidate the free edge boundary conditions which apply with respect to thin elastic plates with a circular geometry, we note that (see

[217], §2.2 ) in the orthogonal $(s, n)$ coordinate system introduced along the boundary $\partial\Omega$ of a domain $\Omega$ in the $x, y$ plane

$$Q_n = \frac{\partial M_n}{\partial n} + 2\frac{\partial M_{tn}}{\partial s} \qquad (6.86a)$$

Thus, along a free edge we must have $M_n = 0$ as well as

$$\frac{\partial M_n}{\partial n} + 2\frac{\partial M_{tn}}{\partial s} = 0 \qquad (6.86b)$$

Along the edge at $r = R$ of a circular plate, the condition (6.86b) assumes the form

$$M_{r,r} + \frac{2}{r}M_{r\theta,\theta} = 0 \qquad (6.87)$$

For an annular plate with edges at $r = R_i, i = 1, 2 (R_1 = a, R_2 = b > a)$ which exhibits isotropic material symmetry, the condition corresponding to (6.87) becomes

$$K\{(w_{,rr} + \frac{1}{r}w_{,r} + \frac{1}{r^2}w_{,\theta\theta})_{,r}$$

$$+ (1 - \nu)(\frac{1}{r^2}w_{,\theta\theta r} - \frac{1}{r^3}w_{,\theta\theta})\} + M^*_{HT,r} = 0, r = R_i, i = 1, 2 \qquad (6.88)$$

where $M^*_{HT}(r, \theta)$ is given by (6.81). For the same annular plate, this time exhibiting cylindrically orthotropic behavior, we have as a consequence of (6.62)

$$D_r[w_{,rrr} + \nu_\theta(\frac{1}{r}w_{,r} + \frac{1}{r^2}w_{,\theta\theta})_{,r}]$$

$$+ \frac{4}{r}\tilde{D}_{r\theta}(\frac{w}{r})_{,r\theta\theta} + M^r_{HT,r} = 0, \qquad (6.89)$$

$$r = R_i, i = 1, 2$$

where $D_r = E_r h^3/12(1 - \nu_r\nu_\theta)$, $\tilde{D}_{r\theta} = G_{r\theta}h^3/12$, and $M^r_{HT}$ is given by (6.53) and (6.63). Of course, for both the isotropic and cylindrically orthotropic cases we must have $M_r = 0$ along $r = R_i, i = 1, 2$, with $M_r$ given by (6.80) in the isotropic case and by (6.62) in the cylindrically orthotropic case. If $R_1 = a = 0$ the annular case reduces to the case of a disk (circular plate of radius $R_2 = b$).

## 6.4 Thermal Bending and Buckling Equations and Boundary Conditions

Almost all of the literature on hygrothermal bending and buckling of plates has focused on purely thermal problems; in order to survey some of that literature, therefore, we will now specialize some of the equations and boundary conditions specified above for the mixed hygrothermal situation, to the thermal case only. We will also look at the reductions that occur in the thermal buckling and bending equations when special forms of the temperature distribution in the plate are considered or when we restrict ourselves to the small deflection case or ignore middle surface forces (acting in the plane of the plate); at this point we will also append to the first of the relevant von Karman equations, in each case, a distributed force $t = t(x, y)$ normal to the middle surface of the plate. In all the cases to be considered in this section an equivalent hygroexpansive (or hygrocontractive) problem results by replacing the thermal expansion coefficients by hygroscopic coefficients and the plate temperature distribution by an equivalent distribution of moisture in the plate.

We begin by specializing the equations and boundary conditions derived in rectilinear coordinates so as to cover the specific case of thermal bending and buckling. Thus, in (6.2), $\beta \equiv 0$ so that

$$\epsilon_{HT} \equiv \epsilon_T = \alpha \delta T(x, y, z) \tag{6.90}$$

where it is assumed that the thermal expansion coefficient $\alpha$ is constant. For isotropic response, the constitutive relations (6.9) then reduce to

$$\begin{cases} N_x &= \dfrac{Eh}{1-\nu^2}(\epsilon_{xx}^o + \nu\epsilon_{yy}^o) - N^T \\[2mm] N_y &= \dfrac{Eh}{1-\nu^2}(\epsilon_{yy}^o + \nu\epsilon_{xx}^o) - N^T \\[2mm] N_{xy} &= 2Gh\epsilon_{xy}^o \end{cases} \tag{6.91}$$

where $\epsilon_{xx}^o, \epsilon_{xy}^o, \epsilon_{yy}^o$ are the middle surface strains as defined by (6.7) and $N^T$ is given by (6.11a). The bending moments are given as

$$\begin{cases} M_x &= -K(w_{,xx} + \nu w_{,yy}) - M^T \\[2mm] M_y &= -K(w_{,yy} + \nu w_{,xx}) - M^T \\[2mm] M_{xy} &= -(1-\nu)Kw_{,xy} \end{cases} \tag{6.92}$$

with the thermal moment $M^T$ given by (6.14).

The generalized von Karman equations for the hygrothermal case now reduce to (see (6.21), (6.24)):

$$K\triangle^2 w = \Phi_{,yy}w_{,xx} - 2\Phi_{xy}w_{,xy} + \Phi_{,xx}w_{,yy} - \triangle M^T + t \quad (6.93a)$$

and

$$\triangle^2\Phi = -\frac{1}{2}Eh[w,w] - (1-\nu)\triangle N^T \quad (6.93b)$$

where $t = t(x,y)$ is the applied transverse force. In considering small deflection theory one ignores the bracket operator on the right-hand side of (6.93b), in which case, the Airy function, as given by (6.20), satisfies

$$\triangle^2\Phi = -(1-\nu)\triangle N^T \text{(small deflection theory)} \quad (\overline{6.93b})$$

In a purely (thermal) bending problem, middle surface forces do not come into play, in which case, the system (6.93a), reduces to

$$K\triangle^2 w = -\triangle M^T + t \text{ (thermal bending)}. \quad (\overline{6.93a})$$

In many places in the literature, a temperature distribution of the form

$$\delta T(x,y,z) = T_o(x,y) + zT_1(x,y) \quad (6.94)$$

has been considered; for this specific type of distribution it is easy to see that, as a direct consequence of (6.11a) and (6.14),

$$\begin{cases} N^T = \dfrac{\alpha E H}{1-\nu}T_o(x,y) \\ M^T = \dfrac{\alpha E}{1-\nu} \cdot \dfrac{h^3}{12}T_1(x,y) \end{cases} \quad (6.95)$$

For such a temperature distribution within the context of the small deflection equations, it will generally be the case that one is dealing with a thermal bending problem when $\triangle N^T = \triangle T_o = 0$ and a thermal buckling problem when $\triangle M^T = \triangle T_1 = 0$. The boundary conditions associated with the thermal bending and buckling of, say, a rectangular plate of width $a$ and length $b$ exhibiting isotropic material symmetry are as follows:

(i) If all the edges are clamped, then the conditions coincide with (1.130a, b).

(ii) If all the edges are simply supported, then $w = 0$ along each edge and, in addition,

$$\begin{cases} K(w_{,xx} + \nu w_{,yy}) + M^T = 0 \\ \quad\quad x = 0, a; 0 \leq y \leq b \\ K(w_{,yy} + \nu w_{,xx}) + M^T = 0 \\ \quad\quad y = 0, b; 0 \leq x \leq a \end{cases} \quad (6.96)$$

(iii) If the edges are all free, then the conditions in (6.96) hold as well as

$$\begin{cases} K[w_{,yyy} + (2 - \nu)w_{,xxy}] + M^T_{,y} = 0 \\ \quad\quad y = 0, b; 0 \leq x \leq a \\ K[w_{,xxx} + (2 - \nu)w_{,xyy}] + M^T_{,x} = 0 \\ \quad\quad x = 0, a; 0 \leq y \leq b \end{cases} \quad (6.97)$$

In (6.96), (6.97), $M^T$ is given by (6.14). Of course, cases where, e.g., one pair of (parallel) edges is simply supported, while the other pair of edges is free can be considered by combining the conditions in (i)–(iii), above.

For a plate exhibiting rectilinearly orthotropic material symmetry the thermal strains assume the form (see (6.26)):

$$\epsilon_T^1 = \alpha_1 \delta T(x, y, z), \epsilon_T^2 = \alpha_2 \delta T(x, y, z) \quad (6.98)$$

with the coefficients of linear thermal expansion along the $x$ and $y$ axes, $\alpha_1$ and $\alpha_2$, respectively, taken to be constants. The constitutive relations in this situation are given as follows (where $\epsilon_{xx}^o, \epsilon_{xy}^o, \epsilon_{yy}^o$ are, once again, the middle surface strains):

$$\begin{cases} N_x = \{\dfrac{E_1 h}{1 - \nu_{12}\nu_{21}}\}\epsilon_{xx}^o + \{\dfrac{E_2\nu_{21}H}{1 - \nu_{12}\nu_{21}}\}\epsilon_{yy}^o - \tilde{N}_T^1 \\[2mm] N_y = \{\dfrac{E_1\nu_{12}h}{1 - \nu_{12}\nu_{21}}\}\epsilon_{xx}^o + \{\dfrac{E_2 h}{1 - \nu_{12}\nu_{21}}\}\epsilon_{yy}^o - \tilde{N}_T^2 \\[2mm] N_{xy} = (2G_{12}h)\epsilon_{xy}^o \end{cases} \quad (6.99)$$

where (see (6.29)):

$$\begin{cases} \tilde{N}_T^1 = (\dfrac{E_1\alpha_1 + E_2\alpha_2\nu_{21}}{1 - \nu_{12}\nu_{21}}) \displaystyle\int_{-h/2}^{h/2} \delta T(x,y,z)dz \\[4mm] \tilde{N}_T^2 = (\dfrac{E_1\alpha_1\nu_{12} + E_2\alpha_2}{1 - \nu_{12}\nu_{21}}) \displaystyle\int_{-h/2}^{h/2} \delta T(x,y,z)dz \end{cases} \tag{6.100}$$

The expressions for the bending moments, in the thermal bending/buckling problem for a rectilinearly orthotropic elastic plate, assume the form (see (6.35)):

$$\begin{cases} M_x = -(D_{11}w_{,xx} + D_{12}w_{,yy}) - \tilde{M}_T^1 \\[2mm] M_y = -(D_{21}w_{,xx} + D_{22}w_{,yy}) - \tilde{M}_T^2 \\[2mm] M_{xy} = -2D_{66}w_{,xy} \end{cases} \tag{6.101}$$

where

$$\begin{cases} \tilde{M}_T^1 = (c_{11}\alpha_1 + c_{12}\alpha_2) \displaystyle\int_{-h/2}^{h/2} \delta T(x,y,z)zdz \\[4mm] \tilde{M}_T^2 = (c_{21}\alpha_1 + c_{22}\alpha_2) \displaystyle\int_{-h/2}^{h/2} \delta T(x,y,z)zdz \end{cases} \tag{6.102}$$

and the $c_{ij}$ are given by (1.61). The relevant von Karman equations for thermal bending/buckling become (see (6.36), (6.39)):

$$D_{11}w_{,xxxx} + \{D_{12} + 4D_{66} + D_{21}\}w_{,xxyy}$$

$$+D_{22}w_{,yyyy} = [\Phi, w] - \tilde{M}_{T,xx}^1 - \tilde{M}_{T,yy}^1 + t \tag{6.103a}$$

and

$$\begin{aligned} \frac{1}{E_1h}\Phi_{,yyyy} + \frac{1}{h}(\frac{1}{G_{12}} - \frac{2\nu_{12}}{E_2})\Phi_{,xxyy} \\ + \frac{1}{E_2h}\Phi_{,xxxx} = -\frac{1}{2}[w,w] \\ - \frac{1}{E_1h}(\tilde{N}_T^1 - \nu_{21}\tilde{N}_T^2)_{,yy} \\ - \frac{1}{E_2h}(\tilde{N}_T^2 - \nu_{12}\tilde{N}_T^1)_{,xx} \end{aligned} \tag{6.103b}$$

As for the boundary data in this case, with respect, e.g., to a rectangular plate of width $a$ and length $b$, we have the following:

(i) If all four edges are clamped, then conditions (1.130a, b) still apply.

(ii) If the four edges are simply supported, then $w = 0$ along each edge and, in addition,

$$
\begin{cases}
(D_{11}w_{,xx} + D_{12}w_{,yy}) + \tilde{M}_T^1 = 0, \\
\qquad x = 0, a; 0 \le y \le b \\[2mm]
(D_{21}w_{,xx} + D_{22}w_{,yy}) + \tilde{M}_T^2 = 0, \\
\qquad y = 0, b; 0 \le x \le a
\end{cases}
\qquad (6.104)
$$

where $\tilde{M}_T^1, \tilde{M}_T^2$ are given by (6.102).

(iii) If all four edges are free, then the conditions in (6.104) apply as well as

$$
\begin{cases}
D_{21}w_{,xxy} + D_{22}w_{,yyy} + 4D_{66}w_{,xxy} + \tilde{M}_{T,y}^2 = 0, \\
\qquad y = 0, b; 0 \le x \le a \\[2mm]
D_{11}w_{,xxx} + D_{12}w_{,xyy} + 4D_{66}w_{,xyy} = \tilde{M}_{T,x}^1 = 0 \\
\qquad x = 0, a; 0 \le y \le b
\end{cases}
\qquad (6.105)
$$

The same comments apply as in the isotropic case with regard to different types of boundary data holding along pairs of parallel edges of the plate; in a small deflection situation, the "bracket" term $-\frac{1}{2}[w, w]$ would be deleted from the right-hand side of (6.103b) while for purely thermal buckling equation (6.103b) would be deleted in its entirety and (6.103a) would be employed with $\Phi \equiv 0$.

When the temperature difference $\delta T$ varies linearly through the thickness of the plate, as in (4.94), the thermal stress resultants $\tilde{N}_T^i, i = 1, 2$ and the thermal moments $\tilde{M}_T^i, i = 1, 2$, as given by (6.100) and (6.102), respectively, reduce to

$$
\begin{cases}
\tilde{N}_T^1 = (\dfrac{E_1\alpha_1 + E_2\alpha_2\nu_{21}}{1 - \nu_{12}\nu_{21}})hT_o(x, y) \\[4mm]
\tilde{N}_T^2 = (\dfrac{E_1\alpha_1\nu_{12} + E_2\alpha_2}{1 - \nu_{12}\nu_{21}})hT_o(x, y)
\end{cases}
\qquad (6.106a)
$$

and

$$
\left\{ \tilde{M}_T^1 = \dfrac{(c_{21}\alpha_1 + c_{22}\alpha_2)h^3}{12}T_1(x, y) \right.
\qquad (6.106b)
$$

For a circular (or annular) plate exhibiting isotropic material symmetry, the thermal strain (see (6.40)) is given by

$$e_T^* = \alpha \delta T(r, \theta, z) \tag{6.107}$$

For the isotropic case, the constitutive relations (6.44) reduce to

$$
\begin{cases}
N_r &= \dfrac{Eh}{1 - \nu^2}(e_{rr}^o + \nu e_{\theta\theta}^o) - N_T^* \\[2mm]
N_\theta &= \dfrac{Eh}{1 - \nu^2}(e_{\theta\theta}^o + \nu e_{rr}^o) - N_T^* \\[2mm]
N_{r\theta} &= 2Gh e_{r\theta}^o
\end{cases} \tag{6.108}
$$

where the $e_{rr}^o, e_{r\theta}^o, e_{\theta\theta}^o$ are the middle surface strains as given by (1.71) with $\zeta = 0$, while

$$N_T^* = \frac{E\alpha}{1 - \nu} \int_{-h/2}^{h/2} \delta T(r, \theta, z) dz \tag{6.109}$$

The bending moments for the isotropic case in polar coordinates are given by (see [217], §4):

$$
\begin{cases}
M_r &= -K[w_{,rr} + \nu(\dfrac{1}{r}w_{,r} + \dfrac{1}{r^2}w_{,\theta\theta})] - M_T^* \\[2mm]
M_\theta &= -K[\nu w_{,rr} + (\dfrac{1}{r}w_{,r} + \dfrac{1}{r^2}w_{,\theta\theta})] - M_T^* \\[2mm]
M_{r\theta} &= -(1 - \nu)K(w_{,r\theta} - \dfrac{1}{r^2}w_{,\theta})
\end{cases} \tag{6.110}
$$

where

$$M_T^* = \frac{E\alpha}{1 - \nu} \int_{-h/2}^{h/2} \delta T(r, \theta, z) z \, dz \tag{6.111}$$

The von Karman system for this case now assumes the form

$$
\begin{cases}
K\Delta^2 w = [\Phi, w] - \Delta M_T^* + t(r, \theta) \\[2mm]
\Delta^2 \Phi = -\dfrac{1}{2}Eh[w, w] - (1 - \nu)\Delta N_T^*
\end{cases} \tag{6.112}
$$

where $[\Phi, w]$ and $[w, w]$, as well as $\Delta^2 w$, are given by the expressions directly following (6.48) and $\Delta = \dfrac{\partial^2}{\partial r^2} + \dfrac{1}{r}\dfrac{\partial}{\partial r} + \dfrac{1}{r^2}\dfrac{\partial^2}{\partial \theta^2}$; for the small deflection case one again deletes the "bracket" $[w, w]$ in the second equation in (6.112). Thermal bending alone for the circular (or annular) isotropic plate is governed by the first of the two equations in (6.112) with $\Phi \equiv 0$.

For a temperature distribution varying linearly through the thickness of the plate we have, in lieu of (6.94),

$$\delta T(r, \theta, z) = T_o(r, \theta) + zT_1(r, \theta) \tag{6.113}$$

In this special case (6.109) and (6.111) become, respectively,

$$
\begin{cases}
N_T^* = \dfrac{E\alpha h}{1 - \nu} T_o(r, \theta) \\[2mm]
M_T^* = \dfrac{E\alpha h^3}{12(1 - \nu)} T_1(r, \theta)
\end{cases}
\tag{6.114}
$$

We now delineate the boundary conditions that are associated with the system (6.112), or a specialization thereof, for the case of an annular plate with edges at $r = R_i, i = 1, 2, R_1 = a, R_2 = b > a$.

(i) If both edges are clamped, then $w = 0$ and $\dfrac{\partial w}{\partial r} = 0$, along $r = R_i, i = 1, 2$.

(ii) If the plate edges at $r = R_i, i = 1, 2$ are simply supported, then $w = 0$, for $r = R_i, i = 1, 2$, and, in addition,

$$K[w_{,rr} + \nu(\frac{1}{r^2}w_{,\theta\theta} + \frac{1}{r}w_{,r})] + M_T^* = 0 \tag{6.115}$$
$$\text{for } r = R_i, i = 1, 2$$

(iii) When the edges at $r = R_i, i = 1, 2$ are free, we must use (6.115) as well as the condition

$$
K\left\{ (w_{,rr} + \frac{1}{r}w_{,r} + \frac{1}{r^2}w_{,\theta\theta})_{,r} \right.
$$
$$
\left. +(1 - \nu)(\frac{1}{r^2}w_{,\theta\theta r} - \frac{1}{r^3}w_{,\theta\theta}) \right\}
\tag{6.116}
$$
$$
+M_{T,r}^* = 0, r = R_i, i = 1, 2
$$

Our last case in this sequence concerns the thermal bending and/or buckling of thin elastic circular (or annular) plates exhibiting cylindrically orthotropic behavior. The thermal strains in the radial and angular directions are given by (see (6.53))

$$e_T^r = \alpha_r \delta T(r, \theta, Z), e_T^\theta = \alpha_\theta \delta T(r, \theta, z) \tag{6.117}$$

In lieu of (6.108)–(6.111) for the isotropic case, we now have the following sets of expressions for the resultant forces and bending moments:

$$
\begin{cases}
N_r &= \dfrac{E_r h}{1 - \nu_r \nu_\theta} e^o_{rr} + \dfrac{\nu_r E_\theta h}{1 - \nu_r \nu_\theta} e^o_{\theta\theta} - N^r_T \\[2mm]
N_\theta &= \dfrac{\nu_\theta E_r h}{1 - \nu_r \nu_\theta} e^o_{rr} + \dfrac{E_\theta h}{1 - \nu_r \nu_\theta} e^o_{\theta\theta} N^\theta_T \\[2mm]
N_{r\theta} &= 2 G_{r\theta} h e^o_{r\theta}
\end{cases}
\qquad (6.118)
$$

with

$$
\begin{cases}
N^r_T &= \left( \dfrac{E_r \alpha_r + \nu_r E_\theta \alpha_\theta}{1 - \nu_r \nu_\theta} \right) \displaystyle\int_{-h/2}^{h/2} \delta T(r, \theta, z) dz \\[4mm]
N^\theta_T &= \left( \dfrac{\nu_\theta E_r \alpha_r + E_\theta \alpha_\theta}{1 - \nu_r \nu_\theta} \right) \displaystyle\int_{-h/2}^{h/2} \delta T(r, \theta, z) dz
\end{cases}
\qquad (6.119)
$$

and

$$
\begin{cases}
M_r &= -D_r[w_{,rr} + \nu_\theta(\tfrac{1}{r} w_{,r} + \tfrac{1}{r^2} w_{,\theta\theta})] - M^r_T \\[3mm]
M_\theta &= -D_\theta[\nu_r w_{,rr} + (\tfrac{1}{r} w_{,r} + \tfrac{1}{r^2} w_{,\theta\theta})] - M^\theta_T \\[3mm]
M_{r\theta} &= -2\tilde{D}_{r\theta}(\dfrac{w}{r})_{,r\theta}
\end{cases}
\qquad (6.120)
$$

with

$$
\begin{cases}
M^r_T &= \left( \dfrac{E_r \alpha_r + \nu_r E_\theta \alpha_\theta}{1 - \nu_r \nu_\theta} \right) \displaystyle\int_{-h/2}^{h/2} \delta T(r, \theta, z) z dz \\[4mm]
M^\theta_T &= \left( \dfrac{\nu_\theta E_r \alpha_r + E_\theta \alpha_\theta}{1 - \nu_r \nu_\theta} \right) \displaystyle\int_{-h/2}^{h/2} \delta T(r, \theta, z) z dz
\end{cases}
\qquad (6.121)
$$

Employing (6.120) in (6.64), and using (6.46), we obtain the first of the von Karman equations for thermal bending and/or buckling of a thin, elastic, cylindrically orthotropic plate, i.e.,

$$
D_r w_{,rrrr} + 2D_{r\theta} \cdot \frac{1}{r^2} w_{,rr\theta\theta} + D_\theta \cdot \frac{1}{r^4} w_{,\theta\theta\theta\theta}
$$

$$
+2D_r \cdot \frac{1}{r} w_{,rrr} - 2D_{r\theta} \frac{1}{r^3} w_{,r\theta\theta} - D_\theta \frac{1}{r^2} w_{,rr}
$$

$$
+2(D_\theta + D_{r\theta}) \frac{1}{r^4} w_{,\theta\theta} + D_\theta \frac{1}{r^3} w_{,r}
$$

$$
= (\frac{1}{r}\Phi_{,r} + \frac{1}{r^2}\Phi_{,\theta\theta}) w_{,rr}
$$

$$
+\Phi_{,rr}(\frac{1}{r} w_{,r} + \frac{1}{r^2} w_{,\theta\theta})
$$

$$
+2(\frac{1}{r^2}\Phi_{,\theta} - \frac{1}{r}\Phi_{,r\theta})(\frac{1}{r} w_{,r\theta} - \frac{1}{r^2} w_{,\theta})
$$

$$
+\frac{1}{r}(r M_T^r)_{,rr} + \frac{1}{r^2}(M_T^\theta)_{,\theta\theta}
$$

$$
-\frac{1}{r}(M_T^\theta)_{,r} + t(r,\theta)
$$

(6.122)

while the second of the relevant von Karman equations for this case becomes

$$
\frac{1}{E_\theta}\Phi_{,rrrr} + (\frac{1}{G_{r\theta}} - \frac{2\nu_r}{E_r})\frac{1}{r^2}\Phi_{,rr\theta\theta}
$$

$$
+\frac{1}{E_r} \cdot \frac{1}{r^4}\Phi_{,\theta\theta\theta\theta} + \frac{2}{E_\theta} \cdot \frac{1}{r}\Phi_{,rrr}
$$

$$
-(\frac{1}{G_{r\theta}} - \frac{2\nu_r}{E_r})\frac{1}{r^3}\Phi_{,r\theta\theta} - \frac{1}{E_r} \cdot \frac{1}{r^2}\Phi_{,rr}
$$

$$
+(2\frac{1-\nu_r}{E_r} + \frac{1}{G_{r\theta}})\frac{1}{r^4}\Phi_{,\theta\theta} + \frac{1}{E_r} \cdot \frac{1}{r^3}\Phi_{,r}
$$

(6.123)

$$
= -h[w_{,rr}(\frac{1}{r} w_{,r} + \frac{1}{r^2} w_{,\theta\theta}) - (\frac{1}{r} w_{,r\theta} - \frac{1}{r^2} w_{,\theta})^2]
$$

$$
+\frac{1}{E_r}\{\frac{1}{r^2}(N_T^r - \nu_r N_T^\theta)_{,\theta\theta} + \frac{1}{r}(N_{T,r}^r - \nu_r N_{T,r}^\theta)\}
$$

$$
+\frac{1}{E_\theta}\{(N_{T,r}^\theta - \nu_\theta N_{T,rr}^r) + \frac{2}{r}(N_{T,r\theta}^\theta - \nu_\theta N_{T,r}^r)\}
$$

with $N_T^r$ and $N_T^\theta$ as given by (6.119).

**Remarks:** Various combinations of terms on the right-hand side of

(6.123) may be simplified somewhat, e.g.,

$$N_T^r - \nu_r N_T^\theta = E_r \alpha_r \int_{-h/2}^{h/2} \delta T(r, \theta, z) dz$$

but there appears to be little value in carrying out such an exercise except within the context of an application to a specific problem.

Among the special cases of the von Karman system (6.122), (6.123) that are of particular interest are the following:

(i) when the temperature distribution varies linearly through the plate, as in (6.113), the thermal moments and stress resultants in (6.122) and (6.123) reduce to the following expressions:

$$\begin{cases} N_T^r = (\dfrac{E_r \alpha_r + \nu_r E_\theta \alpha_\theta}{1 - \nu_r \nu_\theta}) h T_o(r, \theta) \\[4mm] N_T^\theta = (\dfrac{\nu_\theta E_r \alpha_r + E_\theta \alpha_\theta}{1 - \nu_r \nu_\theta}) h T_o(r, \theta) \end{cases} \qquad (6.124)$$

and

$$\begin{cases} M_T^r = (\dfrac{E_r \alpha_r + \nu_r E_\theta \alpha_\theta}{1 - \nu_r \nu_\theta}) \dfrac{h^3}{12} \cdot T_1(r, \theta) \\[4mm] M_T^\theta = (\dfrac{\nu_\theta E_r \alpha_r + E_\theta \alpha_\theta}{1 - \nu_r \nu_\theta}) \dfrac{h^3}{12} \cdot T_1(r, \theta) \end{cases} \qquad (6.125)$$

(ii) For small deflections, the first term on the right-hand side of (6.123), in square brackets, is deleted.

(iii) The (purely) thermal bending problem is governed by (6.122) with $\Phi = 0$.

(iv) For an axially symmetric problem, both (6.122) and (6.123) reduce to (variable coefficient) ordinary differential equations in the radial variable $r$; we have $w = w(r)$, $\Phi = \Phi(r)$ and $T = T(r, z)$ so that the resultants $N_T^r$, $N_T^\theta$ and moments $M_T^r$, $M_T^\theta$ are functions only of $r$. In (6.122) and (6.123), therefore, derivatives of all quantities, of any order, with respect to $\theta$ vanish.

The boundary data for an annular elastic plate of inner radius $R_1 = a$ and outer radius $R_2 = b > a$, exhibiting cylindrically orthotropic symmetry, may be specified as follows:

(i) If both edges are clamped, then $w = 0$ and $\dfrac{\partial w}{\partial r} = 0$, along $r = R_i, i = 1, 2$.

(ii) If the plate edges are simply supported, then $w = 0$, for $r = R_i, i = 1, 2$ and, additionally,

$$D_r[w_{,rr} + \nu_\theta(\frac{1}{r}w_{,r} + \frac{1}{r^2}w_{,\theta\theta})] + M_T^r = 0 \tag{6.126}$$

$$\text{for } r = R_i, i = 1, 2$$

where $M_T^r$ is given by the first of the relations in (6.121).

(iii) For free edges at $r = R_i, i = 1, 2$, (6.126) applies and, in addition,

$$D_r[w_{,rrr} + \nu_\theta(\frac{1}{r}w_{,r} + \frac{1}{r^2}w_{,\theta\theta})_{,r}]$$

$$+ \frac{4}{r}\tilde{D}_{r\theta}(\frac{w}{r})_{,r\theta} + M_{T,r}^r = 0, \tag{6.127}$$

$$\text{for } r = R_i, i = 1, 2$$

The usual considerations apply if one edge is, e.g., clamped while the other is simply supported, or if one edge is simply supported while the other is free, etc.; for a circular plate of radius $R = b, R_1 = a = 0$.

**Remarks:** All of the problems considered above may be posed in terms of the middle surface displacements $u, v$ and the out-of-plane deflection $w$ in lieu of $w$ and the Airy function $\Phi$; the idea is most feasible within the context of small deflection theory. For an isotropic rectangular plate, small deflection theory corresponds to the substitution of the expressions for the middle surface strains $\epsilon_{xx}^o, \epsilon_{xy}^o$, and $\epsilon_{yy}^o$ from (6.7) into the constitutive relations (6.91), suppression of all those terms involving the out-of-plane displacement $w$, and then substitution of the resultant expressions for $N_x, N_y$, and $N_{xy}$ into the in-plane equilibrium equations (6.17a); this process leads to the pair of equations

$$\begin{cases} \frac{Eh}{1-\nu^2}(u_{,xx} + \nu v_{,xy}) + \frac{Eh}{2(1+\nu)}(u_{,yy} + v_{,xy}) = N_{,x}^T \\ \frac{Eh}{2(1+\nu)}(u_{,xy} + v_{,xx}) + \frac{Eh}{1-\nu^2}(\nu u_{,xy} + v_{,yy}) = N_{,y}^T \end{cases} \tag{6.128}$$

which must be solved in conjunction with (6.93a), which we repeat here as

$$K\Delta^2 w = N_x w_{,xx} + N_{xy} w_{,xy} + N_y w_{,yy} - \Delta M^T + t \tag{6.129}$$

For the small deflection case considered above, (6.128) and (6.129) are decoupled, just as (6.93a), (6.93b) are if, in (6.93b), we delete the bracket

$[w, w]$. Thus (6.128) must be solved subject to appropriate boundary conditions for the middle surface displacements $u, v$ which, in turn, are used to compute $N_x, N_{xy}$, and $N_y$ for subsequent substitution in (6.129). Little use of the displacement formulation of the thermal bending/buckling problem will be made in the work presented below and we will, therefore, not pursue the issue further with respect to other geometries or other classes of material symmetry.

All of the thermal bending/buckling equations and corresponding boundary conditions considered in this section may be derived from energy principles, i.e., from the principle of minimum potential energy in conjunction with elementary techniques in the calculus of variations; we will illustrate the general idea for an isotropic plate in rectangular Cartesian coordinates. Energy principles also serve as the basis for various approximate methods of analysis, including the Rayleigh-Ritz method (for computing critical (buckling) temperatures and the corresponding (initial) buckling modes) and finite element methods. In what follows, we will consider only the thermal bending problem for the sake of simplifying, somewhat, the presentation.

The standard descriptions of the principle of minimum potential energy within the context of structural mechanics is as follows: of all displacement fields which satisfy the prescribed constraint conditions, the state assumed by the structure is the one which makes the total potential energy a minimum.

For an elastic plate, the total potential energy $\Pi$ is the sum of the strain energy $U$ and the potential of any (conservative) applied forces. For the case of an isotropic linearly elastic plate, the strain energy $U$ within the context of small deflection (classical plate) theory assumes the form

$$U = \int_{\mathcal{A}} \int \{ \frac{Eh}{2(1-\nu^2)}(u_{,x} + v_{,y})^2 + \frac{Eh}{4(1+\nu)}[(u_{,y} + v_{,x})^2$$

$$-4u_{,x}v_{,y}] + \frac{K}{2}(w_{,xx} + w_{,yy})^2$$

$$+(1-\nu)K[w_{,xy}^2 - w_{,xx}w_{,yy}]$$

$$-N^T(u_{,x} + v_{,y}) + M^T(w_{,xx} + w_{,yy})\}dxdy,$$

where $\mathcal{A}$ is the area of the middle surface of the plate, while the potential of the transverse loading $t(x,y)$ is

$$V = -\int_A \int twdxdy$$

Therefore, for a rectangular plate $(0 \le x \le a, 0 \le y \le b)$ which is subject to a temperature variation $\delta T(x,y,z)$ and a transverse loading $t(x,y)$, but no applied edge loads, the total potential energy $\Pi$ assumes the form:

$$\Pi = \int_o^b \int_o^a \{ \frac{Eh}{2(1-\nu^2)}(u_{,x} + v_{,y})^2 + \frac{Eh}{4(1+\nu)}[(u_{,y} + v_{,x})^2$$

$$-4u_{,x}v_{,y}] + \frac{K}{2}(w_{,xx} + w_{,yy})^2 \qquad (6.130)$$

$$+(1-\nu)K[w_{,xy}^2 - w_{,xx}w_{,yy}]$$

$$-N^T(u_{,x} + v_{,y}) + M^T(w_{,xx} + w_{,yy}) - tw\}dxdy$$

To apply the principle of minimum potential energy, we note that a necessary condition for $\Pi$ to have a minimum is that the first variation $\delta\Pi = 0$. Using (6.130) to compute $\delta\Pi$, integrating by parts, and employing (6.91) and (6.92), with $\epsilon_{xx}^o, \epsilon_{xy}^o, \epsilon_{yy}^o$ as given by classical plate theory, we find that

$$\delta\Pi = \int_o^b \int_o^a \{-[\frac{Eh}{1-\nu^2}(u_{,xx} + \nu v_{,xy})$$

$$+\frac{Eh}{2(1+\nu)}(u_{,yy} + v_{,xy}) - N_{,x}^T]\delta u$$

$$-[\frac{Eh}{2(1+\nu)}(u_{,xy} + v_{,xx}) + \frac{Eh}{1-\nu^2}(\nu u_{,xy} + v_{,yy}) - N_{,y}^t]\delta v \qquad (6.131)$$

$$+[K(w_{,xxxx} + 2w_{,xxyy} + w_{,yyyy})$$

$$+M_{,xx}^T + M_{,yy}^T - t]\delta w\}dxdy$$

$$+ \int_o^b \{[N_x \delta u]_o^a + [N_{xy} \delta v]_o^a$$

$$+ [(M_{x,x} + 2M_{xy,y}) \delta w]_o^a - [M_x \delta(w_{,x})]_o^a \} dy$$

$$+ \int_o^a \{[N_x \delta u]_o^b + [N_y \delta v]_o^b + [(M_{y,y} + 2M_{y,x}) \delta w]_o^b$$

$$- [M_y \delta(w_{,y})]_o^b \} dx$$

$$- [2M_{xy} \delta w]_{(o,b)}^{(a,b)} + [2M_{xy} \delta w]_{(o,o)}^{(a,o)}$$

In order that (6.131) be satisfied for arbitrary variations $\delta u, \delta v, \delta w,$ $\delta(w_{,x})$, and $\delta(w_{,y})$, all the expressions within the square brackets in (6.131) must vanish identically, which leads to (6.128) and (6.129), with $N_x = N_{xy} = N_y = 0$; the resultant system governs the thermal bending, under an arbitrary temperature distribution $\delta T(x, y, z)$, of a linearly elastic isotropic rectangular plate. From (6.131) it is readily deduced that the following (natural) boundary conditions must be satisfied:

(i) On the edges $x = 0, a$, for $0 \le y \le b$

$$\begin{cases} u \text{ is prescribed or } N_x = 0 \\ v \text{ is prescribed or } N_{xy} = 0 \\ w \text{ is prescribed or } M_{x,x} + 2M_{xy,y} = 0 \\ w_{,x} \text{ is prescribed or } M_x = 0 \end{cases} \quad \text{(6.132a)}$$

(ii) On the edges $y = 0, b$, for $0 \le x \le a$

$$\begin{aligned} u \text{ is prescribed or } N_{xy} &= 0 \\ v \text{ is prescribed or } N_y &= 0 \\ w \text{ is prescribed or } M_{y,y} + 2M_{xy,x} &= 0 \end{aligned} \quad \text{(6.132b)}$$

(iii) At the corners $(0, 0), (a, 0), (0, b),$ and $(a, b)$

$$w \text{ is prescribed or } M_{xy} = 0 \quad \text{(6.132c)}$$

Some consideration of the calculation of thermal stress distributions will be made in § 6.4 of this chapter, while problems of buckling, bending, and postbuckling for rectangular plates and plates with circular symmetry will be treated at length in Chapter 7; however, it is feasible to present here the simple, but important problem of a thin plate (of arbitrary contour) which is subjected to a temperature distribution that varies only through the thickness of the plate, i.e., $T = T(z)$. In this case, we clearly have that $N^T$ and $M^T$ are constants. We will restrict the discussion to the case of an isotropic linearly elastic plate (in rectangular Cartesian coordinates) which is either free or has clamped edges.

For the case of a free plate, a solution to (6.93b), with $[w, w] = 0$, which yields zero force resultants on the boundary, is given by $\Phi = 0$. It then follows that $N_x = N_y = N_{xy} = 0$ throughout the plate. Inversion of (6.91) yields

$$
\begin{cases}
\epsilon_{xx}^o = \dfrac{1}{Eh}[N_x - \nu N_y + (1 - \nu)N^T] \\[2mm]
\epsilon_{yy}^o = \dfrac{1}{Eh}[N_y - \nu N_x + (1 - \nu)N^T] \\[2mm]
\epsilon_{xy}^o = \dfrac{1 + \nu}{Eh}N_{xy}
\end{cases}
\tag{6.133}
$$

Integration of (6.133) yields (recall that $\epsilon_{xx}^o, \epsilon_{xy}^o$, and $\epsilon_{yy}^o$ are the classical plate theory middle surface strains) the in-plane displacements:

$$
\begin{cases}
u = \dfrac{(1 - \nu)N^T}{Eh}x + a + cy \\[2mm]
v = \dfrac{(1 - \nu)N^T}{Eh}y + b - cx
\end{cases}
\tag{6.134}
$$

with $a, b, c$ arbitrary constants of integration. In a similar vein, if we take $M_x = M_y = M_{xy} = 0$ throughout the plate, then the boundary conditions for a free edge are automatically satisfied and, with $\Phi = 0$ and $t = 0$, (6.93a) yields

$$
w = -\frac{M^T}{2(1 + \nu)K}(x^2 + y^2) + d + ex + fy
\tag{6.135}
$$

with $d, e$, and $f$ constants of integration. The resulting thermal stresses for this case are easily computed to be (e.g., [217], § 2.4)

$$
\sigma_{xx} = \sigma_{yy} = \frac{1}{h}N^T + \frac{12}{h^3}M^T z - \frac{E\alpha T(z)}{1 - \nu}, \quad \sigma_{xy} = 0
\tag{6.136}
$$

For the case where the plate has clamped edges, instead of free edges, it again follows that a simple solution exists. With constant $M^T$, and

$t \equiv 0$, (6.93a) and the boundary conditions are satisfied by taking $w = 0$. Then, by virtue of (6.92),

$$M_x = M_y = -M^T, \; M_{xy} = 0 \qquad (6.137)$$

If in-plane edge displacements are prevented, then equations (6.128) and the boundary conditions yield $u = v = 0$ so that, as a consequence of (6.91),

$$N_x = N_y = -N^T, \; N_{xy} = 0 \qquad (6.138)$$

In this situation, it is easily computed that

$$\sigma_{xx} = \sigma_{yy} = -\frac{E\alpha T(z)}{1 - \nu}, \; \sigma_{xy} = 0 \qquad (6.139)$$

If, on the other hand, the middle surface of the plate is free of in-plane tractions, then $N_x = N_y = N_{xy} = 0$ and

$$\sigma_{xx} = \sigma_{yy} = \frac{1}{h} N^T - \frac{E\alpha T(z)}{1 - \nu}, \; \sigma_{xy} = 0 \qquad (6.140)$$

# CHAPTER 7

# Thermal Bending, Buckling, and Postbuckling
# of Rectangular and Circular Plates

There are many excellent surveys of thermoelastic problems in the mechanics literature (e.g., Boley and Weiner [221], Nowacki [222], Hetnarski [223], and Kovalenko [224]) to which the reader may be referred. In this chapter, we present a few thermal stress distribution solutions which have been considered in conjunction with problems involving the thermal bending and/or buckling of thin plates within the context of both small and large deflection theory; the associated bending, buckling and postbuckling solutions are also presented and analyzed.

## 7.1 Small Deflection Theory

Within the context of small-deflection theory, two distinct types of problems may be considered: those in which the effect on the deflections of loads in the plane of the plate is neglected, thus leading to a thermal bending problem, and those in which the effect of such loads is taken into account, thereby leading to a buckling problem; postbuckling behavior, of course, cannot be adequately accounted for in the context of small deflection theory.

If the effect of loads in the plane of the plate on deflections is ignored, then for the simplest case of an isotropic rectangular plate subjected to a transverse loading $t = t(x, y)$ and a general three-dimensional temperature variation $\delta T(x, y, z)$, the pertinent equation is (6.93a) with $\Phi \equiv 0$, i.e.,

$$K(w_{,xxxx} + 2w_{,xxyy} + w_{,yyyy}) = t - M_{,xx}^t - M_{,yy}^T \qquad (7.1)$$

Equation (7.1) is to hold for $0 < x < a, 0 < y < b$. For illustrative purposes, we will assume that the plate is simply supported on all four edges, in which case $w = 0$ for $x = 0, a, 0 \leq y \leq b, w = 0$ for $y = 0, b, 0 \leq x \leq a$, and the conditions (6.96) apply as well. The analysis proceeds by expressing the thermal moment and transverse load as double Fourier sine series of the form

$$
\begin{cases}
M^T = \displaystyle\sum_{m=1}^{\infty}\sum_{n=1}^{\infty} M_{mn} \sin \alpha_m \sin \beta_n y \\[2mm]
t \;\; = \displaystyle\sum_{m=1}^{\infty}\sum_{n=1}^{\infty} t_{mn} \sin \alpha_m \sin \beta_n y
\end{cases}
\tag{7.2}
$$

with $\alpha_m = \dfrac{m\pi}{a}, \beta_n = \dfrac{n\pi}{b}$ and

$$
(M_{mn}, t_{mn}) = \frac{4}{ab} \int_o^b \int_o^a (M^T, t) \sin \alpha_m \sin \beta_n y \, dx dy
$$

To satisfy the boundary conditions we take the deflection to have the form

$$
w = \sum_{m=1}^{\infty}\sum_{n=1}^{\infty} \gamma_{mn} \sin \alpha_m \sin \beta_n y
\tag{7.3}
$$

Thus, in accord with (6.96) it must be assumed that the temperature distribution is such that $M^T = 0$, for $x = 0, a, 0 \leq y \leq b$ and $M^T = 0$ for $y = 0, b, 0 \leq x \leq a$.

Substituting (7.2) and (7.3) into (7.1) and solving for the coefficients $\gamma_{mn}$ we have

$$
\gamma_{mn} = \frac{t_{mn} + (\alpha_m^2 + \beta_n^2) M_{mn}}{K(\alpha_m^2 + \beta_n^2)^2}
\tag{7.4}
$$

so that the resultant moments, as a consequence of (6.92), (7.2)–(7.4), become

$$
\begin{cases}
M_x = \displaystyle\sum_{m=1}^{\infty}\sum_{n=1}^{\infty} \{K(\alpha_m^2 + \nu\beta_n^2)\gamma_{mn} - M_{mn}\} \sin \alpha_m \sin \beta_n y \\[2mm]
M_y = \displaystyle\sum_{m=1}^{\infty}\sum_{n=1}^{\infty} \{K(\nu\alpha_m^2 + \beta_n^2)\gamma_{mn} - M_{mn}\} \sin \alpha_m \sin \beta_n y \\[2mm]
M_{xy} = -\displaystyle\sum_{m=1}^{\infty}\sum_{n=1}^{\infty} (1 - \nu)K\alpha_m \beta_n \gamma_{mn} \cos \alpha_m \cos \beta_n y
\end{cases}
\tag{7.5}
$$

To resolve the in-plane stretching aspect of this problem, we note that the displacements $u, v$ are governed by the differential equations (6.128) with which we may associate two types of boundary conditions: in the first case it may be assumed that normal components of displacements along each edge are permitted while tangential components are not, while, in the second case, the opposite situation would prevail. For the first case alluded to, the boundary data takes on the form

$$
\begin{cases}
N_x = \dfrac{Eh}{1-\nu^2}(u_{,x}+\nu v_{,y}) - N^T = 0, v = 0 \\
\qquad\qquad \text{for } x = 0, a; 0 \le y \le 6 \\
N_y = \dfrac{Eh}{1-\nu^2}(\nu u_{,x}+v_{,y}) - N^T = 0, u = 0 \\
\qquad\qquad \text{for } y = 0, b; 0 \le x \le a
\end{cases}
\tag{7.6}
$$

Of course, small deflection theory has been assumed in writing down (7.6). We now express the thermal force $N^T$ as

$$
N^T = \sum_{m=1}^{\infty}\sum_{n=1}^{\infty} N_{mn} \sin\alpha_m \sin\beta_n y
\tag{7.7}
$$

with

$$
N_{mn} = \frac{4}{ab}\int_o^b \int_o^a N^T \sin\alpha_m \sin\beta_n y\, dx\, dy,
$$

$N^T$ being given by (6.11a) for a general variation $\delta T(x, y, z)$.

To satisfy the boundary conditions (7.6) we seek solutions of (6.128) in the form

$$
\begin{cases}
u = \displaystyle\sum_{m=1}^{\infty}\sum_{n=1}^{\infty} a_{mn}\cos\alpha_m \sin\beta_n y \\
v = \displaystyle\sum_{m=1}^{\infty}\sum_{n=1}^{\infty} b_{mn}\sin\alpha_m \cos\beta_n y
\end{cases}
\tag{7.8}
$$

The expressions in (7.8) are now substituted into (6.128) with the result that

$$
\begin{cases}
a_{mn} = \dfrac{-\alpha_m N_{mn}(1-\nu^2)}{Eh(\alpha_m^2 + \beta_n^2)} \\
b_{mn} = \dfrac{-\beta_n N_{mn}(1-\nu^2)}{Eh(\alpha_m^2 + \beta_n^2)}
\end{cases}
\tag{7.9}
$$

Employing (7.7)–(7.8) in (7.6) we compute, for the stress resultants

$$
\begin{cases}
N_x = -\sum_{m=1}^{\infty}\sum_{n=1}^{\infty}\{\frac{Eh}{1-\nu^2}(\alpha_m a_{mn} + \nu\beta_n b_{mn}) \\
\quad + N_{mn}\}\sin\alpha_m\sin\beta_n y \\[2ex]
N_y = -\sum_{m=1}^{\infty}\sum_{n=1}^{\infty}\{\frac{Eh}{1-\nu^2}(\nu\alpha_m a_{mn} + \beta_n b_{mn}) \\
\quad + N_{mn}\}\sin\alpha_m\sin\beta_n y \\[2ex]
N_{xy} = \sum_{m=1}^{\infty}\sum_{n=1}^{\infty}\frac{1}{2}\frac{Eh}{1+\nu}(\beta_n a_{mn} + \alpha_{nm} b_{mn})\cos\alpha_m\cos\beta_n y
\end{cases}
\tag{7.10}
$$

If, in lieu of the boundary data (7.6), one assumes that tangential displacements are allowed along each edge, but that normal displacements are prevented, the relevant boundary conditions are

$$
N_{xy} = \frac{1}{2}\frac{Eh}{1+\nu}(u_{,y} + v_{,x}) = 0
\tag{7.11a}
$$
$$
\text{for } x = 0, a; 0 \le y \le b, \text{ and } y = 0, b; 0 \le x \le a
$$

and

$$
u = 0, \text{for } x = 0, a; 0 \le y \le b
\tag{7.11b}
$$
$$
v = 0, \text{ for } y = 0, b; 0 \le x \le a
$$

In this case, $N^T$ may be expressed as

$$
N^T = \sum_{m=0}^{\infty}\sum_{n=0}^{\infty}\bar{N}_{mn}\cos\alpha_m\cos\beta_n y
\tag{7.12}
$$

with

$$
\begin{cases}
\bar{N}_{mn} = \frac{\zeta_{mn}}{ab}\int_o^b\int_o^a N^T\cos\alpha_m\cos\beta_n y\,dx\,dy \\[2ex]
\zeta_{mn} = \begin{cases} 4, m > 0, n > 0 \\ 2, m > 0, n = 0 \text{ or } m = 0, n > 0 \\ 1, m = n = 0 \end{cases}
\end{cases}
$$

while the displacements, chosen so as to identically satisfy the boundary conditions in (7.11a, b), have the form

$$
\begin{cases}
u = \displaystyle\sum_{m=1}^{\infty}\sum_{n=0}^{\infty} \bar{a}_{mn} \sin \alpha_m \cos \beta_n y \\[2em]
v = \displaystyle\sum_{m=0}^{\infty}\sum_{n=1}^{\infty} \bar{b}_{mn} \cos \alpha_m \sin \beta_n y
\end{cases}
\tag{7.13}
$$

The same procedure described above for the first set of boundary conditions now leads to

$$
\bar{a}_{mn} = \frac{\alpha_m \bar{N}_{mn}(1-\nu^2)}{Eh(\alpha_m^2 + \beta_n^2)}, \quad \bar{b}_{mn} = \frac{\beta_n \bar{N}_{mn}(1-\nu^2)}{Eh(\alpha_m^2 + \beta_n^2)}
\tag{7.14}
$$

and

$$
\begin{cases}
N_x = \displaystyle\sum_{m=0}^{\infty}\sum_{n=0}^{\infty}\{\frac{Eh}{1-\nu^2}(\alpha_m\bar{a}_{mn} + \nu\beta_n\bar{b}_{mn}) \\[1em]
\qquad\quad -\bar{N}_{mn}\} \cos \alpha_m \cos \beta_n y \\[2em]
N_y = \displaystyle\sum_{m=0}^{\infty}\sum_{n=0}^{\infty}\{\frac{Eh}{1-\nu^2}(\nu\alpha_m\bar{a}_{mn} + \beta_n\bar{b}_{mn}) \\[1em]
\qquad\quad -\bar{N}_{mn}\} \cos \alpha_m \cos \beta_n y \\[2em]
N_{xy} = -\displaystyle\sum_{m=1}^{\infty}\sum_{n=1}^{\infty}\frac{1}{2}\frac{Eh}{1+\nu}(\beta_n\bar{a}_{mn} + \alpha_m\bar{b}_{mn}) \sin \alpha_m \sin \beta_n y
\end{cases}
\tag{7.15}
$$

The thermoelastic stress distributions for either of the two bending problems considered above may be computed by substitution for $N_x$, $N_{xy}$, $N_y$, $N^T$, $M_x$, $M_y$, $M_{xy}$, and $M^T$ in the relations

$$
\begin{cases}
\sigma_{xx} = \dfrac{1}{h}(N_x + N^T) + \dfrac{12z}{h^3}(M_x + M^T) - \dfrac{E\alpha}{1-\nu}T \\[1em]
\sigma_{yy} = \dfrac{1}{h}(N_y + N^T) + \dfrac{12z}{h^3}(M_y + M^T) - \dfrac{E\alpha}{1-\nu}T \\[1em]
\sigma_{xy} = \dfrac{1}{h}N_{xy} + \dfrac{12z}{h^3}M_{xy}
\end{cases}
\tag{7.16}
$$

An alternative solution to the flexure problem for the rectangular plate discussed above has been described in Tauchert [217] and is now described below; we will begin, as in [217], by assuming that the edges

$x = 0$ and $x = a$ are simply supported, that the plate is symmetric with respect to the $x$ axis, so that $-\dfrac{b}{2} \le y \le \dfrac{b}{2}$, and that, for now, the boundary conditions along $y = \pm \dfrac{1}{2} b$ are arbitrary. As $w = 0$ along $x = 0$ and $x = a$ it follows that $w_{,yy} = 0$ along these edges as well. The conditions of simple support of the plate along $x = 0$ and $x = a$ may, therefore, be expressed as

$$w = 0, w_{,xx} = -\frac{1}{K} M^T$$
$$\text{for } x = 0, a; 0 \le y \le b \tag{7.17}$$

We look for a solution of (7.1) satisfying the non-homogeneous boundary conditions (7.17) in the form

$$w = W(x, y) + M^T(0, y) H_o(x) + M^T(a, y) H_a(x) \tag{7.18}$$

with

$$\begin{cases} H_o(x) = \dfrac{a^2}{6K} \{ \dfrac{x}{a} - 3(\dfrac{x}{a})^2 + (\dfrac{x}{a})^3 \} \\[4mm] H_a(x) = \dfrac{a^2}{6K} \{ \dfrac{x}{a} - (\dfrac{x}{a})^3 \} \end{cases} \tag{7.19}$$

Using (7.19) in (7.18) and substituting the resultant expression for $w(x, y)$ into (7.1), it follows that

$$\nabla^4 w(x, y) = F(x, y) \tag{7.20}$$

where

$$F(x, y) = \frac{t(x, y)}{K} - \frac{\nabla^2 M^T}{K}$$
$$- \nabla^4 \{ M^T(0, y) H_o(x) + M^T(a, y) H_a(x) \} \tag{7.21}$$

and

$$w = \frac{\partial^2 w}{\partial x^2} = 0, \text{ for } x = 0, a; 0 \le y \le b \tag{7.22}$$

We express $F$ in terms of the Fourier series

$$F(x, y) = \sum_{m=1}^{\infty} f_m(y) \sin \alpha_m x; \quad \alpha_m = \frac{m\pi}{\alpha} \tag{7.23}$$

with

$$f_m(y) = \frac{2}{a} \int_o^a F(x, y) \sin \alpha_m x \, dx$$

and take $w$ in the form

$$w(x, y) = \sum_{m=1}^{\infty} Y_m(y) \sin \alpha_m x \qquad (7.24)$$

so that $w$ automatically satisfies the edge conditions in (7.22). It is easily shown that $w$, as given by (7.24), satisfies (7.20) provided the $Y_m$ satisfy the ordinary differential equations

$$Y_m^{(iv)}(y) - 2\alpha_m^2 Y_m''(y) + \alpha_m^4 Y_m(y) = f_m(y) \qquad (7.25)$$

whose general solution has the form

$$Y_m = (A_m + B_m y) \cosh \alpha_m y + (C_m + D_m y) \sinh \alpha_m y$$

$$+ e^{-\alpha_m y} \int e^{2\alpha_m y} \left( \int e^{-\alpha_m y} f_m(y) dy \right) dy \qquad (7.26)$$

The constants of integration $A_m, B_m, C_m$, and $D_m$ in (7.26) are to be determined from the boundary data on the edges $y = \pm \frac{1}{2} b$. Suppose, e.g., that the edges $y = \pm \frac{1}{2} b$ are also simply supported and that the thermal moment $M^T$ is constant—as it would be, say, for $\delta T = \delta T(z)$. As the support conditions and loading are both symmetric with respect to the $x$-axis, the deformation must also be symmetric and, thus, $B_m = C_m = 0$ in (7.26). Substituting $Y_m$ from (7.26) into (7.24); the resultant expression for $w$ into (7.18) yields

$$w = \sum_{m=1}^{\infty} (A_m \cosh \alpha_m y + D_m y \sinh \alpha_m y) \sin \alpha_m x$$

$$+ M^T(o, y) H_o(x) + M^T(a, y) H_a(x) \qquad (7.27)$$

The sum of the last two terms on the right-hand side of (7.27) may be expressed as a Fourier series, i.e.,

$$M^T(0, y) H_o(x) + M^T(a, y) H_a(x)$$

$$\equiv \frac{M^T a^2}{2K} \left( \frac{x}{a} - \left( \frac{x}{a} \right)^2 \right) \equiv \sum_{m=1}^{\infty} k_m \sin \alpha_m x \qquad (7.28)$$

where

$$k_m = \begin{cases} 0, & m \text{ even} \\ \dfrac{4M^T}{aK\alpha_m^3}, & m \text{ odd} \end{cases}$$

in which case,

$$w = \sum_{m=1,3,\cdots}^{\infty} (A_m \cosh \alpha_m y + D_m y \sinh \alpha_m y + k_m) \sin \alpha_m x \qquad (7.29)$$

The constants $A_m$ and $D_m$ in (7.29) are to be determined from the boundary conditions

$$w = 0, w_{,yy} = -\frac{1}{K} M^T$$

$$\text{for } y = \pm \frac{1}{2}b; 0 \le x \le a \qquad (7.30)$$

We write

$$M^T = \sum_{m=1}^{\infty} M_m \sin \alpha_m x; \quad m_m = K\alpha_m^2 k_m \qquad (7.31)$$

and substitute (7.29) into (7.30); after solving for $A_m$ and $D_m$ we obtain

$$w(x,y) = \frac{4M^T}{aK} \sum_{m=1,3,\cdots}^{\infty} \frac{1}{\alpha_m^3} \left(1 - \frac{\cosh \alpha_m y}{\cosh \frac{1}{2}\alpha_m b}\right) \sin \alpha_m x \qquad (7.32)$$

for which the corresponding moment resultants are given by

$$\begin{cases} M_x = \frac{-4M^T(1-\nu)}{a} \sum_{m=1,3,\cdots}^{\infty} \frac{\cosh \alpha_m y}{\alpha_m \cosh \frac{1}{2}\alpha_m b} \sin \alpha_m x \\[4mm] M_y = \frac{-4M^T(1-\nu)}{a} \sum_{m=1,3,\cdots}^{\infty} \frac{1}{\alpha_m}\left(1 - \frac{\cosh \alpha_m y}{\cosh \frac{1}{2}\alpha_m b}\right) \sin \alpha_m x \\[4mm] M_{xy} = \frac{4M^T(1-\nu)}{a} \sum_{m=1,3,\cdots}^{\infty} \frac{1}{\alpha_m}\left(\frac{\sinh \alpha_m y}{\cosh \frac{1}{2}\alpha_m b}\right) \cos \alpha_m x \end{cases} \qquad (7.33)$$

For the case in which the plate is clamped along the edges at $y = \pm\frac{1}{2}b$, and subject to a constant thermal moment $M^T$, it has been noted in [217] that the deflection, once again, assumes the form in (7.29), but now with

$$
\begin{cases}
A_m = -k_m(\frac{1}{2}\alpha_m b \cosh \frac{1}{2}\alpha_m b \\
\qquad + \sinh \frac{1}{2}\alpha_m b)/\triangle_m \\
B_m = \frac{1}{2}\alpha_m b + \sinh \frac{1}{2}\alpha_m b \cosh \frac{1}{2}\alpha_m b \\
\triangle_m = \frac{1}{2}\alpha_m b + \sinh \frac{1}{2}\alpha_m b \cosh \frac{1}{2}\alpha_m b
\end{cases}
\tag{7.34}
$$

References for the thermal bending of an isotropic, elastic rectangular plate, under other combinations of edge conditions, may be found in [217].

Next, we consider the problem of thermal bending of an isotropic annular plate; we assume that the plate is subjected to a transverse loading $t = t(r, \theta)$ and a general temperature variation $\delta T(r, \theta, z)$. For this situation, ignoring for now the effect on deflections of loads in the plane of the plate, the relevant equation is the first partial differential equation in (6.112) with $\Phi \equiv 0$, i.e.,

$$
\begin{aligned}
K(w_{,rrrr} & + \frac{2}{r}w_{,rrr} - \frac{1}{r^2}w_{,rr} \\
& + \frac{2}{r^2}w_{,rr\theta\theta} + \frac{1}{r^3}w_{,r} - \frac{2}{r^3}w_{,r\theta\theta} \\
& + \frac{1}{r^4}w_{,\theta\theta\theta\theta} + \frac{4}{r^4}w_{,\theta\theta}) \\
= t & - (M^*_{T,rr} + \frac{1}{r}M^*_{T,r} + \frac{1}{r^2}M^*_{T,\theta\theta})
\end{aligned}
\tag{7.35}
$$

with $M^*_T = \dfrac{E\alpha}{1-\nu} \displaystyle\int_{-h/2}^{h/2} \delta T(r, \theta, z)z\,dz$ being the thermal moment.

Equation (7.35) holds for $a \le r < b, 0 \le \theta < 2\pi$. The associated clamped, simply supported, and free edge boundary conditions are given, respectively, by

(i) $w = 0$ and $\dfrac{\partial w}{\partial r} = 0$, at $r = a, b$ if the edges are clamped

(ii) $w = 0$ and $K[w_{,rr} + \nu(\dfrac{1}{r^2}w_{,\theta\theta} + \dfrac{1}{r}w_{,r})] + M^*_T = 0$, at $r = a, b$ if the edges are simply supported

(iii) $K[w_{,rr} + \nu(\frac{1}{r^2}w_{,\theta\theta} + \frac{1}{r}w_{,r})] + M_T^* = 0$ and

$$K[(w_{,rr} + \frac{1}{r}w_{,r} + \frac{1}{r^2}w_{,\theta\theta})_{,r} + (1-\nu)(\frac{1}{r^2}w_{,\theta\theta r} - \frac{1}{r^3}w_{,\theta\theta})] + M_{T,r}^* = 0,$$

at $r = a, b$ if both edges are free.

Also, for isotropic response, the bending moments in polar coordinates are given by (6.110), the resultant forces by (6.108), with $N_T^* = \dfrac{E\alpha}{1-\nu}$

$\displaystyle\int_{-h/2}^{h/2} \delta T(r,\theta,z)dz$ and the stresses may be expressed by

$$\begin{cases} \sigma_{rr} = \dfrac{1}{h}(N_r + N_T^*) + \dfrac{12z}{h^3}(M_r + M_T^*) - \dfrac{E\alpha}{1-\nu}\delta T \\[2mm] \sigma_{\theta\theta} = \dfrac{1}{h}(N_\theta + N_T^*) + \dfrac{12z}{h^3}(M_\theta + M_T^*) - \dfrac{E\alpha}{1-\nu}\delta T \qquad (7.36) \\[2mm] \sigma_{r\theta} = \dfrac{1}{h}N_{r\theta} + \dfrac{12z}{h^3}M_{r\theta} \end{cases}$$

The simplest case of thermal bending with respect to an annular plate is that of axisymmetric bending in which it is assumed that the loading and boundary conditions are independent of the angular coordinate $\theta$. If, in addition, $t \equiv 0$, then (7.35) reduces to

$$\nabla^4 w = -\frac{1}{K}\nabla^2 M_T^*, \quad a < r < b \qquad (7.37)$$

where $w = w(r)$, $M_T^* = \dfrac{E\alpha}{1-\nu}\displaystyle\int_{-h/2}^{h/2}\delta T(r,z)zdz,$

$$\begin{cases} \nabla^4 = \dfrac{d^4}{dr^4} + \dfrac{2}{r}\dfrac{d^3}{dr^3} - \dfrac{1}{r^2}\dfrac{d^2}{dr^2} + \dfrac{1}{r^3}\dfrac{d}{dr} \\[2mm] \nabla^2 = \dfrac{d^2}{dr^2} + \dfrac{1}{r}\dfrac{d}{dr} \equiv \dfrac{1}{r}\dfrac{d}{dr}(r\dfrac{d}{dr}) \end{cases}$$

The general solution of (7.37) is easily computed to be

$$w = C_1 + C_2 r^2 + C_3 \ln\frac{r}{a} + C_4 r^2 \ln\frac{r}{a}$$

$$+ \int_r^b (\frac{1}{r}\int_a^r \frac{1}{K}M_T^*(r)rdr)dr \qquad (7.38)$$

with the $C_i, i = 1, \cdots, 4$, arbitrary constants of integration. For the problem at hand, a straightforward computation based on (7.38) yields the following expressions for the relevant moments and shear force resultant:

$$
\begin{cases}
M_r &= -K\{2(1+\nu)C_2 - (1-\nu)\dfrac{C_3}{r^2} + (3+\nu)C_4 \\[2ex]
&\quad +2(1+\nu)C_4 \ln\dfrac{r}{a}\} - \dfrac{1-\nu}{r^2}\displaystyle\int_a^r M_T^*(r)r\,dr \\[2ex]
M_{r\theta} &= 0 \\[2ex]
Q_r &\equiv M_{r,r} + \dfrac{M_r - M_\theta}{r} = -4K\dfrac{C_4}{4}
\end{cases}
\tag{7.39}
$$

For the case of a solid plate, in which $a = 0$, the constants $C_3$ and $C_4$ in (7.38) must vanish so that $M_r$ and $Q_r$ remain finite at $r = 0$; if the solid plate is clamped along its edge at $r = b$, then it follows from (7.38) and the fact that $w = w_{,r} = 0$ at $r = b$ that

$$
C_1 = -b^2 C_2 = -\frac{1}{2K}\int_0^b M_T^*(r)r\,dr
\tag{7.40}
$$

while, if the edge at $r = b$ is simply supported,

$$
C_1 = -b^2 C_2 = \frac{1-\nu}{2(1+\nu)K}\int_0^b M_T^*(r)r\,dr
\tag{7.41}
$$

When $(\delta T)_{,\theta} \neq 0$ a solution $w = w(r,\theta)$ must be obtained for (7.35); for simplicity we again set $t \equiv 0$; such problems have, e.g., been considered by Forray and Newman [225] for the special case in which the thermal gradient is assumed to vary linearly through the thickness of the plate. Specifically, it is assumed in [225] that the thermal moment $M_T^*$ may be expressed in the form

$$
M_T^* = \sum_{m=0}^{\infty}\sum_{k=0}^{\infty} A_{km}r^k \cos m\theta + \sum_{m=1}^{\infty}\sum_{k=0}^{\infty} B_{km}r^k \sin m\theta
\tag{7.42}
$$

This form for the thermal moment is a consequence of the assumption that

$$
\delta T(r,\theta,z) = T_0(r,\theta) + zT_1(r,\theta)
$$

with

$$
T_1(r,\theta)z = \frac{-z}{h}T_d(r,\theta)
\tag{7.43}
$$

and $T_d$ the temperature difference between the upper and lower faces of the plate. Using the definition of $M_T^*$ we then easily compute that

$$M_T^* = \frac{1+\nu}{hK}\alpha T_d(r,\theta) \tag{7.44}$$

while (7.35) becomes (with $t \equiv 0$)

$$\nabla^4 w = \frac{1+\nu}{hK}\alpha\,\nabla^2 T_d \tag{7.45}$$

The solution to (7.45) consists of the sum of the general solution of $\nabla^4 w_g = 0$ and a particular solution of $\nabla^2 w_p = \frac{1+\nu}{hK}\alpha T_d$.

In fact, the general solution to (7.45) can be shown, as in [225], to have the form

$$w = a_0 + b_0 r^2 + c_0 r^2 lnr + d_0 lnr \tag{7.46}$$
$$+ (a_1 r + b_1 r^3 + \frac{c_1}{r} + d_1 rlnr)\cos\theta$$
$$+ (a_1' r + b_1' r^3 + \frac{c_1'}{r} + d_1' rlnr)\sin\theta$$
$$+ \sum_{n=2}^{\infty}\left(a_n r^n + b_n r^{n+2} + \frac{c_n}{r^n} + \frac{d_n}{r^{n-2}}\right)\cos n\theta$$
$$+ (a_n' r^n + b_n' r^{n+2} + \frac{c_n'}{r_n} + \frac{d_n'}{r^{n-2}})\sin n\theta$$
$$+ \sum_{m=0}^{\infty} q_m(r)\cos m\theta + \sum_{m=1}^{\infty} h_m(r)\sin m\theta$$

with the $a_n, a_n', b_n, ...(n = 0, 1, ...)$ arbitrary constants and

$$(g_m, h_m) = -\frac{1}{Kr^m}\int\left(r^{2m-1}\int(A_{km}, b_{km})r^{k+1-m}dr\right)dr \tag{7.47}$$

For the special case in which we are dealing with a solid plate, so that $a = 0$, we must set $c_n = c_n' = d_n = d_n' = 0$ so as to avoid singularities at $r = 0$.

**Remarks:** The last two sums on the right-hand side of (7.46) constitute the particular solution $w_p$ of $\nabla^2 w_p = \frac{1+\nu}{Kh}\alpha T_d$; more specifically, if

$$\frac{1+\nu}{hk}\alpha T_d = \begin{cases} A_{km}r^k\cos m\theta \\ B_{km}r^k\sin m\theta \end{cases} \tag{7.48a}$$

then $w_p(r, \theta)$ is given by

$$w_p(r, \theta) = \begin{cases} g_m(r) \cos m\theta \\ h_m(r) \sin m\theta \end{cases} \tag{7.48b}$$

with $g_m(r)$, $h_m(r)$ as defined in (7.47). By carrying out the integrations in (7.47) and using the results in (7.48b) it can be shown that

$$w_p(r_1\theta) = \begin{cases} \dfrac{A_{km} r^{k+2} \cos m\theta}{(k+2)^2 - m^2} \\ \dfrac{B_{km} r^{k+2} \sin m\theta}{(k+2)^2 - m^2} \end{cases} \tag{7.49a}$$

when $k + 2 - m \neq 0$, and (7.48a) applies, while for the case in which $k + 2 - m = 0$,

$$w_p(r, \theta) = \begin{cases} A_{km} \left\{ \dfrac{lnr}{2m} - \dfrac{1}{(2m)^2} \right\} r^m \cos m\theta \\ B_{km} \left\{ \dfrac{lnr}{2m} - \dfrac{1}{(2m)^2} \right\} r^m \sin m\theta \end{cases} \tag{7.49b}$$

Returning to the case of a solid plate, for which $c_n = c_n' = d_n = d_n' = 0$, we note that the constants $a_n$ and $b_n$ must be determined from the boundary conditions. For the case of a clamped plate, in which $w(b, \theta) = \dfrac{\partial w}{\partial r}(b, \theta) = 0$, the (rather complex) expressions for the deflection, moments, and shears in nondimensional form are given in [225]; these results, for $m = 0, 1, 2, 3$ are depicted in Fig. 7.1.

The problem of bending of a rectangular orthotropic plate (which has two opposite edges simply supported and the other two clamped) due to different temperature distributions on the plate surfaces, has been considered by Misra [226]. It is assumed in [226] that the plate occupies the region

$$0 \leq x \leq a, \quad -\frac{b}{2} \leq y \leq \frac{b}{2}, \quad -\frac{h}{2} \leq z \leq \frac{h}{2}$$

so that the opposite faces are defined by $z = \pm\dfrac{h}{2}$. It is also assumed that the two parallel edges at $x = 0$ and $x = a$ are simply supported while the edges $y = \pm\dfrac{b}{2}$ are clamped. The temperature distribution is taken as having the form

$$\delta T(x, y, z) = \frac{T_1 + T_2}{2} + \frac{T_1 - T_2}{h} z \tag{7.50}$$

so that $T(x, y, \frac{h}{2}) = T_1$, $T(x, y, -\frac{h}{2}) = T_2$, for $0 \le x \le a$, $-\frac{b}{d} \le y \le \frac{b}{2}$, where $T_1$ and $T_2$ are, respectively, the constant temperatures at the top and bottom of the plate; thus the temperature is assumed to remain constant in any plane which is parallel to the $x, y$ plane. The edge conditions are given by

$$w = 0, \; M_x = 0; \quad \text{on } x = 0, a, \; \text{for} \; -\frac{b}{2} \le y \le \frac{b}{2} \qquad (7.51a)$$

and

$$w = 0, \; w_{,y} = 0; \quad \text{on } y = \pm\frac{b}{2}, \; \text{for } 0 \le x \le a \qquad (7.51b)$$

where $M_x$ is given by (6.101). Actually, Misra [226] writes the term $\tilde{M}_T^1$ in (6.102) in the form $\tilde{M}_T^1 = \bar{\beta}_1 M_T$, with $M_T = \int_{-h/2}^{h/2} \delta T z dz$, so that

$$\bar{\beta}_1 = c_{11}\alpha_1 + c_{12}\alpha_2 \qquad (7.52)$$

The superposed bar over the $\beta_1$ in (7.52) does not appear in [226] and has been placed there so as not to confuse this parameter with a hygroscopic coefficient. With the definition of $M_T$, as given above, and (7.50) it is easily seen that

$$M_T = \frac{1}{12}h^2(T_1 - T_2) \equiv k \qquad (7.53)$$

which has the Fourier representation

$$M_T = \frac{4k}{\pi} \sum_{m=1,3,5,\ldots}^{\infty} \frac{1}{m} \sin\frac{m\pi}{a}x, \qquad (7.54)$$

for $0 < x < a$. Using (7.54) in (6.103a), setting $t \equiv 0$ and $\Phi \equiv 0$, and replacing $D_{11} + 4D_{66} + D_{21}$ by $2H$ we obtain, as in [226], the following equation for the bending of the heated, rectangular, orthotropic plate:

$$D_{11}w_{,xxxx} + 2Hw_{,xxyy} + D_{22}w_{,yyyy} = P \sum_{m=1,3,\ldots}^{\infty} m\sin\frac{m\pi}{a}x \qquad (7.55)$$

where

$$P = \frac{4k\pi\bar{\beta}_1}{a^2} \qquad (7.56)$$

From (7.54) it follows that $M_T = 0$ along the edges at $x = 0$ and $x = a$. Then, by virtue of (7.51a), it follows that both $w$ and $w_{,yy}$ must vanish along $x = 0$ and $x = a$. However, as $M_x = 0$ along $x = 0$ and $x = a$, it would follow from (6.101) that $w_{,xx} = 0$ along $x = 0$ and $x = a$ only if $M_T = 0$ along these edges, which it does not—the Fourier representation not withstanding! Thus, the edge conditions in (7.51a), which in [226] are now written in the form

$$w = 0, w_{,xx} = 0, \text{ on } x = 0, a$$
$$\text{for } -\frac{b}{2} \le y \le \frac{b}{2} \tag{7.57}$$

are open to suspicion, as is the remainder of the solution presented below. A solution of the homogeneous equation associated with (7.55) which is compatible with the edge conditions (7.57) is sought in [226] in the form

$$w = \sum_{m=1,3,..}^{\infty} Y_m(y) \sin \frac{m\pi}{a} x \tag{7.58}$$

Substituting this expansion into the relevant homogeneous partial differential equation we are led to the following homogeneous fourth-order ordinary differential equation for the functions $Y_m(y)$:

$$D_{22} Y_m'''' - 2H\alpha_m^2 Y_m'' + D_{11}\alpha_m^4 Y_m = 0 \tag{7.59}$$

where $\alpha_m = \dfrac{m\pi}{a}$. Noting that, because of symmetry, $Y_m$ must be an even function of $y$, a solution of (7.59) is sought in [226] in the form

$$Y_m(y) = A_m \cosh \ p_m y \cos q_m y$$
$$+ B_m \sinh \ p_m y \sin q_m y \tag{7.60}$$

where

$$\begin{cases} p_m^2 = \dfrac{\alpha_m^2 (H + \sqrt{H^2 - D_{11}D_{22}})}{D_{22}} \\ \\ q_m^2 = \dfrac{\alpha_m^2 (H - \sqrt{H^2 - D_{11}D_{22}})}{D_{22}} \end{cases} \tag{7.61}$$

The $A_m, B_m$ are, at this junction, arbitrary functions of $m$. For a particular integral of (7.55) Misra [226] chooses

$$w = \sum_{m=1,3,5...}^{\infty} E_m \sin \alpha_m x \tag{7.62}$$

Substitution of (7.62) into (7.55) then yields

$$E_m = \frac{mP}{\alpha_m^4 D_{11}} \tag{7.63}$$

in which case, the complete solution of (7.55) assumes the form

$$w = \sum_{m=1,3,5...}^{\infty} \{ \frac{mP}{\alpha_m^4 D_{11}} + A_m \cos hp_m y \cos q_m y$$
$$+ B_m \sinh \ p_m y \sin q_m y \} \sin \alpha_m x \qquad (7.64)$$

The edge conditions (7.57) are automatically satisfied by (7.64), while those in (7.51b) are satisfied if and only if $A_m, B_m$ are connected by the relations

$$\frac{mp}{\alpha_m^4 D_{11}} + A_m \cosh \frac{bpm}{2} \cos \frac{bqm}{2}$$
$$+ B_m \sinh \frac{bpm}{2} \sin \frac{bqm}{2} = 0 \qquad (7.65a)$$

$$A_m (p_m \tan h \frac{bpm}{2} - q_m \tan \frac{bqm}{2})$$
$$+ B_m (p_m \tan \frac{bqm}{2} + q_m \tanh \frac{bpm}{2}) = 0 \qquad (7.65b)$$

These relations may be solved for $A_m, B_m$ (see [226] for the details), which are then substituted back into (7.64) so as to yield the deflection at any point of the plate. The expression obtained for $w(x, y)$ can also be employed in (6.101) so as to compute the thermally induced moments at any point in the plate, e.g.,

$$M_x = D_{11} \sum_{m=1,3,5,..}^{\infty} \alpha_m^2 \{ A_m \cosh p_m y \cos q_m y$$
$$+ B_m \sinh p_m y \sin q_m y \} \sin \alpha_m x$$

$$- D_{12} \sum_{m=1,3,5,...}^{\infty} [A_m \{ (p_m^2 - q_m^2) \cosh p_m y \cos q_m y$$
$$- 2 p_m q_m \sinh p_m y \sin q_m y \} \qquad (7.66)$$

$$+ B_m \{ (p_m^2 - q_m^2) \sinh p_m y \sin q_m y$$
$$+ 2 p_m q_m \cosh p_m y \cos q_m y \}] \sin \alpha_m x$$

$$- \frac{4 \bar{\beta}_1 k}{a} \sum_{m=1,3,5,...}^{\infty} \frac{1}{\alpha_m} \sin \alpha_m x$$

with analogous expressions for $M_y$ and $M_{xy}$. Finally, the deflection at the center of the plate, i.e., at $x = \dfrac{a}{2}, y = 0$, is computed in [226] to be

$$w = \sum_{m=1,3,5\ldots}^{\infty} \left( \frac{mP}{\alpha_m^4 D_{11}} + A_m \right) \sin \frac{m\pi}{2} \qquad (7.67)$$

The expressions for the moments, e.g., (7.66) and the final result (7.67) for the deflection at the center of the plate are subject to the criticism (because of the relevance of the boundary conditions (7.57)) which has been levied above.

An alternative approach to solving problems of thermal deflection for plates that has been used extensively in the literature is based on the concept of an influence function and usually goes under the title of Maysel's method; this approach is actually an extension of Betti's reciprocal theorem to thermoelastic problems, and excellent treatments have appeared in several places in the literature, e.g., in Nowacki [222] and in Tauchert [217]. In what follows, we will adhere closely to the presentation in [217] and will assume that the plate exhibits isotropic response; we will also take $t \equiv 0$, so that, in either rectangular or polar coordinates, the relevant partial differential equation is given by

$$K\triangle^2 w = -\triangle M^T \qquad (7.68)$$

If the plate occupies the domain $\mathcal{A}$ in the $x, y$ plane when in its undeflected state, and $w^*(\xi, y; x, y)$ is the Green's function for the operator $K\triangle^2$ then it is easily shown (i.e. [217] or [222]) that

$$w(x, y) = - \iint_{\mathcal{A}} M^T(\xi, \eta) \, \nabla^2 \, w^*(\xi, \eta; x, y) d\mathcal{A}(\xi, \eta) \qquad (7.69)$$

where, for the sake of convenience, we have initiated the discussion by employing rectangular coordinates. In (7.69),

$$\nabla^2 = \frac{\partial^2}{\partial \xi^2} + \frac{\partial^2}{\partial \eta^2}$$

The Green's function in (7.69), $w^*(\xi, \eta; x, y)$, represents the deflection at the point $(\xi, \eta)$ of the plate middle surface which would be due to a concentrated unit force applied at the point $(x, y)$. Thus, the Maysel relation (7.69) may be used to compute a thermally induced deflection $w(x, y)$ whenever $w^*(\xi, \eta; x, y)$ can be calculated for a plate of given shape and assigned support conditions. As an alternative to (7.69) one may use the form obtained by employing Green's formula, i.e.,

$$w(x,y) = - \iint_{\mathcal{A}} w^*(\xi, \eta; x, y) \, \nabla^2 M^T(\xi, \eta) dA(\xi, \eta)$$
$$- \int_{\partial \mathcal{A}} \left( M^T \frac{\partial w^*}{\partial n} - w^* \frac{\partial M^T}{\partial n} \right) ds \qquad (7.70)$$

where $n, s$ denote, respectively, the directions that are normal and tangential to the plate boundary $\partial \mathcal{A}$. If $\nabla^2 M^T = 0$, such as for the case in which $M^T$ is constant, then (7.70) reduces to

$$w(x,y) = - \int_{\partial \mathcal{A}} \left( M^T \frac{\partial w^*}{\partial n} - w^* \frac{\partial M^T}{\partial n} \right) ds \qquad (7.71)$$

and if the plate is simply supported, so that $w^* = 0$ along $\partial \mathcal{A}$, then

$$w(x,y) = - \int_{\partial \mathcal{A}} M^T(\xi, \eta) \frac{\partial w^*}{\partial n}(\xi, \eta; x, y) ds \qquad (7.72)$$

Of course, if $\nabla^2 M^T = 0$ in $\mathcal{A}$, and the plate is clamped along $\partial \mathcal{A}$, then $w^* = \dfrac{\partial w^*}{\partial n} = 0$ along $\partial \mathcal{A}$ in which case $w \equiv 0$ throughout the plate.

As a first example of the influence function method we consider the simply supported rectangular plate which is depicted in Fig. 7.2. We assume that the thermal moment is nonzero within an arbitrary region $\mathcal{A}^T$ of the plate, while $M^T = 0$ is the complement of this region. It is easily shown that the deflection $w^*$ at an arbitrary point $(\xi, \eta)$ of the same simply supported plate subject to a concentrated unit force at $(x,y)$ is

$$w^*(\xi, \eta; x, y) = \frac{4}{abK} \sum_{m=1}^{\infty} \sum_{n=1}^{\infty} \frac{\sin \alpha_m \xi \sin \beta_n \eta \sin \alpha_m x \sin \beta_n y}{(\alpha_m^2 + \beta_n^2)^2} \qquad (7.73)$$

where $\alpha_m = \dfrac{m\pi}{a}$, $\beta_n = \dfrac{n\pi}{b}$, provided $M^T$ vanishes along the edges of the plate. Substituting (7.73) into (7.69), and carrying out an elementary computation, we are led to the following expression for the deflection:

$$w(x,y) = \frac{4}{abK} \sum_{m=1}^{\infty} \sum_{n=1}^{\infty} \left( \frac{\sin \alpha_m x \sin \beta_n y}{(\alpha_m^2 + \beta_n^2)^2} \right.$$
$$\left. \times \iint_{\mathcal{A}^T} M^T(\xi, \eta) \sin \alpha_m \xi \sin \beta_n y d\xi d\eta \right) \qquad (7.74)$$

If the thermal moment is constant over the entire plate, say, $M^T = M$, then (7.74) formally reduces to

$$w(x, y) = \frac{16M}{abK} \sum_{m=1,3,\ldots}^{\infty} \sum_{n=1,3,\ldots}^{\infty} \frac{\sin \alpha_m x \sin \beta_n y}{\alpha_m \beta_n (\alpha_m^2 + \beta_n^2)} \qquad (7.75)$$

but, once again, such a solution is subject to the criticisms raised earlier as now $M^T$ does not vanish along the edges of the plate.

As a second example, we consider the application of Maysel's relation (7.69) to the thermal deflection of a solid circular plate of radius $b$; in this case (7.69) assumes the following form in terms of polar coordinates:

$$w(r, \theta) = -\int_0^{2\pi} \int_0^b M^T(\rho, \psi) \nabla^2 w^*(\rho, \psi; N, \theta) \rho \, d\rho \, d\psi \qquad (7.76)$$

where

$$\nabla^2 = \frac{\partial^2}{\partial \rho^2} + \frac{1}{\rho} \frac{\partial}{\partial \rho} + \frac{1}{\rho^2} \frac{\partial^2}{\partial \psi^2}$$

In lieu of (7.76) we may write, in analogy with (7.70), that

$$
\begin{aligned}
w(r, \theta) = & -\int_0^{2\pi} \int_0^b w^*(\rho, \psi; r, \theta) \nabla^2 M^T(\rho, \psi) \rho \, d\rho \, d\psi \\
& -\int_0^{2\pi} \left( M^T(b, \psi) \frac{\partial w^*(b, \psi; r, \theta)}{\partial \rho} \right. \\
& \left. -w^*(b, \psi; r, \theta) \frac{\partial M^T(b, \psi)}{\partial \rho} \right) b \, d\psi
\end{aligned}
\qquad (7.77)
$$

If the circular plate is clamped along its edge at $r = b$ then $w^* = \dfrac{\partial w^*}{\partial \rho} = 0$, for $r = b$, $0 \le \theta < 2\pi$, and (7.77) reduces to

$$w(r, \theta) = -\int_0^{2\pi} \int_0^b w^*(\rho, \psi; r, \theta) \nabla^2 M^T(\rho, \psi) \rho \, d\rho \, d\psi \qquad (7.78)$$

The appropriate Green's function $w^*$ in (7.78) for the case of a clamped edge is known to have the form

$$
\begin{aligned}
w^*(\rho, \psi; r, \theta) = \frac{b^2}{16\pi K} \Bigg\{ & (1 - \rho'^2)(1 - r'^2) \\
& \left( \rho'^2 + r'^2 - 2\rho'r'\cos(\theta - \psi) \ln \frac{\rho'^2 + r'^2 - 2\rho'r'\cos(\theta - \psi)}{1 + \rho'^2 r'^2 - 2\rho'r'\cos(\theta - \psi)} \right) \Bigg\}
\end{aligned}
$$

$$(7.79)$$

with $\rho' = \rho/b$ and $r' = r/b$. Various forms of the Green's function are available for the case of the simply supported solid circular isotropic plate but, as noted in [217] the expressions tend to be quite complex.

Before concluding this brief description of thermal bending of plates (and moving on to describe some problems of thermal buckling within the context of small deflection theory) we want to note some approximate methods that have been employed to deal with problems of thermal flexure of plates; two of the better-known techniques are the Rayleigh-Ritz and Galerkin procedures. In the Rayleigh-Ritz method, the displacement field $w$ is approximated by functions which contain a finite number of independent coefficients. The functions employed are chosen so as to satisfy the kinematic boundary conditions, but they do not have to satisfy the static boundary conditions. The unknown coefficients in the assumed solution are then determined by employing the principle of minimum potential energy. For the problem of thermal flexure we may, in particular, represent the transverse displacement $w(x, y)$ in the form

$$w(x, y) = \sum_{m=1}^{M} \sum_{n=1}^{N} c_{mn} \phi_{mn}(x, y) \tag{7.80}$$

It is assumed here that the $\phi_{mn}(x, y)$ satisfy the boundary conditions which involve $w$, $w_{,x}$, and $w_{,y}$. The assumed form of the solution (7.80) is then substituted into the expression for the potential energy $\Pi$ which, for a problem of (purely) thermal flexure of a homogeneous isotropic plate, is given by the following reduced form of (6.130):

$$\Pi = \int_0^b \int_0^a \left\{ \frac{K}{2} (w_{,xx} + w_{,yy})^2 \right.$$
$$+ (1 - \nu) K (w_{,xy}^2 - w_{,xx} w_{,yy}) \tag{7.81}$$
$$\left. + M^T (w_{,xx} + w_{,yy}) - tw \right\} dx dy$$

Setting $\delta \Pi = 0$, after substituting (7.80) into (7.81), yields a system of $M + N$ simultaneous algebraic equations, i.e.,

$$\frac{\partial \Pi}{\partial c_{mm}} = 0; \; m = 1, 2, ..., M; \; n = 1, 2, ..., N \tag{7.82}$$

which are then employed so as to compute the $c_{mn}$. To illustrate the use of the Rayleigh-Ritz procedure, we may consider the simple example of a square plate of side length $a$, which is simply supported along the edges at $x = 0$ and $x = a$, clamped along the edges at $y = 0$ and $y = a$, and subjected to a uniform thermal moment $M^T$. If we use the representation

$$w = \sum_{m=1}^{M} \sum_{n=1}^{N} c_{mn} \sin \frac{m\pi x}{a} \left( 1 - \cos \frac{2n\pi y}{a} \right) \tag{7.83}$$

for the transverse deflection, then we satisfy the kinematic boundary conditions for this problem but not the static boundary condition $M_x = 0$ along the edges $x = 0$ and $x = a$, $0 \le y \le a$. Retaining only the term corresponding to $m = 1$, $n = 1$ in (7.83), it is easily verified that the Rayleigh-Ritz method yields an approximation to $w(x, y)$ in which the maximum deflection, which occurs at $x = y = \frac{1}{2}a$, is given by $0.0191a^2 M^T/K$. As noted in [217], two-and three-term approximations using (7.83), yield maximum deflections of $0.0144a^2 M^T/K$ and $0.0157a^2 M^T/k$, respectively, while the "exact" value of the maximum deflection in this case is given (approximately) by $0.0158a^2 M^T/K$.

To implement the Galerkin procedure, we work directly with the relevant differential equation instead of with the associated potential energy; the equation, for the problem of thermal flexure of an isotropic, homogeneous thin plate is just (6.93a) with $\Phi \equiv 0$, i.e.,

$$K\Delta^2 w + \Delta M^T - t = 0 \tag{7.84}$$

An approximate solution of the form (7.80) is again sought, the difference being that the $\phi_{mn}(x, y)$ must satisfy all the pertinent boundary conditions. If we substitute (7.80) into (7.84) we will obtain an error (or residual) $e(x, y)$, which is given by

$$e(x, y) = K\Delta^2 w + \nabla^2 M^T - t \tag{7.85}$$

and in the Galerkin method it is required that $e(x, y)$ be orthogonal to each of the $\phi_{mn}(x, y)$, i.e., that (assume a rectangular plate, $0 \le x \le a$, $0 \le y \le b$)

$$\begin{cases} \displaystyle\int_0^b \int_0^a e(x, y)\phi_{mn}(x, y)dxdy = 0, \\ m = 1, 2, ..., M \\ n = 1, 2, ...N \end{cases} \tag{7.86}$$

By computing the integrals in (7.86) we are led to a system of $M + N$ algebraic equations for the coefficients $c_{mn}$.

**Remarks:** If one incorporates boundary residuals into the Galerkin procedure it is possible to relax the constraint that the $\phi_{mn}(x,y)$ satisfy the static as well as the kinematic boundary conditions. The first variation $\delta\Pi$ of the total potential energy $\Pi$ is, for purely thermal flexure problems, given by the following reduced (and modified) form of (6.131)

$$\delta\Pi = \int_0^b \int_0^a (K\Delta^2 w + \Delta M^T - t)\delta w\,dx\,dy \qquad (7.87)$$

$$+ \int_0^b \left\{ \left[(M_{x,x} + 2M_{xy,y} - \bar{K}_x)\delta w\right]_{x=0}^{x=a} \right.$$

$$\left. - \left[(M_x - \bar{M}_x)\delta\left(\frac{\partial w}{\partial x}\right)\right]_{x=0}^{x=a} \right\} dy$$

$$+ \int_0^a \left\{ \left[(M_{y,y} + 2M_{xy,x} - \bar{K}_y)\delta w\right]_{y=0}^{y=b} \right.$$

$$\left. - \left[(M_y - \bar{M}_y)\delta\left(\frac{\partial w}{\partial w}\right)\right]_{y=0}^{y=b} \right\} dx$$

$$- \left[(2M_{xy} - \bar{R}_{xy})\delta w\right]_{x=0,y=b}^{x=a,y=b}$$

$$+ \left[(2M_{xy} - \bar{R}_{xy})\delta w\right]_{x=0,y=0}^{x=a,y=0} = 0$$

Equation (7.87) includes the possibility of nonzero prescribed edge and corner loads $\bar{K}_x$, $\bar{K}_y$, $\bar{M}_x$, $\bar{M}_y$, and $\bar{R}_{xy}$. The variation $\delta w$ is, by virtue of (7.80), computed as

$$\delta w = \sum_{m=1}^{M} \sum_{n=1}^{N} \delta c_{mn}\phi_{mn}(x,y) \qquad (7.88)$$

If all of the boundary conditions are of kinematic type, then substitution of (7.80) and (7.88) into (7.87) yields

$$\int_0^b \int_0^a (K\Delta^2 w + \Delta M^T - t)\phi_{mn}(x,y)\,dx\,dy = 0 \qquad (7.89)$$
$$(m = 1,2,...,M; n = 1,2,...,N)$$

which is, of course, equivalent to (7.85). One also obtains (7.89) if certain of the boundary conditions, as noted in [217], are static; however, these static conditions must be satisfied identically by (7.80). Suppose we consider, as an example, the case treated earlier in this section by the Raleigh-Ritz method, i.e., a square plate clamped along two parallel edges and simply supported along the other two, and subjected to a uniform thermal moment $M^T$. For this problem, the static boundary condition $M_x = 0$ is not satisfied, as already noted, by the assumed form (7.83) of the solution. The condition (7.87) leads, in this case, to the following system of equations for the coefficients $c_{mn}$:

$$
\int_0^b \int_0^a K\Delta^2 w \cdot \phi_{mn}(x,y)dxdy
$$
$$
+ \int_0^b \left[(Kw_{,xx} + \nu Kw_{,yy} + M^T)\phi_{mn,x}\right]_{x=0}^{x=a} dy = 0 \tag{7.90}
$$
$$
(m = 1,2,...,M; \; n = 1,2,...,N)
$$

It is easily demonstrated that the coefficients $c_{mn}$ which are determined by solving (7.90) are, in fact, identical to those that are obtained by applying the Rayleigh-Ritz procedure.

In all of the work discussed to this point in this section, not only have we assumed that we are working within the domain of small deflection theory but also that the stress resultants in the plane of the plate were small enough so as to not materially influence the transverse deformations of the plate; if such is not the case then, e.g., for an isotropic thin elastic plate in rectangular coordinates, the basic equations governing the flexure and buckling of the plate are (see (6.93a,b))

$$
K\Delta^2 w = t - \Delta M^T + N_x w_{,xx} + N_y w_{,yy} + 2N_{xy} w_{,xy} \tag{7.91a}
$$

$$
\Delta^2 \Phi = -(1-\nu)\Delta N^T \tag{7.91b}
$$

where $\Phi$ is given by (6.20), $N^T$ by (6.11a), $M^T$ by (6.14), and the small deflection assumption has been enforced in writing down (7.91b). For a given temperature distribution $\delta T(x,y,z)$, and given boundary conditions along the edge of the plate, one would first compute $\Delta N^T$ and then solve (7.91b) for $\Phi \equiv \Phi_0(x,y)$; the Airy function $\Phi_0$ is then used to compute the in-plane, prebuckling stress resultants $N_x^0, N_y^0, N_{xy}^0$, which are substituted into (7.91a), along with $t$ and $\Delta M^T$. Finally (7.91a), together with appropriate support conditions with respect to $w$ along the edges of the plate, is treated as an eigenvalue–eigenfunction problem with the first eigenvalue (for a purely thermal problem) corresponding to the (smallest) critical temperature and the corresponding eigenfunction representing the first buckling mode. To illustrate the procedure delineated above, we will begin our discussion by presenting three examples that have been highlighted in Boley and Weiner [221] for isotropic plates and a rectilinear geometry; we will then proceed to examples involving circular plates as well as problems for plates with orthotropic material symmetry.

The first case treated in [221] concerns the buckling of plates subjected to heat conduction (but no transverse loads) with their edges unrestrained in the plane. We are reminded in [221] of the basic fact that if the ends of a column are free to displace axially, and the column is free from axial loads, then the column cannot buckle no matter what the temperature distribution may be; this is clearly not the case with plates. Because the plate is assumed to be free of external tractions in its plane, equilibrium relations of the form

$$\int N_x dy = 0 \qquad\qquad (7.92)$$

must be satisfied in which the integration extends across the entire plate along a line given by $x = $ const; a relation such as (7.92) cannot hold unless $N_x > 0$ along part of this line while $N_x < 0$ along its complement, thus leading to the conclusion that for this class of problems, compressive stresses will always occur in the plane of the plate. A very well known example of the type referenced above occurs in the often quoted paper of Gossard, Seide, and Roberts [216], which will be discussed in some detail later in this chapter; although our focus, with respect to the discussion of the work in [216], will be on postbuckling behavior, it should be clear that the buckling problem described, e.g., by the system (6.93a,b), within the context of small deflection theory is mathematically isomorphic to the initial buckling problem for the full non-linear system. Indeed, some specific initial buckling problems for such systems will be discussed at the end of this section.

A second class of thermal buckling problems in the realm of small deflection theory, which is discussed in [221] and which is mathematically similar to the first class of problems, concerns the buckling of plates which are subjected to heat and loads in their plane with, once again, their edges unrestrained in the plane of the plate. As an example, we consider the plate strip of Fig. 7.3, which is loaded at its ends by a uniformly distributed stress $\sigma_0$; the strip, of width $b$, is reinforced along its edges at $y = 0$, $y = b$ by longitudinals of area $A$ which act as a heat sink, thus causing the temperature to be higher along the center of the plate than near its edges. For illustration purposes, the temperature will be assumed to be uniform across the thickness of the plate and of the form

$$\delta T(x, y) = c_0 - c_1 \cos\left(\frac{2\pi y}{b}\right) \tag{7.93}$$

in the plane of the plate where $c_0, c_1$ are constants which may be chosen so as to fit empirical data. We consider a single panel of the strip, as depicted in Fig. 7.3., which extends from $x = 0$ to $x = a$; it is assumed that this panel is at a large enough distance from the ends of the strip so that the stresses can be taken to be independent of $x$. Also, we assume that $w = 0$ along the line segments $x = 0, x = a$, for $0 \le y \le b$. As we have already indicated in the discussion of the procedure for solving (7.91a,b), the first step in the solution of the problem at hand consists of determining a stress function $\Phi$ from (7.91b) and the pertinent boundary conditions; these boundary conditions are as follows:

$$\begin{cases} u(0, y) = 0, \ u(a, y) = u_0, \ 0 \le y \le b \\ v(x, 0) = 0, \ v(x, b) = v_0, \ 0 \le x \le a \end{cases} \tag{7.94}$$

where $u_0$ and $v_0$ are constants which are chosen so that

$$\int_0^b N_x(0, y)dy = \int_0^b N_x(a, y)dy = bh\sigma_0 \tag{7.95}$$

For the temperature distribution (7.93), (6.11a) and (7.91b) yield.

$$\Delta^2\Phi = -4\left(\frac{\pi}{b}\right)^2 \alpha E c_1 \cos\frac{2\pi y}{b} \tag{7.96}$$

a solution of which is

$$\Phi \equiv \Phi_0(x, y) = \frac{\sigma_0 y^2}{2} - \frac{\alpha E c_1 b^2}{4\pi^2}\cos\frac{2\pi y}{b} \tag{7.97}$$

It is easily computed that corresponding to $\Phi_0$, as given by (7.97), we have the following expressions, modulo rigid-body motions, for the stress, strain, and displacement components:

$$\begin{cases} \sigma_{yy} = N_y = \sigma_{xy} = N_{xy} = \epsilon_{xy} = 0 \\ h\sigma_{xx} = N_x = \Phi_{,yy} = h\left\{\sigma_0 + \alpha E c_1 \cos\frac{2\pi y}{b}\right\} \end{cases} \tag{7.98a}$$

$$\begin{cases} \epsilon_{xx} = \frac{1}{E}\sigma_0 + \alpha c_0 \\ \epsilon_{yy} = -\frac{\nu\sigma_0}{E} + \alpha c_0 - (1+\nu)\alpha c_1 \cos\frac{2\pi y}{b} \end{cases} \tag{7.98b}$$

$$\begin{cases} u = (\sigma_0 + \alpha E c_0)\dfrac{x}{E} \\ v = (-\nu\sigma_0 + \alpha E c_0)\dfrac{y}{E} - \dfrac{(1+\nu)\alpha c_1 b}{2\pi}\sin\dfrac{2\pi y}{b} \end{cases} \qquad (7.98c)$$

It is easily seen that (7.98a,b,c) satisfy all the boundary conditions delineated above provided

$$\begin{cases} u_0 = (\sigma_0 + \alpha E c_0)\dfrac{a}{E} \\ v_0 = (-\nu\sigma_0 + \alpha E c_0)\dfrac{b}{E} \end{cases} \qquad (7.99)$$

The next step for the problem at hand involves the computation of the transverse deflection $w(x, y)$ and the corresponding critical combination of the temperature levels and the applied load. Using the fact that $t \equiv 0$ in (7.91a), and that $M^T = 0$ for the temperature distribution defined by (7.93), it is easily seen that the use of (7.98a) in (7.91a) reduces this equation to

$$\Delta^2 w = \dfrac{h}{K}\left\{\sigma_0 + \alpha E c_1 \dfrac{\cos 2\pi y}{b}\right\} w_{,xx} \qquad (7.100)$$

We consider (7.100) with the conditions relevant for simply supported edges, namely,

$$\begin{cases} w = w_{,xx} = 0; \ x = 0, a; \ 0 \le y \le b \\ w = w_{,yy} = 0; \ y = 0, b; \ 0 \le x \le a \end{cases} \qquad (7.101)$$

and thus seek a solution of the form

$$w(x, y) = \sum_{m=1}^{\infty}\sum_{n=1}^{\infty} a_{mn} \sin\left(\dfrac{m\pi x}{a}\right)\sin\left(\dfrac{n\pi y}{b}\right) \qquad (7.102)$$

By substituting (7.102) into (7.100) and then comparing the coefficients of like terms we are led to the following system of algebraic equations for the coefficients $a_{mn}$:

$$\begin{cases} \left[k_{m1} + \sigma_0 - \dfrac{\alpha E c_1}{2}\right] a_{m1} + \dfrac{\alpha E c_1}{2} a_{m3} = 0 \\ \left[k_{m2} + \sigma_0\right] a_{m2} + \dfrac{\alpha E c_1}{2} a_{m4} = 0 \\ \left[k_{mn} + \sigma_0\right] a_{mn} + \dfrac{\alpha E c_1}{2}(a_{m,n+2} + a_{m,n-2}) = 0, \ (n > 2) \end{cases} \qquad (7.103)$$

where

$$k_{mn} = \frac{K}{h}\left(\frac{m\pi}{a}\right)^2\left\{1 + \left(\frac{na}{mb}\right)^2\right\}^2 \qquad (7.104)$$

The critical combination of $\sigma_0$ and $c_1$ is obtained by setting the determinant of the homogeneous system (7.103) equal to zero. As noted in [221] no coupling exists between coefficients with different values of $m$ or between coefficients with even and odd values of $n$. Thus, a single value of $m$ may be employed in the series (7.102), i.e., the one which yields the lowest critical combination of load and temperature levels for the loading and geometry being considered. Furthermore, as also noted in [221], one may set up two independent determinants, one with only odd values of $n$ and one with even values only. It may be shown that the symmetric case corresponding to the determinant involving only odd values of $n$ corresponds to the lower "buckling" load; the symmetric determinant has the form

$$\begin{vmatrix} \{k_{m1} + \sigma_o - \dfrac{\alpha E c_1}{2}\} & \dfrac{\alpha E c_1}{2} & 0 \cdots \\[2ex] \dfrac{\alpha E c_1}{2} & k_{m3} + \sigma_o & \dfrac{\alpha E c_1}{2} \cdots \\[2ex] 0 & \dfrac{\alpha E c_1}{2} & (k_{m5} + \sigma_o)\cdots \\[2ex] \cdots & \cdots & \cdots \end{vmatrix} = 0 \qquad (7.105)$$

for which two special cases are of interest: If $c_1 = 0$ then only the edge stress distribution $\sigma_o$ acts to buckle the panel and only the diagonal terms in (7.105) survive. In this case, the critical value of $\sigma_o$ is given by the same expression that has already been noted in Chapter 2, namely,

$$(\sigma_o)_{cr}|_{c_1=0} = -K\frac{\pi^2 E}{12(1-\nu^2)}\left(\frac{h}{b}\right)^2 \qquad (7.106a)$$

in which $n = 1$ (so as to obtain the lowest possible critical stress) and

$$k = \left(\frac{bm}{a} + \frac{a}{bm}\right)^2 \qquad (7.106b)$$

is computed, for a given aspect ratio $a/b$, by choosing the integral value of $m$ for which it is a minimum; a full discussion may be found in Chapter 2. The more interesting special case of (7.105), from the viewpoint of (purely) thermal buckling, corresponds to taking $\sigma_o = 0$ in (7.105) and seeking the smallest root $T_1 = T_{cr}$ of the resulting infinite determinant; approximations to $T_{cr}$ may be obtained from (7.105), with $\sigma_o = 0$, by retaining only a finite number of rows and columns of the determinant. By retaining only the element in the first row and the first column of (7.105), and setting $\sigma_o = 0$, we obtain

$$(\frac{\alpha E}{2})T_{cr}|_{\sigma_0=0} \approx k\frac{\pi^2 E}{12(1-\nu^2)}(\frac{h}{b})^2 = -(\sigma_0)_{cr}|_{c_1=0} \qquad (7.107)$$

with $k$ again given by (7.106b). If the first $2 \times 2$ block in (7.105) is retained and $\sigma_0$ is set equal to zero, it is possible to show that

$$(\frac{\alpha E}{2})T_{cr}|_{\sigma_0=0} = k_1\frac{\pi^2 E}{12(1-\nu^2)}(\frac{h}{b})^2 \qquad (7.108)$$

with the coefficient $k_1$ given by

$$k_1 = \frac{1}{2}(\frac{a}{mb})^2 \left\{ \sqrt{[(\frac{mb}{a})^2 + 9]^4 + 4[(\frac{mb}{a})^2 + 1]^2[(\frac{mb}{a})^2 + 9]^2} \right.$$
$$\left. -[(\frac{mb}{a})^2 + 9]^2 \right\} \qquad (7.109)$$

As indicated in [221], computations performed using larger subdeterminants of (7.105) with $\sigma_o = 0$, yield results which are very close to those presented above. In Fig. 7.4 we show a plot of $k_1$ versus the aspect ratio $a/b$ for various values of $m$; the graphs indicate that for $\frac{a}{b} \gg 1$ a good approximation to $k_1$ is given by $k_1 \approx 3.848$, which is the value that corresponds to $m = \frac{a}{b}$, while for $\frac{a}{b} < 1$ the curve for $m = 1$ in Fig. 7.4 should be used. The more general case in which both heat and applied edge loads act on the panel can be treated in a manner similar to that for the case in which $\sigma_o = 0$. Retention of the first $2 \times 2$ block in (7.105), with $\sigma_0 \neq 0, c_1 \neq 0$, leads to the results depicted in Fig. 7.7 which are interpolated quite accurately by the equation

$$\frac{T_{cr}}{T_{cr}|_{\sigma_o=0}} + \frac{(\sigma_o)_{cr}}{(\sigma_o)_{cr}|_{c_1=0}} = 1 \qquad (7.110)$$

for all combinations of heat and edge stress and all aspect ratios.

The last basic example within the context of small deflection theory that is presented in [221] concerns plates whose edges are restrained in the plane of the plate. Consider, e.g., the case of a simply supported rectangular plate whose edges are fixed in the plane of the plate and which is subjected to a temperature distribution which varies through the thickness of the plate, i.e. $\delta T = \delta T(z)$, in such a way as to cause bending; no external loads act on the plate. The displacement boundary conditions when the plate occupies the domain $0 \le x \le a, 0 \le y \le b, -\dfrac{h}{2} \le z \le \dfrac{h}{2}$ are

$$u(0, y) = u(a, y) = v(x, 0) = v(x, b) = 0 \qquad (7.111)$$

for $0 \le y \le b$ and $0 \le x \le a$. Under these conditions, the solution for the displacement components in the plane of the plate is $u = v = 0$, which implies that (see (6.91))

$$N_x = N_y = -N^T; N_{xy} = 0 \qquad (7.112)$$

so that the in-plane equilibrium equations are automatically satisfied. As $t \equiv 0$, and $\triangle M^T = 0$, equation (6.93a) reduces, in view of (7.112), to

$$K\triangle^2 w + N^T \triangle w = 0 \qquad (7.113)$$

If we associate with (7.113) the boundary conditions corresponding to simply supported edges along $x = 0, x = a$ and $y = 0, y = b$ then the pertinent boundary value problem can be shown to be equivalent to

$$\begin{cases} K \bigtriangledown^2 w + N^T w = -M^T; 0 < x < a, 0 < y < b \\[2mm] w = 0; \begin{cases} x = 0, a; \ 0 \le y \le b \\ y = 0, b; \ 0 \le x \le a \end{cases} \end{cases} \qquad (7.114)$$

Taking $w(x, y)$ is the form

$$w = \sum_{m=1}^{\infty} y_m(y) \sin\left(\frac{m\pi x}{a}\right) \qquad (7.115)$$

and expanding the constant $M^T$ as

$$M^T = \left(\frac{4M^T}{\pi}\right) \sum_{m=1,3,5...}^{\infty} \left(\frac{1}{m}\right) \sin\left(\frac{m\pi x}{a}\right) \qquad (7.116)$$

the authors [221] obtain the following equation for the $y_m$:

$$\begin{cases} \dfrac{d^2 y_m}{dy^2} - \beta_m^2 y_m = -\dfrac{4M^T}{km\pi}; m = 1, 3, 5... \\[3mm] \beta_m = \sqrt{\left(\dfrac{m\pi}{a}\right)^2 - \dfrac{1}{K}N^T} \end{cases} \qquad (7.117)$$

Once again, as $M^T$ does not vanish along the edges of the plate, such a solution technique is subject to the criticisms raised earlier. In Fig. 7.6, we show a nondimensional plot of the deflection (as computed in [221] at the center of the rectangular plate) against the temperature parameter $N^T/(N^T)_{cr}$ for various aspect ratios $\dfrac{a}{b}$; here $(N^T)_{cr}$, the value of $N^T$ at which buckling occurs is computed to be

$$(N^T)_{cr} = (1 + \frac{a^2}{b^2})\frac{\pi^2}{a^2} \tag{7.118}$$

The (nondimensional) variation of the bending moment $M_x$ in a square plate is shown in Figs. 7.7 and 7.8 for two different values of the "temperature level" $N^T/(N^T)_{cr}$; such plots are sufficient to determine both $M_x$ and $M_y$ throughout the entire plate because of the double symmetry exhibited by the plates; the results depicted in Figs. 7.7 and 7.8 indicate that the maximum bending moment occurs at the center of the plate. Finally, in Fig. 7.9 we show, for various aspect ratios, the variations in $M_x$ at the center of the plate, with the temperature parameter $N^T/(N^T)_{cr}$. All of the results depicted in Figs. 7.6–7.9 are valid only for values of $N^T$ that are sufficiently close to $(N^T)_{cr}$ because of the small deflection assumption employed in their derivation.

For the special case (in rectangular coordinates) in which (6.94) reduces to

$$\delta T(x, y, z) = T_o(x, y) \tag{7.119}$$

so that, by virtue of (6.95),

$$N^T = \frac{\alpha E h}{1 - \nu} T_o(x, y); \quad M^T = 0 \tag{7.120}$$

Equations (6.93a) and (6.93b) for the case of isotropic response, reduce to

$$K\triangle^2 w = \Phi_{,yy} w_{,xx} - 2\Phi_{,xy} w_{,xy} + \Phi_{,xx} w_{,yy} \tag{7.121a}$$

$$\triangle^2 \Phi + \alpha E h \triangle T_o = 0 \tag{7.121b}$$

provided $t \equiv 0$ and small deflection theory is assumed. Equations (7.121a, b) appear, e.g., in Nowacki [222] with $F = \dfrac{1}{h}\Phi$, the Airy function associated with the stress components, in lieu of $\Phi$. If we assume that, for a given temperature field (7.119), equation (7.121b) has been solved for $\Phi$, then the stress resultants $N_x, N_y, N_{xy}$ are known and equation (7.121a), namely,

$$\Delta^2 w = \frac{1}{K}(N_x w_{,xx} + 2N_{xy} w_{,xy} + N_y w_{,yy}) \equiv \frac{1}{K} N_{ij} w_{,ij} \qquad (7.122)$$

must be solved subject to the specification of boundary conditions on $w$. As an example, we consider a rectangular plate with $0 \leq x \leq a, 0 \leq y \leq b$, which is simply supported along its edges; for this case $M^T = 0$ along the edges of the plate so our previous criticisms do not apply.

One approach to dealing with (7.22) is similar to Maysel's method, which was discussed in connection with the thermal bending of plates. We introduce the Green's function $w^*$, which satisfies the equation

$$\Delta w^*(x, y; \xi_1, \xi_2) = \frac{1}{K} \delta(x - \xi_1) \delta(y - \xi_2) \qquad (7.123)$$

in the sense of distributions and the same boundary conditions as $w$. Combining (7.122) and (7.123) we obtain

$$w(x, y) = \int_o^a \int_o^b w^*(x, y; \xi_1, \xi_2) N_{ij}(\xi_1, \xi_2) w_{,ij}(\xi_1, \xi_2) d\xi_1 d\xi_2 \qquad (7.124)$$

If we apply the Green's transformation to the right-hand side of (7.124), assume that the plate is simply supported along its edges (or clamped), and use the planar equilibrium equations $N_{ij,ij} = 0$, we find that (7.124) yields the following Fredholm integral equation of the second kind for $w(x, y)$:

$$w(x, y) = \int_o^a \int_o^b w(\xi_1, \xi_2) N_{ij}(\xi_1, \xi_2) \frac{\partial^2 w^*}{\partial \xi_i \partial \xi_j} d\xi_1 d\xi_2 \qquad (7.125)$$

Now, for the rectangular plate described above which is simply supported on all four edges, $w^*(x, y, \xi_1, \xi_2)$ is given by (7.73) with $\xi \to \xi_1, \eta \to \xi_2$. If we assume that the solution of the integral equation (7.125) can be represented in series form as

$$w(x, y) = \sum_{i,k}^{\infty} A_{ik} \sin \alpha_i \sin \beta_k y, \qquad (7.126)$$

$\alpha_n = \dfrac{n\pi}{a}, \beta_m = \dfrac{m\pi}{b}$ (thus automatically satisfying the boundary conditions of simple support) then by substituting the series representations for $w$ and $w^*$ into (7.125), and performing elementary computations, we are led to an infinite system of linear equations for the coefficients $A_{ik}$ in (7.126) of the form

$$A_{ik} + \frac{4}{abk(\alpha_i^2 + \beta_k^2)} \sum_{n,m} A_{mn}G_{nimk} = 0$$

$$(i, k = 1, 2, \cdots, \infty)$$

(7.127)

where

$$G_{nimk} = \alpha_i^2 a_{nimk} + 2\alpha_i\beta_k c_{nimk} + \beta_k^2 b_{nimk}$$

with

$$
\begin{cases}
a_{nimk} = \displaystyle\int_0^a \int_0^b N_{11}(\xi_1, \xi_2) \sin \alpha_i\xi_1 \sin \alpha_n\xi_1 \sin \beta_k\xi_2 \sin \beta_m\xi_2 d\xi_1 d\xi_2 \\[2ex]
b_{nimk} = \displaystyle\int_0^a \int_0^b N_{22}(\xi_1, \xi_2)) \sin \alpha_i\xi_1 \sin \alpha_n\xi_1 \sin \beta_k\xi_2 \sin \beta_m\xi_2 d\xi_1 d\xi_2 \\[2ex]
c_{nimk} = \displaystyle\int_0^a \int_0^b N_{12}(\xi_1, \xi_2) \cos \alpha_k\xi_1 \sin \alpha_n\xi_1 \cos \beta_k\xi_2 \sin \beta_m\xi_2 d\xi_1 d\xi_2
\end{cases}
$$

The condition for buckling of the plate is, of course, that the (infinite) determinant of the system of equations (7.117) be zero. A related (classical) treatment of the thermal buckling of a simply supported isotropic rectangular plate which employs the Rayleigh-Ritz procedure may be found in Klosner and Forray [227]; in [227], the temperature distribution is of the form (7.119) and, in fact, is assumed to be symmetrical about the centerlines of the plate so that it is representable in the form of a double Fourier series in the functions $\cos \dfrac{n\pi x}{a} \cos \dfrac{m\pi y}{b}$ (the plate in [227] has length $2a$ and width $2b$).

There are many excellent treatments of the problem of general thermal deflections of an isotropic elastic circular plate, one of which may be found in [228]. We assume that the plate has radius $b > 0$ and is subjected to the radially symmetric subcase of (6.113), namely,

$$\delta T = T_o(r) + zT_1(r)$$

(7.128)

It is also assumed in [228] that the edge of the plate is subjected to a uniform force $P$ per unit length of the arc parameter $s$ along the edge; for problems involving deflections due to temperature variations only, we may set $P = 0$ in the results which follow below. Within the scope of

small deflection theory it is easily seen that (7.128) and (6.109) combine
with the second equation in (6.112) so as to yield

$$\triangle^2 \Phi = -\alpha E h \triangle T_o \qquad (7.129)$$

Clearly, if $\Phi_p$ is a solution of

$$\triangle \Phi_p = -\alpha E h T_o \qquad (7.130)$$

it is also a solution of (7.129). A particular solution of (7.129) is thus
obtained by using potential theory to integrate the Poisson's equation
(7.130), where $\triangle = \dfrac{1}{r}\dfrac{d}{dr}(r\dfrac{d}{dr})$ in view of the radial symmetry of $T_o$; we
obtain

$$\Phi_p(r) = -\alpha E h \int_o^r (\int_o^\xi T_o(\lambda)\lambda d\lambda)\frac{1}{\xi}d\xi \qquad (7.131)$$

The general solution of (7.129) which satisfies the edge conditions delineated above may then be shown [228] to have the form

$$\Phi(r) = \frac{1}{2}\{-P + \frac{\alpha E h}{r^2}\int_o^r T_o(\lambda)\lambda d\lambda\}r^2$$

$$\qquad (7.132)$$

$$-\alpha E h \int_o^r (\int_o^\xi T_o(\lambda)\lambda d\lambda)\frac{1}{\xi}d\xi$$

The first equation in the set (6.112) may now be written in the following
form when $t \equiv 0$:

$$\triangle^2 w = -\Gamma \triangle T_1 + \frac{1}{K}\{N_r w_{,rr} + 2N_{r\theta}(\frac{1}{r}w_{,\theta})_{,r}$$

$$\qquad (7.133)$$

$$+N_\theta(\frac{1}{r}w_{,r} + \frac{1}{r^2}w_{,\theta\theta})\}$$

where $N_r, N_\theta, N_{r\theta}$ are given in terms of $\Phi$ by (6.46) and

$$\Gamma = E\alpha h^3/12K(1-\nu)$$

In fact, with $\Phi$ as given by (7.132), we have

$$\begin{cases} \sigma_{rr}^o = -P + \dfrac{\alpha E}{b^2}\displaystyle\int_o^b T_o(\lambda)\lambda d\lambda - \dfrac{\alpha E}{r^2}\displaystyle\int_o^r T_o(\lambda)\lambda d\lambda \\[4mm] \sigma_{\theta\theta}^o = -P + \dfrac{\alpha E}{b^2}\displaystyle\int_o^b T_o(\lambda)\lambda d\lambda + \dfrac{\alpha E}{r^2}\displaystyle\int_o^r T_o(\lambda)\lambda d\lambda \\[4mm] \qquad\qquad -\alpha E T_o(r) \\[4mm] \sigma_{r\theta}^o = 0 \end{cases} \qquad (7.134)$$

the superscripted $o$ denoting, of course, that we are computing the middle surface stress field distribution. Because $N_r, N_\theta, N_{r\theta}$, and $T_1$ are independent of $\theta$, it is easily seen that (7.133) for the transverse deflection $w = w(r)$ may be rewritten in the form:

$$\frac{d}{dr}(r\frac{d}{dr}[\frac{1}{r}\frac{d}{dr}\{r\frac{dw}{dr}\}])$$

$$= -\Gamma\frac{d}{dr}(r\frac{dT_1}{dr})$$

$$+ \frac{h}{K}\left[\left(P_0 r - \frac{\alpha E}{r}\int_0^r T_0(\lambda)\lambda d\lambda\right)\frac{d^2 w}{dr^2}\right.$$

$$\left. + \left(P_0 + \frac{\alpha E}{r^2}\int_0^r T_0(\lambda)\lambda d\lambda - \alpha E T_0\right)\frac{dw}{dr}\right] \tag{7.135}$$

with the constant $P_0$ given by

$$P_0 = \frac{\alpha E}{b^2}\int_0^b T_0(\lambda)\lambda d\lambda - P \tag{7.136}$$

An integration of (7.135) now yields

$$\frac{d}{dr}\left\{\frac{1}{r}\frac{d}{dr}\left(r\frac{dw}{dr}\right)\right\} = -\Gamma\frac{dT_1}{dr}$$

$$+ \frac{h}{K}\left(P_0 - \frac{\alpha E}{r^2}\int_0^r T_0(\lambda)\lambda d\lambda\right)\frac{dw}{dr} + \frac{k_1}{r} \tag{7.137}$$

with $k_1$ a constant of integration. Thus, our problem has been reduced to that of solving the third-order ordinary differential equation (7.137) with appropriate boundary conditions. It is noted in [228] that further integrations of (7.137) are not possible unless specific forms for $T_0(r)$ and $T_1(r)$ are assumed; in [228] these are taken as the truncated power series expansions

$$\begin{cases} T_0(r) = \dfrac{K}{\alpha E h}\displaystyle\sum_{j=0}^n t_{0j}r^j \\[3mm] T_1(r) = -\dfrac{1}{\Gamma}\displaystyle\sum_{j=0}^n t_{1j}r^j \end{cases} \tag{7.138}$$

with the $t_{0j}, t_{1j}$ real constants and $m, n$ arbitrary positive integers. A solution of (7.137) is now sought in the form of a power series. Substituting (7.138) in (7.137) and setting

$$c_j = \frac{t_{0j}}{j+2}, \quad d_j = jt_{1j}$$

we obtain

$$\frac{d^3w}{dr^3} + \frac{d^2w}{dr^2} + \left\{ \sum_{j=0}^{m} c_j(r^j - b^j) \right.$$

$$+ \frac{hP}{K} - \frac{1}{r^2} \right\} \frac{dw}{dr} = \sum_{j=1}^{n} d_j r^{j-1}$$

(7.139)

Setting

$$\begin{cases} u(r) = \dfrac{dw}{dr} \\ b_0' = \dfrac{hP}{K} - \displaystyle\sum_{j=1}^{m} c_j b^j \end{cases}$$

then yields the equation

$$\frac{d^2u}{dr^2} + \frac{1}{r}\frac{du}{dr} + \left( \sum_{j=1}^{m} c_j r^j + b_0' - \frac{1}{r^2} \right) u = \sum_{j=0}^{n-1} d_{j+1} r^j \qquad (7.140)$$

for $u = u(r)$. Relative to (7.140) we now consider solutions of the form

$$u(r) = r \sum_{i=0}^{\infty} \lambda_i r^i \qquad (7.141)$$

Inserting (7.141) in (7.140) a recurrence relation of the form

$$(j+4)(j+2)\lambda_{j+2} + b_0'\lambda_j + \sum_{i=0}^{m} c_j\lambda_{j-1}$$

$$= \begin{cases} d_{j+2}, \ j = -1, 0, ..., n-2 \\ 0, \ j = n-1, n, ... \end{cases} \qquad (7.142)$$

is generated for the $\lambda_j$ with $\lambda_j = 0$ if $j < 0$. A careful discussion of the convergence of the series (7.141) with the $\lambda_j$ as given by (7.142) may be found in [228]. Inserting (7.141) into the equation $\dfrac{dw}{dr} = u(r)$ and integrating, we find that the transverse deflection may be represented in the form

$$w(r) = r^2 \sum_{i=0}^{\infty} \kappa_i r^i + \kappa \qquad (7.143)$$

with $\kappa$ a constant of integration and $\kappa_i = \lambda_i/(i+2)$. By referring to (7.142) it may be deduced that

$$\kappa_i = \lambda_0 \xi_i + \delta_i$$

where $\xi_i$ contains the parameters $P, K, h, b$ and some (or all) of the $t_{0j}$, while $\delta_i$ contains these parameters as well as some (or all) of the $t_{1j}$. Thus, (7.143) may be rewritten in the form

$$w(r) = \lambda_0 w_0(r) + w_1(r) + r \qquad (7.144)$$

where

$$w_0(r) = \sum_{j=0}^{\infty} \xi_j r^j, \quad w_1(r) = \sum_{j=0}^{\infty} \delta_j r^j \qquad (7.145)$$

It is easily seen that $w_1(r)$ is a particular solution of (7.139), while $w_0(r)$ is the solution of the corresponding homogeneous equation which is bounded at $r = 0$.

The constants $\lambda_0$ and $r$ in (7.144) are determined by the support conditions along the edge of the plate at $r = b$; these support conditions for the clamped edge and the simply supported edge are, respectively, in view of (6.115)

$$w'(b) = 0, \ w(b) = 0 \qquad (7.146a)$$

and

$$\begin{cases} K[w''(b) + \dfrac{\nu}{b} w'(b)] + \Gamma T_1(b) = 0 \\ w(b) = 0 \end{cases} \qquad (7.146b)$$

We will proceed by considering the clamped edge conditions only; the analysis corresponding to the simply supported conditions in [228] would appear to be correct only if $T_1$ vanishes along the edge of the plate at $r = b$.

Substituting (7.144) into (7.146a) yields, for the clamped plate,

$$\lambda_0 = -\frac{w_1'(b)}{w_0'(b)}, \ r = \frac{w_1'(b)}{w_0'(b)} w_0(b) - w_1(b) \qquad (7.147)$$

Inserting the values of $\lambda_0$, $\kappa$ in (7.137) into (7.144) we obtain for the plate which is clamped along the edge at $r = b$

$$w(r) = \frac{w_1'(b)}{w_0'(b)} \{w_0(b) - w_0(r)\} + w_1(r) - w_1(b) \qquad (7.148)$$

It will be assumed that $w_0'(b) \neq 0$; as noted in [228], if $w_0'(b) = 0$ then $b$ is the radius of the clamped plate with given temperature distribution $t_0(r)$ for which $P$ is a critical buckling load.

In the special case in which $m = 0$, $T_0(r)$ reduces, in light of (7.138), to a constant while (7.140) becomes

$$\frac{d^2u}{dr^2} + \frac{1}{r}\frac{du}{dr} + \left(\frac{hP}{K} - \frac{1}{r^2}\right)u = \sum_{j=0}^{n-1} d_{j+1}r^j \qquad (7.149)$$

Using (7.141) again, and working with the special case of the recurrence relation (7.142) for $m = 0$, it is not difficult to show that the transverse deflection assumes the form

$$w(r) = \lambda_0 J_0\left(r\sqrt{\frac{hP}{K}}\right) + w_{10}(r) + \kappa \qquad (7.150)$$

with $J_0$ the Bessel function of the first kind of order zero and $w_{10}(r)$ the form assumed by $w_1(r)$ for $m = 0$. Applying the boundary conditions (7.146a) we now find as the expression for the transverse deflection of the circular, isotropic plate clamped along its edge at $r = b$

$$w(r) = \frac{w'_{10}(b)}{\sqrt{\frac{h}{K}}J_0\left(b\sqrt{\frac{hP}{K}}\right)}\left\{J_0\left(r\sqrt{\frac{hP}{K}}\right)\right.$$
$$\left. - J_0\left(b\sqrt{\frac{hP}{K}}\right)\right\} + w_{10}(r) - w_{10}(b) \qquad (7.151)$$

The analysis of the isotropic circular plate presented above involved no considerations of plate stability; as has already been indicated, the thermal buckling problem within the context of small deflection theory is mathematically equivalent to the initial (thermal) buckling problem without the small deflection assumption. For an isotropic plate, the thermal stress resultant $N^T$ as given by (6.11a) may be written in the form

$$N^T = \frac{\alpha Eh}{1 - \nu}T_m(x, y) \qquad (7.152)$$

where

$$T_m(x, y) = \frac{1}{h}\int_{-h/2}^{h/2} \delta T(x, y, z)dz \qquad (7.153)$$

may be thought of as the medium temperature in the plate. Suppose that

$$\int_{-h/2}^{h/2} \delta T(x, y, z)zdz = 0$$

so that $M^T = 0$. The buckling problem (in rectangular coordinates), either assuming small deflection theory or focusing on the initial buckling problem, then takes the form (see (6.93a,b))

$$\Delta^2 \Phi = -\alpha E h \Delta T_m$$
$$\Delta^2 w = \frac{1}{K}(\Phi_{,yy} w_{,xx} - 2\Phi_{,xy} w_{,xy} + \Phi_{,xx} w_{,yy})$$

(7.154)

if we again assume that $t \equiv 0$. Following the stability analysis in [229], we let $T_0$ represent the maximum value (upper bound) of $T_m$ on the domain of the plate so that $0 \leq T_m \leq T_0$ and we introduce the dimensionless temperature parameter $\mathcal{T} = T_m/T_0$ so that $0 \leq \mathcal{T} \leq 1$. If $b$ represents a characteristic length associated with the plate, then in rectangular coordinates, we may introduce the dimensionless variables $\xi = x/b$, $\eta = y/b$, in which case (7.154) would assume the form

$$\Delta^2 \phi = -\Delta \mathcal{T}$$
$$\Delta^2 w = \lambda(\phi_{,\eta\eta} w_{,\xi\xi} - 2\phi_{,\xi\eta} w_{,\xi\eta} + \phi_{,\xi\xi} w_{,\eta\eta})$$

(7.155)

where

$$\phi = \Phi/b^2 \alpha E h T_0$$

(7.156)

while, by virtue of the definition of $K$,

$$\lambda = 12(1 - \nu^2)\left(\frac{b}{h}\right)^2 \alpha T_0$$

(7.157)

and, of course, $\Delta = \dfrac{\partial^2}{\partial \xi^2} + \dfrac{\partial^2}{\partial \eta^2}$. Following the discussion in [229], the edge(s) of the plate are assumed to be free of applied forces and moments so that the buckling (stability) problem consists of finding the temperature parameter $\mathcal{T}$, which minimizes the value of $\lambda$ in (7.155); this may be achieved, e.g., by following the Raleigh-Ritz procedure and considering the expression

$$\Lambda = \Lambda_N/\Lambda_D$$

(7.158)

with

$$\begin{cases} \Lambda_N = \displaystyle\iint_A \left\{(w_{,\xi\xi} + w_{,\eta\eta})^2 - 2(1 - \nu)[w_{,\xi\xi} w_{,\eta\eta} - w_{,\xi\eta}^2]\right\} d\xi d\eta \\[6mm] \Lambda_D = \displaystyle\iint_A \left\{2\phi_{,\xi\eta} w_{,\xi} w_{,\eta} - \phi_{,\eta\eta} w_{,\xi}^2 - \phi_{,\xi\xi} w_{,\eta}^2\right\} d\xi d\eta \end{cases}$$

where $\mathcal{A}$ is the domain occupied by the midplane of the plate. As $\lambda \leq \Lambda$ for all admissible functions, $w = w^*(\xi, \eta)$, which satisfy the edge support conditions, and, for any such $w^*$, $\Lambda_N$ remains constant while $\Lambda_D$ is a functional of $\phi$, the buckling problem may be cast in the following form: compute the

$$\max_{w^*} \iint_{\mathcal{A}} \{2\phi_{,\xi\eta} w^*_{,\xi} w^*_{,\eta} - \phi_{,\eta\eta} w^{*2}_{,\xi} - \phi_{,\xi\xi} w^{*2}_{,\eta} - \phi_{,\xi\xi} w^{*2}_{,\eta}\} \, d\xi d\eta$$

for $\phi(\xi, \eta)$ satisfying (7.155), subject to $0 \leq \mathcal{T} \leq 1$.

We now return to the problem of thermal buckling of an isotropic elastic circular plate and apply the general methodology elucidated above to the special case of a circular plate subjected to a radially symmetric (nondimensional) temperature field $\mathcal{T}$; as in [229], however, the buckling mode will be allowed to depend on the angular coordinate $\theta$. As the Airy function $\Phi$ is also independent of $\theta$ the membrane equation (7.154) reduces to an ordinary differential equation in the radial coordinate $r$; setting $G = F'(r)$ this equation becomes

$$G''(r) + \frac{1}{r}G'(r) - \frac{1}{r^2}G = -E\alpha h \frac{dT_m}{dr} \tag{7.159}$$

and

$$N_r = \frac{1}{r}G, \quad N_\theta = G'(r) \tag{7.160}$$

Using the plate radius $b$ as a reference length and introducing the nondimensional coordinate $\rho = r/b$ we may rewrite (7.159) in the form

$$g''(\rho) + \frac{1}{\rho}g'(\rho) - \frac{1}{\rho^2}g = -\frac{d\mathcal{T}}{d\rho} \tag{7.161}$$

with

$$g(\rho) = G(\rho)/E\alpha h b T_0 \tag{7.162}$$

The solution of (7.161) which satisfies the boundary condition $N_r(b) = 0$ and also satisfies the condition that $N_r(0)$ and $N_\theta(0)$ remain bounded, is

$$g(\rho) = -\frac{1}{\rho}\int_0^\rho \mathcal{T}(\lambda)\lambda d\lambda + \rho \int_0^1 \mathcal{T}(\lambda)\lambda d\lambda \tag{7.163}$$

As noted in [229], it may be demonstrated that the solution defined by (7.163) remains valid for a noncontinuous temperature distribution. For the problem at hand, (7.158) can be written in the following form

$$\Lambda = \frac{\displaystyle\int_0^{l'} \int_0^{2\pi} [(\Delta w)^2 - (1-\nu)L(w,w)]\rho\, d\rho\, d\theta}{-\displaystyle\int_0^{l'} \int_0^{2\pi} \left[\frac{1}{\rho}g(\rho)\left(\frac{\partial w}{\partial \rho}\right)^2 + \frac{1}{\rho^2}\frac{dg}{d\rho}\left(\frac{\partial w}{\partial \theta}\right)^2\right]\rho\, d\rho\, d\theta} \tag{7.164}$$

where

$$\Delta w = \frac{\partial^2 w}{\partial \rho^2} + \frac{1}{\rho}\frac{\partial w}{\partial \rho} + \frac{1}{\rho^2}\frac{\partial^2 w}{\partial \theta^2}$$

and

$$L(w,w) = 2\left(\frac{1}{\rho}\frac{\partial w}{\partial \rho} + \frac{1}{\rho^2}\frac{\partial^2 w}{\partial \theta^2}\right)\frac{\partial^2 w}{\partial \rho^2}$$
$$-2\left(\frac{1}{\rho}\frac{\partial^2 w}{\partial \rho\partial \theta} - \frac{1}{\rho^2}\frac{\partial w}{\partial \theta}\right)^2 \tag{7.165}$$

Following the analysis in [229] we now introduce the admissible functions

$$w = w^*(\rho,\theta) = w_m(\rho)\cos m\theta, \quad m = 0,1,2... \tag{7.166}$$

which are subject to the boundary conditions

$$m_r(b) = 0, \quad \left[q_r + \frac{1}{b}\frac{\partial m_{r\theta}}{\partial \theta}\right]_{r=b} = 0 \tag{7.167}$$

where $q_r, m_r$, and $m_{r\theta}$ are (in $r,\theta$ coordinates) the $z$-direction shear force, bending moment, and twisting moment, respectively, per unit length along the edge at $r = b$. With respect to the set of admissible functions in (7.166), the boundary conditions (7.167) become

$$\left[\frac{d^2 w_m}{d\rho^2} + \nu\left(\frac{dw_m}{d\rho} - m^2 w_m\right)\right]_{\rho=1} = 0 \tag{7.168a}$$

$$\left\{\frac{d^3 w_m}{d\rho^3} + 2\frac{d^2 w_m}{d\rho^2} + [1 + m^2(2-\nu)]\frac{dw_m}{d\rho}\right.$$
$$\left. + m^3(3-\nu)w_m\right\}_{\rho=1} = 0 \tag{7.168b}$$

To the conditions (7.167) we append

or

$$\begin{cases} w(0) = 0, \; \dfrac{dw}{dr}\bigg|_{r=0} = 0 \\[4mm] w_m\bigg|_{\rho=0} = 0, \; \dfrac{dw_m}{d\rho}\bigg|_{\rho=0} = 0 \end{cases} \tag{7.168c}$$

which has the effect of eliminating undetermined rigid body displacements of the plate.

By introducing the $w^*(\rho, \theta)$ in (7.166) into (7.164) we obtain for $\Lambda$ the expression

$$\Lambda = \frac{\int_0^1 \left\{ [\Delta_m(w_m)]^2 - (1-\nu)L_m(w_m, w_m) \right\} \rho \, d\rho}{-\int_0^1 \left[ g(\rho) \left( \frac{dw_m}{d\rho} \right)^2 + \frac{m^2}{\rho} \frac{dg}{d\rho} w_m^2 \right] d\rho} \tag{7.169}$$

where

$$\Delta_m(w_m) = \frac{d^2 w_m}{d\rho^2} + \frac{1}{\rho} \frac{dw_m}{d\rho} - \frac{m^2}{\rho^2} w_m \tag{7.170a}$$

and

$$L_m(w_m, w_m) = 2 \left[ \left( \frac{1}{\rho} \frac{dw_m}{d\rho} - \frac{m^2}{\rho^2} w_m \right) \frac{d^2 w_m}{d\rho^2} \right.$$
$$\left. -m^2 \left( \frac{1}{\rho} \frac{dw_m}{d\rho} - \frac{1}{\rho^2} w_m \right)^2 \right] \tag{7.170b}$$

Therefore, for any $w_m(\rho)$ satisfying (7.168a,b,c) and for any given wave number $m$ of the buckling mode, $\Lambda$ is minimized by requiring that the denominator of (7.169) be a maximum over all functions $g(\rho)$ of the form (7.163) for $0 \leq T \leq 1$. We recall that the circumferential membrane force $N_\theta(\rho)$ must, because of the planar equilibrium conditions, change its sign at least once in the interval $0 \leq \rho \leq 1$, i.e., ([221], §13), $\int_0^1 N_\theta(\rho) d\rho = 0$. If, as in [229], we assume that there exists but one change of sign of $N_\theta(\rho)$ at the value $\rho = \mu$ then, using the relations (7.160) as applied to $g$ in lieu of $G$, we have the following:
1) For a plate heated near its center at $\rho = 0$

$$\begin{cases} \dfrac{dg}{d\rho} \leq 0, \ g \leq 0, \text{ for } 0 \leq \rho \leq \mu \\ \dfrac{dg}{d\rho} \geq 0, \ g \leq 0, \text{ for } \mu \leq \rho \leq 1 \end{cases} \tag{7.171a}$$

2) For a plate heated near its edge at $\rho = 1$

$$\begin{cases} \dfrac{dg}{d\rho} \geq 0, \ g \geq 0, \text{ for } 0 \leq \rho \leq \mu \\ \dfrac{dg}{d\rho} \leq 0, \ g \geq 0, \text{ for } \mu \leq \rho \leq 1 \end{cases} \tag{7.171b}$$

From (7.163) we compute

$$\frac{dg}{d\rho} = \frac{1}{\rho^2}\int_0^\rho T(\lambda)\lambda d\lambda - T(\rho) + \int_0^1 T(\lambda)\lambda d\lambda \qquad (7.172)$$

so that, by virtue of (7.171a,b)

$$\frac{1}{\mu^2}\int_0^\mu T(\lambda)\lambda d\lambda - T(\mu) + \int_0^1 T(\lambda)\lambda d\lambda = 0 \qquad (7.173)$$

Equation (7.173) serves to determine all points $\rho = \mu$ at which a change of sign of $N_\theta(\rho)$, i.e., $g'(\rho)$, may occur, it being understood that the assumption that $N_\theta(\rho)$ can change sign at only one point in the interval $0 \le \rho \le 1$ places a restriction on the temperature distribution.

We now write the denominator $\Lambda_0$ in (7.169) in the form

$$\Lambda_0 = \int_0^\mu \left[ g(\rho)\left(\frac{dw_m}{d\rho}\right)^2 + \frac{m^2}{\rho}\frac{dg}{d\rho}w_m^2 \right] d\rho$$
$$- \int_\mu^1 \left[ g(\rho)\left(\frac{dw_m}{d\rho}\right)^2 + \frac{m^2}{\rho}\frac{dg}{d\rho}w_m^2 \right] d\rho$$

and effect the same splitting in both (7.163) and (7.172), i.e., if we set

$$\begin{cases} I = \displaystyle\int_0^\mu T(\lambda)\lambda d\lambda \\[2mm] II = \displaystyle\int_\mu^1 T(\lambda)\lambda d\lambda \end{cases} \qquad (7.174)$$

then

$$\begin{cases} g = -\dfrac{1}{\rho}\displaystyle\int_0^\rho T(\lambda)\lambda d\lambda + \rho I + \rho II \\[3mm] \dfrac{dg}{d\rho} = \dfrac{1}{\rho^2}\displaystyle\int_0^\rho T(\lambda)\lambda d\lambda - T(\rho) + I + II \end{cases}, \quad 0 \le \rho \le \mu \qquad (7.175)$$

and

$$\begin{cases} g = -\dfrac{1}{\rho}\displaystyle\int_\mu^\rho T(\lambda)\lambda d\lambda + \rho II - \left(\dfrac{1}{\rho} - \rho\right)I \\[3mm] \dfrac{dg}{d\rho} = \dfrac{1}{\rho^2}\displaystyle\int_\mu^\rho T(\lambda)\lambda d\lambda - T(\rho) + \left(\dfrac{1}{\rho^2} + 1\right)I + II \end{cases}, \quad \mu \le \rho \le 1$$

$$(7.176)$$

If we begin by first considering a radially symmetric buckling mode then, by (7.166), $m = 0$ and $\Lambda_D$ reduces to

$$\left.\Lambda_D\right|_{m=0} = -\int_0^\mu g(\rho)\left(\frac{dw_0}{d\rho}\right)^2 d\rho - \int_\mu^1 g(\rho)$$
$$- \int_\mu^1 g(\rho)\left(\frac{dw_0}{d\rho}\right)^2 d\rho \qquad (7.177)$$

where, for a plate heated near its center at $\rho = 0$, $g \leq 0$. As in [229], we assume that the temperature field $\mathcal{T}$ is restricted so as to satisfy

$$-\frac{1}{\rho}\int_\mu^\rho \mathcal{T}(\lambda)\lambda d\lambda + \rho II \geq 0 \qquad (7.178)$$

which prohibits the presence of strong oscillation in $\mathcal{T}$. Using (7.178) it follows from the form of $g(\rho)$ in (7.175), (7.176) that $-g(\rho)$ is maximized for $0 \leq \rho \leq 1$ by setting $\mathcal{T}(\rho) \equiv 0$ for $\mu \leq \rho \leq 1$; this latter condition now reduces $g(\rho)$ to the form

$$g(\rho) = \begin{cases} -\dfrac{1}{\rho}\displaystyle\int_0^\rho \mathcal{T}(\lambda)\lambda d\lambda + \rho I, \ 0 \leq \rho \leq \mu \\ -\left(\dfrac{1}{\rho} - \rho\right) I, \ \mu \leq \rho \leq 1 \end{cases} \qquad (7.179)$$

However, for $\mu \leq \rho \leq 1$, it is clear that $-g(\rho)$ is maximized by maximizing $I$ and, in view of (7.174), we obtain (set $\mathcal{T}(\lambda) \equiv 1$, $0 \leq \lambda \leq \mu$)

$$g(\rho) = -\left(\frac{1}{\rho} - \rho\right)\frac{\mu^2}{2}, \ \mu \leq \rho \leq 1 \qquad (7.180)$$

An analogous result holds for $-g(\rho)$ with $\rho$ in the interval $[0, \mu]$, namely, we have

$$g(\rho) = -\frac{\rho}{2}(1 - \mu^2), \ 0 \leq \rho \leq \mu \qquad (7.181)$$

By introducing (7.180) and (7.181) into (7.177) we see that the function to be maximized becomes

$$\left.\Lambda_D\right|_{m=0} = \frac{1}{2}(1 - \mu^2)\int_0^\mu \rho\left(\frac{dw_m}{d\rho}\right)^2 d\rho$$
$$+ \frac{1}{2}\mu^2 \int_\mu^1 \left(\frac{1}{\rho} - \rho\right)\left(\frac{dw_m}{d\rho}\right)^2 d\rho \qquad (7.182)$$

As $0 \le \rho \le 1$, and $0 < \mu < 1$, the expression on the right-hand side of (7.182) is positive; for any admissible function $\Lambda_D\big|_{m=0}$ may be computed and, moreover, the value of $\mu$ can be determined which maximizes $\Lambda_D\big|_{m=0}$.

If we retain the same temperature field, i.e., $\mathcal{T} = 1$ for $0 \le \rho \le \mu$ and $\mathcal{T} \equiv 0$ for $\mu \le \rho \le 1$ but consider buckling modes which are not radially symmetric ($m \neq 0$) the situation described above changes. In this case, it follows from (7.175), (7.176) that

$$
\frac{dg}{d\rho} = 
\begin{cases}
-\dfrac{1}{2}(1 - \mu^2), \ 0 \le \rho \le \mu \\[2mm]
\dfrac{1}{2}\mu^2 \left( \dfrac{1}{\rho^2} + 1 \right), \ \mu \le \rho \le 1
\end{cases}
\tag{7.183}
$$

so that, as per (7.171a), $\dfrac{dg}{d\rho} \ge 0$, for $\mu \le \rho \le 1$ while $\dfrac{dg}{d\rho} \le 0$, for $0 \le \rho \le \mu$. Using (7.180), (7.181), and (7.183), it follows that the function to be maximized is given by

$$
\Lambda_D = \frac{1}{2}(1 - \mu^2) \left[ \int_0^\mu \rho \left( \frac{dw_m}{d\rho} \right)^2 d\rho + m^2 \int_0^\mu \frac{w_m^2}{\rho} d\rho \right]
$$
$$
+ \frac{1}{2}\mu^2 \left[ \int_\mu^1 \left( \frac{1}{\rho} - \rho \right) \left( \frac{dw_m}{d\rho} \right)^2 d\rho \right.
\tag{7.184}
$$
$$
\left. - m^2 \int_\mu^1 \left( \frac{1}{\rho^2} + 1 \right) \frac{w_m^2}{\rho} d\rho \right]
$$

which, for arbitrary $m$, must be a positive quantity. However, if we consider admissible functions $w_m(\rho)$ of the (polynomial) form

$$
w_m(\rho) = C\rho^2 (1 + c_1\rho + c_2\rho^2)
\tag{7.185}
$$

satisfying all the boundary conditions in (7.168a,b,c,d) then the right-hand side of (7.184) will be negative for $m \ge 2$ and, thus, no buckling mode with $m \ge 2$ is possible if $\mathcal{T} \equiv 1$, for $0 \le \rho \le \mu$, while $\mathcal{T} \equiv 0$, for $\mu \le \rho \le 1$; rather, buckling modes corresponding to wave numbers $m \ge 2$ are caused by circumferential compressive stresses in a neighborhood of the edge of the plate at $r = b$ (i.e., the relevant conditions with respect to $g(\rho)$ are those in (7.171b)). It is, in fact, shown in [229] that corresponding to the temperature field $\mathcal{T}$ (within the domain of piecewise constant temperature fields) $-\dfrac{dg}{d\rho}$ is maximized by the field

$\mathcal{T} \equiv 0$, for $0 \le \rho \le \mu$, $\mathcal{T} \equiv 1$, for $\mu \le \rho \le 1$, and

$$
g(\rho) = \begin{cases} \dfrac{1}{2}\rho(1 - \mu^2),\ 0 \le \rho \le \mu \\[2mm] \left(\dfrac{1}{\rho} - \rho\right)\dfrac{\mu^2}{2},\ \mu \le \rho \le 1 \end{cases} \tag{7.186a}
$$

and

$$
\frac{dg}{d\rho} = \begin{cases} \dfrac{1}{2}(1 - \mu^2),\ 0 \le \rho \le \mu \\[2mm] -\left(\dfrac{1}{\mu^2} + 1\right)\dfrac{\mu^2}{2},\ \mu \le \rho \le 1 \end{cases} \tag{7.186b}
$$

In lieu of (7.184) we now obtain

$$
\begin{aligned}
\Lambda_D = -\frac{1}{2}(1 - \mu^2)\left[\int_0^\mu \rho\left(\frac{dw_m}{d\rho}\right)^2 d\rho + m^2 \int_0^\mu \frac{w_m^2}{\rho} d\rho\right] \\
+ \frac{1}{2}\mu^2\left[-\int_\mu^1 \left(\frac{1}{\rho} - \rho\right)\left(\frac{dw_m}{d\rho}\right)^2 d\rho\right. \\
\left. + m^2 \int_\mu^1 \left(\frac{1}{\rho^2} + 1\right)\frac{w_m^2}{\rho} d\rho\right]
\end{aligned} \tag{7.187}
$$

as the function to be maximized. The expression in (7.187) can now be used, in conjunction with a class of admissible $w_m(\rho)$, to determine $\mu$ for $m \ge 2$; if we use that class of admissible functions $w_m(\rho)$ which is defined by (7.185), we obtain the results shown in Table 7.1. A related treatment of the thermal buckling of isotropic, circular elastic plates may be found in [224] and further discussion of several aspects of initial buckling will be presented in the next section. An excellent discussion of closed-form representations of the stability boundary associated with two-dimensional temperature fields (in simply supported elastic rectangular plates) that produce combined compression, tension, and shear, temperature fields without shear, and one-dimensional temperature fields, may be found in Bargmann [230]. We will return to the problem of initial buckling of a thermally loaded elastic circular plate in §7.2 of this chapter.

## 7.2 Large Deflection Theory

In this section, we relinquish the small deflection assumption that was imposed in §7.1; we will work within the context of large deflection theory and will study the buckling and postbuckling behavior of isotropic and orthotropic elastic plates. For the case of a rectangular plate, our focus will be on the postbuckling behavior in the isotropic case and on large thermal deflections in the orthotropic case; postbuckling problems for rectangular orthotropic plates will be considered in §7.3 within the context of Berger's approximation. For the case of an isotropic circular plate, the emphasis in this section will be on computing the critical buckling temperature and the corresponding buckling mode; a discussion of initial buckling, within the context of thermoelasticity theory, for cylindrically orthotropic circular plates may be found in the paper by Stavsky [231].

The solution of a large deflection problem for an isotropic rectangular plate involves the determination of the two unknown functions $w(x, y)$ and $\Phi(x, y)$ from equations (6.93a,b) and suitable boundary support conditions. In the small deflection case, the buckling problem could be solved in two steps, namely, one would solve (6.93b) with $[w, w] = 0$ for $\Phi$, and then employ that solution in (6.93a) to independently determine $w$. In the present situation, a simultaneous solution for both $\Phi$ and $w$ is needed and, in most situations, one must resort to (approximate) series solution, to iteration schemes, or to numerical methods. An iterative procedure due to S.R. Boley [232], which has been described in [221], will be presented for the case of thermal buckling of a rectangular isotropic plate; in fact, the temperature distribution and applied boundary conditions on the edges are identical with those in (7.93)–(7.95), respectively, so that for edges at $x = 0, a(0 \leq y \leq b)$ and $y = 0, b(0 \leq x \leq a)$ simply supported, i.e. (7.101), the critical combination of applied edge stress $\sigma_0$ and "temperature" $c_1$ is obtained by setting the determinant of the system of homogeneous linear algebraic equations (7.103) equal to zero, where $k_{mn}$ is given by (7.104).

The iterative procedure consists of determining a set of successive approximations where the first approximation is the linear solution, i.e., those expressions for $\Phi$ and $w$ which correspond to the loading that initiates buckling; these expressions are then employed in the nonlinear terms of (6.93a,b) in the manner described below, with $t \equiv 0$, to obtain new expressions for $\Phi$ and $w$ and successive iteratives are computed in the same fashion. Provided that the actual solution is close to the initial iterative (i.e., the loading does not greatly exceed the critical load) the convergence of the iterative scheme may be shown to be rapid [232]; in fact, as little as two iterations may suffice to produce reasonable results. Suppose that we retain only two terms of the infinite series for $w$, i.e., of (7.102); then, as shown in [221] the deflected shape at this level of approximation assumes the form

$$w = \sin \frac{m\pi x}{a} \left\{ a_1 \frac{\sin m\pi y}{a} + a_3 \frac{\sin 3m\pi y}{a} \right\} \qquad (7.188)$$

As a first approximation we take $\left( \dfrac{a_3}{a_1} \right) = L_3$ with $L_3$ the solution of the equation which corresponds to either the first or second line of (7.105); thus

$$L_3 = 1 - 2 \left( \frac{k_{m1} + \sigma_0}{\alpha E c_1} \right) \qquad (7.189)$$

The first approximation for $w$ in the iterative scheme say, $w^{(1)}$, is then given by (7.188), with $a_3/a_1$ given by (7.189); it is, therefore, expressed in terms of an amplitude $a_1$, which is still indeterminate at this point. To obtain the second iteration, we substitute the first set of iteratives, i.e $w^{(1)}(x, y)$, as determined above, and $\Phi^{(1)}(x, y)$, as given by (7.97) into the system (6.93a,b) as follows; first $w^{(1)}$ is used in (6.93b) so as to produce the equation

$$\triangle^2 \Phi^{(2)} = -\frac{1}{2} Eh[w^{(1)}, w^{(1)}] - (1 - \nu)N^T \qquad (7.190)$$

for $\Phi^{(2)}(x, y)$; the solution of (7.190) which satisfies the same boundary conditions as $\Phi^{(1)}(x, y)$ is given by

$$\Phi^{(2)} = \frac{\sigma_0 y^2}{2} - \sum_{n=2,4,6} r_n \cos \left( \frac{nm\pi y}{a} \right)$$
$$+ \cos \left( \frac{2m\pi x}{a} \right) \sum_{n=0,2,4} s_n \cos \left( \frac{nm\pi y}{a} \right) \qquad (7.191)$$

with

$$\begin{cases} r_2 = \dfrac{-\alpha E c_1 a^2}{2m^2 \pi^2} + \dfrac{E a_1^2}{32}(1 - 2L_3) \\[2mm] r_4 = \dfrac{E a_1^2}{64} L_3 \\[2mm] r_6 = \dfrac{E a_1^2}{288} L_3^2 \end{cases} \qquad (7.192a)$$

and

$$
\begin{cases}
s_0 = \dfrac{Ea_1^2}{32}(1 + 9L_3^2) \\[2mm]
s_2 = \dfrac{Ea_1^2}{16}L_3 \\[2mm]
s_4 = -\dfrac{Ea_1^2}{400}L_3
\end{cases}
\tag{7.192b}
$$

Next, $w^{(1)}$ and $\Phi^{(2)}$ are substituted into the right-hand side of (6.93a) so as to produce for $w^{(2)}(x,y)$ the equation

$$
K \triangle^2 w^{(2)} = \Phi^{(2)}_{,yy} w^{(1)}_{,xx} - 2\Phi^{(2)}_{,xy} w^{(1)}_{,xy}
$$
$$
+ \Phi^{(2)}_{,xx} w^{(1)}_{,yy} - \triangle M^T
\tag{7.193}
$$

Solving (7.193), subject to the support conditions, we now obtain $w^{(2)}$ as an infinite series involving the constants $a_1$ and $a_3$, which are now subject to the conditions

$$
\left( k_{m1} + \sigma_0 - \frac{\alpha E c_1}{2} \right) + \frac{\alpha E c_1}{2}\left( \frac{a_3}{a_1} \right)
$$
$$
= 3(1 - \nu^2)k_{m1}\left( \frac{a_1}{h} \right)^2 \left( 2 - 3L_3 + \frac{3801}{400}L_3^2 \right)
\tag{7.194a}
$$

and

$$
\left( \frac{\alpha E c_1}{2} \right) + (k_{m3} + \sigma_0)\left( \frac{a_3}{a_1} \right)
$$
$$
= 3(1 - \nu^2)k_{m1}\left( \frac{a_1}{h} \right)^2 \left( -1 + \frac{426}{25}L_3 \right)
\tag{7.194b}
$$

By solving the (simultaneous) system (7.194a, b) for $a_1, a_3$ we obtain $w^{(2)}$, thus producing, for the problem at hand, the second iteration $(w^{(2)}, \Phi^{(2)})$ based on retaining just two terms in the series for $w(x,y)$. In Fig. 7.10 we show some of the corresponding numerical results for the variation of the deflection at the center point of the panel with increasing thermal load. It is indicated in [221] that if one applies the scheme described above, but retains three terms in the series for the deflection $w(x,y)$ and proceeds through the third iteration $(w^{(3)}, \Phi^{(3)})$, then for the range of thermal loading displayed in Fig. 7.10 the results are practically identical to those determined above.

One of the earliest and most frequently referenced studies of thermal buckling and postbuckling of rectangular isotropic plates is the NACA Technical Note [216] by Gossard, Seide, and Roberts which treats buckling of a simply supported plate that is subjected to a tentlike temperature distribution; this paper also deals with the effects of initial imperfections on the buckling analysis, a subject that we will cover, briefly, in Chapter 8. The rectangular plate in [216] is heated along its longitudinal center line by a uniform line source of heat and cooled along its edges by two uniform equal line sinks of heat; such an arrangement yields a temperature distribution in the plate which is constant through the thickness and which varies in a tentlike manner over the face of the plate (Figs. 7.11a, 7.11b). All edges of the plate are restrained in a direction normal to the plane of the plate by simple rigid supports, but are free to move in the plane of the plate.

The analysis in [216] proceeds by first computing the thermal stresses at temperature levels below the critical level, after which the critical (buckling) temperature is determined; then the (postbuckling) behavior of the plate at temperature levels above the critical level is analyzed. The actual details of the calculation of the thermal stress distribution in the plate of reference [216] are given in [233] and are obtained by employing the first-order approximation that on any cross section normal to the $x$-axis the stress component $\widetilde{\sigma}_{xx}$ is distributed as in Fig. 7.12. The (prebuckling) Airy function $\widetilde{\Phi}_0$ based on the stress distribution (instead of on the resultant or averaged stresses) may, in this case, be expressed as the product of a known function of $y$ and an arbitrary function of $x$; its approximate form, for $0 \leq y \leq b$, is

$$
\widetilde{\Phi}_0(x, y) = \frac{1}{12} b^2 E \alpha T_0 \left( 1 - 3\frac{y^2}{b^2} + 2\frac{y^3}{b^3} \right)
$$
$$
x \left( B_1 \sin hR_1 \frac{x}{a} \sin R_2 \frac{x}{a} + \right.
$$
$$
\left. B_2 \cos hR_1 \frac{x}{a} \cos R_2 \frac{x}{a} + 1 \right)
$$

(7.195)

with the constants $B_1, B_2, R_1, R_2$ as defined in appendix A of [216], i.e.,

$$
R_1 = k_1 \frac{a}{b}, \quad R_2 = k_2 \frac{a}{b}
$$

(7.196a)

$$\begin{cases} k_1 = \sqrt[4]{\dfrac{105}{13}} \cdot \sqrt{1 + \sqrt{\dfrac{21}{65}}} \\[4mm] k_2 = \sqrt[4]{\dfrac{105}{13}} \cdot \sqrt{1 - \sqrt{\dfrac{21}{65}}} \end{cases} \tag{7.196b}$$

$$\begin{cases} B_1 = \dfrac{k_1 \sin hR_1 \cos R_2 - k_2 \cos hR_1 \sin R_2}{k_1 \sin R_2 \cos R_2 + k_2 \sin hR_1 \cos hR_2} \\[4mm] B_2 = -\dfrac{k_1 \cos hR_1 \sin R_2 - k_2 \sin hR_1 \cos R_2}{k_1 \sin R_2 \cos R_2 + k_2 \sin hR_1 \cos hR_2} \end{cases} \tag{7.196c}$$

Corresponding to $\tilde{\Phi}_0(x, y)$ in (7.195), the prebuckling stress distribution in the plate is given for $0 \le y \le b$ by

$$\begin{aligned} \sigma_{xx}^0 \equiv \tilde{\Phi}_{0,yy} &= E\alpha T_0 \left( \frac{y}{b} - \frac{1}{2} \right) \\ &\times \left( B_1 \sin hR_1 \frac{x}{a} \sin R_2 \frac{x}{a} + B_2 \cos hR_1 \frac{x}{a} \cos R_2 \frac{x}{a} + 1 \right) \end{aligned} \tag{7.197a}$$

$$\begin{aligned} \sigma_{yy}^0 \equiv \tilde{\Phi}_{0,xx} &= \frac{1}{12} E\alpha T_0 \left( 1 - 3\frac{y^2}{b^2} + 2\frac{y^3}{b^3} \right) \\ &\times \left( D_1 \sin hR_1 \frac{x}{a} \sin R_2 \frac{x}{a} + D_2 \cos hR_1 \frac{x}{a} \cos R_2 \frac{x}{a} \right) \end{aligned} \tag{7.197b}$$

and

$$\begin{aligned} \sigma_{xy}^0 \equiv -\tilde{\Phi}_{0,xy} &= \frac{1}{2} E\alpha T_0 \left( 1 - \frac{y}{b} \right) \frac{y}{b} \\ &\times \left( D_3 \sin hR_1 \frac{x}{a} \cos R_2 \frac{x}{a} + D_4 \cos hR_1 \frac{x}{a} \sin R_2 \frac{x}{a} \right) \end{aligned} \tag{7.197c}$$

with the stress components in the domain $-b \le y \le 0$ identical with those given by (7.197a,b,c). In (7.197a,b,c) the $D_i's$ are given by

$$\begin{cases} D_1 = B_1(k_1^2 - K_2^2) - 2B_2 k_1 k_2 \\ D_2 = B_2(k_1^2 - k_2^2) + 2B_1 k_1 k_2 \\ D_3 = B_1 k_2 + B_2 k_1 \\ D_4 = B_1 k_1 - B_2 k_2 \end{cases} \tag{7.198}$$

The stresses in the plate are a function of the temperature differential $T_0$ for the tentlike temperature distribution depicted in Fig. 7.11b and are independent of the edge temperature $T_1$; as this temperature differential $T_0$ increases a value $(T_0)_{cr}$ or will be achieved, at which the plate will buckle under the action of the induced thermal stresses. When only small deflection theory is used it may be assumed that the middle surface of the plate does not stretch and thus the stress distribution in the plate does not change after the onset of buckling; this stress distribution is given by (7.197a,b,c), while the corresponding plate deflection is governed by (6.21), with $\Delta M_{HT} = 0$ (as the temperature distribution is constant through the plate thickness) and $\Phi = \Phi_0$, i.e., by

$$K\Delta^2 w = h \left( \sigma_{xx}^0 w_{,xx} + 2\sigma_{xy}^0 w_{,xy} + \sigma_{yy}^0 w_{,yy} \right) \tag{7.199}$$

Of course, (6.21) also serves to determine the critical temperature differential and the corresponding first buckling mode within the context of large deflection theory; such a determination, in this case, may be achieved through the Rayleigh-Ritz method. Assuming a buckle pattern which is symmetrical about the center of the plate of the form

$$w = \sum_{m=1,3,5}^{\infty} \sum_{n=1,3,5}^{\infty} a_{mn} \cos \frac{m\pi x}{2a} \cos \frac{n\pi y}{2b}, \tag{7.200}$$

substituting (7.200) and (7.197a,b,c) into the potential energy expression

$$U = \frac{1}{2} K \int_{-b}^{b} \int_{-a}^{a} \left\{ (\Delta w)^2 - 2(1-\nu)[w_{,xx} w_{,yy} - w_{,xy}^2] \right\} dx dy$$
$$+ \frac{1}{2} h \int_{-b}^{b} \int_{-a}^{a} \left[ \sigma_{xx}^0 w_{,x}^2 + \sigma_{yy}^0 w_{,y}^2 + 2\sigma_{xy}^0 w_{,x} w_{,y} \right] dx dy$$

and then minimizing with respect to the unknown coefficients $a_{mn}$, leads to a set of simultaneous equations constituting a characteristic value problem of the form

$$\frac{\pi^2 K}{b^2 E \alpha (T_0)_{cr} h} K_{pq} a_{pq} + \sum_{m=1,3,5}^{\infty} \sum_{n=1,3,5}^{\infty} K_{pqmn} a_{mn} = 0 \tag{7.201}$$

$$p = 1, 3, 5...., q = 1, 3, 5, ...$$

The solutions of (7.201) yield sets of relative values of the coefficients $a_{mn}$, and associated values of the critical temperature $(T_0)_{cr}$; the expressions for the coefficients $k_{pq}$, $k_{pqmn}$ in (7.201) are quite involved and are delineated in Appendix A of [216].

If only the terms $a_{11}$, $a_{13}$, $a_{31}$, and $a_{33}$ are retained in the deflection function (7.200) then equations (7.201) can be written, for a plate having an aspect ratio of 1.57, in the matrix form

$$
\begin{bmatrix}
15.73 & 22.52 & 14.88 & -5.96 \\
0.504 & 1.426 & 0.871 & 0.377 \\
1.35 & 3.54 & 7.42 & 7.79 \\
-0.0735 & 0.208 & 1.043 & 0.437
\end{bmatrix}
\begin{bmatrix}
a_{11} \\
a_{13} \\
a_{31} \\
a_{33}
\end{bmatrix}
=
\frac{100\pi^2 K}{b^2 E\alpha(T_0)_{cr} h}
\begin{bmatrix}
a_{11} \\
a_{13} \\
a_{31} \\
a_{33}
\end{bmatrix}
\quad (7.202)
$$

The solution of (7.202), for the smallest value of

$$
\lambda = \frac{\pi^2 K}{b^2 E\alpha(T_0)_{cr} h},
$$

is obtained by matrix iteration and yields $\lambda_{cr} = 5.39$. The relative values of the four coefficients which have been retained in the deflection function $w$ are given by

$$
\begin{bmatrix}
a_{11} \\
a_{13} \\
a_{31} \\
a_{33}
\end{bmatrix}
=
\begin{bmatrix}
1 \\
0.0365 \\
0.1360 \\
0.0042
\end{bmatrix}
$$

in which case

$$
w_0 = a_{11} \left( \cos\frac{\pi x}{2a}\cos\frac{\pi y}{2b} + 0.0365\cos\frac{\pi x}{2a}\cos\frac{3\pi y}{2b} \right.
$$
$$
\left. +0.1360\cos\frac{3\pi x}{2a}\cos\frac{\pi y}{2b} + 0.0042\cos\frac{3\pi x}{2a}\cos\frac{3\pi y}{2b} \right)
\quad (7.203)
$$

It is easily computed, based on (7.203) that the deflection $w_c$ at the center of the rectangular plate is $w_c = 1.1767\, a_{11}$ in which case

$$
w_0 = \frac{w_c}{1.1767} \left( \cos\frac{\pi x}{2a}\cos\frac{\pi y}{2b} + 0.0365\cos\frac{\pi x}{2a}\cos\frac{3\pi y}{2b} \right.
$$
$$
\left. +0.1360\cos\frac{3\pi x}{2a}\cos\frac{\pi y}{2b} + 0.0042\cos\frac{3\pi x}{2a}\cos\frac{3\pi y}{2b} \right)
\quad (7.204)
$$

In Table 7.2, we have indicated the convergence of the critical temperature parameter as additional terms are included in the deflection function; as indicated in [216], retaining further terms in the deflection function beyond those four already chosen above has a negligible effect on both the critical temperature and the buckle pattern. The initial buckling pattern ($\equiv$ small deflection buckle pattern) in one quadrant of the plate is depicted in Fig. 7.13 for a plate with an aspect ratio of 1.57.

The postbuckling behavior of the heated plate which is analyzed in [216] takes into account, as all analyses of postbuckling behavior must, the stretching of the plate middle surface due to bending and the corresponding changes in the plate stress distribution as the plate deflects; thus, this analysis is based on the generalized von Karman equations for the hygrothermal case, i.e., (6.93a,b) with $\Delta M^T = 0$, $t = 0$. Actually, since the analysis in [216] uses the Airy function $\tilde{\Phi}$ based on the local stress distribution, the relevant form of (6.93a,b) in this case is

$$
\begin{cases}
K\Delta^2 w = h \left( \tilde{\Phi}_{,yy} w_{,xx} + \tilde{\Phi}_{,xx} w_{,yy} - 2\tilde{\Phi}_{,xy} w_{,xy} \right) \\
\Delta^2 \tilde{\Phi} = -E\alpha \, \nabla^2 T + E \left[ w_{,xy}^2 - w_{,xx} w_{,yy} \right]
\end{cases}
\tag{7.205}
$$

where the lateral loading $t \equiv 0$.

The system (7.205) is solved (approximately) in [216] by using a procedure based on the Galerkin method. The stress function $\tilde{\Phi}$ is decomposed as the sum

$$
\tilde{\Phi} = \tilde{\Phi}_0 + \tilde{\Phi}_1
\tag{7.206}
$$

where $\tilde{\Phi}_0$ is the thermal stress (Airy) function for the unbuckled plate; $\tilde{\Phi}_0$ satisfies

$$
\Delta^2 \tilde{\Phi}_0 = -E\alpha \, \nabla^2 T
\tag{7.207}
$$

and the stress boundary conditions and is given by (7.195), (7.196a,b,c). The function $\tilde{\Phi}_1$ is taken to be the solution of

$$
\Delta^2 \tilde{\Phi}_1 = E \left[ w_{,xy}^2 - w_{,xx} w_{,yy} \right],
\tag{7.208}
$$

with $w$ the buckle pattern determined by (7.200) and the boundary conditions on the stresses. For $\tilde{\Phi}_1$ a series of the form

$$
\tilde{\Phi}_1 = (x^2 - a^2)^2 (y^2 - b^2)^2 (c_1 + c_2 x^2 + c_3 y^2 + ...)
\tag{7.209}
$$

is chosen and the coefficients $e_i$, $i = 1, 2, ...$, are then determined in terms of the coefficients $a_{mn}$ in (7.200) by the equations

$$
\int_{-a}^{a} \int_{-b}^{b} \frac{\partial \tilde{\Phi}_1}{\partial c_i} \left\{ \Delta^2 \tilde{\Phi}_1 - E \left[ w_{,xy}^2 - w_{,xx} w_{,yy} \right] \right\} dx dy = 0
\tag{7.210}
$$

The resulting stress function $\tilde{\Phi}$ is now substituted into the first equation in (7.205) and the Galerkin approach is again used so as to determine the values of the coefficients $a_{mn}$ of the deflection function $w$; as shown in [216], this leads to the set of simultaneous equations

$$\int_{-a}^{a}\int_{-b}^{b}\cos\frac{m\pi x}{a}\cos\frac{n\pi y}{2b}$$

$$\times\left(\frac{K}{n}\Delta^2 w - \tilde{\Phi}_{,yy}w_{,xx} - \tilde{\Phi}_{,xx}w_{,yy} + 2\tilde{\Phi}_{,xy}w_{,xy}\right)dx\,dy = 0 \qquad (7.211)$$

$$(m = 1, 3, 5, ..., \quad n = 1, 3, 5, ...)$$

for the $a_{mn}$. The equations (7.211) are, of course, nonlinear, and their solution becomes more difficult as the number of terms retained in the deflection function increases. In [216], therefore, it is assumed that the shape of the deflected surface of the plate for large deflections may be taken to be the one the plate has at the onset of buckling; with such an assumption only the coefficient $a_{11}$ remains arbitrary, while the ratios $a_{mn}/a_{11}$ are taken to be those which are given by the initial buckling solution described above. In lieu of (7.211) we have, therefore, as the Galerkin-type equation, from which the coefficient $a_{11}$ may be determined, the relation

$$\int_{-a}^{a}\int_{-b}^{b}w_s\left(\frac{K}{h}\Delta^2 w_s - \tilde{\Phi}_{,yy}w_{s,xx} - \tilde{\Phi}_{,xx}w_{s,yy}\right.$$

$$\left. +2\tilde{\Phi}_{,xy}w_{s,xy}\right)dx\,dy = 0 \qquad (7.212)$$

where

$$w_s = \sum_{m=1,3,5}^{\infty}\sum_{n=1,3,5}^{\infty}\frac{a_{mn}}{a_{11}}\cos\frac{m\pi x}{2a}\cos\frac{n\pi y}{2b} \qquad (7.213)$$

In (7.212) the ratios $a_{mn}/a_{11}$, which were obtained from the initial buckling solution, must also be substituted into the stress function $\tilde{\Phi}$.

For a plate with aspect ratio $a/b = 1.57$ it has already been determined that

$$w_s = \cos\frac{\pi x}{2a}\cos\frac{\pi y}{2b} + 0.0365\cos\frac{\pi x}{2a}\cos\frac{3\pi y}{2b}$$

$$+0.1360\cos\frac{3\pi x}{2a}\cos\frac{\pi y}{2b} + 0.0042\cos\frac{3\pi x}{2a}\cos\frac{3\pi y}{2b}$$

in which case (7.212) yields the following relation between the temperature differential $T_0$ and the center deflection $w_c$:

$$\frac{b^2 E\alpha T_0 h}{\pi^2 K} = 5.39 + 1.12(1 - \nu^2)\frac{w_c^2}{h^2} \qquad (7.214)$$

At any other point of the plate (assumed to have stress-free edges and an aspect ratio $a/b = 1.57$) the deflection assumes the form

$$\frac{w}{h} = 0.723 \sqrt{\frac{\frac{b^2 E\alpha T_0 h}{\pi^2 K} - 5.39}{1.12(1 - \nu^2)}} \; w_s \qquad (7.215)$$

A comparison of the calculated deflections at the plate center with some experimental data is shown in Fig. 7.14, while, in Figs. 7.15 a,b, respectively, we depict the predicted growth of the deflections, with increasing temperature differential $T_0$, along the longitudinal center line and the transverse center line of the rectangular plate. We will return to the problem treated in [216] when we discuss hygrothermal buckling in the presence of imperfections in Chapter 8.

As a final example of hygrothermal buckling of heated isotropic rectangular plates, we consider the large deflection analysis presented in [234]; in this paper, the plate is subjected to both heating and resultant edge loads due to an elastic edge restraint in the plane of the plate. A major result of the analysis presented in [234] is that a decreased buckling temperature results from a lowered edge flexibility. The temperature gradient in the $z$-direction through the thickness of the plate is assumed, in [234], to be negligible and the edges of the plate are assumed to remain straight during deformation. The rectangular plate, as depicted in Fig. 7.16, is simply supported along all four edges and free to rotate at the edges but translation of the edges in the plane of the plate is resisted by spring forces of magnitude $k_y$ (force/cubic volume) along $y = 0$ and $y = b$ and $k_x$ along $x = 0$ and $x = a$.

The relevant form of the large deflection equations for the problem at hand is, precisely, (7.205) where $\tilde{\Phi}$ is the Airy function associated with the local stress field $(\sigma_{xx}, \sigma_{xy}, \sigma_{yy})$ as opposed to the resultant stresses $(N_x, N_{xy}, N_y)$. The boundary conditions require that the deflection $w$ and the edge bending moments per unit length be zero along the edges of the plate; as $M^T = 0$, because of the assumed constancy of the temperature distribution through the thickness of the plate, these boundary conditions (see (6.96)) assume the form $w = 0$ along $x = 0, a$, for $0 \le y \le b$, $w = 0$ along $y = 0, b$, for $0 \le x \le a$, and

$$\begin{cases} w_{,xx} + \nu w_{,yy} = 0, \; x = 0, a; \; 0 \le y \le b \\ w_{,yy} + \nu w_{,xx} = 0, \; y = 0, b; \; 0 \le x \le a \end{cases} \qquad (7.216)$$

The boundary conditions are satisfied by choosing

$$w = \sum_{r=1}^{\infty} \sum_{s=1}^{\infty} a_{rs} \sin \frac{r\pi x}{a} \sin \frac{s\pi y}{b} \tag{7.217}$$

The lateral loading $t(x, y)$ is represented as a Fourier series in the form

$$t = \sum_{r=1}^{\infty} \sum_{s=1}^{\infty} b_{rs} \sin \frac{r\pi x}{a} \sin \frac{s\pi y}{b} \tag{7.218}$$

while the prescribed temperature distribution is represented as

$$T = T_0 + \sum_{p=1}^{\infty} \sum_{q=1}^{\infty} T_{pq} \cos \frac{p\pi x}{a} \cos \frac{q\pi y}{b} \tag{7.219}$$

with $T_0$ the average temperature over the surface of the plate. As the plate is subjected to the temperature distribution given by (7.219), the plate will expand and this expansion will result in a distribution of average edge thrusts which are caused by the restraining spring forces acting along the edges of the plate; we denote these average edge stresses by $\rho$ and $\xi$, respectively, along the edges $x = 0, a$ and $y = 0, b$. The magnitudes of $\rho$ and $\xi$ depend on the temperature distribution, the plate dimensions, and the elastic moduli of the plate. Employing (7.217) and (7.219) it may be shown that the second of the large deflection equations in (7.205) is satisfied if $\tilde{\Phi}$ has the form

$$\tilde{\Phi} = \frac{\xi x^2}{2} + \frac{\rho y^2}{2} + \sum_{p=0}^{\infty} \sum_{q=0}^{\infty} c_{pq} \cos \frac{p\pi x}{a} \cos \frac{q\pi y}{b} \tag{7.220}$$

in which

$$c_{pq} = E \left\{ \sum_{i=1}^{9} B_i + A_{pq} \right\} \Big/ 4 \left( \frac{p^2 b}{a} + \frac{q^2 a}{b} \right)^2 \tag{7.221}$$

where

$$A_{pq} = \left( \frac{4\alpha}{\pi^2} \right) T_{pq} ab \left( \frac{p^2 b}{a} + \frac{a^2 a}{b} \right) \tag{7.222}$$

while the $B_i, i = 1, 2, ...9$ are given in [235]; the work in [235] covers the case of a laterally loaded plate subject to edge thrusts in the plane of the plate with lateral displacements in the large deflection range but does not consider temperature effects. Specific expressions for the coefficients $b_{rs}$, in terms of the coefficients $a_{rs}$, $c_{rs}$, so that the first equation in (7.205) is satisfied are quite complex; they are delineated in [234] and will not be replicated here.

For the plate to be in equilibrium it is necessary that

$$\int_0^a \left[ \tilde{\Phi}_{,xx} - k_y v(x, y) \right] dx = 0 \text{ at } y = 0, b; \text{ for } 0 \le x \le a \quad (7.223a)$$

and

$$\int_0^b \left[ \tilde{\Phi}_{,yy} - k_x u(x, y) \right] dy = 0 \text{ at } x = 0, a; \text{ for } 0 \le y \le b \quad (7.223b)$$

We recall the constitutive relations in the form

$$\begin{cases} \epsilon_{xx}^0 \equiv u_{,x} + \dfrac{1}{2} w_{,x}^2 = \dfrac{1}{E} \left( \tilde{\Phi}_{,yy} - \nu \tilde{\Phi}_{,xx} + \alpha ET \right) \\[2mm] \epsilon_{yy}^0 \equiv v_{,y} + \dfrac{1}{2} w_{,y}^2 = \dfrac{1}{E} \left( \tilde{\Phi}_{,xx} - \nu \tilde{\Phi}_{,yy} + \alpha ET \right) \end{cases} \quad (7.224)$$

As the spring constants on opposite edges of the plate are equal, it follows that, for plate equilibrium to hold, the displacements along opposite edges must also be equal in magnitude, i.e.,

$$\begin{cases} u(a, y) = -u(0, y), \ 0 \le y \le b \\ v(x, b) = -v(x, 0), \ 0 \le x \le a \end{cases} \quad (7.225)$$

Employing (7.225) in conjunction with (7.224) we obtain

$$\begin{cases} u(0, y) = -\dfrac{1}{2E} \int_0^a \left[ \tilde{\Phi}_{,yy} - \nu \tilde{\Phi}_{,xx} + \alpha ET - \dfrac{E}{2} w_{,x}^2 \right] dx \\[3mm] v(x, 0) = -\dfrac{1}{2E} \int_0^b \left[ \tilde{\Phi}_{,xx} - \nu \tilde{\Phi}_{,yy} + \alpha ET - \dfrac{E}{2} w_{,y}^2 \right] dy \end{cases} \quad (7.226)$$

Carrying out the integrations in (7.226), and making use of (7.217)-(7.219), we find for the edge displacements in the plane of the plate

$$u(0, y) = -\dfrac{1}{2E} \left[ \rho a - \nu \xi a + \alpha E T_0 a \right.$$

$$\left. -\dfrac{aE}{4} \sum_{r=1}^{\infty} \sum_{s=1}^{\infty} \sum_{s'=1}^{\infty} a_{rs} a_{rs'} \left( \dfrac{r\pi}{a} \right)^2 \sin \dfrac{s\pi y}{b} \sin \dfrac{s'\pi y}{b} \right] \quad (7.227a)$$

and

$$v(x,0) = -\frac{1}{2E}\left[\xi b - \nu\rho b + \alpha E T_0 b\right.$$

$$\left. -\frac{bE}{4}\sum_{r=1}^{\infty}\sum_{r'=1}^{\infty}\sum_{s=1}^{\infty}a_{rs}a_{rs'}\left(\frac{s\pi}{b}\right)^2\sin\frac{r\pi x}{a}\sin\frac{r'\pi x}{a}\right] \tag{7.227b}$$

Substituting (7.220) and (7.227a, b) into (7.223a, b) and carrying out the indicated integrations we obtain

$$\begin{cases} \rho\lambda_{11} + \xi\lambda_{12} = \mu_{11} \\ \rho\lambda_{21} + \xi\lambda_{22} = \mu_{22} \end{cases} \tag{7.228}$$

where

$$\begin{cases} \lambda_{11} = -\dfrac{\nu k_y b}{2E}, \quad \lambda_{12} = 1 + \dfrac{k_y b}{2E} \\[2mm] \lambda_{22} = -\dfrac{\nu k_x a}{2E}, \quad \lambda_{22} = 1 + \dfrac{k_x a}{2E} \end{cases} \tag{7.229a}$$

$$\begin{cases} \mu_{11} = -\dfrac{k_y b}{2E}\left[\alpha E T_0 - \dfrac{1}{8}E\sum_{r=1}^{\infty}\sum_{s=1}^{\infty}a_{rs}^2\left(\dfrac{s\pi}{b}\right)^2\right] \\[4mm] \mu_{22} = -\dfrac{k_x a}{2E}\left[\alpha E T_0 - \dfrac{1}{8}E\sum_{r=1}^{\infty}\sum_{s=1}^{\infty}a_{rs}^2\left(\dfrac{r\pi}{a}\right)^2\right] \end{cases} \tag{7.229b}$$

From (7.228) we obtain the average edge (thrust) stresses in the form

$$\rho = (\lambda_{22}\mu_{11} - \lambda_{12}\mu_{22})/(\lambda_{11}\lambda_{22} - \lambda_{12}\lambda_{21}) \tag{7.230a}$$

and

$$\xi = (\lambda_{11}\mu_{22} - \lambda_{21}\mu_{11})/(\lambda_{11}\lambda_{22} - \lambda_{12}\lambda_{21}) \tag{7.230b}$$

An expression for the (critical) buckling temperature can now be derived in terms of the physical properties of the plate, the plate dimensions, and the spring constants $k_x$ and $k_y$. In [234], only the case of a uniform temperature distribution $T = T_0$ is considered; we assume, also, that the lateral loading $t \equiv 0$ so that $b_{rs} \equiv 0$. As $(T_0)_{cr}$ corresponds to the first buckling mode, the summation indices $r, s$ in (7.217) are both equal to one when $T_0 = (T_0)_{cr}$. For initial buckling, $w = 0$ in the second equation in (7.205) in which case, $\tilde{\Phi}$ as defined by (7.220), reduces to

$$\tilde{\Phi}_0 = \frac{\xi x^2}{2} + \frac{\rho y^2}{2} \tag{7.231}$$

and the first equation in (7.205) reduces to the statement that

$$\frac{E\pi^2}{12(1-\nu^2)} \left(\frac{h}{a}\right)^2 \left[1 + \left(\frac{a}{b}\right)^2\right] = -\rho - \left(\frac{a}{b}\right)^2 \xi \tag{7.232}$$

Using (7.229a,b), (7.230a,b) in (7.232), with the summations extending only over $r = s = 1$, then yields the following expression for the buckling temperature

$$\begin{cases} (\alpha T_0)_{cr} = \hat{P}/\hat{Q} \\ \hat{P} = \dfrac{\pi^2}{12(1-\nu^2)} \left(\dfrac{h}{a}\right)^2 \left[1 + \left(\dfrac{a}{b}\right)^2\right] \left[\left(1 + \dfrac{2E}{k_x a}\right) \left(1 + \dfrac{2E}{k_y b}\right) - \nu^2\right] \\ \hat{Q} = \left[1 + \left(\dfrac{a}{b}\right)^2\right](1+\nu) + \left(\dfrac{2E}{k_y b}\right) + \left(\dfrac{a}{b}\right)^2 \left(\dfrac{2E}{k_x a}\right) \end{cases} \tag{7.233}$$

If $k_x = k_y = \infty$ then (7.233) yields the expression for the buckling temperature for the case of a nonflexible edge restraint in the form

$$(\alpha T_0)_{cr}^\infty = \frac{\pi^2}{12(1+\nu)} \left(\frac{h}{a}\right)^2 \left[1 + \left(\frac{a}{b}\right)^2\right] \tag{7.234}$$

in which case, for $a = b$ and $k_x = k_y$

$$\frac{(\alpha T_0)_{cr}}{(\alpha T_0)_{cr}^\infty} = \frac{1}{1-\nu} \cdot \frac{\left(1 + \dfrac{2E}{k_x a}\right)^2 - \nu^2}{1 + \nu + \dfrac{2E}{k_x a}} \tag{7.235}$$

Numerical results have been generated by the authors in [234] for the case of a uniform temperature distribution $T = T_0$ and a sinusoidal lateral loading of the form

$$t = b_{11} \sin \frac{\pi x}{a} \sin \frac{\pi y}{b} \tag{7.236}$$

Plots of the center deflection of the plate versus a lateral load parameter based on $q_{11}$ are depicted in Fig. 7.17 for various values of $T_0$, the aspect ratio $a/b$, and the spring constant $k_x$ ($k_x = k_y$ in the application). For a uniform lateral loading, the variation of the buckling temperature ratio $(\alpha T_0)_{cr}/(\alpha T_0)_{cr}^\infty$ with the edge spring rate parameter $k_x a/2E$ is depicted in Fig. 7.18 ($a = b$, $k_x = k_y$ in the application) while for the same loading condition the effect of the edge fixity on the critical temperature is displayed in Fig. 7.19; finally, the center deflection versus temperature ratio (postbuckling) curves for this case are depicted in Fig. 7.20 for several different values of $k_x a/2E$.

Having discussed several examples of buckling and postbuckling behavior for heated isotropic rectangular plates, we now turn to the rectilinear orthotropic case for such a geometry. For this problem, the relevant generalized von Karman equations are given by (6.103a,b) where we set the applied transverse loading $t(x, y) \equiv 0$. Athough many of the papers in the literature, which deal with the buckling behavior of heated, rectangular, orthotropic plates do so within the context of the Berger's approximation, which will be discussed in the next section, there are a few analyses which deal with the full system (6.103a,b) in the context of plate bending, e.g., the work of Biswas [236]; this work appears to contain a serious flaw common to much of the literature on thermal bending and buckling of simply supported plates, as we shall indicate below.

In [236], the author writes (6.103a), with $t \equiv 0$, in the form

$$
\begin{aligned}
D_x w_{,xxxx} + 2H w_{,xxyy} + D_y w_{,yyyy} \\
+ \tilde{\beta}_1 M_{T,xx} + \tilde{\beta}_2 M_{T,yy} = \\
F_{,xx} w_{,yy} - 2F_{,xy} w_{,xy} + F_{,yy} w_{,xx}
\end{aligned}
\tag{7.237}
$$

$$
M_T = \int_{-h/2}^{h/2} \delta T(x, y, z) z \, dz
$$

and it is easy to see that, with respect to the notation employed in the present work

$$
D_x = D_{11}, \ 2H = D_{12} + 4D_{66} + D_{21}, D_y = D_{22}
\tag{7.238}
$$

while $F = \Phi$ and

$$
\begin{cases}
\tilde{\beta}_1 = c_{11}\alpha_1 + c_{12}\alpha_2 \\
\tilde{\beta}_2 = c_{21}\alpha_1 + c_{22}\alpha_2
\end{cases}
\tag{7.239}
$$

The $\tilde{\beta}_i$ in (7.239) are not to be confused with their previous interpretation as hygroscopic expansion coefficients (where we denoted them as $\beta_i$). To bring the second von Karman equation for this case into line with the form employed in [236], we multiply (6.103b) through by $E_2 h$ thus producing

$$\Phi_{,xxxx} + \left(\frac{E_2}{G_{12}} - 2\nu_{12}\right)\Phi_{,xxyy} + \left(\frac{E_2}{E_1}\right)\Phi_{,yyyy}$$
$$= -\frac{1}{2}E_2 h[w, w] - (\tilde{N}_T^2 - \nu_{12}\tilde{N}_T^1)_{,xx} \qquad (7.240)$$
$$- \left(\frac{E_2}{E_1}\right)(\tilde{N}_T^1 - \nu_{21}\tilde{N}_T^2)_{,yy}$$

Next, we note that, by virtue of (6.100),

$$\tilde{N}_T^2 - \nu_{12}\tilde{N}_T^1 = \left[\left(\frac{E_1\alpha_1\nu_{12} + E_2\alpha_2\nu_{21}}{1 - \nu_{12}\nu_{21}}\right)\right.$$
$$\left. -\nu_{21}\left(\frac{E_1\alpha_1 + E_2\alpha_2\nu_{21}}{1 - \nu_{12}\nu_{21}}\right)\right] N_T \equiv E_2\alpha_2 N_T \qquad (7.241a)$$

while

$$\tilde{N}_T^2 - \nu_{12}\tilde{N}_T^2 = \left[\left(\frac{E_1\alpha_1 + E_2\alpha_2\nu_{21}}{1 - \nu_{12}\nu_{21}}\right)\right.$$
$$\left. -\nu_{21}\left(\frac{E_1\alpha_1\nu_{12} + E_2\alpha_2}{1 - \nu_{12}\nu_{21}}\right)\right] N_T \equiv E_1\alpha_1 N_T \qquad (7.241b)$$

where

$$N_T = \int_{-h/2}^{h/2} \delta T(x, y, z)dz$$

In view of (7.241a,b), (7.240) becomes

$$\Phi_{,xxxx} + \left(\frac{E_2}{G_{12}} - 2\nu_{12}\right)\Phi_{,xxyy} + \left(\frac{E_2}{E_1}\right)\Phi_{,yyyy}$$
$$= E_2 h(w_{,xy}^2 - w_{,xx}w_{,yy}) - E_2\alpha_2 N_{T,xx} - E_2\alpha_1 N_{T,yy} \qquad (7.242)$$

Equation (7.242) corresponds to the second generalized von Karman equation (for this case) in [236], i.e., to

$$F_{,xxxx} + p^2 F_{,xxyy} + q^2 F_{,yyyy} + \lambda_1 N_{T,xx}$$
$$+ \lambda_2 N_{T,yy} = E_2 h(w_{,xy}^2 - w_{,xx}w_{,xx}w_{,yy}) \qquad (7.243)$$

if we again identify $F = \Phi$ and take

$$
\begin{cases}
p^2 = \dfrac{E_2}{G_{12}} - 2\nu_{12} \\
q^2 = E_2/E_1 \\
\lambda_1 = E_2\alpha_2 \\
\lambda_2 = E_2\alpha_1
\end{cases}
\tag{7.244}
$$

With the correlations $F = \Phi$, (7.238), (7.244)), and the definitions of $N_T, M_T$ given above, (7.237) and (7.243) are identical with our earlier equations for the heated rectangular orthotropic plate, namely, (6.103a,b). **Remarks:** Biswas [236] writes the $\lambda_i$ of (7.244) in the form

$$
\begin{cases}
\lambda_1 = \dfrac{E_2}{E_1}\tilde{\beta}_1 - \nu_{12}\tilde{\beta}_2 \\
\lambda_2 = \tilde{\beta}_2 - \nu_{12}\tilde{\beta}_1
\end{cases}
\tag{7.245}
$$

It is easily verified by using (7.239) that these expressions are equivalent to those in (7.244). We also note that Biswas [236] works with the inverse constitutive relations, i.e., with the matrix $s_{ij} = c_{ij}^{-1}$, whose components are related to the $c_{ij}$ by

$$
\begin{cases}
c_{11} = s_{22}/\Delta \\
c_{12} = -s_{12}/\Delta \\
c_{21} = s_{11}/\Delta \\
c_{22} = -s_{12}/\Delta \\
c_{66} = 1/s_{66}
\end{cases}
\tag{7.246}
$$

with $\Delta = s_{11}s_{22} - s_{12}^2$. It is any easy task to show that, based on (1.61) and (7.246)

$$
\begin{cases}
E_1 = \dfrac{1}{s_{11}}, \quad E_2 = \dfrac{1}{s_{22}}, \quad \nu_{12} = -\dfrac{s_{12}}{s_{11}} \\
\nu_{21} = -\dfrac{s_{12}}{s_{22}}, \quad G_{12} = \dfrac{1}{s_{66}}
\end{cases}
\tag{7.247}
$$

We now consider a simply supported orthotropic plate which occupies the region $0 \le x \le a$, $0 \le y \le b$, $-\frac{1}{2}h \le z \le \frac{1}{2}h$ in the three-space and assume a temperature distribution which varies linearly through the plate of the form

$$
\delta T(x,y,z) = \frac{1}{2}(T_1 + T_2) + z\left(\frac{T_1 - T_2}{h}\right)
\tag{7.248}
$$

so that

$$T(x, y, \frac{1}{2}h) = T_1, \quad T(x, y, -\frac{1}{2}h) = T_2 \tag{7.249}$$

Corresponding to the simple-support conditions, along the edges of the plate, the author, in [236] looks for plate deflections of the form

$$w(x, y) = w_0 \sin \frac{\pi x}{a} \sin \frac{\pi y}{b} \tag{7.250}$$

However, by virtue of (6.104), the assumption (7.250) cannot possibly be valid unless $\tilde{M}_T^1 = \tilde{M}_T^2 = 0$, i.e., unless $M_T = 0$ (which is not true in view of (7.248)). For the temperature distribution (7.248), $N_{T,x} = N_{T,y} = 0$; Using (7.250) in (7.243) it may be shown that the general solution of (7.243) is given by an Airy function $F(x, y)$ of the form

$$F = \frac{1}{2}Ax^2 + \frac{1}{2}By^2 + \frac{1}{32}E_2 h w_0^2 \left( \frac{a^2}{b^2} \cos \frac{2\pi x}{a} + \frac{b^2}{a^2 q^2} \cos \frac{2\pi y}{b} \right) \tag{7.251}$$

with $A$ and $B$ arbitrary constants that must be determined from the in-plane boundary conditions. We assume that all edges of the plate remain straight after deformation; then, using the strain-displacement equations, the constitutive relations (6.99) for the resultant stresses, (7.251), and the definitions (1.61), (7.246) of the constitutive constants one obtains for the in-plane displacements

$$u = \int_0^a \left[ \frac{s_{11}}{h} \left( B - \frac{E_2 h w_0^2 \pi^2}{8a^2 q^2} \cos \frac{2\pi y}{b} \right) \right.$$

$$+ \frac{s_{12}}{h} \left( A - \frac{E_2 h w_0^2 \pi^2}{8b^2} \cos \frac{2\pi x}{a} \right) \tag{7.252a}$$

$$\left. - \frac{1}{2} \left( \frac{\partial w}{\partial x} \right)^2 + \frac{\tilde{\beta}_1 s_{11} + \tilde{\beta}_2 s_{12}}{h} N_T \right]_{y=0} dx$$

$$v = \int_0^b \left[ \frac{s_{22}}{h} \left( A - \frac{E_2 h w_0^2 \pi^2}{8b^2} \cos \frac{2\pi x}{a} \right) \right.$$

$$+ \frac{s_{12}}{h} \left( B - \frac{E_2 h w_0^2 \pi^2}{8a^2 q^2} \cos \frac{2\pi y}{b} \right) \tag{7.252b}$$

$$\left. - \frac{1}{2} \left( \frac{\partial w}{\partial y} \right)^2 + \frac{\tilde{\beta}_1 s_{12} + \tilde{\beta}_2 s_{22}}{h} N_T \right]_{x=0} dy$$

For immovable plate edges (in the plane of the plate) $u = v = 0$; combining this latter assumption with (7.252a,b) we are led to

$$A = C_1 w_0^2 - \tilde{\beta}_2 N_T, \quad B = C_2 w_0^2 \tilde{\beta}_1 N_T \tag{7.253a}$$

with

$$\begin{cases} C_1 = \dfrac{E_2 h \pi^2}{8\Delta} \left( \dfrac{s_{11} s_{22}}{a^2 q^2} - \dfrac{s_{22} s_{11}}{b^2} \right) \\[3mm] C_2 = \dfrac{E_2 h \pi^2}{8\Delta} \left( \dfrac{s_{11} s_{22}}{a^2 q^2} - \dfrac{s_{22} s_{12}}{b^2} \right) \end{cases} \tag{7.253b}$$

Finally, an algebraic equation for $w_0/h$ is determined in [236] by using (7.250), (7.251), (7.253a,b) and the Fourier expansion

$$M_T = \sum_{m=1,3,\dots}^{\infty} \sum_{n=1,3,\dots}^{\infty} \frac{16 M_T}{mn\pi^2} \sin \frac{m\pi x}{a} \sin \frac{n\pi y}{b} \tag{7.254}$$

in (7.237); this procedure yields the following first-order approximation

$$\frac{w_0}{h} = \frac{16(M_T/h)(\tilde{\beta}_1 a^{-2} + \tilde{\beta}_2 b^{-2})}{\pi^4 [(D_x/a^4) + (2H/a^2 b^2) + (D_y/b^4)]} \tag{7.255}$$

We note, however, the objection to the analysis which has been raised following (7.250) as well as the observation that the double Fourier series for $M_T$, as given by (7.254), formally vanishes for $x = 0, a$ $(0 \le y \le b)$ and $y = 0, b$ $(0 \le x \le a)$ whereas, in point of fact

$$M_T \equiv \int_{-n/2}^{h/2} \delta T(x, y, z) z \, dz = \frac{1}{12}(T_1 - T_2)h^2 \ne 0$$

if $T_1 \ne T_2$. Variations of the non-dimensional (bending) deflections as a function of the temperature parameter $\dfrac{\tilde{\beta}_2}{10\tilde{\beta}_1}(T_1 - T_2)$, which are based on (7.255), are depicted in Fig. 7.21. We now comment on what appears to be the central shortcoming in the work presented in [236]: For a simply supported rectangular orthotropic elastic plate occupying the domain $0 \le x \le a$, $0 \le y \le b$, $-\dfrac{h}{2} \le z \le \dfrac{h}{2}$, the relevant boundary conditions are given by (see (6.102), (6.104)):

$$\begin{cases} D_{11} w_{,xx} + D_{12} w_{,yy} + \tilde{M}_T^1 = 0 \\ w = 0 \end{cases} \tag{7.256a}$$

for $x = 0, a, \ 0 \le y \le b$, and

$$\begin{cases} D_{21}w_{,xx} + D_{22}w_{,yy} + \tilde{M}_T^2 = 0 \\ w = 0 \end{cases} \qquad (7.256b)$$

for $y = 0, b, 0 \le x \le a$, where

$$\tilde{M}_T^1 = (c_{11}\alpha_1 + c_{12}\alpha_2) \int_{-h/2}^{h/2} \delta T(x, y, z)dz \qquad (7.257a)$$

$$\tilde{M}_T^2 = (c_{21}\alpha_1 + c_{22}\alpha_2) \int_{-h/2}^{h/2} \delta T(x, y, z)dz \qquad (7.257b)$$

From the assumed form of the temperature distribution in Biswas [236], i.e.,

$$\delta T(x, y, z) = \frac{1}{2}(T_1 + T_2) + z\left(\frac{T_1 - T_2}{h}\right),$$

$T_1$ and $T_2$ being, respectively, the constant temperatures at the top and bottom plate surfaces,

$$\int_{-h/2}^{h/2} \delta T(x, y, z)dz = \frac{1}{2}h(T_1 + T_2)$$

so that

$$\begin{cases} \tilde{M}_T^1 = \dfrac{(c_{11}\alpha_1 + c_{12}\alpha_2)}{2}h(T_1 + T_2) \equiv A^* \\ \tilde{M}_T^2 = \dfrac{(c_{21}\alpha_1 + c_{22}\alpha_2)}{2}h(T_1 + T_2) \equiv B^* \end{cases} \qquad (7.258)$$

Setting $D_{11} = \lambda_1$, $D_{12} = \lambda_2$, $D_{21} = \gamma_1$, $D_{22} = \gamma_2$, the boundary conditions of simple support which are represented by (7.256a,b), (7.257a,b) become

$$\begin{cases} w(0, y) = w(a, y) = 0 \\ \lambda_1 w_{,xx}(0, y) + \lambda_2 w_{,yy}(0, y) = A^* \\ \lambda_1 w_{,xx}(a, y) + \lambda_2 w_{,yy}(a, y) = A^* \end{cases} \qquad (7.259a)$$

for $0 \le y \le b$ and

$$\begin{cases} w(x, 0) = w(x, b) = 0 \\ \gamma_1 w_{,xx}(x, 0) + \gamma_2 w_{,yy}(x, 0) = B^* \\ \gamma_1 w_{,xx}(x, b) + \gamma_2 w_{,yy}(x, b) = B^* \end{cases} \qquad (7.259b)$$

for $0 \le x \le a$. However, as $w(0, y) = 0$, $0 \le y \le b$, it follows that $w_{,yy}(0, y) = 0$, $0 \le y \le b$, and, similarly, $w(a, y) = 0$, $0 \le y \le b$, implies that $w_{,yy}(a, y) = 0$, $0 \le y \le b$. In an analogous manner, we obtain from $w(x, 0) = 0$, $0 \le x \le a$, the fact that $w_{,xx}(x, 0) = 0$ $0 \le x \le a$, and from $w(x, b) = 0$, $0 \le x \le a$, the conclusion that $w_{,xx}(x, b) = 0, 0 \le x \le a$. Thus, (7.259a,b) imply that along the edges of the plate we must have

$$\begin{cases} w(0, y) = w(a, y) = 0, \ 0 \le y \le b \\ w(x, 0) = w(x, b) = 0, \ 0 \le x \le a \end{cases} \tag{7.260a}$$

$$\begin{cases} w_{,xx}(0, y) = w_{,xx}(a, y) = A^*/\lambda_1, \ 0 \le y \le b \\ w_{,yy}(x, 0) = w_{,yy}(x, b) = B^*/\gamma_2, \ 0 \le x \le a \end{cases} \tag{7.260b}$$

The situation is depicted in Fig. 7.22. In [236], Biswas, as we have already indicated, looks for plate deflections of the form specified in (7.250). However, such deflections can satisfy (7.260a,b) if and only if $A^* = B^* = 0$, i.e., if and only if $T_1 = T_2 = 0$. If, on the other hand, the boundary conditions of simple support are to be compatible with an applied temperature distribution which is independent of position in the middle surface of the plate, and varies linearly through the plate thickness, trial (bending) deflections would have to satisfy the inhomogeneous boundary data (7.260a,b); this is, clearly, not possible, as should be evident from Fig. 7.22, because at each of the four corners of the plate the boundary data is inconsistent, e.g., $w_{,xx} = A^*/\lambda_1 \ne 0$, for $x = 0$, $0 \le y \le b$, which implies that $w_{,xx}(0, 0) = A^*/\lambda_1$. However, $w = 0$, for $y = 0$, $0 \le x \le a$, so that $w_{,xx}(x, 0) = 0$, $0 \le x \le a$, in which case $w_{,xx}(0, 0) = 0$. Thus any (trial) deflection which satisfied the inhomogeneous boundary conditions (7.260a,b) would have to exhibit jump discontinuities in its second derivatives at each of the four corners of the rectangular plate. Such a problem would not exist for an applied temperature distribution $\delta T(x, y, z)$ which produced nontrivial thermal moments $\tilde{M}_T^1$, $\tilde{M}_T^2$ satisfying $\tilde{M}_T^1(0, 0) = \tilde{M}_T^1(0, b) = \tilde{M}_T^1(a, 0) = \tilde{M}_T^1(a, b) = 0$ and $\tilde{M}_T^2(0, 0) = \tilde{M}_T^2(a, 0) = \tilde{M}_T^2(0, b) = \tilde{M}_T^2(a, b) = 0$; such thermal moments could still satisfy

$$\begin{cases} \dfrac{\partial \tilde{M}_T^1}{\partial y}(0, y) \ne 0, \ \dfrac{\partial \tilde{M}_T^1}{\partial y}(a, y) \ne 0, \ 0 \le y \le b \\ \dfrac{\partial \tilde{M}_T^2}{\partial x}(x, 0) \ne 0, \ \dfrac{\partial \tilde{M}_T^2}{\partial x}(x, b) \ne 0, \ 0 \le x \le a \end{cases} \tag{7.261}$$

as long as each of $\tilde{M}_T^1$, $\tilde{M}_T^2$ vanished at the four corners of the plate.

We now turn to a discussion of the initial buckling behavior of circular plates within the context of large deflection theory. Although the interpretation is somewhat different, the analysis of initial buckling for an isotropic circular plate subject to thermal stresses, within the context of large deflection theory, mathematically parallels the small deflection analysis presented in the previous section. Klosner and Forray [227] studied the buckling of simply supported isotropic circular plates subjected to a symmetrical temperature distribution by using the Rayleigh-Ritz method. For the case of a circular plate of variable thickness, Mansfield [237] has analyzed the buckling, curling, and postbuckling behavior when the temperature varies through the thickness of the plate and the edge of the plate is restrained in its plane; we will discuss Mansfield's work in Chapter 8 and other discussions of the postbuckling behavior of isotropic circular plates will appear in section 7.3 of the present chapter within the context of Berger's approximation. In this section, we will content ourselves with discussions of the work of Sarkar [238], who treats the case of a heated thin circular plate of isotropic material under uniform thermal compression in the plane of the plate, and the analysis in Nowacki [222], which is based on the work of Vinokurov [239]; in [238], the critical buckling temperature is calculated for plates under different edge conditions and different temperature distributions as well. The edge of the plate is restrained in the plane of the plate.

The solid circular isotropic plate considered by Sarkar [238] has uniform thickness, occupies the domain $0 \leq r \leq a$, $-\dfrac{h}{2} \leq z \leq \dfrac{h}{2}$, and is subjected to an arbitrary temperature distribution (which, however, must be symmetrical for the author's [238] equations to apply). As the edge of the plate is assumed to be restrained in the plane of the plate, the displacement $u(r) = 0$. Setting $\phi = -\dfrac{dw}{dr}$ we have, as a direct consequence of (6.108), (6.110), the assumed radial symmetry of $w$, and the fact that $u(r) \equiv 0$, $0 \leq r \leq a$,

$$N_r = -N_T^* = -\frac{E\alpha}{1-\nu} \int_{-h/2}^{h/2} T(r,z)dz \qquad (7.262)$$

and

$$\begin{cases} M_r = K\left[\dfrac{d\phi}{dr} + \dfrac{\nu}{r}\phi\right] - M_T^* \\[2mm] M_\theta = K\left[\nu\dfrac{d\phi}{dr} + \dfrac{1}{r}\phi\right] - M_T^* \end{cases} \qquad (7.263)$$

where $M_T^*$ is given by (6.111). With the assumptions given above, the initial buckling equation that follows from the system (6.112) by setting $w = 0$ in the second equation of this set, and substituting the resultant Airy function in the first equation, is

$$r^2\frac{d^2\phi}{dr^2} + r\frac{d\phi}{dr} + \left(\frac{N_r}{K}r^2 - 1\right)\phi = r^2\frac{dM_T^*}{dr} \qquad (7.264)$$

Replacing $\dfrac{N_r}{K} \equiv -\dfrac{1}{K}N_T^*$ in (7.264) by $\gamma^2$ and then setting $v(r) = \gamma r$ we obtain the equation

$$\frac{d^2\phi}{dv^2} + \frac{1}{v}\frac{d\phi}{dv} + \left(1 - \frac{1}{v^2}\right)\phi = \frac{1}{\gamma}\frac{dM_T^*}{dv} \qquad (7.265)$$

As a simple example in [238] the author considers the case of a temperature distribution which varies only through the thickness of the plate, i.e., $\delta T = \delta T(z)$. In this case, $M_T^*$ is clearly independent of $r$, and hence of $v$, so that $\dfrac{dM_T^*}{dv} = 0$. Equation (7.265) then reduces to

$$\frac{d^2\phi}{dv^2} + \frac{1}{v}\frac{d\phi}{dv} + \left(1 - \frac{1}{v^2}\right)\phi = 0 \qquad (7.266)$$

for $0 \leq v \leq \gamma a$. The solution of (7.266) which is finite at $v = 0$ has the form

$$\phi = CJ_1(v) \equiv CJ_1(\gamma r) \qquad (7.267)$$

in which case,

$$w(r) = C\gamma J_0(\gamma r) + C_1 \qquad (7.268)$$

with $C_1$ an arbitrary constant of integration.

If the plate is clamped along its edge at $r = a$ so that $w = \phi = 0$ for $r = a$, then, by virtue of (7.267), $J_1(\gamma a) = 0$; the smallest (approximate) root of this equation is $\gamma = 3.832/a$, in which case,

$$(N_r)_{cr} = \frac{K}{a^2}(3.832)^2 \qquad (7.269a)$$

so that

$$(N_T)_{cr} = -\frac{K(1-\nu)}{a^2}(3.832)^2 \qquad (7.269b)$$

A second case considered in [238] is one for which the temperature distribution is such that

$$M_T^* = \frac{M_0}{1 - \nu}\left(1 - \frac{v^2}{\gamma^2 a^2}\right) \tag{7.270}$$

In this case, (7.265) becomes

$$\frac{d^2\phi}{dv^2} + \frac{1}{v}\frac{d\phi}{dv} + \left(1 - \frac{1}{v^2}\right)\phi = -\frac{2M_0}{\gamma^3 a^2(1 - \nu)}v \tag{7.271}$$

whose solution is given by

$$\phi = C_2 J_1(v) - \frac{2M_0}{\gamma^3 a^2(1 - \nu)}v \tag{7.272}$$

From (7.272) it follows that

$$w = C_3 + \frac{M_0 r^2}{\gamma^2 a^2(1 - \nu)} - \frac{C_2}{\gamma}J_0(\gamma r) \tag{7.273}$$

Imposing upon (7.272), (7.273) the conditions at $r = a$ for a clamped plate we are led to the solution

$$w(r) = \frac{2M_0[J_0(\gamma a) - J_0(\gamma r)]}{a^2\gamma^3(1 - \nu)J_1(\gamma a)} - \frac{M_0}{\gamma^2(1 - \nu)} + \frac{M_0 r^2}{\gamma^2 a^2(1 - \nu)} \tag{7.274}$$

The deflection in (7.274) becomes infinite when $J_1(\gamma a) = 0$, which again leads to the critical values reflected in (7.269a,b); the absolute value of $(N_T)_{cr}$ in (7.269b) yields the critical buckling temperature.

In [222] a somewhat different presentation of buckling of an elastic isotropic circular plate subjected to an axisymmetric temperature distribution is given; the temperature distribution is taken in the form

$$\delta T(r) = T_0(r) + zT_1(r) \tag{7.275}$$

which is, of course, a special case of (6.113). Thus,

$$\begin{cases} N_T^* = \dfrac{E\alpha h}{1 - \nu}T_0(r) \\[4mm] M_T^* = \dfrac{E\alpha h^3}{12(1 - \nu)}T_1(r) \end{cases} \tag{7.276}$$

The radial and circumferential stress fields in this situation are given, respectively, by

$$\begin{cases} \sigma_{rr} = \dfrac{E}{1-\nu^2}(e_{rr} + \nu e_{\theta\theta}) - \dfrac{E\alpha}{1-\nu}(T_0 + zT_1) \\ \sigma_{\theta\theta} = \dfrac{E}{1-\nu^2}(e_{\theta\theta} + \nu e_{rr}) - \dfrac{E\alpha}{1-\nu}(T_0 + zT_1) \end{cases} \qquad (7.277)$$

We note that (6.108) follows from (7.276), (7.277) if (7.277) is integrated through the thickness of the plate. In (7.277), of course

$$\begin{cases} e_{rr} = v_{,r} + \dfrac{(w_{,r})^2}{2} - zw_{,rr} \\ e_{\theta\theta} = \dfrac{1}{r}v - \dfrac{z}{r}w_{,r} \end{cases} \qquad (7.278)$$

where $v = v(r)$ again denotes the radial displacement of the plate. Combining (7.277) and (7.278) the resultant forces and moments are easily shown to have the form

$$\begin{cases} N_r = \dfrac{E}{1-\nu^2}\left[v_{,r} + \dfrac{1}{2}(w_{,r})^2 + \dfrac{\nu}{r}v - (1+\nu)\alpha T_0\right] \\ N_\theta = \dfrac{Eh}{1-\nu^2}\left[\dfrac{1}{r}v + \nu(v_{,r} + \dfrac{1}{2}(w_{,r}^2)) - (1+\nu)\alpha T_0\right] \end{cases} \qquad (7.279)$$

and

$$M_r = -K\left[w_{,rr} + \dfrac{\nu}{r}w_{,r} + (1+\nu)\alpha T_1\right]$$

$$M_\theta = -K\left[\dfrac{1}{r}w_{,r} + \nu w_{,rr} + (1+\nu)\alpha T_1\right] \qquad (7.280)$$

In [222], Nowacki works directly with the equilibrium equations, for the symmetric situation, in the form

$$\begin{cases} (rN_r)_{,r} - N_\theta = 0 \\ (rM_{rr})_{,r} - M_\theta - Q_r = 0 \\ (rQ)_{,r} - (rN_r w_{,r})_{,r} - tr = 0 \end{cases} \qquad (7.281)$$

with $Q$ the shear force and $t$ the external loading of the plate. By inserting $Q$ from the second equation in (7.281) into the third equation in this set, and then employing the relations (7.280), one obtains

$$K\nabla_r^4 w + K(1+\nu)\alpha \nabla_r^2 T_1 = \dfrac{1}{r}(rN_r w_{,r})_{,r} + t \qquad (7.282)$$

with $\nabla_r^4 = \nabla_r^2 \nabla_r^2$ and $\nabla_r^2 = \partial_r^2 + \frac{1}{r}\partial_r$. Equation (7.282) is, of course, the form assumed by the first of the von Karman equations in (6.112) under the assumption of radial symmetry.

From the relations (7.279) one obtains the radial displacement $v(r)$ in the form

$$v(r) = \frac{r}{Eh}(N_\theta - \nu N_r) + \alpha r T_0 \qquad (7.283)$$

or, if we use the first of the relations in (7.281) to eliminate the resultant stress $N_\theta$ in (7.283)

$$v(r) = \frac{r}{Eh}[(rN_r)_{,r} - \nu N_r] + \alpha r T_0 \qquad (7.284)$$

Finally, inserting (7.284) into the first of the relations in (7.279) we obtain

$$r\left[\frac{1}{r}(r^2 N_r)_{,r}\right]_{,r} + \frac{1}{2}Eh(w_{,r})^2 = 0 \qquad (7.285)$$

Equation (7.286) is just the form assumed by the condition of geometric compatability of the strains for the special problem under consideration. To illustrate the utility of the relations (7.282), (7.283) with respect to axially symmetric temperature distributions of the form (7.275), Nowacki [222] presents, following Vinokurov [239], the solution of the clamped circular plate problem for which the boundary conditions are

$$\begin{cases} w(a) = 0, \ w_{,r}(a) = 0 \\ v(a) = \dfrac{a}{Eh}[(rN_r)_{,r} - \nu N_r]_{r=a} + \alpha a T_0 = 0 \end{cases} \qquad (7.286)$$

At $r = 0$ it is required, as usual, that $r^{-1}w_{,r}$ and $N_r$ be finite. Assuming that $T_1 = $ const. and $t = $ const. simplifies (7.282), after an integration, to the form

$$K\left(\frac{1}{r}(rw_{,r})_{,r}\right)_{,r} = N_r w_{,r} + \frac{tr}{2} \qquad (7.287)$$

We now introduce the dimensionless variables

$$\begin{cases} \rho = \dfrac{r}{a}, \ \phi = w_{,r} = \dfrac{1}{a}w_{,\rho} \\ \psi = \dfrac{rN_r}{Eah}, \ k = 12(1-\nu^2)\dfrac{a^2}{h^2}, \ m = 6(1-\nu^2)\dfrac{ta^2}{Eh^3} \end{cases} \qquad (7.288)$$

in terms of which (7.285) and (7.287) assume the respective forms

$$
\begin{cases}
\rho\psi_{,\rho\rho} + \psi_{,\rho} - \dfrac{1}{\rho}\psi = \dfrac{1}{2}\phi^2 \\[2mm]
\rho\phi_{,\rho\rho} + \phi_{,\rho} - \dfrac{1}{\rho}\phi = -k\phi\psi + m\rho^2
\end{cases}
\tag{7.289}
$$

while the boundary conditions (7.286) become

$$
\phi(0) = 0, \quad \phi(1) = 0, \quad [\psi_{,\rho} - \nu\psi - \alpha T_0]_{\rho=1} = 0
\tag{7.290}
$$

with $\dfrac{1}{\rho}\psi$ and $v$ bounded at $\rho = 0$.

The system (7.289) may be solved (approximately) by applying Galerkin's method: as a first approximation, we use the classical solution for the deflection of an isotropic elastic circular plate loaded by the uniform loading $t = \text{const.}$, namely,

$$
\begin{cases}
w^{(1)}(\rho) = w_0[2(1 - \rho^2) - (1 - \rho^4)] \\[2mm]
w_0 = ta^4/64K
\end{cases}
\tag{7.291}
$$

Then, as $\phi = \dfrac{1}{a}w_{,\rho}$

$$
\phi^{(1)} = c(\rho^3 - \rho), \quad c = 4w_0/a
\tag{7.292}
$$

Substituting $\phi^{(1)}$ for $\phi$ in the first equation in (7.289) and using the boundary conditions, yields

$$
\begin{cases}
\psi^{(1)} = \dfrac{c^2}{96}(\rho^7 - 4\rho^5 + 6\rho^3 - b\rho) \\[2mm]
b = \dfrac{5 - 3\nu}{1 - \nu} - \dfrac{96}{c^2}\dfrac{\alpha T_0}{1 - \nu}
\end{cases}
\tag{7.293}
$$

If we now substitute $\psi^{(1)}$, from (7.293), for $\psi$ in the second equation in (7.289), multiply this equation by $\rho^3 - \rho$, and integrate from $\rho = 0$ to $\rho = 1$, we obtain

$$
c + c^3\dfrac{k}{64}\left(\dfrac{b}{24} - \dfrac{1}{14}\right) = \dfrac{1}{8}m
\tag{7.294}
$$

In (7.294) we recall that $c = \dfrac{4}{a}w_0$, while the presence of $b = \dfrac{5 - 3\nu}{1 - \nu} -$ $\dfrac{96}{c^2}\dfrac{\alpha T_0}{1 - \nu}$ accounts for the influence of the temperature distribution on

the plate deflection (which is seen to depend only on the function $T_0(r)$).
For the special case in which $t = 0$ (so that $m = 0$ in (7.294)) it is possible
to obtain the critical buckling temperature in the form

$$(T_0)_{cr} \simeq \frac{4}{3(1 + \nu)} \cdot \frac{h^2}{\alpha a^2} \tag{7.295}$$

Note is made, in [222], of the fact that (7.295) is independent of the
Young's modulus $E$ of the plate.

For results on the thermoelastic stability of cylindrically orthotropic
(laminated) circular plates, the reader is referred to the paper of Stavsky
[231] in which axisymmetric stability and thermal buckling equations are
established for circular plates; the plates in [231] are composed of polar
orthotropic layers subjected to mechanical loads depending only on the
radial variable $r$ and thermal fields of the form $\delta T = \delta T(r, z)$. For the
initial buckling (eigenvalue) problems in [231], closed form solutions are
presented in terms of Bessel functions of the first kind (and of fractional
order) and the variation of the critical loads with the anisotropy parameters is analyzed. It is worth noting that, unlike the situation for isotropic
circular plates, where (7.295) holds for the specific problem considered
above (so that $(T_0)_{cr}$ is independent of $E$), for the same problem for a
polar orthotropic circular plate Stavsky [231] obtains the result

$$(T_0)_{cr} \simeq \frac{4}{3} \cdot \frac{D_{rr}}{A_r ha^2} \tag{7.296}$$

where

$$\begin{cases} D_{rr} = \int_{-h/2}^{h/2} E_{rr} z \, dz \\ A_r = -(\alpha_r E_{rr} + \alpha_\theta E_{r\theta}) \end{cases} \tag{7.297}$$

---

## 7.3 Applications of the Berger's Approximation

For the case of isotropic, linearly elastic response, we have shown that
the von Karman theory for large thermal deflections of a thin plate
leads to the coupled system of nonlinear partial differential equations
(6.93a,b), when rectangular Cartesian coordinates are used, and to the
system (6.112) when polar coordinates are employed; these systems,
together with associated boundary conditions cannot be expected to

yield exact solutions. Among the many approximate methods which
have been devised to treat large deflection von Karman systems is the
technique first promulgated by Berger in [240] for thin plates subjected
to mechanical loading. As Tauchert [217] notes, "The method is based on
neglecting the strain energy associated with the second invariant of the
strains in the plate middle surface. Although there does not appear to
be any physical justification for this approximation, comparisons of the
results with known solutions indicate that for a broad range of problems
Berger's approach yields sufficiently accurate results." In fact, all of the
applications of Berger's approximation in the literature, where accurate
results appear to be obtained, are associated with problems in which the
edges of the plate are constrained against in-plane movements; Nowinski
and Ohnabe [241] have shown that, for the case of mechanical loading,
the accuracy of Berger's method is closely tied to this specific type of
boundary condition. In fact, by applying Galerkin's method to both
the usual von Karman large deflection equations, and the approximate
equations governing large deflections which are obtained from Berger's
approach, it was determined in [241] that Berger's method is susceptible
to serious errors if the edge (or edges) of the plate are free to move. It has
been surmised in [217] that the conclusions reached in [241] are almost
certainly extendible to the case of thermal deflections. The accuracy of
Berger's method was also assessed by Jones, Mazumdar, and Cheung in
[242] for the special case in which $t \equiv 0$, $M^T \equiv 0$ in (6.93a,b), and a
perturbation technique was proposed for use in those situations where
Berger's technique may not be sufficiently accurate. We will comment
below on the work in [242] as well as on the work of Banerjee and Datta
[243], who proposed a modified energy expression which leads, as in
the case of Berger's method, to large deflection plate equations which
are decoupled. Although only mechanical loadings are considered in
[243], the results given for transverse loading of a circular plate are in
good agreement with prior results for this problem for both movable and
immovable edges and an extension of the method to the case of thermal
loading has been indicated.

Berger's method was originally formulated, as we have already indicated, for isothermal plate problems; it was extended to non-isothermal
problems in [244] by Basuli, who used the technique to treat large deflection problems for rectangular and circular isotropic plates subjected to
normal pressure and heating. We will begin our study in this section of
the applications of Berger's method to large thermal deflection problems
for thin plates by reviewing the work in [244] and the related work of

Pal [245], which also assumes isotropic response. Then we will consider in some detail the application of Berger's technique to large thermal deflection problems for heated rectilinearly orthotropic rectangular plates as presented in Biswas [246]; this is followed by a brief description of the similar results for polar orthotropic circular plates, which were described by Pal [247].

The starting point for the application of Berger's approximation to large deflection problems for isotropic rectangular plates is the observation that the total potential energy of the plate, which is given by (6.130), may be rewritten in the form

$$
\Pi = \int_0^b \int_0^a \left\{ \frac{Eh}{2(1 - \nu^2)} [\epsilon_1 - 2(1 - \nu)\epsilon_2] \right.
$$
$$
+ \frac{1}{2} K \left[ (\nabla^2 w)^2 - 2(1 - \nu) \left\{ (w_{,xy}^2 - w_{,xx} w_{,yy}) \right\} \right] \qquad (7.298)
$$
$$
\left. - N^T \epsilon_1 + M^T \nabla^2 w - tw \right\} dx dy
$$

where $\epsilon_1$ and $\epsilon_2$ are, respectively, the first-and second-strain invariants of the middle surface of the plate, i.e.,

$$
\epsilon_1 = \epsilon_{xx}^0 + \epsilon_{yy}^0 = u_{,x} + v_{,y} + \frac{1}{2} \left( w_{,x}^2 + w_{,y}^2 \right) \qquad (7.299a)
$$

$$
\epsilon_2 = \epsilon_{xx}^0 \epsilon_{yy}^0 - \frac{1}{4} (\gamma_{xy}^0)^2
$$
$$
= u_{,x} v_{,y} + \frac{1}{2} u_{,x} w_{,y}^2 + \frac{1}{2} v_{,y} w_{,x}^2 \qquad (7.299b)
$$
$$
- \frac{1}{4} \left\{ u_{,y}^2 + v_{,x}^2 + 2u_{,y} v_{,x} + 2u_{,y} w_{,x} w_{,y} + 2v_{,x} w_{,x} w_{,y} \right\}
$$

Assuming a temperature distribution $\delta T$ of the form

$$
\delta T(x, y, z) = T_0(x, y) + g(z) T_1(x, y) \qquad (7.300)
$$

Basuli [244] writes (7.298) in the form

$$
\Pi = \int_0^b \int_0^a \left[ \frac{1}{2} K \left\{ (\nabla w)^2 + \frac{12}{h^2} \epsilon_1^2 \right. \right.
$$
$$
- 2(1 - \nu) \left[ \frac{12}{h^2} \epsilon_2 + w_{,xx} w_{,yy} - w_{,xy}^2 \right] \right\} - tw \qquad (7.301)
$$
$$
\left. - \frac{E\alpha}{1 - \nu} \left\{ \epsilon_1 h T_0 - f(h) T_1 \nabla^2 w \right\} \right] dx dy
$$

where

$$f(h) = \int_{-h/2}^{h/2} g(z)z\,dz$$

The functional $\Pi$ is now minimized using the Euler-Lagrange variational equations

$$\frac{\partial \Pi}{\partial u} - \frac{\partial}{\partial x}\frac{\partial \Pi}{\partial u_{,x}} - \frac{\partial}{\partial y}\frac{\partial \Pi}{\partial u_{,y}} = 0 \qquad (7.302a)$$

$$\frac{\partial \Pi}{\partial v} - \frac{\partial}{\partial x}\frac{\partial \Pi}{\partial v_{,x}} - \frac{\partial}{\partial y}\frac{\partial \Pi}{\partial v_{,y}} = 0 \qquad (7.302b)$$

$$\frac{\partial \Pi}{\partial w} - \frac{\partial}{\partial x}\frac{\partial \Pi}{\partial w_{,x}} - \frac{\partial}{\partial y}\frac{\partial \Pi}{\partial w_{,y}} + \frac{\partial^2}{\partial x^2}\frac{\partial \Pi}{\partial w_{,xx}}$$
$$+ \frac{\partial^2}{\partial x \partial y}\frac{\partial \Pi}{\partial w_{,xy}} + \frac{\partial^2}{\partial y^2}\frac{\partial \Pi}{\partial w_{,yy}} = 0 \qquad (7.302c)$$

which yields (upon following Berger [240] and setting $\epsilon_2 = 0$) the relations

$$\frac{\partial}{\partial x}\left(\frac{Eh}{1-\nu^2}\epsilon_1 - N^T\right) = 0, \quad \frac{\partial}{\partial y}\left(\frac{Eh}{1-\nu^2}\epsilon_1 - N^T\right) = 0 \quad (7.303a)$$

and

$$\frac{\partial^2}{\partial x^2}(K\,\nabla^2\,w + M^T) + \frac{\partial^2}{\partial y^2}(K\,\nabla^2\,w + M^T)$$
$$- \frac{\partial}{\partial x}\left\{\left(\frac{Eh}{1-\nu^2}\epsilon_1 - N^T\right)w_{,x}\right\} \qquad (7.303b)$$
$$- \frac{\partial}{\partial y}\left\{\left(\frac{Eh}{1-\nu^2}\epsilon_1 - N^T\right)w_{,y}\right\} - t = 0$$

in the general case and

$$\frac{\partial}{\partial x}\{\epsilon_1 - (1+\nu)\alpha T_0\} = 0, \quad \frac{\partial}{\partial y}\{\epsilon_1 - (1+\nu)\alpha T_0\} = 0 \quad (7.304a)$$

$$\frac{\partial^2}{\partial x^2}\left\{K\,\nabla^2\,w + \frac{E\alpha}{1-\nu}f(h)T_1\right\} + \frac{\partial^2}{\partial y^2}\left\{K\,\nabla^2\,w + \frac{E\alpha}{1-\nu}f(h)T_1\right\}$$
$$- \frac{E}{1-\nu}\frac{\partial}{\partial x}\{[\epsilon_1 - (1+\nu)\alpha T_0]w_{,x}\} \qquad (7.304b)$$
$$- \frac{E}{1-\nu}\frac{\partial}{\partial y}\{[\epsilon_1 - (1+\nu)\alpha T_0]w_{,y}\} - t = 0$$

in the special case in which $\delta T(x, y, z)$ is given by (7.300).
From (7.303a) one easily obtains

$$\frac{Eh}{1 - \nu^2}\epsilon_1 - N^T = \lambda^2, \ (\lambda^2 \ \text{const.}) \tag{7.305}$$

which, upon substitution into (7.303b), yields

$$K \nabla^2 (\nabla^2 - \lambda^2) w = t - \nabla^2 M^T \tag{7.306}$$

As Tauchert [217] has noted, the system consisting of (7.305), (7.306) is much simpler than the corresponding Von Karman system (6.93a,b); the equations (7.305), (7.306) are quasi-linear and decoupled in the following sense: (7.306) is linear in $w$ and may be solved independently of (7.305). Equation (7.305) is linear in $u_{,x}$ and $v_{,y}$ and may be solved once $w$ has been determined. The only coupling that remains in (7.305), (7.306) is through the parameter $\lambda$. In a similar manner, for the special case governed by (7.300), (7.304a) yields

$$\epsilon_1 - (1 + \nu)\alpha T_0 = \frac{\mu^2 h^2}{12}, \ (\mu^2 \ \text{const.}) \tag{7.307}$$

whose substitution into (7.304b) produces

$$K \nabla^2 (\nabla^2 - \mu^2) w = t - \frac{\alpha E}{1 - \nu} f(h) \nabla^2 T_1 \tag{7.308}$$

We now consider the application of the system (7.307), (7.308) to the thermal buckling of a simply supported rectangular plate which occupies the domain $0 \leq x \leq a, 0 \leq y \leq b$ in the $x, y$ plane; in this case, the assumed temperature distribution has the form given by (7.300).
**Remarks:** In [244] Basuli writes the conditions of simple support along the edges of the plate in the form

$$\begin{cases} w = w_{,xx} = 0, \ x = 0, a; \ 0 \leq y \leq b \\ w = w_{,yy} = 0, \ y = 0, b; \ 0 \leq x \leq a \end{cases} \tag{7.309}$$

However, for the simply supported plate

$$K(w_{,xx} + \nu w_{,yy}) + M^T = 0, \ x = 0, a; \ 0 \leq y \leq b$$
$$K(w_{,yy} + \nu w_{,xx}) + M^T = 0, \ y = 0, b; \ 0 \leq x \leq a$$

with $w = 0$ along each of the four edges; thus (7.309) holds, in the case of simply supported edges if and only if $M^T(0, y) = M^T(a, y) = 0$, for

$0 \leq y \leq b$, and $M^T(x,0) = M^T(x,b) = 0$, for $0 \leq x \leq a$. For the temperature distribution represented by (7.300)

$$M^T = \frac{\alpha E}{1-\nu} T_1(x,y) \int_{-h/2}^{h/2} g(z)z\,dz \equiv \frac{\alpha E}{1-\nu} f(h)T_1(x,y) \qquad (7.310)$$

and thus (7.309) represents the conditions of simple support along the edges of the plate if and only if

$$\begin{cases} T_1(0,y) = T_1(a,y) = 0; \ 0 \leq y \leq b \\ T_1(x,0) = T_1(x,b) = 0; \ 0 \leq x \leq a \end{cases} \qquad (7.311)$$

To maintain the validity of the analysis in [244] we will assume, below, that the function $T_1$ satisfies (7.311).

Proceeding with the analysis, we assume that each of $t(x,y)$, $T_1(x,y)$ may be expanded in a double Fourier series in the products $\sin \dfrac{m\pi x}{a}$ $\sin \dfrac{n\pi y}{b}$ so that, in particular, on the right-hand side of (7.308)

$$t - \frac{\alpha E}{1-\nu} f(h) \nabla^2 T_1 = \sum_{m=1}^{\infty} \sum_{n=1}^{\infty} q_{mn} \sin \frac{m\pi x}{a} \sin \frac{n\pi y}{b} \qquad (7.312)$$

Also, in view of (7.309), $w(x,y)$ is sought in the form

$$w = \sum_{m=1}^{\infty} \sum_{n=1}^{\infty} w_{mn} \sin \frac{m\pi x}{a} \sin \frac{n\pi y}{b} \qquad (7.313)$$

Assuming the edges of the plate are constrained against in-plane motion is equivalent to requiring that

$$\begin{cases} u(0,y) = u(a,y) = 0; \ 0 \leq y \leq b \\ v(x,0) = v(x,b) = 0; \ 0 \leq x \leq a \end{cases} \qquad (7.314)$$

By substituting (7.312) and (7.313) into (7.308) we obtain, in the usual manner, the relations

$$w_{mn} = \frac{q_{mn}}{K\left[\left(\frac{m\pi}{a}\right)^2 + \left(\frac{n\pi}{b}\right)^2\right]\left[\left(\frac{m\pi}{a}\right)^2 + \left(\frac{n\pi}{b}\right)^2 + \mu^2\right]} \qquad (7.315)$$

so that the deflection has been determined up to the constant $\mu^2$. We now write out (7.307) in the form

$$u_{,x} + v_{,y} + \frac{1}{2}w_{,x}^2 + \frac{1}{2}w_{,y}^2 - (1+\nu)\alpha T_0 = \frac{\mu^2 h^2}{12} \qquad (7.316)$$

Integration of (7.316) over the domain of the plate and use of the boundary conditions (7.314), which express the constraint of the plate against in-plane motion, yields the condition

$$\int_0^a \int_0^b \left\{ \frac{1}{2} (w_{,x}^2 + w_{,y}^2) - (1+\nu)\alpha T_0 \right\} dx\,dy = \frac{\mu^2 h^2}{12} ab \qquad (7.317)$$

We now substitute the expansion (7.313) into (7.317) and use standard results involving trigonometric integrals, e.g.,

$$\int_0^a \int_0^b \cos^2 \frac{m\pi x}{b} \sin^2 \frac{n\pi y}{b} dx\,dy = \frac{1}{4} ab$$

so as to obtain from (7.317) the equation

$$\frac{ab}{8} \sum_{m=1}^{\infty} \sum_{n=1}^{\infty} w_{mn}^2 \left[ \left(\frac{m\pi}{a}\right)^2 + \left(\frac{n\pi}{b}\right)^2 \right]$$
$$-(1+\nu)\alpha \int_0^a \int_0^b T_0(x,y)dx\,dy = \frac{\mu^2 h^2}{12} ab \qquad (7.318)$$

The further substitution of the expressions for the $w_{mn}$ in (7.315) into (7.318) then produces an algebraic relation from which $\mu^2$ may be determined; once $\mu^2$ has been computed (7.313), (7.315) serve to completely determine the deflection $w(x, y)$.

**Remarks:** Basuli [244] considers a specific example for which

$$T_0(x,y) = T_0 = const., \quad T_1(x,y) = T_1 \cos\left(\frac{2\pi y}{b}\right) \qquad (7.319)$$

however, the subsequent results are specious inasmuch as $T_1(x, y)$ does not satisfy (7.311). The source of confusion in much of the literature may be traced to the following elementary fact: any piecewise continuous function on the domain $0 \le x \le a$, $0 \le y \le b$, say, $M^T(x, y)$, can be extended to the entire $x, y$ plane in such a manner that the extended function may be represented by the Fourier series

$$M^T(x,y) = \sum_{m=1}^{\infty} \sum_{n=1}^{\infty} a_{mn} \sin \frac{m\pi x}{a} \sin \frac{n\pi y}{b} \qquad (7.320)$$

However, while the series on the right hand side of (7.320) converges to the value of $M^T(x, y)$, at each point of continuity of $M^T$ in the domain $0 < x < a$, $0 < y < b$, it will not converge to the values of $M^T$ along the boundary of the domain; it is the function $M^T(x, y)$, as defined by the temperature distribution $\delta T(x, y)$, and not its Fourier series representation, which must satisfy the conditions (7.311) so the criteria for a simply supported rectangular plate, under thermal loading, reduce to (7.309). In a one-dimensional setting, this phenomenon is easily represented by the Fourier sine series expansion of $f(x) = 1$ on $(0, \pi)$, i.e.

$$1 = \frac{4}{\pi} \sum_{k=1}^{\infty} \sin(2k - 1)x, \ 0 < x < \pi \tag{7.321}$$

For all $x$, $0 < x < 1$, the series on the right-hand side of (7.321) converges to 1; however, convergence breaks down at $x = 0$, $x = \pi$ where the series converges, by the Fourier theorem, to the average value of the left-hand and right-hand limits (of the periodic extension, with period $2\pi$, of the odd extension of $f(x)$ to $(-\pi, \pi)$). The situation is depicted below:

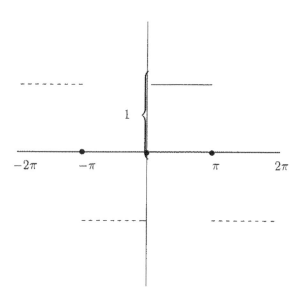

The second application of Berger's method that we want to consider is to the problem of axisymmetric deformations of a solid circular elastic plate subjected to a uniform load $t$ and a stationary temperature field

$\delta T(r, z)$. In this case, equations (7.305), (7.306) assume the form

$$\frac{Eh}{1-\nu^2}\left[u_{,r} + \frac{1}{r}u + \frac{1}{2}w_{,r}^2\right] - N^T = K\lambda^2 \tag{7.322}$$

$$K\left(\frac{d^2}{dr^2} + \frac{1}{r}\frac{d}{dr}\right)\left(\frac{d^2}{dr^2} + \frac{1}{r}\frac{d}{dr} - \lambda^2\right)w = t - \nabla^2 M^T \tag{7.323}$$

Under the assumption that $t' = t - \nabla^2 M^T$ is a constant, Tauchert [217] obtains a solution to (7.323) in the form

$$w = C_1 I_0(\lambda r) + C_2 - \frac{t'r^2}{4K\lambda^2} \tag{7.324}$$

with $\Phi_0$ the modified Bessel function of the first kind of order zero and $C_1, C_2$ arbitrary constants. If the edge of the plate at $r = b$ is constrained against in-plane motion so that $u(b) = 0$, then integration of (7.322) over the domain $0 \leq r \leq b$, $0 \leq \theta < 2\pi$, yields the relation

$$\frac{1}{2}C_1^2\left\{\frac{1}{2}\lambda^2 b^2\left[I_1^2(\lambda b) - I_0^2(\lambda b)\right] + \lambda b I_1(\lambda b)I_0(\lambda b)\right\}$$

$$-\frac{C_1 t'b^2 I_2(\lambda b)}{K\lambda^2} + \frac{t'b^4}{16K^2\lambda^4} - \frac{1-\nu^2}{Eh}\int_0^b N^T r \, dr = b^2 t^2 \lambda^2/24 \tag{7.325}$$

Assuming the plate to be clamped so that $w(b) = \dfrac{dw}{dr}(b) = 0$, we obtain from (7.324)

$$w = -\frac{t'b^2}{4K\lambda^2}\left\{\frac{2[I_0(\lambda b) - I_0(\lambda r)]}{\lambda b I_1(\lambda b)} - 1 + \frac{r^2}{b^2}\right\} \tag{7.326}$$

while the relation (7.325) for $\lambda$ assumes the form

$$\left(\frac{t'b^4}{Kt}\right)^2 = \frac{\left\{\dfrac{1}{3}(\lambda b)^6 + \dfrac{8\displaystyle\int_0^b N^T r \, dr}{\left(\dfrac{Eh}{1-\nu^2}\right)t^2}(\lambda b)^4\right\}}{\left\{1 + \dfrac{I_0(\lambda b) - 4I_2(\lambda b)}{\lambda b I_1(\lambda b)} - \dfrac{I_0^2(\lambda b)}{2I_1^2(\lambda b)}\right\}} \tag{7.327}$$

With the same assumption of circular symmetry and a temperature distribution of the form given by

$$\delta T(r, z) = T_0(r) + g(z)T_1(r),$$

Basuli [244] writes the system (7.322), (7.323) in the form

$$\frac{du}{dr} + \frac{v}{r} + \frac{1}{2}w_{,r}^2 - (1+\nu)\alpha T_0 = \frac{\beta^2 h^2}{12} \qquad (7.328)$$

$$\left(\frac{d^2}{dr^2} + \frac{1}{2}\frac{d}{dr}\right)\left(\frac{d^2}{dr^2} + \frac{1}{r}\frac{d}{dr} - \beta^2\right)w$$
$$= \frac{1}{K}\left[t - \frac{\alpha E}{1-\nu}f(h)\left(\frac{d^2}{dr^2} + \frac{1}{r}\frac{d}{dr}\right)T_1\right] \qquad (7.329)$$

and obtains, for a uniform load distribution $t$, that special case of (7.324) which has the form

$$w = AI_0(\beta r) + B - \frac{t'}{4K\beta^2}r^2 \qquad (7.330)$$

with $t' = t - \dfrac{\alpha E}{1-\nu}f(h)\left(\dfrac{d^2}{dr^2} + \dfrac{1}{r}\dfrac{d}{dr}\right)T_1$. Clearly, $t' = t$ if either $f(h) = 0$ or if $T_1$ is harmonic. With $u(b) = 0$, and $w(b) = \dfrac{dw}{dr}(b) = 0$, the solution of the system (7.328), (7.329) follows the same pattern as that outlined above for temperature distributions more general than (7.300) and will, therefore, not be repeated here. Somewhat more problematic, however, is the solution of (7.328), (7.329), which is presented in [244] for the case of a simply supported circular plate subject to the temperature distribution (7.300) and the assumption of constrained in-plane motion; in [244] Basuli takes the conditions of simple support along the edge of the plate at $r = b$ to have the form $w = 0$, at $r = b$, and

$$\frac{d^2w}{dr^2} + \frac{w}{r}\frac{dw}{dr} = 0, \quad \text{at } r = b \qquad (7.331)$$

However, by virtue of (7.323), with $w_{,\theta} = 0$, and $R_1 = 0$, $R_2 = b$, we have, along the edge at $r = b$, $w(b) = 0$ and

$$K\left(\frac{d^2w}{dr^2} + \frac{1}{r}\frac{dw}{dr}\right)\Bigg|_{r=b} + M_T^*\big|_{r=b} = 0 \qquad (7.332)$$

The problem that arises now is the same one which has been discussed at length earlier in this section, namely, in view of (6.111)

$$M_T^*\big|_{r=b} = \frac{E\alpha}{1-\nu}\int_{-h/2}^{h/2} \delta T(b,z)z\,dz = \frac{E\alpha}{1-\nu}f(h)T_1(b) \qquad (7.333)$$

so that (7.331) follows from (7.333) if and only if $f(h) = 0$ or $T_1(b) = 0$; under this latter assumption, the results in [244] are valid and the full solution of the boundary-value problem associated with (7.331), (7.329), when the edge at $r = b$ is simply supported, assumes the form

$$w = \frac{-t'b^4}{4K\beta^3 b^3 H} \left\{ 2(1+\nu)[I_0(\beta b) - I_0(\beta r)] - H\beta b \left( 1 - \left(\frac{r}{b}\right)^2 \right) \right\}$$

(7.334)

$$\left(\frac{t'b^4}{Kh}\right)^2$$

$$= \frac{\frac{1}{3}(\beta b)^6 + \frac{(1+\nu)\alpha F(b)}{h^2}\beta^4 b^4}{\frac{1}{2} + \left\{ \frac{1}{2}[I_1^2(\beta b) - I_0^2(\beta b)] + \frac{I_0(\beta b)I_1(\beta b)}{\beta b} \right\} - \frac{4(1+\nu)I_2(\beta b)}{H\beta b}}$$

(7.335)

where

$$\begin{cases} H = \beta b I_0(\beta b) - (1-\nu)I_1(\beta b) \\ \\ F(b) = \int_0^b T_0(r) r \, dr \end{cases}$$

(7.336)

Although Basuli's work [244] extending Berger's method [240] to non-isothermal plate deflection problems preceded the work of Pal [247], the latter author appears not to have been aware of the earlier research in this direction. Thus, Pal [247] rederives the Berger's type equations by minimizing the total potential energy associated with a circular plate under the assumption of a radially symmetric temperature; these decoupled large deflection equations are then solved in [247], for the case in which $t \equiv 0$, by using the method of successive approximations. Included in the analysis in [247] are results for annular plates, i.e., circular plates with a concentric circular hole. In line with our related discussions in this section, the applicability of the results in [247] for circular and annular plates with simply supported edges is subject to the same constraints on the behavior of $T_1(r)$ along the plate edges, which have been delineated previously. In Pal [247], the temperature distribution is taken in the form

$$\delta T(\zeta, z) = \{\bar{T}_0 + \bar{T}_1(1 - \zeta^2)\} \left( 1 + \frac{2z}{3h} \right)$$

(7.337)

where $\zeta = r/a$. Thus, for $\delta T(r, z)$ we have

$$\delta T = T_0(r) + g(z)T_1(r)$$

with

$$\begin{cases} T_0(r) = \bar{T}_0 + \bar{T}_1 \left(1 - \left(\frac{r}{a}\right)^2\right) = T_1(r) \\ g(z) = \frac{2}{3} \cdot \frac{z}{h} \end{cases} \tag{7.338}$$

It is clear from (7.337) that $T_1(a) = 0$ if and only if $\bar{T}_0 = 0$; in [247] the outer edge of an annular plate is located at $r = a$ while the inner edge is at $r = b$ (with $b = 0$ for a solid circular plate). Therefore, in Pal's work [247], (7.331) is equivalent to (7.332), with $b$ replaced by $a$ provided $\bar{T}_0 = 0$. In Figs. 7.23 and 7.24, which are taken from [247], we show the relation between the temperature rise and the normalized maximal deflection $\delta/h(\delta = w(0))$ at the center of the plate for the simply supported and clamped edge conditions, respectively; in these figures

$$\begin{cases} \tilde{T} = \int_{-h/2}^{h/2} \delta T(r, z) dz \\ \bar{T} = \int_{-h/2}^{h/2} \delta T(r, z) z dz \end{cases} \tag{7.339}$$

Fig. 7.25 depicts, for an annular plate with aspect ratio $\dfrac{b}{a} = 0.4$, the relation between the temperature rise and the normalized deflection $\delta/h$, for the same temperature distribution (7.337); the plate in Fig. 7.25 has both inner and outer edges simply supported. For the same plate and the same temperature distribution, the analogous results for an annular plate with both inner and outer edges clamped are depicted in Fig. 7.26. The results in [247] for the case where the same annual plate has its inner edge clamped and its outer edge simply supported is shown in Fig. 7.27. The results shown in Fig. 7.25 (the annular plate with both edges simply supported) are open to the criticism that even if $\bar{T}_0 = 0$, so that $T_1(a) = 0$ is a direct consequence of (7.338), i.e., it does not necessarily follow that $T_1(b) = 0$; in fact $T_1(b) = 0$, when $\bar{T}_0 = 0$, if and only if $a = b$ which reduces the problem to one for a solid circular plate of radius $a$.

Both Biswas [248] and Pal [249] have applied Berger's approximation method to linearly elastic thin plates exhibiting anisotropic response. In [248], Biswas solves the large-deflection problem for a rectilinearly orthotropic rectangular plate which is simply supported and subjected to a stationary temperature distribution, while in [249], Pal obtains solutions for cylindrically orthotropic circular plates using the method of successive approximations; it is also demonstrated in [249] that the solutions for a polar orthotropic circular plate reduce to those for an isotropic circular plate as a special case. We now briefly review the analysis in [248] and [249] beginning with the work of Biswas [248], which is, unfortunately, subject, once again, to the same criticism concerning the interpretation of simple support boundary conditions in thermal deflection problems. The notation employed by Biswas in [248] parallels that of his earlier paper [236]; thus, in particular, (7.238) applies, i.e.

$$D_x = D_{11}, \ H = D_1 + 2D_{66}, \ D_y = D_{22} \qquad (7.340)$$

with $D_1 = \dfrac{1}{2}(D_{12} + D_{21})$, while the $\tilde{\beta}_i$, $i = 1, 2$, are given by (7.239). The first strain invariant $\epsilon_1$ is given in this situation by

$$\epsilon_1 = \epsilon_{xx}^0 + \kappa\epsilon_{yy}^0, \ \kappa = \sqrt{\frac{D_y}{D_x}} \qquad (7.341)$$

and $M_T$, as in (7.237), is given by $M_T = \displaystyle\int_{-h/2}^{h/2} \delta T z dz$. The full set of von Karman large deflection equations has the form (7.237), (7.240) where the $E_1, E_2, \nu_{12}, \nu_{21}, G_{12}$, and the resultant (thermal) stresses $N_T, \tilde{N}'_T$, $\tilde{N}^2_T$ have been defined in Chapter 6. Neglecting the second strain invariant $\epsilon_2$, Biswas [248] writes the total potential energy for the rectilinearly orthotropic rectangular plate in the form

$$\begin{aligned}
\Pi = \frac{1}{2} \int_0^b \int_0^a & \left\{ D_x w_{,xx}^2 + 2D_1 w_{,xx} w_{,yy} \right. \\
& \left. + D_y w_{,yy}^2 + 4D_{66} w_{,xy}^2 + D_x \frac{12}{h^2} \epsilon_1^2 \right\} dxdy \\
+ \int_{-h/2}^{h/2} \int_0^b \int_0^a & \left\{ \tilde{\beta}_1 \epsilon_{xx} \delta T(x, y, z) + \tilde{\beta}_2 \epsilon_{yy} \delta T(x, y, z) \right\} dxdydz
\end{aligned}$$
$$(7.342)$$

The temperature variation in [248] is chosen so as to have the form

$$\delta T(x, y, z) = 2\theta \left(\frac{z}{h}\right), \ -\frac{h}{2} \le z \le \frac{h}{2} \qquad (7.343)$$

in which case

$$T\left(x, y, \frac{h}{2}\right) = \theta, \ T\left(x, y, -\frac{h}{2}\right) = -\theta \qquad (7.344)$$

Combining (7.342) with (7.343), and using Euler's equations (7.302a, b, c), one obtains

$$\frac{\partial \epsilon_1}{\partial x} = 0, \ \frac{\partial \epsilon_1}{\partial y} = 0 \qquad (7.345a)$$

and

$$\frac{\partial^2}{\partial x^2}\left\{D_x w_{,xx} + D_1 w_{,yy} + \tilde{\beta}_1 M_T\right\}$$
$$+\frac{\partial}{\partial y^2}\left\{D_y w_{,yy} + D_1 w_{,xx} + \tilde{\beta}_2 M_T\right\}$$
$$+4\frac{\partial^2}{\partial x^2}(D_{66}w_{,xy}) - \frac{\partial}{\partial x}\left\{D_x \frac{12}{h^2}\epsilon_1 w_{,x}\right\} \quad\quad (7.345b)$$
$$-\frac{\partial}{\partial y}\left\{D_y \frac{12}{h^2}\epsilon_1 \kappa w_{,y}\right\} = 0$$

Clearly, (7.345a) implies that

$$\epsilon_1 = \text{const.} \equiv \frac{\alpha^2 h^2}{12} \quad\quad (7.346)$$

while (7.345b), in conjunction with (7.346), yields

$$D_x w_{,xxxx} + 2H w_{,xxyy} + D_y w_{,yy}$$
$$- D_x \alpha^2 (w_{,xx} + \kappa w_{,yy}) + \tilde{\beta}_1 M_{T,xx} + \tilde{\beta}_2 M_{T,yy} = 0 \quad\quad (7.347)$$

For the temperature variation given by (7.343) we have

$$M_T \equiv \int_{-h/2}^{h/2} \delta T(x,y,z)dz = \frac{1}{6}\theta h^2 = \text{const.}$$

which Biswas [248] expands in the double Fourier series

$$M_T = \sum_{m=1,3,\dots}^{\infty} \sum_{n=1,3,\dots}^{\infty} \frac{16 M_T}{mn\pi^2} \sin\frac{m\pi x}{a} \sin\frac{n\pi y}{b} \quad\quad (7.348)$$

While (7.348) is correct as a Fourier representation of $M_T$, convergence of the series on the right-hand side of (7.348) to $M_T$ holds only on the open domain $0 < x < a$, $0 < y < b$; in particular, $M_T$ does not vanish along the edges of the plate for the temperature variation chosen. For this reason, $\tilde{M}_T^1$ and $\tilde{M}_T^2$, as given by (6.102), do not vanish along the edges of the plate; therefore, satisfaction of the conditions expressing the fact that the plate is simply supported, i.e., (6.104) does not follow as a consequence of choosing a trial deflection $w(x,y)$ with the property that both $w_{,xx}$ and $w_{,yy}$ vanish along the edges at $x = 0, a$, for $0 \le y \le b$, and along the edges $y = 0, b$, for $0 \le x \le a$. However, Biswas [248] chooses a trial deflection function of the form

$$w = \sum_{m=1,3,\dots}^{\infty} \sum_{n=1,3,\dots}^{\infty} A_{mn} \sin\frac{m\pi x}{a} \sin\frac{n\pi y}{b} \quad\quad (7.349)$$

so that substitution of (7.348) and (7.349) into (7.347) yields

$$
A_{mn} = \frac{\dfrac{16M_T}{mn\pi^2}\left(\tilde{\beta}_1\dfrac{m^2\pi^2}{a^2}+\tilde{\beta}_2\dfrac{n^2\pi^2}{b^2}\right)}{D_x\dfrac{m^4\pi^4}{a^4}+2H\dfrac{m^2n^2\pi^2}{a^2b^2}+D_y\dfrac{n^4\pi^4}{b^4}+D_x\alpha^2\left(\dfrac{m^2\pi^2}{a^2}+\dfrac{n^2\pi^2}{b^2}\sqrt{\dfrac{D_y}{D_x}}\right)}
$$

(7.350)

An additional criticism of the procedure followed in [248] would focus on the fact that, as $M_T$ is constant on the open domain $0 < x < a$, $0 < y < b$, both $M_{T,xx}$ and $M_{T,yy}$ vanish in (7.347).

The relation (7.350) for the coefficients $A_{mn}$ still contains the unknown constant $\alpha^2$, which arises in (7.346). Assuming that

$$
\begin{cases}
u = 0, & \text{on } x = 0, a, \text{ for } 0 \leq y \leq b \\
v = 0, & \text{on } y = 0, b, \text{ for } 0 \leq x \leq a
\end{cases}
$$

(7.351)

the displacements $u, v$ in [248] are taken to have the form

$$
\begin{cases}
u = \displaystyle\sum_{m=1,3,\ldots}^{\infty}\sum_{n=1,3,\ldots}^{\infty} X_{mn}\sin\dfrac{m\pi x}{a}\cos\dfrac{n\pi y}{b} \\
v = \displaystyle\sum_{m=1,3,\ldots}^{\infty}\sum_{n=1,3,\ldots}^{\infty} X_{mn}\cos\dfrac{m\pi x}{a}\sin\dfrac{n\pi y}{b}
\end{cases}
$$

(7.352)

in which case, substitution of (7.349) and (7.352) into (7.346) and integration over the domain $0 \leq x \leq a$, $0 \leq y \leq b$, yields

$$
\sum_{m=1,3,\ldots}^{\infty}\sum_{n=1,3,\ldots}^{\infty}\frac{1}{8}A_{mn}^2\left(\frac{m^2\pi^2}{a^2}+\frac{n^2\pi^2}{b^2}\right)=\frac{\alpha^2h^2}{12}
$$

(7.353)

Substitution of (7.350) into (7.353) now serves to determine $\alpha^2$ and, hence, $w(x, y)$; for the data

$$
\frac{D_y}{D_x}=0.32,\quad \frac{H}{D_x}=0.2,\quad \frac{a}{h}=20,\quad \frac{\tilde{\beta}_2}{\tilde{\beta}_1}=0.5
$$

Fig. 7.28 depicts the predictions for the theory delineated above of the variation of the central deflection of the plate with changes in the temperature parameter $\theta\left(\dfrac{a}{h}\right)^2\dfrac{\tilde{\beta}_1 h^3}{D_x}$.

The counterpart of [248] for the case of heated cylindrically orthotropic circular plates is the work of Pal [249] who treats both dynamic and static behavior; Pal [249], as in [247], again uses, as an example, the temperature variation (7.337). In [249], however, the *correct* form of the simply supported boundary condition is used and the resultant problem is treated by employing a series expansion of $w$ in the variable $\zeta = \dfrac{r}{a}$. For the case of polar orthotropic response, the total potential energy in the static case assumes the form

$$
\begin{aligned}
\Pi = {} & \frac{E_r h}{2(1 - \nu_r \nu_\theta)} \int_0^{2\pi} \int_0^a [\tilde{e}_1^2 - 2(k - \nu_\theta)\tilde{e}_2] r \, dr \, d\theta \\
& + \frac{h^2}{12} \int_0^{2\pi} \int_0^a \left\{ \left[ w_{,rr}^2 + \left( \frac{1}{r} w_{,r} + \frac{1}{r^2} w_{,\theta\theta} \right) \right]^2 \right. \\
& \quad - 2(k - \nu_\theta) \left[ w_{,rr} \left( \frac{1}{r} w_{,r} + \frac{1}{r^2} w_{,\theta\theta} \right) \right. \\
& \quad \left. - \frac{2 G_{r\theta}(1 - \nu_r \nu_\theta)}{E_r(k - \nu_\theta)} \left( \frac{1}{r} w_{,r\theta} - \frac{1}{r^2} w_{,\theta} \right)^2 \right] \bigg\} r \, dr \, d\theta \\
& - \frac{2}{h} \left\{ \int_0^{2\pi} \int_0^a \left[ (\alpha_r + \nu_\theta \alpha_\theta) e_{rr}^0 + \frac{\nu_\theta}{\nu_r} (\alpha_\theta + \nu_r \alpha_r) \epsilon_{\theta\theta}^0 \right] \bar{T} r \, dr \, d\theta \right\}
\end{aligned}
\tag{7.354}
$$

with

$$
\begin{cases}
\tilde{e}_1 = e_{rr}^0 + k e_{\theta\theta}^0 \\[4pt]
\tilde{e}_2 = e_{rr}^0 e_{\theta\theta}^0 - \dfrac{G_{r\theta}(1 - \nu_r \nu_\theta)}{2 E_r(k - \nu_\theta)} (e_{r\theta}^0)^2 \\[6pt]
k = (\nu_\theta / \nu_r)^{1/2}
\end{cases}
\tag{7.355a}
$$

and

$$
\bar{T} = \int_{-h/2}^{h/2} \delta T(r, z) \, dz, \quad \tilde{T} = \int_{-h/2}^{h/2} \delta T(r, z) z \, dz
\tag{7.355b}
$$

The functions $\tilde{e}_1$ and $\tilde{e}_2$ are, respectively, the first-and second-strain invariants of the middle surface of the polar orthotropic plate. Introducing the dimensionless variables

$$
\tilde{w} = \frac{w}{h}, \quad \zeta = \frac{r}{a}, \quad T^* = \frac{\bar{T}}{h}, \quad \hat{T} = \frac{\tilde{T}}{h^2}
\tag{7.356}
$$

and deleting the second strain invariant $\tilde{e}_2$ from (7.354), the Euler-Langrange equations associated with (7.354) assume the form

$$
\frac{1}{\zeta}\frac{d}{d\zeta}\left\{\zeta\frac{d}{d\zeta}\left[\frac{1}{\zeta}\left(\zeta\frac{d\tilde{w}}{d\zeta}\right)\right]\right\}
$$

$$
= \frac{k^2-1}{\zeta^2}\left(\frac{d^2\tilde{w}}{d\zeta^2} - \frac{1}{\zeta}\frac{d\tilde{w}}{d\zeta}\right)
$$

$$
+12\left(\frac{a}{h}\right)^2\left\{\frac{1}{\zeta}\frac{d}{d\zeta}\left(\tilde{e}_1\zeta\frac{d\tilde{w}}{d\zeta}\right)\right. \tag{7.357}
$$

$$
-\frac{1}{\zeta}\left[(\alpha_r + \nu_\theta\alpha_\theta)\frac{d}{d\zeta}\left(\zeta T^*\frac{d\tilde{w}}{d\zeta}\right) + (\alpha_r + \nu_\theta\alpha_\theta)\frac{d^2}{d\zeta^2}(\hat{T}\zeta)\right.
$$

$$
\left.\left.-\left(\frac{\nu_\theta}{\nu_r}\right)(\alpha_\theta + \nu_r\alpha_r)\frac{d\hat{T}}{d\zeta}\right]\right\}
$$

$$
\frac{d\tilde{e}_1}{d\zeta} + \frac{1-k}{\zeta}\tilde{e}_1 = \left[\frac{(1-\nu_\theta)\alpha_r\nu_r - (1-\nu_r)\alpha_\theta\nu_\theta}{\nu_r\zeta}\right]T^*
$$

$$
+(\alpha_r + \nu_\theta\alpha_\theta)\frac{dT^*}{d\zeta} \tag{7.358}
$$

Equations (7.357), (7.358) are, therefore, the forms assumed by the Berger's equations for the polar orthotropic elastic plate. In terms of the current set of variables, the boundary conditions along the edge of the plate at $\zeta = 1$ assume the form
(i) Clamped Plate:

$$
\tilde{w} = 0, \ \tilde{w}_{,\zeta} = 0, \ \text{at} \ \zeta = 1 \tag{7.359a}
$$

(ii) Simply Supported Plate:

$$
\begin{cases}
\tilde{w} = 0 \\
\tilde{w}_{,\zeta\zeta} + \dfrac{\nu_\theta}{\zeta}\tilde{w}_{,\zeta} + 12(\alpha_r + \nu_\theta\alpha_\theta)\left(\dfrac{a}{h}\right)^2\hat{T} = 0 \ \text{at} \ \zeta = 1
\end{cases} \tag{7.359b}
$$

The general temperature variations $\delta T(r, z)$ in [249] are assumed to be such that the dimensionless forms of (7.355b), namely $T^*$ and $\hat{T}$, have the representations

$$
\begin{cases}
T^* = \displaystyle\sum_{j=0,2,4,\ldots}^{\infty} c_j^*\zeta^j \\
\hat{T} = \displaystyle\sum_{j=0,2,4,\ldots}^{\infty} d_j^*\zeta^j
\end{cases} \tag{7.360}
$$

and solutions generated by of the system of Berger's type equations (7.357), (7.358) are then sought in the form of a series expansion

$$\tilde{w} = 1 + A_2 \zeta^2 + A_4 \zeta^4 + \dots. \tag{7.361}$$

with the solutions generated by (7.361) subject to either (7.359a) or (7.359b); the results are also specialized to the case of an isotropic circular elastic plate, i.e., $\nu_r = \nu_\theta = \nu$, $\alpha_r = \alpha_\theta = \alpha$, $E_r = E_\theta = E$ and for both types of symmetry computations are effected in [249] for the temperature variation given by (7.337). While the associated analytical results are too complex to reproduce here, we do depict in Fig. 7.29, for both the isotropic and orthotropic cases, some of the conclusions relating temperature increases to deflections at the center of the plate for various ratios $\dfrac{\alpha_\theta}{\alpha_r}$, $\dfrac{\nu_\theta}{\nu_r}$, and $\bar{T}_0/\bar{T}_1$. It may be observed that, in the case when the temperature gradient through the thickness of the plate is taken into account, the plate starts to deflect at the beginning of heating without exhibiting the Euler buckling phenomenon at a critical temperature. The case of no temperature gradient through the thickness of the plate, which corresponds to $\hat{T} = 0$, is shown in Fig. 7.29 by the sets of broken curves (depicting the deflection after buckling).

As we have already indicated, critiques and extensions of Berger's approximate method have appeared in several places in the thermal (and load) buckling literature; we content ourselves here with reviewing just two such pieces of work, that of Jones, Mazumdar, and Cheung [242], and that of Banerjee and Datta [243]. The results in [248] cover the case of dynamic (as well as static) behavior of plates at elevated temperatures, but we will confine our attention to static situations only; the behavior of viscoelastic plates within the context of Berger's approximate scheme is also discussed in [242] and this analysis will be briefly reviewed in Chapter 8.

In [242], the authors begin by reviewing the system of Berger's approximate equations, which are written in the form (compare with (7.305), (7.306))

$$K \nabla^4 w + \lambda^2 \nabla^2 w = t - \nabla^2 \frac{M_T}{1 - \nu} \tag{7.362a}$$

$$\frac{N_T}{1 - \nu} - \frac{12K}{h^2} \epsilon_1 = \lambda^2 \tag{7.362b}$$

for isotropic response in rectangular Cartesian coordinates where $M_T = (1 - \nu)M^T = \int_{-h/2}^{h/2} \alpha E \delta T z \, dz$, $N_T = (1 - \nu)N^T = \int_{-h/2}^{h/2} \alpha E \delta T \, dz$, and

$\epsilon_1$ is given by (7.299a); pure thermal buckling corresponds, of course, to the case in which $t \equiv 0$, $M_T = 0$. For pure thermal buckling, therefore, (7.362a) reduces to

$$K \triangledown^4 w + \lambda^2 \triangledown^2 w = 0 \qquad (7.363)$$

Equation (7.363) is formally identical to the equation governing the mechanical buckling of a rectangular plate subject to uniform compression, i.e., $N_x = N_y = N$, $N_{xy} = 0$. In order, therefore, for a nontrivial solution to exist, we must have $\lambda^2 = N_{cr}$, $N_{cr}$ being the critical buckling load. Taking the corresponding solution for $w$ in the form $w = \Lambda w_b(x, y)$, with $\Lambda$ an arbitrary parameter, we note that $w_b$ is the deformation due to biaxial loading and, thus, satisfies

$$K \triangledown^4 w_b + \frac{N_{cr}}{K} \triangledown^2 w_b = 0 \qquad (7.364)$$

By nondimensionalizing $w_b$ so that its maximum value is one, we may interpret $\Lambda$ as being the maximum plate deflection. By substituting for $w$ in (7.362b) the temperature deflection curve may now be obtained from

$$N_{cr} = \frac{N_T}{1 - \nu} - \frac{12K}{h^2}(u_{,x} + v_{,y} + \frac{1}{2}\Lambda^2| \triangledown w_b|^2) \qquad (7.365)$$

In fact, for a plate whose edges are restrained in the plane of the plate (7.365) yields

$$\left(\frac{\Lambda}{h}\right)^2 = \frac{\int_0^{2b} \int_0^{2a} \left\{\frac{N_T}{1-\nu} - N_{cr}\right\} dxdy}{6K \int_0^{2b} \int_0^{2a} | \triangledown w_b|^2 dxdy} \qquad (7.366)$$

for a plate occupying the domain $0 \le x \le 2a$, $0 \le y \le 2b$. From (7.366) it follows that the critical buckling temperature corresponds to

$$\int_0^{2b} \int_0^{2a} \left\{\frac{N_T}{1-\nu} - N_{cr}\right\} dxdy = 0 \qquad (7.367)$$

By denoting the value of $N_T$ which corresponds to the critical temperature $T_{cr}$ as $(N_T)_{cr}$, and employing (7.367), we may rewrite (7.366) in the form

$$\left(\frac{\Lambda}{h}\right)^2 = \frac{\int_0^{2b} \int_0^{2a} (N_T - (N_T)_{cr})dxdy}{(1-\nu)6K \int_0^{2b} \int_0^{2a} | \triangledown w_b|^2 dxdy} \qquad (7.368)$$

To assess the accuracy of the Berger's approximation technique, the authors [242] consider the rectangular plate described above subject to a temperature variation of parabolic type, i.e.,

$$\delta T(x,y,z) = T_0 + T_1 \left[ 1 - \left( \frac{x-a}{a} \right)^2 \right] \left[ 1 - \left( \frac{y-b}{b} \right)^2 \right] \tag{7.369}$$

The plate is simply supported (note that (7.369) yields $M^T = 0$) and $T_0$, $T_1$ in (7.369) are constants; for this case

$$w_b(x,y) = \sin \frac{\pi x}{2a} \sin \frac{\pi y}{2b}$$

Comparative values of $T_{cr}$ are taken from the work of Forray and Newman [225]. In Fig. 7.30, we depict (for aspect ratios 1 and 3, with $T_0/T_1$ as parameter along the curves) the temperature–deflection relation which is obtained from the definition of $N_T$ and (7.368) along with the corresponding results in [225]; there appears to be excellent agreement between the results obtained by the two different approaches. Comparisons of the results generated by Berger's method with standard results available for clamped and simply supported circular plates subject to uniform temperature fields, and for long narrow plate strips subjected to an arbitrary temperature variation, are also made in [242]. In fact, for the latter problem, assuming the $x$-axis to be in the direction of the longest side, and neglecting the component of the deflection in the $y$-direction in comparison with the components in the $x$ and $z$-directions, (so that the displacements $u$, $w$ are regarded as functions of $x$ alone), equations (7.362a) and (7.362b) become,

$$\begin{cases} K \dfrac{d^4 w}{dx^4} + \lambda^2 \dfrac{d^2 w}{dx^2} = 0 \\ \lambda^2 = \dfrac{N_T}{1-\nu} - \dfrac{12K}{h^2} \left( \dfrac{1}{2} \left( \dfrac{dw}{dx} \right)^2 + \dfrac{du}{dx} \right) \end{cases} \tag{7.370}$$

which are precisely the same as the equations for thermal deflections of a plate strip as derived by Williams [250].

A highlight of the work presented in [242] is the conclusion that the solution of Berger's equations, for the response of thermally heated plates, represents the first term in a perturbation solution of the generalized von Karman equations; this conclusion is arrived at in the following manner: We write (in rectangular Cartesian coordinates) the governing equations for the classical analysis of thermally heated isotropic plates in the form

$$N_{x,x} + N_{xy,y} = 0 \tag{7.371a}$$

$$N_{y,y} + N_{xy,y} = 0 \qquad\qquad (7.371\text{b})$$

$$K \nabla^4 w - N_x w_{,xx} - 2N_{xy} w_{,xy}$$
$$-N_y w_{,yy} = t - \nabla^2 \left( \frac{M_T}{1 - \nu} \right) \qquad (7.371\text{c})$$

with (compare with (6.91))

$$\begin{cases} N_x = h\hat{K}(e - \mu e^0_{yy}) - \dfrac{N_T}{1 - \nu} \\[2mm] N_y = k\hat{K}(e - \mu e^0_{xx}) - \dfrac{N_T}{1 - \nu} \\[2mm] N_{xy} = \dfrac{1}{2} \mu h \hat{K} e^0_{xy} \end{cases} \qquad (7.372)$$

where $e = e^0_{xx} + e^0_{yy}$, $\mu = 1 - \nu$, and $e^0_{xx}, e^0_{xy}, e^0_{yy}$ are the usual middle surface strain components, and $\hat{K} = E/(1 - \nu^2)$ is the plane stress plate stiffness. Following Schmidt [251] and Schmidt and DaDeppo [252] a solution of the equations (7.371a,b,c), (7.372) is sought in the form

$$\begin{cases} u = \displaystyle\sum_{n=0}^{\infty} \mu^n u_n(x, y) \\[2mm] v = \displaystyle\sum_{n=0}^{\infty} \mu^n v_n(x, y) \\[2mm] w = \displaystyle\sum_{n=0}^{\infty} \mu^n w_n(x, y) \end{cases} \qquad (7.373)$$

By substituting (7.373) into (7.371a,b,c) and equating, in the usual manner, the coefficients of equal powers of $\mu$, one obtains the following systems of equations for $w_n$, $u_n$, and $v_n$:
(i) For $n = 0$:

$$\begin{cases} K \nabla^4 w_0 + \left( \dfrac{N_T}{1 - \nu} - \hat{K} h e_0 \right) \nabla^2 w_0 = t - \nabla^2 \left( \dfrac{M_T}{1 - \nu} \right) \\[3mm] \dfrac{\partial}{\partial x} \left[ \hat{K} h e_0 - \dfrac{N_T}{1 - \nu} \right] = 0 \\[3mm] \dfrac{\partial}{\partial y} \left[ \hat{K} h e_0 - \dfrac{N_T}{1 - \nu} \right] = 0 \end{cases} \qquad (7.374\text{a})$$

where $e_0$ is $e = \epsilon^0_{xx} + \epsilon^0_{yy}$, based on the functions $u_0(x, y)$ and $v_0(x, y)$.

(ii) For $n \neq 0$:

$$
\begin{cases}
K \nabla^4 w_n - \hat{K} h e_n \nabla^2 w_n \\
\quad + \hat{K} h [\epsilon_{(n-1)xx} w_{n,xx} + \epsilon_{(n-1)yy} w_{n,yy} \\
\quad + 2\epsilon_{(n-1)xy} w_{n,xy}] = 0 \\[6pt]
\dfrac{\partial e_n}{\partial x} - \dfrac{\partial}{\partial x} \epsilon_{(n-1)xx} + \dfrac{\partial}{\partial y} \epsilon_{(n-1)xy} = 0 \\[6pt]
\dfrac{\partial e_n}{\partial y} - \dfrac{\partial}{\partial y} \epsilon_{(n-1)yy} + \dfrac{\partial}{\partial x} \epsilon_{(n-1)xy} = 0
\end{cases}
\tag{7.374b}
$$

where $\epsilon_{(n)xx}$ is $\epsilon_{xx}^0$ based on $u = u_n$ and $w = w_n$, etc., while $e_n$ is $\epsilon_{xx}^0 + \epsilon_{yy}^0$ based on $u = u_n$, $v = v_n$.

The relations (7.374a) for $n = 0$ may be integrated so as to yield

$$
\hat{K} h e_0 - \frac{N_T}{1 - \nu} = \lambda^2 (\lambda^2 \text{ const.})
\tag{7.375}
$$

If the edges of the plate are restrained in the plane of the plate, then the first equation in (7.374a) becomes

$$
K \nabla^4 w_0 + \frac{E\alpha T}{1 - \nu} h \nabla^2 w_0 = t - \nabla^2 \left( \frac{M_T}{1 - \nu} \right)
\tag{7.376}
$$

We note that the first relation in (7.374a) and (7.375) are formally equivalent to (7.305), (7.306), i.e., the solution generated by Berger's approximation method is formally the same as that generated by the zeroth order perturbation solution of the generalized von Karman system for thermal deflection of an isotropic plate.

Finally, we consider the work of Banerjee and Datta [243]; in this paper, a modified energy expression is developed and a new system of governing differential equations, in decoupled form, is obtained for thin plates undergoing large deflections. The equations obtained are tested for circular and square plates with immovable as well as movable edge conditions under uniform static loads, and an extension to cover the case of thermal loading is outlined. For both movable and immovable edges the results obtained appear to be in excellent agreement with other known results for the same test problems considered. The basic motivation for the work presented in [243] appears to be the analysis reported in Nowinski and Ohnabe [241], who establish that the Berger's approximation leads to meaningless results for problems involving movable edge conditions.

For a thin isotropic circular plate of radius $b$ the total potential energy $\Pi$ under isothermal conditions is given by

$$\Pi = \frac{K}{2} \int_0^b \left[ \left( \frac{d^2 w}{dr^2} \right)^2 + \frac{2\nu}{r} \frac{dw}{dr} + \frac{1}{r^2} \left( \frac{dw}{dr} \right)^2 \right.$$
$$\left. + \frac{12}{h^2} \left\{ e_1^2 + 2(\nu - 1)e_2 \right\} \right] r\, dr - \int_0^b twr\, dr \quad (7.377)$$

where $e_1$ and $e_2$ are, respectively, the first and second invariants of the middle surface strains, i.e.,

$$\begin{cases} e_1 = \dfrac{du}{dr} + \dfrac{1}{2}\left( \dfrac{dw}{dr} \right)^2 + \dfrac{u}{r} \\[2mm] e_2 = \dfrac{u}{r}\left[ \dfrac{du}{dr} + \dfrac{1}{2}\left( \dfrac{dw}{dr} \right)^2 \right] \end{cases} \quad (7.378)$$

and where an axisymmetric deformation has been assumed. It is easily seen that (7.377) may be rewritten in the form

$$\Pi = \frac{K}{2} \int_0^b \left[ \left( \frac{d^2 w}{dr^2} \right)^2 + \frac{2\nu}{r} \frac{dw}{dr} \frac{d^2 w}{dr^2} + \left( \frac{1}{r} \frac{dw}{dr} \right)^2 \right.$$
$$\left. \frac{12}{h^2} \left\{ \bar{e}_1^2 + (1 - \nu^2) \frac{u^2}{r^2} \right\} \right] r\, dr - \int_0^b twr\, dr \quad (7.379)$$

with

$$\bar{e}_1 = \frac{du}{dr} + \nu \frac{u}{r} + \frac{1}{2}\left( \frac{dw}{dr} \right)^2 \quad (7.380)$$

It is now noted that if the expression $(1 - \nu^2)\dfrac{u^2}{r^2}$ in (7.379) is replaced by $\dfrac{1}{4}\lambda \left( \dfrac{dw}{dr} \right)^4$, with $\lambda$ a factor depending on $\nu$, decoupling of (7.379) is possible; carrying out this replacement and using the usual ideas of the variational calculus, one obtains the following system of decoupled differential equations from the energy functional $\Pi$:

$$\nabla^4 w - \frac{12}{h^2} \left( \frac{d^2 w}{dr^2} + \frac{\nu}{r} \frac{dw}{dr} \right) Ar^{\nu - 1}$$
$$- \frac{6\lambda}{h^2} \left( \frac{dw}{dr} \right)^2 \left( \nabla^2 w + 2 \frac{d^2 w}{dr^2} \right) = \frac{t}{K} \quad (7.381a)$$

with $A$ a constant of integration to be determined from

$$\frac{du}{dr} + \nu\frac{u}{r} + \frac{1}{2}\left(\frac{dw}{dr}\right)^2 = Ar^{\nu-1} \qquad (7.381b)$$

The decoupled system (7.381a,b), in rectangular Cartesian coordinates, reads as follows:

$$\nabla^4 w - \frac{12}{h^2}A\left(\frac{\partial^2 w}{\partial x^2} + \nu\frac{\partial^2 w}{\partial y^2}\right)$$
$$-\frac{6\lambda}{h^2}\left[\nabla^2 w\left\{\left(\frac{\partial w}{\partial x}\right)^2 + \left(\frac{\partial w}{\partial y}\right)^2\right\}\right.$$
$$\left.+2\left\{\frac{\partial^2 w}{\partial x^2}\left(\frac{\partial w}{\partial x}\right)^2 + \frac{\partial^2 w}{\partial y^2}\left(\frac{\partial w}{\partial y}\right)^2\right\}\right] = \frac{t}{K} \qquad (7.382a)$$

and

$$A = \frac{\partial u}{\partial x} + \nu\frac{\partial v}{\partial y} + \frac{1}{2}\left(\frac{\partial w}{\partial x}\right)^2 + \frac{\nu}{2}\left(\frac{\partial w}{\partial y}\right)^2 \qquad (7.382b)$$

To gauge the accuracy of the systems (7.381a,b) and (7.382a,b) various sample problems are studied in [243] within the context of an isothermal situation. We will confine our remarks to that case involving the circular plate. For the circular plate, the usual boundary conditions are, of course,

$$\begin{cases} w = 0, \quad \dfrac{dw}{dr} = 0, \text{ at } r = b \text{ (clamped)} \\[2ex] w = 0, \quad \dfrac{d^2 w}{dr^2} + \dfrac{\nu}{r}\dfrac{dw}{dr} = 0, \text{ at } r = b \text{ (simply supported)} \end{cases}$$

For movable edge conditions, we have $A = 0$ in (7.381b) while, for an immovable edge, $u = 0$ at $r = b$; the deflection function which conforms to these conditions may be taken in the form

$$w = w_0\left[1 - 2\gamma\frac{r^2}{b^2} + \delta\frac{r^4}{b^4}\right] \qquad (7.383)$$

with $\gamma = \delta = 1$ if the edge at $r = b$ is clamped, while

$$\gamma = \frac{3+\nu}{5+\nu}, \quad \delta = \frac{1+\nu}{5+\nu}$$

if the edge at $r = b$ is simply supported. By substituting (7.383) into (7.381a,b) an error function $E(r)$ is obtained. The Galerkin procedure then requires that

$$\int_0^b E(r)w(r)rdr = 0 \qquad (7.384)$$

and (7.384) may be used to obtain the central deflection $w_0$. Once $w(r)$ has been determined, the constant $A$ may be obtained from (7.381b) by substituting the expression for $w(r)$ into this relation and then integrating over the area of the plate; the terms which involve the in-plane displacements $u$ and $v$ may be eliminated by employing appropriate expressions for these displacements, which are compatible with the boundary conditions. While Berger's method leads to meaningless results when the edge of the plate at $r = b$ is free to move in the plane of the plate, for both the clamped and simple support boundary conditions, numerical data presented in [243] seems to indicate that very accurate results may be obtained through the use of (7.381a,b) for both a movable and an immovable edge.

For a (rectangular plate) subject to a stationary temperature variation of the form

$$\delta T(x, y, z) = T_0(x, y) + g(z)T_1(x, y)$$

(7.382a) may be easily shown to generalize to

$$\nabla^4 w - \frac{12}{h^2}A\left(\frac{\partial^2 w}{\partial x^2} + \frac{\partial^2 w}{\partial y^2}\right) - \frac{6\lambda}{h^2}\left[\nabla^2 w\left\{\left(\frac{\partial w}{\partial x}\right)^2 + \left(\frac{\partial w}{\partial y^2}\right)\right\}\right.$$
$$\left. +2\left\{\frac{\partial^2 w}{\partial x^2}\left(\frac{\partial w}{\partial x}\right)^2 + \frac{\partial^2 w}{\partial y^2}\left(\frac{\partial w}{\partial y}\right)^2\right\}\right]$$
$$+12T_0\alpha\sqrt{\lambda(1 - \nu^2)}\frac{\nabla^2 w}{h^2} = -\frac{E\alpha f(h)}{K(1 - \nu)}\nabla^2 T_1$$

$$(7.385)$$

where $f(h) = \displaystyle\int_{-h/2}^{h/2} zg(z)dz$ and it is assumed that $\displaystyle\int_{-h/2}^{h/2} g(z)dz = 0$. The constant of integration $A$ is determined by the following thermal analogue of (7.382b):

$$\frac{\partial u}{\partial x} + \nu\frac{\partial v}{\partial y} + \frac{1}{2}\left(\frac{\partial w}{\partial x}\right)^2 + \frac{\nu}{2}\left(\frac{\partial w}{\partial y}\right)^2 - (1 + \nu)\alpha T_0 = A \qquad (7.386)$$

It does not appear that the equations (7.385), (7.386) have been applied to any sample problems involving thermal buckling of isotropic rectangular plates; nor does it appear that the analogous equations have been developed for the case of rectilinear orthotropic symmetry.

## 7.4 Thermal Bending, Buckling, and Postbuckling Figures, Graphs, and Tables I

**Table 7.1** **Buckling Loads for the Circular Plate (with permission of the Taylor & Francis Group plc)**

| I1 | I2 | I3 | Temperature Distribution |
|---|---|---|---|
| 0 | 0.63 | 26.6 | |
| 1 | 0.59 | 99.2 | |
| 2 | 0.67 | 20.0 | |
| 3 | 0.69 | 47.1 | |

**Table 7.2**  Critical Buckling Temperature for a Rectangular Plate (Adopted, in modified form, from [216])

| Terms retained | $\dfrac{b^2 EaT_{0_{cr}} t}{\pi^2 D}$ |
|:---:|:---:|
| $a_{11}$ | 6.35 |
| $a_{11}, a_{31}$ | 5.65 |
| $a_{11}, a_{31}, a_{13}$ | 5.40 |
| $a_{11}, a_{31}, a_{13}, a_{33}$ | 5.39 |

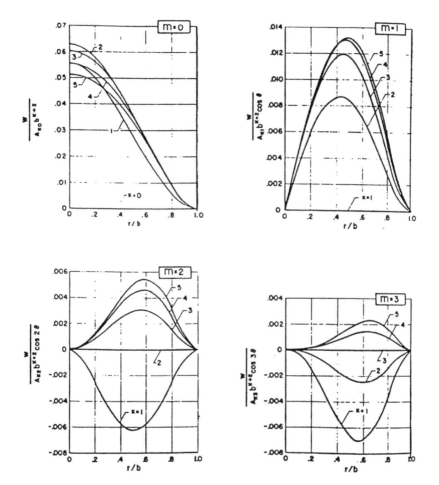

**FIGURE 7.1**
Nondimensional deflections as a function of distance from the
center of the plate. (Adopted, in modified form, from [225].)

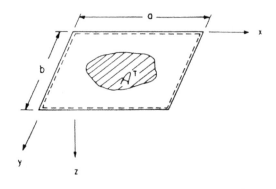

**FIGURE 7.2**
Plate containing a heated region $\mathcal{A}^T$. (Adopted, in modified form, from [217].)

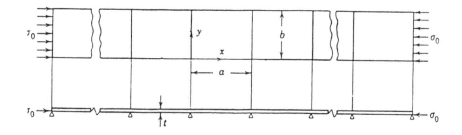

**FIGURE 7.3**
Buckling of a heated plate strip. (From Boley, B.A. and Weiner, J.H., *Theory of Thermal Stresses*, John Wiley & Sons, Inc., N.Y., 1960. With permission of Elsevier Science.)

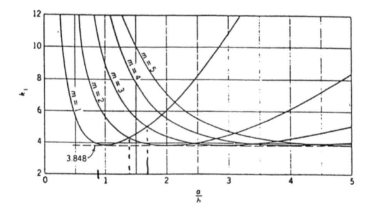

**FIGURE 7.4**
Critical buckling temperature coefficients
$\frac{\alpha E}{2} T_{cr} = \frac{k\pi^2 E}{12(1-\nu^2)} \left(\frac{h}{b}\right)^2$ [221]. (From Boley, B.A. and Weiner, J.H., Theory of Thermal Stresses, John Wiley & Sons, Inc., N.Y., 1960. With per-
mission of Elsevier Science.)

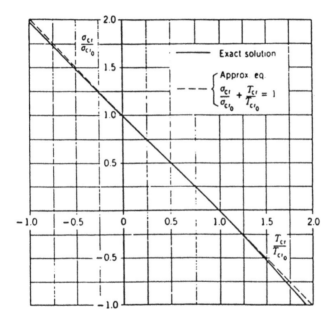

**FIGURE 7.5**
Interaction of temperature level and applied stress in plate buckling.
(From Boley, B.A. and Weiner, J.H., *Theory of Thermal Stresses*,
John Wiley & Sons, Inc., N.Y., 1960. With permission of Elsevier Science.)

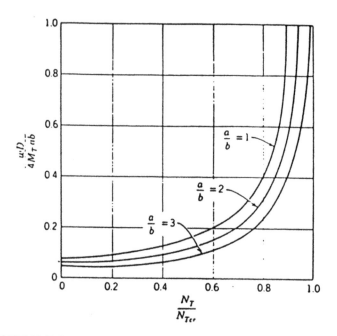

**FIGURE 7.6**
Deflection $w$ at the center of the rectangular plate; poisson ratio $\nu = 1/4$. (From Boley, B.A. and Weiner, J.H., *Theory of Thermal Stresses*, John Wiley & Sons, Inc., N.Y., 1960. With permission of Elsevier Science.)

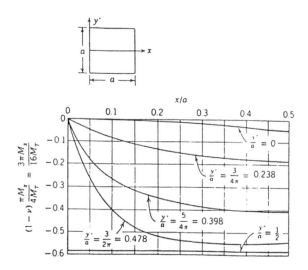

**FIGURE 7.7**
Distribution of the bending moment $M_x$ in a square plate;
$N_T/(N_T)_{cr} = 0.25$, $\nu = 1/4$.    (From Boley, B.A. and Weiner,
J.H., *Theory of Thermal Stresses*, John Wiley & Sons, Inc., N.Y.,
1960. With permission of Elsevier Science.)

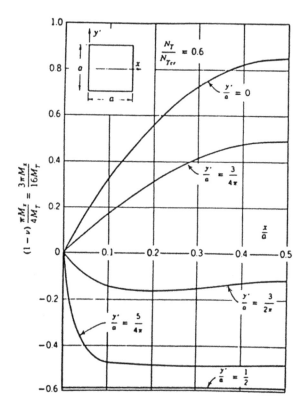

**FIGURE 7.8**
Distribution of the bending moment $M_x$ in a square plate; $NT/(N_T)_{cr} = 0.6$, $\nu = 1/4$. (From Boley, B.A. and Weiner, J.H., *Theory of Thermal Stresses*, John Wiley & Sons, Inc., N.Y., 1960. With permission of Elsevier Science.)

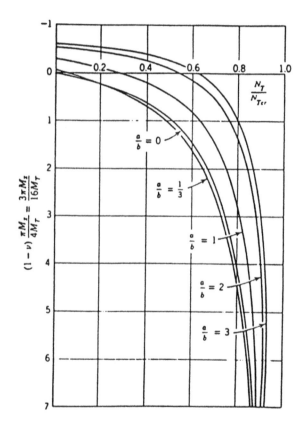

**FIGURE 7.9**
Bending moment $M_x$ at the center of the rectangular plate. (From Boley, B.A. and Weiner, J.H., *Theory of Thermal Stresses*, John Wiley & Sons, Inc., N.Y., 1960. With permission of Elsevier Science.)

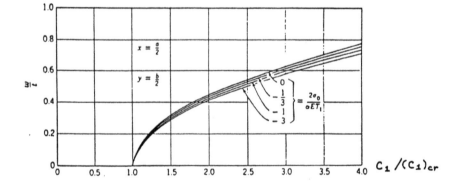

**FIGURE 7.10**
Deflection at the center of the buckled plate. (From Boley, B.A. and Weiner, J.H., *Theory of Thermal Stresses*, John Wiley & Sons, Inc., N.Y., 1960. With permission of Elsevier Science.)

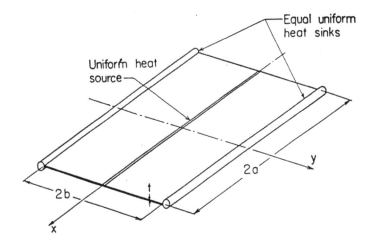

FIGURE 7.11a
The thermal buckling problem of ref. (Adopted, in modified form, from [216].)

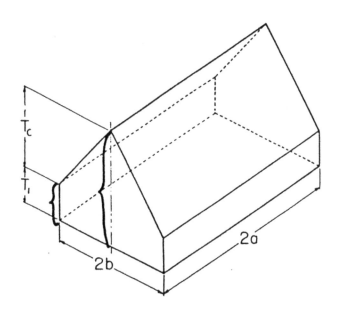

FIGURE 7.11b
The tentlike temperature distribution. (Adopted, in modified form, from [216].)

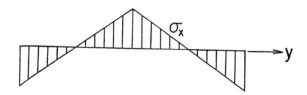

**FIGURE 7.12**
Variation of the normal stress $\sigma_x$. (Adopted, in modified form, from [233].)

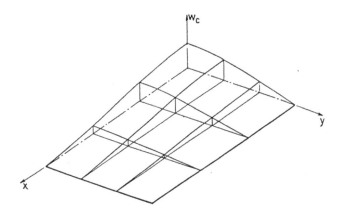

**FIGURE 7.13**
Small-deflection buckle pattern in one quadrant of a plate of
aspect ratio 1.57. (Adopted, in modified form, from [216].)

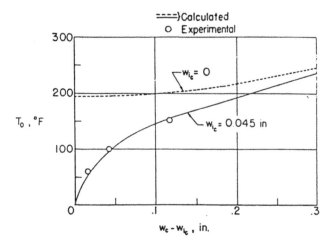

**FIGURE 7.14**
Comparison of calculated and experimental deflections at the plate center. (Adopted, in modified form, from [216].)

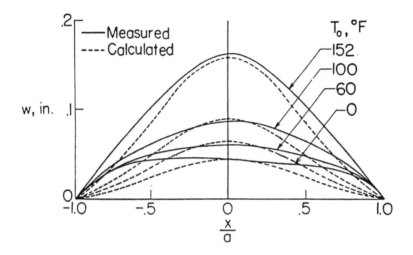

**FIGURE 7.15a**
Longitudinal center line deflection. (Adopted, in modified form, from [216].)

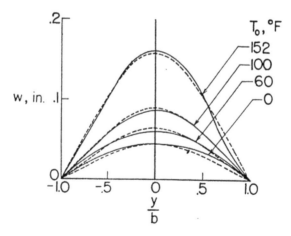

**FIGURE 7.15b**
Growth of deflections with temperature. (Adopted, in modified form, from [216].)

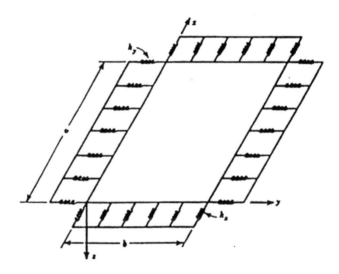

**FIGURE 7.16**
Rectangular plate with edge restraint in the plane of the plate
[234].

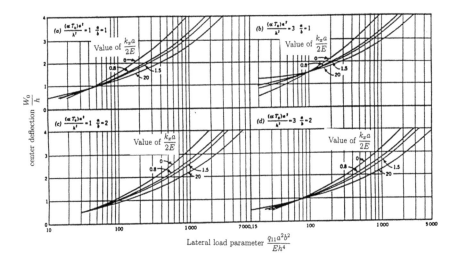

**FIGURE 7.17**
Center deflection versus lateral load parameter [234].

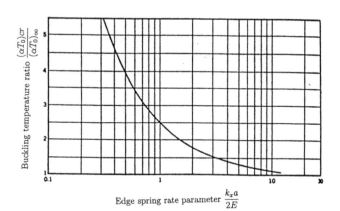

**FIGURE 7.18**
Effect of edge spring rate on buckling temperature ratio [234].

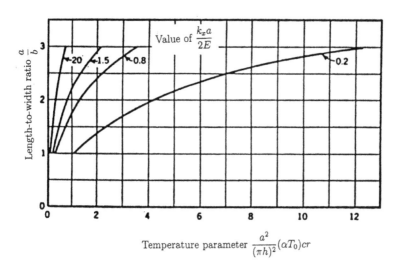

**FIGURE 7.19**
Effect of edge fixity on buckling temperature [234].

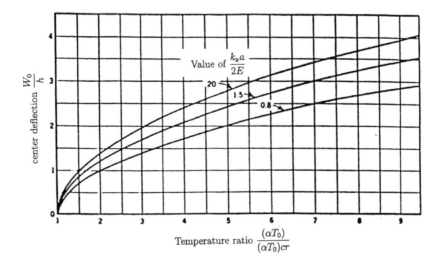

**FIGURE 7.20**
Center deflection vs. temperature Ratio, $\frac{a}{b} = 1$ [234].

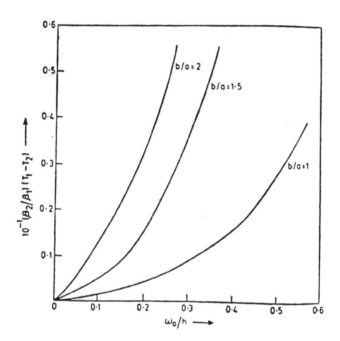

**FIGURE 7.21**
Variation of non-dimensional deflections $w_o/h$ for different values of the temperature parameter [236]. (From Biswas, P., Nonlinear Analysis of Heated Orthotropic Plates, Indian J. Pure Appl. Math, 1380, 12, 1981. By permission of the Indian National Science Academy, New-Delhi)

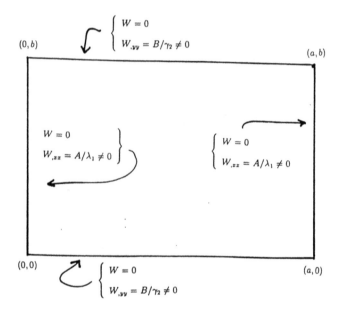

**FIGURE 7.22**
Simply supported rectangular plate with a non-zero thermal moment.

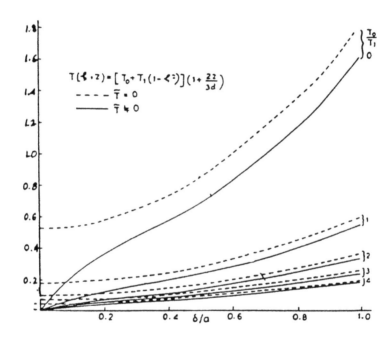

**FIGURE 7.23**
Relation between the temperature rise and the deflection (simply supported edge) [245]. (From Pal, M.C., Acta Mechanica, 8, 82, 1969. With permission.)

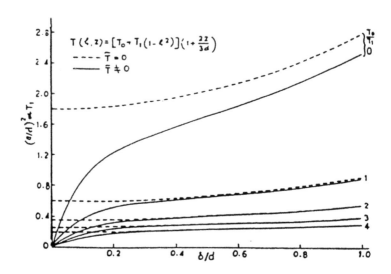

**FIGURE 7.24**
Relation between the temperature rise and the deflection (clamped edge) [245]. (From Pal, M.C., Acta Mechanica, 8, 82, 1969. With permission.)

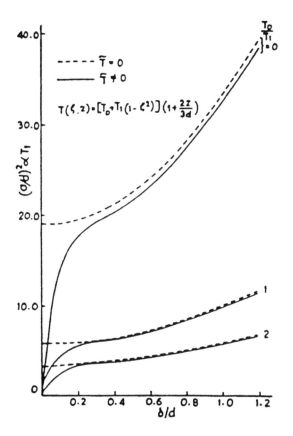

FIGURE 7.25
Relation between the temperature rise and the deflection;
$b/a = 0.4$ with edges simply supported [245]. (From Pal, M.C.,
Acta Mechanica, 8, 82, 1969. With permission.)

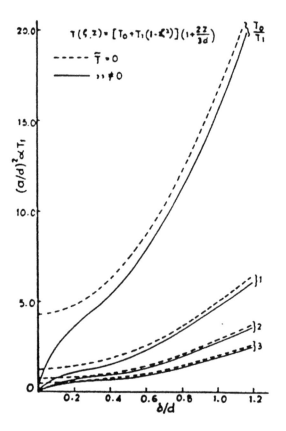

**FIGURE 7.26**
Relation between the temperature rise and the deflection;
$b/a = 0.4$ with both edges clamped [245]. (From Pal, M.C.,
Acta Mechanica, 8, 82, 1969. With permission.)

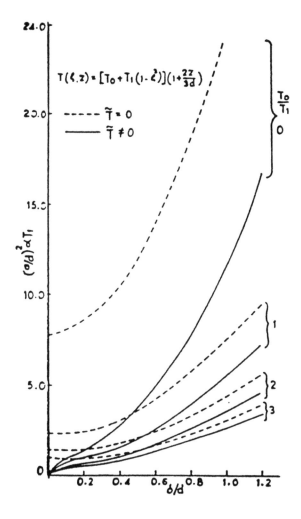

**FIGURE 7.27**
Relation between the temperature rise and the deflection;
$b/a = 0.4$ with the inner edge clamped and the outer edge
simply supported [245]. (From Pal, M.C., Acta Mechanica, 8,
82, 1969. With permission.)

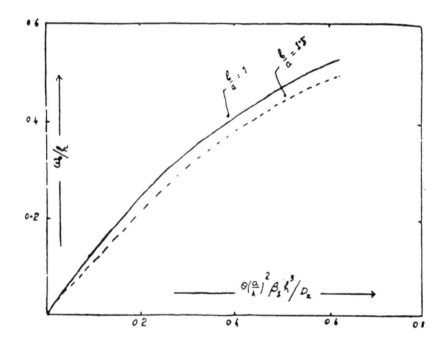

**FIGURE 7.28**
Variation of central deflections with the temperature Parameter [246]. (Biswas, P., Large Deflections of Heated Orthotropic Plates, Indian J. Pure Appl. Math, 1027, 9, 1978. By permission of the Indian National Science Academy, New-Delhi)

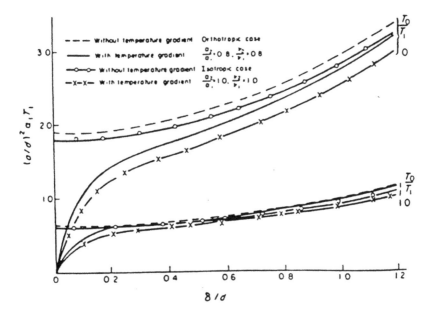

**FIGURE 7.29**
Variation of central deflections with and without a temperature gradient [249]. (Reprinted from the Int. J. Nonlinear Mech., 8, Pal, M.C., Static and Dynamic Non-Linear Behavior of Heated Orthotropic Circular Plates, 489-504, 1973, with permission from Elsevier Science)

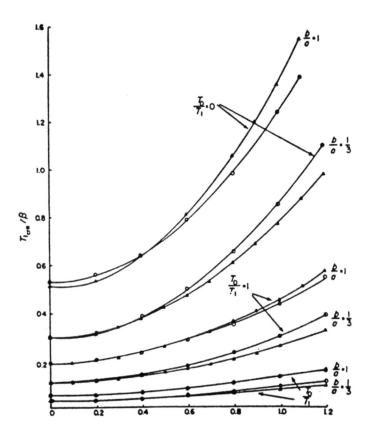

**FIGURE 7.30**
Comparison of results for central deflection vs. temperature rise; $b/a = 1.0$, $1/3$, $\beta = 1/(\alpha(1-\nu^2))(h/b)^2$ [242]. (Reprinted from the Int. J. Solids and Structures, 16, Jones, R., Mazumdar, J., and Y.K. Cheung, Vibration and Buckling of Plates at Elevated Temperatures, 61-70, 1980, with permission from Elsevier Science)

# CHAPTER 8

# Other Aspects of Hygrothermal and Thermal Buckling

In this chapter, we focus our attention on four specific subareas within the domain of hygroexpansive and thermal buckling, namely, buckling in the presence of imperfections, buckling of plates of variable thickness, and the buckling and postbuckling behavior of both heated viscoelastic and heated plastic plates.

## 8.1 Hygroexpansive/Thermal Buckling in the Presence of Imperfections

In the literature, the treatment of hygroexpansive or thermal plate buckling problems in the presence of imperfections, is not very extensive; here we follow the work in Gossard, Seide, and Roberts [216]. We work in rectangular Cartesian coordinates and consider only the case of an isotropic linearly elastic rectangular plate which possesses an (initial) out-of-plane deflection $w_i = w_i(x, y)$. In lieu of (6.22), the condition representing strain compatability for a deflected plate with initial imperfections assumes the form

$$
\begin{aligned}
\frac{\partial^2 \epsilon_{xx}^0}{\partial y^2} + \frac{\partial^2 \epsilon_{yy}^0}{\partial x^2} - 2\frac{\partial^2 \epsilon_{xy}^0}{\partial x \partial y} = & \left(\frac{\partial^2 w}{\partial x \partial y}\right)^2 - \frac{\partial^2 w}{\partial x^2}\frac{\partial^2 w}{\partial y^2} \\
& - \left(\frac{\partial^2 w_i}{\partial x \partial y}\right)^2 + \frac{\partial^2 w_i}{\partial x^2} \cdot \frac{\partial^2 w_i}{\partial y^2}
\end{aligned}
\tag{8.1}
$$

where in the case, e.g., of thermal expansion, the strain tensor components $\epsilon_{xx}, \epsilon_{xy}$, and $\epsilon_{yy}$ are related to the Cartesian stress components $\sigma_{xx}, \sigma_{xy}$, and $\sigma_{yy}$ by that special case of (6.4), which has the following form for isotropic response:

$$
\begin{cases}
\epsilon_{xx} = \dfrac{1}{E}(\sigma_{xx} - \nu\sigma_{yy}) + \alpha T \\[2mm]
\epsilon_{yy} = \dfrac{1}{E}(\sigma_{yy} - \nu\sigma_{xx}) + \alpha T \\[2mm]
\epsilon_{xy} = \dfrac{1+\nu}{E}\sigma_{xy}
\end{cases}
\tag{8.2}
$$

In lieu of (8.2) we have, in terms of the Airy function $\Phi$, as defined by (6.20),

$$
\begin{cases}
\epsilon^0_{xx} = \dfrac{1}{E}(\Phi_{,yy} - \nu\Phi_{,xx}) + \alpha T \\[2mm]
\epsilon^0_{yy} = \dfrac{1}{E}(\Phi_{,xx} - \nu\Phi_{,yy}) + \alpha T \\[2mm]
\epsilon^0_{xy} = -\dfrac{(1+\nu)}{E}\Phi_{,xy}
\end{cases}
\tag{8.3}
$$

Substitution of (8.3) into (8.1) then yields, for the case where an imperfection $w_i = w_i(x, y)$ is present in the plate,

$$
\begin{aligned}
\triangle^2\Phi = Eh\Bigg\{ \left(\frac{\partial^2 w}{\partial x \partial y}\right)^2 - \left(\frac{\partial w}{\partial x}\right)^2\left(\frac{\partial w}{\partial y}\right)^2 \\
-(\frac{\partial^2 w_i}{\partial x \partial y})^2 + \frac{\partial^2 w_i}{\partial x^2}\frac{\partial^2 w_i}{\partial y^2}\Bigg\} - (1-\nu)\Delta N^T
\end{aligned}
\tag{8.4}
$$

where $N^T$ is given by (6.11a). Equation (8.4) represents, for an isotropic linearly elastic, plate in Cartesian coordinates, the first of the generalized von Karman equations modified for the effects of an initial imperfection (deflection). It is easily demonstrated that the second of the pair of generalized von Karman equations in this case has the following form, which extends (6.93a):

$$
K\Delta^2(w - w_i) = \Phi_{,yy}w_{,xx} - 2\Phi_{,xy}w_{,xy} + \Phi_{,xx}w_{,yy} - \Delta M^T + t
\tag{8.5}
$$

with the thermal moment $M^T$ as given by (6.14). In the example to be considered below we will assume that $M^T \equiv 0$, $t \equiv 0$ so that there is neither a thermal moment nor a distributed applied force acting normal to the middle surface of the plate. Furthermore, following the analysis in [216], we will assume that the deflections of the plate are merely a magnification of the initial deflections $w_i$, i.e., if $w_c$ is the net plate center deflection, while $w_{i_c}$ is the plate center deflection, which can be attributed to the imperfection, then

$$\frac{w_i}{w} = \frac{w_{i_c}}{w_c} \qquad (8.6)$$

Substitution of (8.6) into (8.4) and (8.5), with $M^T = t = 0$, yields

$$\Delta^2 \Phi = Eh \left(1 - \frac{w_{i_c}^2}{w_c^2}\right) \left\{\left(\frac{\partial^2 w}{\partial x \partial y}\right)^2 - \frac{\partial^2 w}{\partial x^2}\frac{\partial^2 w}{\partial y^2}\right\} \\ -(1-\nu)\Delta N^T \qquad (8.7a)$$

and

$$K \left(1 - \frac{w_{i_c}}{w_c}\right) \Delta^2 w = \Phi_{,yy} w_{,xx} - 2\Phi_{,xy} w_{,xy} + \Phi_{,xx} w_{,yy} \qquad (8.7b)$$

In (8.7a,b) we now set

$$\begin{cases} E^* = E \left(1 - \dfrac{w_{i_c}^2}{w_c^2}\right) \\[2mm] \alpha^* = \dfrac{\alpha}{1 - (w_{i_c}^2/w_c^2)} \\[2mm] h^* = h \bigg/ \sqrt{1 + \dfrac{w_{i_c}}{w_c}} \end{cases} \qquad (8.8)$$

and

$$\Phi^* = \frac{1}{h}\Phi \qquad (8.9)$$

so that $\Phi^*(x,y)$ is the Airy function associated with the stress tensor components $\sigma_{xx}, \sigma_{xy}$, and $\sigma_{yy}$ instead of the stress resultants $N_x$, $N_y$, and $N_{xy}$. We also assume that $\delta T_{,z} = 0$ so that $N^T = \dfrac{\alpha Eh}{1-\nu}\delta T(x,y)$. Employing this latter assumption and (8.8), (8.9) in (8.7a,b), and noting that $\alpha E = \alpha^* E^*$, it is easily seen that the governing equations may be written in the form

$$\Delta^2 \Phi^* = E^* \left\{\left(\frac{\partial^2 w}{\partial x \partial y}\right)^2 - \frac{\partial^2 w}{\partial x^2}\frac{\partial^2 w}{\partial y^2}\right\} - \alpha^* E^* \nabla^2 (\delta T) \quad (8.10a)$$

$$\left(\frac{K^*}{h^*}\right) \Delta^4 w = \Phi^*_{,yy} w_{,xx} + \Phi^*_{,xx} w_{,yy} - 2\Phi^*_{,xy} w_{,xy} \qquad (8.10b)$$

Equations (8.10a,b) are identical to the large-deflection equations for thermal buckling of a flat plate with Young's modulus $E^*$, thermal expansion coefficient $\alpha^*$, and thickness $h^*$. If the initial deflection $w_i$ satisfies the same homogeneous boundary conditions as would be satisfied by the deflection of an initially flat plate, then the solutions of (8.10a,b) are identical with those for a flat plate of the same aspect ratio but with $E$ replaced by $E^*$, $\alpha$ by $\alpha^*$, and $h$ by $h^*$. Thus, in order to analyze an initially imperfect plate, a flat plate having the same aspect ratio may be analyzed with the quantities $E, \alpha$, and $h$ being left arbitrary; then, everywhere in the resulting expression for the solution, one may replace $E$ by $E(1 - \{w_{i_c}^2/w_c^2\})$, $\alpha$ by $\alpha/(1 - \{w_{i_c}^2/w_c^2\})$, and $h$ by $h\left/\sqrt{1 + \dfrac{w_{i_c}}{w_c}}\right.$.

This relatively simple procedure, introduced in [216], then yields the stresses and deflections for the initially imperfect plate.

For the problem of thermal buckling of a flat isotropic simply supported rectangular plate of aspect ratio 1.57, subjected to a tentlike temperature distribution, the relationship between the temperature differential $T_0$ and the center deflection $w_c$ of the plate was determined to be given by (7.214). Noting that the plate thickness $h$ also appears in the expression for the bending stiffness $K = Eh^2/12(1 - \nu^2)$, and replacing $\alpha$ and $h$, respectively, by $\alpha\left/\left(1 - \dfrac{w_{i_c}^2}{w_2^c}\right)\right.$ and $h\left/\sqrt{1 + (w_{i_c}/w_c)}\right.$ in (7.214) yields the relation

$$\frac{b^2 E \alpha h}{\pi^2 D} T_0 = 5.39 \left(1 - \frac{w_{i_c}}{w_c}\right) + 1.12(1 - \nu^2) \frac{w_c^2 - w_{i_c}^2}{h^2} \qquad (8.11)$$

The Young's modulus $E$ need not be replaced by $E^*$ inasmuch as the relation (7.214) is independent of $E$.

The method described above was employed in [216] to compute curves of center deflection versus average edge compressive stress for the problem of bending of a square isotropic elastic plate with initial imperfections under applied compressive edge stresses; the curves which resulted from this technique turned out to be in excellent agreement with numerical results obtained from an approximate solution of the von Karman large deflection equations for initially imperfect plates, with the agreement existing for all cases in which the initial imperfection was a half-sine wave deflection in both the $x$ and $y$ directions. Experimental and theoretical plate center deflection differences $w_c - w_{i_c}$ have been plotted in Fig. 7.14 as a function of the temperature differential $T_0$ with the theoretical curve based on (8.11); on this graph

$\alpha = 0.127 \times 10^{-4} in./in. - {}^\circ F, \ h = 0.25 in., \ b = 11.25 in., \ w_{i_c} = 0.045 in.,$
and $\nu = 0.33$ so that (8.11) has the explicit form

$$T_0 = 194.1 \left( 1 - \frac{0.045}{w_c} \right) + 573.6(w_c^2 - (0.045)^2) \qquad (8.12)$$

with $T_0$ in degrees Fahrenheit and $w_c$ in inches. We have already referred, in Chapter 7, to Fig. 7.14 in connection with (7.214), which
corresponds to $w_{i_c} = 0$. To the best of the author's knowledge, the
technique described in [216] for approximating the thermal buckling solutions of rectangular isotropic elastic plates has not been extended to
either non-rectangular geometries or to the thermal buckling behavior
of polar orthotropic or rectilinearly orthotropic elastic plates.

---

## 8.2    Buckling of Plates of Variable Thickness

Very little work has appeared in the buckling literature which deals
specifically with the thermal or hygrothermal buckling of thin plates of
variable thickness; the most relevant work in this area would seem to
be the papers of Mansfield [253], [254] and the paper of Biswas [255].
In [253], the author generates the extension and flexual equations which
govern the elastic behavior of a plate of variable thickness, taking into
account temperature variations in the plane of the plate as well as across
the thickness of the plate. Also to be found in [253] are general solutions for a rectangular plate whose thickness varies exponentially along
the length and for a circular or annular plate whose thickness varies
as a power of the plate radius. Mansfield also discusses in [253] the
large deflection plate equations, including the effects of initial irregularities, i.e., imperfections. A companion paper to [253], which appeared
about the same time, is Mansfield's work [254]; this paper discusses the
large deflection analysis of a thin circular lenticular plate whose temperature varies linearly through its thickness. The analysis in [254] covers
both the buckled and the unbuckled regimes. Finally, in [255], Biswas
treats the buckling of a heated annular plate whose thickness varies
as the $m$-th power ($m < 1$) of the radial distance from the center of
the plate; the general stability criterion is derived and expressions are
obtained for the critical buckling compression and the critical buckling
temperature. Other work of note in this area includes that of Mendelson

and Herschberg [256], who develop approximate methods of plane stress analysis for variable thickness plates with temperature variations in the plane, and Thrun [257] who extended the analysis of Olsson [258] for rectangular plates with a linearly varying rigidity, and Olsson [259] for an arbitrarily loaded circular plate with a quadratically varying rigidity, so as to take into account the effect of temperature variations through the thickness of the plate. We now turn to a description of the analysis in [253]–[255].

In [253], Mansfield introduces the invariant differential operator $\Diamond^4$ defined for two sufficiently differentiable functions $f(x, y)$, $g(x, y)$, by

$$\Diamond^4(f, g) \equiv \frac{1}{2} \{ (\nabla^2 f)(\nabla^2 g) + \nabla^2 (f \nabla^2 g + g \nabla^2 f) \}$$
$$- \frac{1}{4} \{ \nabla^4 (fg) + f \nabla^4 g + g \nabla^4 f \}$$

However, for all sufficiently smooth pairs $(f, g)$ it follows that, in fact,

$$\Diamond^4(f, g) = [f, g]$$

where $[\ ,\ ]$ is the usual "bracket" differential operator

$$[f, g] = f_{,xx} g_{,yy} - 2f_{,xy} g_{,xy} + f_{,yy} g_{,xx} \tag{8.13}$$

Thus, to the extent possible in the discussion which follows below, we will use the bracket $[\ ,\ ]$ notation in lieu of Mansfields "die" notation $\Diamond^4$. In particular, in polar coordinates $r, \theta$,

$$\Diamond^4(f, g) \equiv [f, g] = f_{,rr} \left( \frac{1}{r} g_{,r} + \frac{1}{r^2} g_{,\theta\theta} \right)$$
$$- 2 \left( \frac{1}{r} f_{,\theta} \right)_{,r} \left( \frac{1}{r} g_{,\theta} \right)_{,r} \tag{8.14}$$
$$+ g_{,rr} \left( \frac{1}{r} f_{,r} + \frac{1}{r^2} f_{,\theta\theta} \right)$$

and the following special results may be noted:

(i) $[r, g] = \dfrac{1}{r} g_{,rr}$

(ii) $[r^2, g] = 2 \nabla^2 g$

(iii) $[r^n, g] = n(n-1) r^{n-2} \left( \dfrac{1}{r} g_{,r} + \dfrac{1}{r^2} g_{,\theta\theta} \right)$
$\qquad\qquad + n r^{n-2} g_{,rr}$

(iv) $[rh(\theta), g] = \dfrac{1}{r} g_{,rr} (h + h_{,\theta\theta})$

(v) $[\theta, g] = \dfrac{2}{r^2} \left( \dfrac{1}{r} g_{,\theta} \right)_{,r}$

(8.15)

In deriving the small deflection equations for a heated isotropic rectangular plate of variable thickness, Mansfield writes the in-plane equilibrium equations in the form

$$\begin{cases} (h\sigma_{xx})_{,x} + (h\tau_{xy})_{,y} = 0 \\ (h\sigma_{yy})_{,y} + (h\tau_{xy})_{,x} = 0 \end{cases}$$

(8.16)

where $h = h(x, y)$ and introduces the (stress-component based) Airy function $\phi(x, y)$, which satisfies

$$\sigma_{xx} = \dfrac{1}{h}\phi_{,yy}, \sigma_{yy} = \dfrac{1}{h}\phi_{,xx}, \tau_{xy} = -\dfrac{1}{h}\phi_{,xy}$$

(8.17)

For this situation, one could still try to define the usual Airy stress function, i.e., $N_x = \Phi_{,yy}$, etc., where

$$N_x = \int_{-h/2}^{h/2} \sigma_{xx} dz$$

but it is no longer true that $N_x = h\sigma_{xx}$; thus, in particular,

$$\Phi_{,yy} = \int_{-h/2}^{h/2} \dfrac{1}{h}\phi_{,yy} dz \neq \phi_{,yy}$$

(8.18)

Of course, because of the structure of the in-plane equations (8.16), it no longer makes sense to work with the stress resultants $N_x, N_y$, and $N_{xy}$ in the case of a plate with variable thickness $h(x, y)$. The small deflection stress–strain relations are just those given in (6.4) with $\epsilon_{HT} \equiv \epsilon_T = \alpha\delta T$

where $\delta T$ is taken, in [253], to be the temperature variation through the thickness of the plate. The strain compatibility equation (for small deflections) then assumes, in terms of the Airy function $\phi(x,y)$, the form

$$\left\{\frac{1}{h}(\phi_{,xx} - \nu\phi_{,yy})\right\}_{,xx} + \left\{\frac{1}{h}(\phi_{,yy} - \nu\phi_{,xx})\right\}_{,yy}$$
$$+2(1+\nu)\left(\frac{1}{h}\phi_{,xy}\right)_{,xy} + \alpha E \,\nabla^2 \,(\delta T) = 0 \tag{8.19}$$

or, in terms of the bracket differential operator

$$\nabla^2 \left(\frac{1}{h}\nabla^2\phi\right) - (1+\nu)\left[\frac{1}{h},\phi\right] + \alpha E \,\nabla^2 \,(\delta T) = 0 \tag{8.20}$$

With $t(x,y)$ denoting the applied loading, the third equilibrium equation for small deflections assumes the usual form, i.e.,

$$M_{x,xx} + M_{y,yy} + 2M_{xy,xy} = -t \tag{8.21}$$

Using (6.4), with $\epsilon_{HT} \equiv \epsilon_T = \alpha\delta T$, the definitions of the moments $M_x, M_y$, and $M_{xy}$, and assuming that $\delta T(z) = T_0 + T_1 z \equiv T_0 + \dfrac{\partial(\delta T)}{\partial z} \cdot z$, i.e., that the temperature varies linearly through the plate thickness, it is easy to derive the moment-curvature relations

$$\begin{cases} w_{,xx} = -\dfrac{12}{Eh^3}(M_x - \nu M_y) - \kappa_T \\[2mm] w_{,yy} = -\dfrac{12}{Eh^3}(M_y - \nu M_x) - \kappa_T \\[2mm] w_{,xy} = -\dfrac{12(1+\nu)}{Eh^3}M_{xy} \end{cases} \tag{8.22}$$

where

$$\kappa_T = \alpha\frac{\partial(\delta T)}{\partial z} \equiv T_1 \tag{8.23}$$

Although not considered in Mansfield [253], it is an easy exercise to show that, for a general variation $\delta T(z)$ through the plate thickness,

$$\kappa_T = \frac{12\alpha}{h^3}\int_{-h/2}^{h/2} \delta T(z)z\,dz \tag{8.24}$$

with $h = h(x,y)$. By solving (8.22) for $M_x, M_y, M_{xy}$ in terms of $w_{,xx}$, $w_{,yy}$, and $w_{,xy}$, and substituting the results into (8.21), we obtain

$$\nabla^2(K\,\nabla^2 w) - (1-\nu)[K,w] + (1+\nu)\,\nabla^2\,(K\kappa_T) = t \tag{8.25}$$

If, in lieu of (8.21), we include the middle surface forces in the equilibrium equation, i.e., use

$$M_{x,xx} + M_{y,yy} + 2M_{xy,xy}$$
$$= -(t + \phi_{,xx}w_{,yy} - 2\phi_{,xy}w_{,xy} + \phi_{,yy}w_{,xx}) \tag{8.26}$$

then, in place of (8.25) we would clearly obtain

$$\nabla^2(K\,\nabla^2\,w) - (1-\nu)[K,w] + (1+\nu)\,\nabla^2\,(K\kappa_T) = t + [\phi, w] \quad (8.27)$$

To consider large deflections we replace, of course, the small deflection
strain compatibility relation by the relation (6.22) where the middle
surface strains are given by (6.7); in this case, (8.20) is replaced by

$$\nabla^2\left(\frac{1}{h}\nabla^2\,\phi\right) - (1+\nu)\left[\frac{1}{h},\phi\right] + \alpha E\,\nabla^2\,(\delta T) + \frac{1}{2}E[w,w] = 0 \quad (8.28)$$

Finally, for a rectangular plate which is initially stress-free, but which
possesses an initial irregularity (i.e., imperfection) in the sense that the
plate middle surface prior to deflection is given by a relation of the form
$z = w_i(x, y)$, the moment-curvature relations are given as in (8.22) with
$w$ replaced by $w - w_i$, e.g.,

$$(w - w_i)_{,xx} = -\frac{12}{Eh^3}(M_x - \nu M_y) - \kappa_T, \quad \text{etc.,}$$

with $w(x, y)$ the final shape of the deflected middle surface of the plate.
Also, the compatibility equation assumes the form (8.1). The set of
governing equations for large deflections, assuming an initial imperfection $w_i(x, y)$, and also assuming a linear variation of $\delta T$ through the
thickness of the plate, now take on the form

$$\nabla^2\left(\frac{1}{h}\nabla^2\,\phi\right) - (1+\nu)\left[\frac{1}{h},\phi\right] + \alpha E\,\nabla^2\,(\delta T)$$
$$+ \frac{1}{2}E([w,w] - [w_i, w_i]) = 0 \tag{8.29a}$$

and

$$\nabla^2\{K\,\nabla^2\,(w - w_i)\} - (1-\nu)[K, w - w_i]$$
$$+ (1+\nu)\,\nabla^2\,(K\kappa_T) \tag{8.29b}$$
$$= t + [\phi, w]$$

In [253], (8.29a,b) and their analogues in terms of polar coordinates,
are applied to the solution of small deflection problems and plane stress

problems for rectangular plates with an exponential variation in thickness and to circular or annular plates whose thickness varies as a power of the radius.

Consider a rectangular plate such that

$$h = h_0 \exp \left( \gamma \pi y / 3a \right) \tag{8.30}$$

The plate occupies the domain $0 \le x \le a$, $0 \le y \le b$. Assuming, e.g., that the thickness varies from $h_0$ to $h_b$, as $y$ varies from $0$ to $b$,

$$\gamma = \frac{3a}{\pi b} ln \left( \frac{h_b}{h_0} \right) \tag{8.31}$$

The thickness variation (8.30) implies a variation in rigidity which is given by

$$K = K_0 \exp \left( \gamma \pi y / a \right) \tag{8.32}$$

Substitution of (8.32) into the small deflection equilibrium equation (8.25) yields

$$\nabla^4 w + \frac{2\gamma \pi}{a} (w_{,xxy} + w_{,yyy})$$
$$\frac{\gamma^2 \pi^2}{a^2} (w_{,yy} + \nu w_{,xx}) = \frac{\bar{t}}{K_0} \exp \left( -\gamma \pi y / a \right) \tag{8.33}$$

with

$$\bar{t} = t - (1 + \nu) \nabla^2 \left( K \kappa_T \right) \tag{8.34}$$

We suppose that the edges of the plate at $x = 0$ and $x = a$ are simply supported and that the distributed applied loading is such that

$$\bar{t} = e^{\beta \pi y / a} \sum_{m=1}^{\infty} q_m \sin \left( \frac{m \pi x}{a} \right) \tag{8.35}$$

for $0 \le x \le a$, $0 \le y \le b$, with $\beta$ an arbitrary constant. A solution of (8.33), (8.35) is sought in the form

$$w = e^{(\beta - \gamma) \pi y / a} w_1(x) + \sum_{m=1}^{\infty} Y_m(y) \sin \left( \frac{m \pi x}{a} \right)$$

with the first term in (8.35) representing a particular integral of (8.33); it is easily seen that the particular integral must satisfy the ordinary differential equation

$$\frac{d^4 w_1}{dx^4} + \frac{\pi^2}{a^2}\{2\beta(\beta - \gamma) + \nu\gamma^2\}\frac{d^2 w_1}{dx^2}$$
$$+\frac{\pi^4}{a^4}\beta^2(\beta - \gamma)^2 w_1 = \frac{1}{K_0}\sum_{m=1}^{\infty} q_m \sin\left(\frac{m\pi x}{a}\right) \quad (8.36)$$

whose integration leads to

$$w_1(x) = \frac{a^4}{K_0 \pi^4}\sum_{m=1}^{\infty}\frac{q_m}{(m^2 + \beta\gamma - \beta^2)^2 - \nu m^2 \gamma^2}\sin\left(\frac{m\pi x}{a}\right) \quad (8.37)$$

On the other hand, substitution of the sum in (8.35) into (8.33) yields

$$a^4\frac{d^4 Y_m}{dy^4} + 2\gamma\pi a^3\frac{d^3 Y_m}{dy^3} + \pi^2 a^2(\gamma^2 - 2m^2)\frac{d^2 V_m}{dy^2}$$
$$-2\pi^3\gamma m^2 a\frac{dY_m}{dy} + \pi^4 m^2(m^2 - \nu\gamma^2) = 0 \quad (8.38)$$

Taking the $Y_m(y)$ to have the form

$$Y_m = \sum_{j=1}^{4} A_{m,j}\exp\left(r_{m,j}\pi y/a\right) \quad (8.39)$$

and substituting into (8.38) we find that the $r_{m,j}$ must be the roots of the quartic equation

$$r_m^4 + 2\gamma r_m^3 + (\gamma^2 - 2m^2)r_m^2 - 2\gamma m^2 r_m + m^2(m^2 - \nu\gamma^2) = 0 \quad (8.40)$$

while the $A_{m,j}$ are determined by the boundary conditions along the plates edges at $y = 0$ and $y = b$.

An analysis similar to that outlined above may be effected to analyze the plane stress distribution in a rectangular heated plate with an exponential variation in thickness. Indeed, if $h = h_0 \exp(-\lambda\pi y/a)$, a solution of (8.20), with $\nabla^2(\delta T) = 0$, may be obtained in the form

$$\begin{cases} \phi = \sum_{m=1}^{\infty} \tilde{Y}_m \left\{\begin{matrix}\sin\\\cos\end{matrix}\left(\frac{m\pi x}{a}\right)\right\} \\ \tilde{Y}_m = \sum_{j=1}^{4} B_{m,j}e^{s_{m,j}\pi y/a} \end{cases} \quad (8.41)$$

with the $s_{m,j}$ the roots of

$$s_m^4 + 2\lambda s_m^3 + (\lambda^2 - 2m^2)s_m^2)s_m^2 - 2\lambda m^2 s_m + m^2(m^2 + \nu\lambda^2) = 0 \quad (8.42)$$

For $\nabla^2(\delta T) \neq 0$ a particular integral of (8.20) may be obtained, e.g., in the case where $\alpha E \nabla^2 T$ has the form

$$\alpha E \nabla^2 T = -\sum_\beta e^{\beta\pi y/a}\psi_\beta(x)$$

$$= -\sum_\beta \left( e^{\beta\pi y/a} \sum_{m=1}^\infty \psi_{\beta,m} \sin\left(\frac{m\pi x}{a}\right) \right) \quad (8.43)$$

In fact, as a particular solution of (8.20), in this case, Mansfield [253] obtains

$$\phi_p(x,y) = \frac{h_0 a^4 e^{-\lambda\pi y/a}}{\pi 4} \sum_\beta \left( e^{\beta\pi y/a} \right.$$

$$\times \sum_{m=1}^\infty \frac{\psi_{\beta,m}}{(m^2 + \beta\lambda - \beta^2)^2 + \nu m^2\lambda^2} \sin\left(\frac{m\pi x}{a}\right) \Bigg) \quad (8.44)$$

We now consider the small deflection analysis which is presented in [253] for circular or annular plates whose thickness varies as the power of the radial distance from the center of the plate; specifically, we assume that the plate thickness varies like

$$h(r) = h_o\rho^{\beta/3}, \quad \rho = r/r_0 \quad (8.45)$$

where $r_0$ is the plate radius and $\beta$ is an arbitrary constant. The plate rigidity $K$ then varies as $K(r) = K_o\rho^\beta$. Equation (8.25) may now be written in the following form, which is homogeneous in $\rho$:

$$\nabla^2(\rho^\beta \nabla^2 w) - (1-\nu)[\rho^\beta, w] = \bar{t}/K_0 \quad (8.46)$$

where $\bar{t}$ is given by (8.34). The solution of (8.46) corresponding to $\bar{t} = 0$ is given by

$$w = \sum_k R_k \frac{\sin}{\cos}\{k\theta\} \quad (8.47)$$

where

$$R_k = \sum_{i=1}^4 A_{k,i}\rho^{\gamma_{k,i}} \quad (8.48)$$

with the $\gamma_{k,i}$ the roots of the bi-quadratic

$$
\begin{cases}
\Gamma(\gamma_k) = L^2 - L\{2k^2 + \beta(1-\nu)\} \\
\qquad\quad +k^2\{k^2 + \beta(\beta-1)(1-\nu) - (\beta-2)^2\} \\
L = \gamma_k(\gamma_k + \beta - 2)
\end{cases}
\tag{8.49}
$$

For a circular plate $k$ assumes successive integral values, while for a sector plate subtending an angle $\theta_0$ $k = n\pi/\theta_0$. If $\bar{t}$ in (8.46) can be expressed in the form of a double summation with respect to arbitrary parameters $\chi$ and $k$, i.e.,

$$
\bar{t} = \sum_\chi \sum_k \bar{t}_{\chi,k}\rho^x
\begin{Bmatrix} \sin \\ \cos \end{Bmatrix} \{k\theta\}
\tag{8.50}
$$

then it is possible to find a particular integral of (8.46); in fact, a particular integral will be given by

$$
w_1(r,\theta) = \frac{r_0^4}{K_0} \sum_\chi \sum_k \frac{\bar{t}_{\chi,k}\rho^{x+4-\beta}}{\Gamma(\chi+4-\beta)}
\begin{Bmatrix} \sin \\ \cos \end{Bmatrix} \{k\theta\}
\tag{8.51}
$$

As an application (actually, an extension) of the approach delineated above Mansfield [253] considers the symmetrical large deflection bending of a heated circular plate: an unsupported circular plate of lenticular section is subjected to a temperature distribution which varies linearly through the thickness of the plate. For such a plate, we may write

$$
\begin{cases}
h = h_0(1 - \rho^2) \\
K = \dfrac{Eh_0^3}{12(1-\nu^2)}(1 - \rho^2)^3 \\
\rho = r/r_0
\end{cases}
\tag{8.52}
$$

and $\kappa_T = const.$, $w = w(r)$, $\phi = \phi(r)$. The governing differential equations assume the form (8.27), with $t = 0$ and (8.28), with $\nabla^2(\delta T) = 0$. For the small deflection version of these equations we would have $w = -\frac{1}{2}\kappa_T r^2$, with $r$ a constant to be determined, and would substitute this form of $w$, together with (8.52) into (8.27), with $t \equiv 0$, so as to obtain

$$
(1+\nu)(\kappa_T - \kappa)\nabla^2 K = -\kappa\nabla^2\phi
\tag{8.53}
$$

A solution of (8.53) is given by

$$
\phi = -(1+\nu)\left(\frac{\kappa_T - \kappa}{\kappa}\right)K(r)
\tag{8.54}
$$

It may be shown that (8.54) yields a stress distribution which produces a self-equilibrating system of middle surface forces and which satisfies, therefore, the condition that the edges of the plate are free. Substitution of (8.54), (8.52), and $w(r) = -\frac{1}{2}\kappa r^2$ into (8.28), with $\nabla^2(\delta T) = 0$, then shows that $w = -\frac{1}{2}\kappa r^2$, and the expression in (8.54) for $\phi$, yields a solution of the large deflection problem provided

$$\kappa_T - \kappa = \frac{(1-\nu)r_0^4\kappa^3}{2(7+\nu)h_0^3} \qquad (8.55)$$

In [254], the author gives an exact solution for the bending, buckling, and curling of a thin circular plate of lenticular section with a uniform temperature gradient through its thickness; also examined in [254] is the behavior of such a plate with an initial spherical curvature, when the further possibility of snap-through buckling exists. In the work considered, the plate thickness and rigidity once again vary, as in (8.52). With the temperature gradient variation $\frac{\partial}{\partial z}(\delta T)$ through the thickness constant, each unrestrained element of the plate would assume a uniform spherical curvature $\kappa_T = \alpha\frac{\partial(\delta T)}{\partial z} \equiv \frac{\alpha T_1}{h_0}$ where $T_1$ is the temperature difference across the thickness of the plate at the center of the plate; also, for a plate subject to such a temperature distribution the deflection of the plate assumes the form

$$w = ar^2\{1 + b\cos 2(\theta - \theta_0)\} \qquad (8.56)$$

with $a, b$ constants depending on the magnitude of $\kappa_T$ and the initial spherical curvature $\kappa_0$. If the plate deforms with rotational symmetry, then $w$ assumes the form given in [253], i.e., $w = -\frac{1}{2}\kappa r^2$; otherwise (8.56) may be written in the form

$$w = -\frac{1}{2}(\kappa_x x^2 + \kappa_y y^2) \qquad (8.57)$$

where $\kappa_x, \kappa_y$ are independent of $x$ and $y$ and are the curvatures in the $x$ and $y$ directions, respectively. For a suitable choice of the arbitrary angle $\theta_0$ in (8.57), one may achieve $\kappa_x \geq \kappa_y$. Thus, the deflection of the plate is completely determined by the curvatures $\kappa_x$ and $\kappa_y$ with $\kappa_x = \kappa_y = \kappa$ in the case of rotational symmetry. Another feature of the problem under consideration is that while the magnitude of the middle-surface stresses depends on $\kappa_T$ and $\kappa_o$, the distribution of these stresses depends only on $r$; in fact, the middle surface forces may be derived, in the usual way, from an Airy function $\phi$ with

$$
\begin{cases}
h\sigma_{rr} = N_r = \dfrac{1}{r}\phi_{,r} \\[2mm]
h\sigma_{\theta\theta} = N_\theta = \phi_{,rr} \\[2mm]
\sigma_{r\theta} = 0
\end{cases}
\tag{8.58}
$$

where

$$
\phi \sim (1 - \rho^2)^3
\tag{8.59}
$$

As $\phi$ varies with $r$ in the same way that $K$ does, and $\phi$ and $K$ have the same dimensions, we may write that

$$
\phi(r) = \beta K(r)
\tag{8.60}
$$

so that the distribution of middle-surface stresses is determined once the value of $\beta$ is known.

The bending stresses $\sigma_{xx}^b$ and $\sigma_{yy}^b$ are more conveniently referred to Cartesian coordinates; they vary linearly through the thickness of the plate with their values on the surface $z = \dfrac{1}{2}h$ given by

$$
\sigma_{xx}^b = \frac{6}{h^2} M_x \text{ and } \sigma_{yy}^b = \frac{6}{h^2} M_y
\tag{8.61}
$$

The moments $M_x, M_y$ per unit length are given by

$$
\begin{cases}
M_x = -K\{w_{,xx} + \nu w_{,yy} + (1+\nu)\kappa_T\} \\
M_y = -K\{w_{,yy} + \nu w_{,xx} + (1+\nu)\kappa_T\}
\end{cases}
\tag{8.62}
$$

so that by combining (8.61), (8.62) we obtain

$$
\begin{cases}
\sigma_{xx}^b = \dfrac{Eh}{2(1-\nu^2)}[\kappa_x + \nu\kappa_y - (1+\nu)\kappa_T] \\[3mm]
\sigma_{yy}^b = \dfrac{Eh}{2(1-\nu^2)}[\kappa_y + \nu\kappa_x - (1+\nu)\kappa_T]
\end{cases}
\tag{8.63}
$$

At $\rho = 1$ (the edge of the plate) there are no moments or forces applied; these conditions are automatically satisfied because the variation of $K$ (and $\phi$, by virtue of (8.60)) is such that $K = K_{,r} = 0$ at $\rho = 1$.

It is convenient to introduce the following non-dimensional quantities, which are employed in [254]:

$$
\{\hat{\kappa}_T, \hat{\kappa}, \hat{\kappa}_x, \hat{\kappa}_y, \hat{\kappa}_0\} = \frac{r_0^2}{h_0}\{\kappa_T, \kappa, \kappa_x, \kappa_y, \kappa_0\}
$$

$$
\{\hat{\sigma}_{xx}, \hat{\sigma}_{yy}, \hat{\sigma}_{\theta\theta}\} = \frac{r_0^2}{Eh_0^2}\{\sigma_{xx}, \sigma_{yy}, \sigma_{rr}, \sigma_{\theta\theta}\}
\tag{8.64}
$$

With a small temperature gradient through the thickness of the plate, the plate deforms into a shallow "saucer" with constant spherical curvature $\kappa$; this spherical curvature is initially governed by small deflection theory so that $\kappa = \kappa_T$ and the plate is stress free. However, as $\kappa$ increases, middle-surface stresses are developed which stiffen the plate so that the curvature $\kappa$ becomes less than $\kappa_T$, which results in the formation of bending stresses. At a critical value $\kappa_T = \kappa_T^*$ middle-surface stresses dominate the process and for $\kappa_T > \kappa_T^*$ the plate is forced into a shape which is no longer rotationally symmetric. For $\kappa_T \gg \kappa_T^*$ the plate approximates a developable surface with parallel generators, i.e., the plate curls up about a diameter. To illustrate the development of this process, we use the large-deflection equations for a plate of variable thickness, i.e., (8.27), with $t \equiv 0$, and (8.28), with $\nabla^2(\delta T) = 0$, and will, henceforth, refer to these respective specializations of (8.27) and (8.28) as $\overline{(8.27)}$ and $\overline{(8.28)}$; these equations are coupled and nonlinear.

However, by virtue of the ansatz (8.60)

$$\nabla^2\left(\frac{1}{h}\nabla^2\phi\right) = \frac{\beta K_0}{h_0 r_0^4}\left(\frac{\partial^2}{\partial\rho^2} + \frac{1}{\rho}\frac{\partial}{\partial\rho}\right)$$
$$\times \left\{\frac{1}{1-\rho^2}\left(\frac{\partial^2}{\partial\rho^2} + \frac{1}{\rho}\frac{\partial}{\partial\rho}\right)(1-\rho^2)^3\right\} \quad (8.65)$$
$$= 144\frac{\beta K_0}{h_0 r_0^4}$$

In a similar fashion,

$$\left[\frac{1}{h}, \phi\right] = \frac{\beta K_0}{h_0 r_0^4}\frac{1}{\rho}\left\{\frac{\partial^2}{\partial\rho^2}\left(\frac{1}{1-\rho^2}\right)\frac{\partial}{\partial\rho}(1-\rho^2)^3\right.$$
$$\left. +\frac{\partial^2}{\partial\rho^2}(1-\rho^2)^3\frac{\partial}{\partial\rho}\left(\frac{1}{1-\rho^2}\right)\right\} \quad (8.66)$$
$$= -24\frac{\beta K_0}{h_0 r_0^4}$$

Also, as $w = -\dfrac{1}{2}(\kappa_x x^2 + \kappa_y y^2)$, we have

$$\frac{1}{2}[w, w] = \kappa_x \kappa_y (\equiv \text{ const.}) \quad (8.67)$$

With the assumptions that have been made relative to the form of $w$ and $\phi$, $\overline{(8.27)}$ assumes the non-dimensional form

$$\lambda\beta + (1+\nu)\hat{\kappa}_x\hat{\kappa}_y = 0 \quad (8.68)$$

with $\lambda = 2(7 + \nu)/(1 - \nu)$. In [254], the author first treats the case in which $\kappa_x = \kappa_y = \kappa$; for that situation

$$\nabla^2 w = -2\kappa \qquad (8.69)$$

while

$$\begin{cases} [K, w] = -\kappa \nabla^2 K \\ [\Phi, w] = -\beta\kappa \nabla^2 K \end{cases} \qquad (8.70)$$

In this case, $(\overline{8.27})$ is easily seen to reduce to the following non-dimensional equation

$$\hat{\kappa}(1 + \nu - \beta) = \hat{\kappa}_T(1 + \nu) \qquad (8.71)$$

For the situation in which $\kappa_x \neq \kappa_y$, $(\overline{8.27})$ becomes

$$\{(1 + \nu)\kappa_T - (\kappa_x + \kappa_y)\} \nabla^2 K = (\beta + 1 - \nu)[K, w] \qquad (8.72)$$

which is clearly satisfied if

$$\begin{cases} \hat{\kappa}_x + \hat{\kappa}_y = (1 + \nu)\hat{\kappa}_T \\ \beta + 1 - \nu = 0 \end{cases} \qquad (8.73)$$

By taking $\hat{\kappa}_x = \hat{\kappa}_y = \hat{\kappa}$ in (8.69) it may be verified that the solution of (8.68), (8.71) is given by

$$\beta = -(1 + \nu)\hat{\kappa}^2/\lambda \qquad (8.74)$$

with $\hat{\kappa}$ the root of the cubic

$$\hat{\kappa}(1 + \hat{\kappa}^2/\lambda) = \hat{\kappa}_T \qquad (8.75)$$

and $\lambda \simeq 20.9$ when $\nu = 0.3$. It is shown, below, that this solution is valid in the range

$$\begin{cases} |\hat{\kappa}_T| \leq \hat{\kappa}_T^* \\ \hat{\kappa}_T^* = \dfrac{2}{1 + \nu} \left\{ \dfrac{2(7 + \nu)}{1 + \nu} \right\}^{1/2} \simeq 5.15 \end{cases} \qquad (8.76)$$

For $\hat{\kappa}_T = \hat{\kappa}_T^*$ the plate is about to buckle and (8.74)–(8.76) implies that

$$\hat{\kappa} = \left\{ \dfrac{2(7 + \nu)}{1 + \nu} \right\}^{1/2} \simeq 3.35 \qquad (8.77)$$

Equation (8.77) may be written in dimensional form so as to express the deflection at the edge of the plate in terms of $h_0$, i.e.

$$w|_{\rho=1} = -\frac{1}{2}\hat{\kappa}h_0 \simeq -1.67h_0 \qquad (8.78)$$

The prebuckling middle surface stresses are given by

$$\begin{cases} \hat{\sigma}_{rr} = \dfrac{(1-\rho^2)\hat{\kappa}^2}{4(7+\nu)} \\[2mm] \hat{\sigma}_{\theta\theta} = \dfrac{(1-5\rho^2)\hat{\kappa}^2}{4(7+\nu)} \end{cases} \qquad (8.79)$$

while the bending stresses are

$$\hat{\sigma}_{xx}^b = \hat{\sigma}_{yy}^b = -\frac{1-\rho^2)\hat{\kappa}^3}{4(7+\nu)} \qquad (8.80)$$

Thus, the middle surface stresses vary as the square of the plate curvature, while the bending stresses vary as the cube of the plate curvature; the variation of the stress distribution with the magnitude of the temperature gradient is complicated by virtue of the non-linear variation of $\kappa$ with $\kappa_T$. To determine the postbuckling behavior of the plate, we solve (8.68) and (8.73) so as to obtain

$$\begin{cases} \beta = -(1-\nu) \\[2mm] \hat{\kappa}_x = \dfrac{1}{2}(1+\nu)\left\{\hat{\kappa}_T + (\hat{\kappa}_T^2 - \hat{\kappa}_T^{*2})^{1/2}\right\} \\[2mm] \hat{\kappa}_y = \dfrac{1}{2}(1+\nu)\left\{\hat{\kappa}_T + (\hat{\kappa}_T^2 - \hat{\kappa}_T^{*2})^{1/2}\right\} \end{cases} \qquad (8.81)$$

The minimum value of $\hat{\kappa}_T^*$ for which the solution (8.81) exists is $\hat{\kappa}_T^*$, and, at this value, the solutions represented by (8.74), (8.75), and (8.81) coincide. If $|\hat{\kappa}_T| > \hat{\kappa}_T^*$ then the plate strain energy for the asymmetrical mode of deformation is less than that for the rotationally symmetric mode so that the correct solution is given by (8.81) and the solution given by (8.74), (8.75) then represents an unstable state.

For the postbuckling solution given by (8.81), the fact that $\beta$ is constant implies that the middle-surface stresses are independent of $\kappa_T$ and are given by

$$\begin{cases} \hat{\sigma}_{rr} = \dfrac{1-\rho^2}{2(1+\nu)} \\[2mm] \hat{\sigma}_{\theta\theta} = \dfrac{1-5\rho^2}{2(1+\nu)} \end{cases} \qquad (8.82)$$

while the bending stresses assume the form

$$\begin{cases} \hat{\sigma}^b_{xx} = -\dfrac{1}{4}(1-\rho^2)\left\{\hat{\kappa}_T - (\hat{\kappa}_T^2 - \hat{\kappa}_T^{*2})^{1/2}\right\} \\ \hat{\sigma}^b_{yy} = -\dfrac{1}{4}(1-\rho^2)\left\{\hat{\kappa}_T + (\hat{\kappa}_T^2 - \hat{\kappa}_T^{*2})^{1/2}\right\} \end{cases} \qquad (8.83)$$

For $|\hat{\kappa}_T| \gg \hat{\kappa}_T^*$, (8.81) together with (8.83) yield the asymptotic results

$$\begin{cases} \hat{\kappa}_x \to (1+\nu)\hat{\kappa}_T + \mathcal{O}(1/\hat{\kappa}_T) \\ \hat{\kappa}_y \to 0 + \mathcal{O}(1/\hat{\kappa}_T) \end{cases} \qquad (8.84)$$

and

$$\begin{cases} \hat{\sigma}^b_{xx} \to 0 + \mathcal{O}(1/\hat{\kappa}_T) \\ \hat{\sigma}^b_{yy} \to -\dfrac{1}{2}(1-\rho^2)\hat{\kappa}_T + \mathcal{O}(1/\hat{\kappa}_T) \end{cases} \qquad (8.85)$$

The asymptotic results delineated in (8.84) and (8.85) are in agreement with analogous results for inextensional plate theory as given, e.g., in Mansfield [260]. The variation of the principal plate curvatures with the temperature gradient through the thickness of the plate is depicted in Fig. 8.1, while the variation of both the middle surface stresses and the bending stresses with the temperature gradient through the thickness of the plate is shown in Fig. 8.2.

For the case in which the circular lenticular plate has a uniform spherical curvature $\kappa_0$ when $\kappa_T$ is zero, and is initially stress-free, the governing differential equations, assuming an initial deflection $w_i$, are (8.29a), with $\nabla^2(\delta T) = 0$, and (8.29b), with $t \equiv 0$, i.e.

$$\nabla^2\left(\frac{1}{h}\nabla^2\phi\right) - (1+\nu)\left[\frac{1}{h},\phi\right]$$
$$+\frac{1}{2}E([w,w] - [w_i,w_i]) = 0 \qquad (8.86a)$$

and

$$\nabla^2(K\nabla^2(w-w_i)) - (1-\nu)[K,w-w_i]$$
$$+(1+\nu)\nabla^2(K\kappa_T) - [\phi,w] = 0 \qquad (8.86b)$$

In (8.86a,b) we may assume that

$$w_i = -\frac{1}{2}\kappa_0 r_0^2$$

Taking, again, trial solutions $\phi = \beta K$, $w = -\dfrac{1}{2}(\kappa_x x^2 + \kappa_y y^2)$, (8.86a) reduces to the (non-dimensional) form

$$\lambda\beta + (1+\nu)(\hat{\kappa}_x\hat{\kappa}_y - \hat{\kappa}_0^2) = 0 \tag{8.87}$$

With $\kappa_x = \kappa_y$, (8.86b) becomes (for the rotationally symmetric case)

$$\hat{\kappa}(1 + \nu - \beta) = (1+\nu)(\hat{\kappa}_0 + \hat{\kappa}_T) \tag{8.88}$$

while for $\kappa_x \neq \kappa_y$ (8.86b) assumes the form

$$\begin{aligned} \{(1+\nu)(\kappa_0 + \kappa_T) - (\kappa_x + \kappa_y)\}\, \nabla^2 K \\ = (\beta + 1 - \nu)[K, w] \end{aligned} \tag{8.89}$$

which is clearly satisfied provided

$$\begin{cases} \hat{\kappa}_x + \hat{\kappa}_y = (1+\nu)(\hat{\kappa}_0 + \hat{\kappa}_T) \\ \beta = \nu - 1 \end{cases}$$

The solution of (8.87), (8.88) is given by

$$\beta = -(1+\nu)(\hat{\kappa}^2 - \hat{\kappa}_0^2)/\lambda \tag{8.90a}$$

with $\hat{\kappa}$ a root of the cubic

$$(\hat{\kappa} - \hat{\kappa}_0)\{1 + \hat{\kappa}(\hat{\kappa} + \hat{\kappa}_0)\lambda\} = \hat{\kappa}_T \tag{8.90b}$$

depending on the relative magnitudes of $\hat{\kappa}_0$ and $\hat{\kappa}_T$, (8.90b) has either one or three real roots. Following the work in [254], we set

$$J = 27\lambda^2(\hat{\kappa}_0 + \hat{\kappa}_T)^2 - 4(\hat{\kappa}_0^2 - \lambda)^3 \tag{8.91}$$

and note that for $J < 0$, (8.90b) possesses three real roots; the (algebraically) largest and smallest roots correspond to stable configurations, while the middle root is associated with an unstable configuration. When $\hat{\kappa}_T = 0$ there are two stable states for $|\hat{\kappa}_0| > 2\lambda^{1/2}$, which are given by

$$\begin{cases} \hat{\kappa} = \hat{\kappa}_0 \\ \hat{\kappa} = -\dfrac{1}{2}\left\{\hat{\kappa}_0 + (\hat{\kappa}_0^2 - 4\lambda)^{1/2}\right\} \end{cases} \tag{8.92}$$

It is noted, in [254], that when $\hat{\kappa}$ and $\hat{\kappa}_0 + \hat{\kappa}_T$ are of opposite sign, and $J = 0$, snap-through buckling of the plate will occur and the curvature

$\hat{\kappa}$ will jump from the unstable value of $\hat{\kappa} = -\frac{1}{2}\{4\lambda(\hat{\kappa}_0 + \hat{\kappa}_T)\}^{1/2}$ to the stable value of $\hat{\kappa} = \{4\lambda(\hat{\kappa}_0 + \hat{\kappa}_T)\}^{1/2}$. For $J > 0$, (8.90b) has only one real root, but the associated plate configuration is stable only if $K < 0$ where

$$K = \frac{1}{4}(1+\nu)^2(\hat{\kappa}_0 + \hat{\kappa}_T)^2 - \hat{\kappa}_0^2 - \frac{2(7+\nu)}{1+\nu} \qquad (8.93)$$

When $K > 0$, the solution of (8.87), (8.89) is given by

$$\begin{cases} \beta = -(1-\nu) \\ \hat{\kappa}_x = \frac{1}{2}(1+\nu)(\hat{\kappa}_0 + \hat{\kappa}_T) + K^{1/2} \\ \hat{\kappa}_y = \frac{1}{2}(1+\nu)(\hat{\kappa}_0 + \hat{\kappa}_T) - K^{1/2} \end{cases} \qquad (8.94)$$

Therefore, after buckling into a mode which is asymmetrical, the middle-surface forces remain constant, and independent of $\hat{\kappa}_0$; also, for large values of $|\hat{\kappa}_T|$ the plate configuration may be approximated by a developable surface for which

$$\begin{cases} \hat{\kappa}_x \to (1+\nu)(\hat{\kappa}_0 + \hat{\kappa}_T) + \mathcal{O}(1/\hat{\kappa}_T) \\ \hat{\kappa}_y \to 0 + \mathcal{O}(1/\hat{\kappa}_T) \end{cases} \qquad (8.95)$$

The variation of the principal curvatures with the temperature variation through the thickness of the plate is shown in Figs. 8.3–8.6 for, respectively, $\hat{\kappa}_0 = \lambda^{1/2}$ and $\hat{\kappa}_0 = 6, 10, 12$; the graphs in Figs. 8.3–8.6 possess point symmetry with respect to the point $\hat{\kappa} = 0$, $\hat{\kappa}_T = -\hat{\kappa}_0$. For $|\hat{\kappa}_0| \le \lambda^{1/2}$, the possibility of snap-through buckling does not exist; this limiting case is depicted in Fig. 8.3. When $\hat{\kappa}_T = -\hat{\kappa}_0$ the plate is flat but has zero stiffness. If $|\hat{\kappa}_0| > \lambda^{1/2}$, snap-through buckling will occur at appropriate values of $\hat{\kappa}_T$; however, for $|\hat{\kappa}_0| < 2\lambda^{1/2}$ there exists only one stable state when $\hat{\kappa}_T = 0$. Following snap-through buckling, the plate assumes a symmetrical mode in Fig. 8.5 but an asymmetrical mode in Fig. 8.6.

In this section, the last work we have made reference to on the buckling of heated plates with variable thickness is that of Biswas [255]; this work, which is concerned with the buckling of a heated thin annular (circular) plate, appears to possess a serious flaw will be noted below. Biswas [255] considers an annular (circular) plate whose thickness varies as the $m^{th}$ power of the radial distance from the center of the plate. The edges of the plate are restrained from in-plane movement and the plate is subjected to uniform compression along its edges as well as to a stationary temperature distribution of the form

$$T(r, z) = \tau_0(r) + z\tau(r) \tag{8.96}$$

so that (pure) buckling will occur for $\tau \equiv 0$, $\tau_0 \neq 0$. As a consequence of (8.96), $M_T = 0$; therefore, with $\phi(r) = -w'(r)$, we have

$$
\begin{cases}
M_r = K\left(\phi'(r) + \dfrac{\nu}{r}\phi(r)\right) \\[2mm]
M_\theta = K\left(\nu\phi'(r) + \dfrac{1}{r}\phi(r)\right) \\[2mm]
K = K(r) \equiv \dfrac{Eh^3(r)}{12(1 - \nu^2)}
\end{cases}
\tag{8.97}
$$

while

$$N_r = \frac{Eh}{1 - \nu^2}\left(\frac{du}{dr} - \frac{\nu}{r}u\right) - \frac{N_T}{1 - \nu} \equiv \frac{-N_T}{1 - \nu} \tag{8.98}$$

as the radial displacement $u = u(r) \equiv 0$ in the plane of the plate, due to the restraint imposed along the edge of the plate. In (8.98) we have, of course,

$$
\begin{aligned}
N_T &= \alpha E \int_{-h/2}^{h/2} \tau_0(r)dz \\
&\equiv \alpha E \tau_0(r) h(r)
\end{aligned}
\tag{8.99}
$$

The thickness variation for the plate is taken to be of the form

$$h(r) = h_0 r^m, \quad m < 1 \tag{8.100}$$

in which case,

$$K(r) = \frac{Eh_0^3}{12(1 - \nu^2)} r^{3m} \equiv K_0 r^{3m}, \quad m < 1 \tag{8.101}$$

Setting $\beta = \dfrac{b}{a}$, and $\rho = \dfrac{r}{a}$, (so that the outer and inner boundaries of the annular plate correspond, respectively, to $\rho = 1$ and $\rho = \beta$), and employing (8.94) in the equilibrium equation

$$M_r + rM_r' - M_\theta + hN_r\phi = 0 \tag{8.102}$$

Biswas [255] arrives at the following ordinary differential equation for $\phi = \phi(r)$:

$$
\rho^2 \frac{d^2\phi}{d\rho^2} + (3m+1)\rho\frac{d\phi}{d\rho}
$$
$$
+ \left(\frac{h_0 N_r}{K_0}a^{-2m+2}\rho^{-2m+2} + 3m\nu - 1\right)\phi = 0 \tag{8.103}
$$

whose solution (it is claimed in [255]) has the form

$$
\phi(\rho) = \rho^{-3m/2}\left\{AJ_\mu(\lambda\rho^{-m+1}) + BY_\mu(\lambda\rho^{-m+1})\right\} \tag{8.104}
$$

with

$$
\begin{cases}
\lambda = \dfrac{1}{(1-m)}\left(\dfrac{h_0 N_r}{K_0}a^{-2m+2}\right)^{1/2} \\
(1-m)^2\mu^2 = \dfrac{9}{4}m^2 - 3\nu + 1
\end{cases} \tag{8.105}
$$

The specification of the roots $\lambda$ is then sought for the case in which the outer-plate and inner-plate boundary conditions are given by

$$
\phi = 0, \text{ for } \rho = 1, \beta; \; w = 0, \text{ for } \rho = 1 \tag{8.106}
$$

Using the boundary date (8.106) in (8.104), we obtain as the condition from which the critical compression (and thus the critical temperature) can be determined for different values of $m, \nu$, and $\beta$

$$
J_\mu(\lambda)Y_\mu(\lambda\beta^{1-m}) - J_\mu(\lambda\beta^{1-m})Y_\mu(\lambda) = 0 \tag{8.107}
$$

Unfortunately, $\lambda$ as given by (8.105) is a function of $r$ because $N_r \equiv -\dfrac{1}{1-\nu}N_T = \dfrac{-\alpha E}{1-\nu}\tau_0(r)h(r)$. Even if $\tau_0 \equiv$ const., we would still have a situation in which $N_r = N_r(r)$ because of the thickness variation of the plate with radial distance. As $\lambda = \lambda(r)$, (8.103), while an equation of Bessel type, is a variable coefficient (not a constant coefficient) ordinary differential equation for which the standard solution (8.104) does not hold.

---

## 8.3   Viscoelastic and Plastic Buckling

While various papers have appeared in the literature which have investigated the nature of the stress distributions in both viscoelastic and plastic plates at elevated temperatures, few studies exist that are concerned with the deflection or buckling behavior of viscoelastic or plastic plates which are thermally loaded. For viscoelastic plates, or plates subject to temperature-dependent creep buckling, notable contributions have been made by Ross and Berke [261], Jones and Mazumdar [242], and Das [262], which will be summarized below. A significant piece of work in the literature, which is related to the thermal buckling of plates in the plastic regime, is the paper of Williams [263]; in this paper, the field equations associated with Neale's variational theorem [264], [265] are developed and applied to the problem of thermal buckling and post-buckling of constrained rectangular plates; the resulting equations are a generalization of von Karman's equations in rate form, and, among the results, which will be described in this section, is the fact that, in the immediate vicinity of a critical point, the theory predicts a substantial reduction of the buckling temperature due to plasticity effects. We begin our analysis of these problems by first looking at viscoelastic and temperature-dependent creep buckling.

In [261], the authors employ the so-called Norton-Bailey power law for material creep to predict the time-dependent lateral deflection of flat rectangular plates with a through-thickness steady-state temperature distribution. The usual Norton-Bailey or power creep law (e.g., Norton [266]) has the form

$$\dot{\epsilon} = k\sigma^n \qquad (8.108)$$

where $\dot{\epsilon}$ is the creep strain rate, $k$ is the creep constant, and $\sigma$ is the stress. In considering temperature variations, one modifies (8.108) so that it assumes the form given by Maxwell's law [267], i.e.,

$$\dot{\epsilon} = ke^{-H/RT}\sigma^n \qquad (8.109)$$

with $H$ the creep activation energy, $R$ the universal gas constant, and $T$ the absolute temperature. To study the creep buckling of plates, (8.109) is replaced by the two-dimensional equations

$$\dot{\epsilon}_i = \frac{1}{2}k_n e^{-H/RT} J_2^{n-1}(2\sigma_i - \sigma_j) \qquad (8.110)$$

with the subscripts $i, j$ denoting $xx, yy$ in cyclic substitution, $\sigma_{xx}$, $\sigma_{yy}$ being the usual in-plane stresses in the $x$ and $y$ directions, respectively, while $J_2$ is the stress invariant

$$J_2 \equiv \sigma_{xx}^2 + \sigma_{yy}^2 - \sigma_{xx}\sigma_{yy} \qquad (8.111)$$

In [261] a comparison of creep buckling predictions is made for the two cases where the creep exponent $n$ in (8.110) assumes the values $n = 3$ and $n = 5$. For a simply supported flat rectangular plate, such as is depicted in Fig. 8.7, it is assumed that the deflected shape may be adequately represented by a half-wave cosine function in each of two perpendicular directions at all times, in which case, the plate behavior is determined by the plate deflection at the center of the plate. If the plate possesses an initial imperfection given by its value at the plate center, say, $w_i^0$, and a temperature differential is applied between the sides of the plate, in the thickness direction, the center deflection $w^0$ due to that temperature differential may, as in [268], be approximated by

$$w_0^T = \frac{\alpha(T_0 - T_i)(1 + \nu)ab}{\pi^3 h} \tag{8.112}$$

where $T_i, T_0$ are, respectively, the inner-and outer-plate temperatures. The central deflection after the application of a through-thickness temperature differential, but prior to the application of in-plane stresses, is taken to be given by

$$w_0^c = w_i^0 + w_0^T \tag{8.113}$$

In [261] an axial force at time $t = 0$ is applied to the plate. The immediate deflection $w_0^p$, which is attained upon adding the axial load, has been determined, in [269], to have the value

$$w_0^p = w_0^c \left( \frac{\sigma_E}{\sigma_E - \sigma} \right) \tag{8.114}$$

with $\sigma$ the applied axial stress and $\sigma_E$ the Euler buckling stress; for a plate (which is isotropic, as we have assumed here) subject to an in-plane compressive load

$$\sigma_E = \frac{\pi^2 E}{12(1 - \nu^2)} \left[ \frac{b}{a} + \frac{a}{b} \right]^2 \frac{h^2}{a^2} \tag{8.115}$$

If the applied stress $\sigma$ is larger than the material yield stress, that stress is used in (8.114) in lieu of $\sigma_E$. The stresses and lateral displacements in [261] are assumed to be represented as products of $\cos \frac{\pi x}{a} \cos \frac{\pi y}{b}$, e.g.,

$$w(t) = w_0(t) \cos \frac{\pi x}{a} \cos \frac{\pi y}{b} \tag{8.116}$$

where $w_0(t)$, the deflection of the center of the plate at time $t$, is taken to be the sum of a small deflection $w_s(t)$ and a large deflection $w_l(t)$;

it has been shown, in [270], that for the creep exponent $n = 3$ these components of the plate deflection are given by the following expressions in which $\Gamma = b^2/a^2$ and $k_3$ is the material creep constant for $n = 3$:

$$\ln\left[\frac{w_s(t)}{w_0^p}\right] = \frac{36k_3\sigma^3 t e^{-2H/3RT_0}\left[e^{-H/3RT_0} + e^{-H/3RT_i}\right]}{h^2(2\Gamma^2 + 2\Gamma + 1)\left(1 + \exp\left\{\dfrac{-H}{3R}\left(\dfrac{1}{T_0} - \dfrac{1}{T_i}\right)\right\}\right)}\left(\frac{b}{\pi}\right)^2 \qquad (8.117)$$

and

$$w_l(t) =$$

$$\frac{4\pi}{3}\frac{h^2 w_0^p(\Gamma^2 + \Gamma + 1)}{\sqrt{\dfrac{16h^2}{9}\pi^2(\Gamma^2 + \Gamma + 1) - 81b^2 k_3 w_0^{p2}\sigma^3 t\left(e^{-H/RT_0} + e^{-H/RT_i}\right)}} \qquad (8.118)$$

For creep exponent $n = 5$, on the other hand,

$$\ln\left[\frac{w_s(t)}{w_0^p}\right] = \frac{2160k_5\sigma^5 t}{h^2} \cdot \frac{1}{(10\Gamma^2 + 10\Gamma + 4)}$$

$$\times \left\{\frac{e^{-4H/5RT_0}\left(e^{-H/5RT_0} + e^{-H/5RT_i}\right)}{\left[1 + e^{-H/5R\left(\frac{1}{T_0} - \frac{1}{T_i}\right)}\right]^4}\right\}\left(\frac{b}{\pi}\right)^2 \qquad (8.119)$$

and

$$\frac{1}{w_0^{p4}} - \frac{1}{w(t)^4} = \frac{18,225}{32}\cdot\frac{k_5\sigma^5 t}{h^6}\cdot\frac{1}{(\Gamma^2 + \Gamma + 1)^3}$$

$$\times \left\{e^{-H/RT_0} + e^{-H/RT_i}\right\}\left(\frac{b}{\pi}\right)^2 \qquad (8.120)$$

The predictions embodied in (8.117)–(8.120) were studied, numerically, in [261] for specific values of the various parameters involved, e.g., $k_3 = 3.05 \times 10^8/MPa \cdot \sec$ and $k_5 = 5.71 \times 10^5/MPa \cdot \sec$. In Figs. 8.8–8.10 we indicate the type of predictions which follow from the work in [261], for the lateral deflections as functions of time, and for applied axial stresses of 6.9, 13.8, and 34.5 MPa, respectively; these figures, for given values of $\Delta T$ and $\sigma$, compare the predictions with respect to lateral deflection which are made by the creep power law (8.110) when the creep exponent $n$ is taken as either 3 or 5. All the curves in Figs. 8.8–8.10 display the familiar phenomena of increasing strain rate with time.

In Fig. 8.11, we indicate comparative results for the predictions of the time until the creep deflection reaches a specified value for the creep exponents $n = 3$ and $n = 5$; the chosen value of the fixed creep deflection in Fig. 8.11 is $5.08 \times 10^{-3} m$ (which is approximately equal to the plate thickness). The range of applicability of the results in Figs. 8.8–8.11 is subject to question, as they correspond to material properties that were adopted for a constant mean temperature of the plate; in reality, there is ample experimental evidence to suggest that these properties may be strongly dependent on temperature.

In [242], Jones, Mazumdar, and Cheung have considered the small amplitude response of a thermo-rheologically simple viscoelastic plate; in this case, the model is such that (8.27) is replaced by

$$K(p) \bigtriangledown^4 w + \frac{N_T}{1 - \nu(p)} \bigtriangledown^2 w = t - p^2(\rho h w) - \frac{\bigtriangledown^2 M_T}{1 - \nu(p)} \qquad (8.121)$$

with $p \equiv \dfrac{\partial}{\partial t}$ and $K(p), \nu(p)$ the viscoelastic time operators corresponding to the flexual rigidity $K$ and Poisson's ratio $\nu$ for the elastic case. Equation (8.121) is identical to the governing equation for the analysis of a viscoelastic plate with in-plane loads

$$N_x = N_y = N_T / (1 - \nu(p))$$

and transverse load

$$t - \bigtriangledown^2 M_T / (1 - \nu(p)),$$

the derivation following the analysis in Deleeuw and Mase [271]. If $t = \bigtriangledown^2 M_T = 0$, and inertia effects are negligible, then a deflection $w(x, y, t)$ may be sought which is separable in time and space, i.e.,

$$w(x, y, t) = W(x, y)T(t) \qquad (8.122)$$

where $W(x, y)$ satisfies

$$\bigtriangledown^4 W - C \bigtriangledown^2 W = 0 \qquad (8.123)$$

and where $T(x)$, which represents the magnitude of the deflection, satisfies

$$K(\rho)T(t) - T(t)\frac{N_T}{C(1 - \nu(p))} = 0 \qquad (8.124)$$

with $c$ representing a separation constant. Thus, the mode shapes are time independent and only the amplitudes vary with time. If $N_T =$

const. then it may be shown that there exist upper and lower critical temperatures where the lower critical temperature corresponds to a zero deflection rate, while the upper critical temperature corresponds to an infinite deflection. The lower critical temperature is determined by setting $p = 0$ in (8.124), which yields

$$C = \frac{N_T}{(1 - \nu(0))K(0)} \tag{8.125}$$

Using (8.125), (8.123) becomes

$$\nabla^4 W - \frac{N_T}{K(0)(1 - \nu(0))} \nabla^2 W = 0 \tag{8.126}$$

In a similar manner, the upper critical temperature is determined by setting $p = \infty$; this subsequently yields, in lieu of (8.126),

$$\nabla^4 W - \frac{N_T}{K(\infty)(1 - \nu(\infty))} \nabla^2 W = 0 \tag{8.127}$$

with the notation employed being the same as that in [271]. When the boundary conditions for the plate contain the Poisson's ratio $\nu$ (as in the case for a plate with either free or simply supported edges) then in solving (8.126) or (8.127) one simply replaces $\nu$ by $\nu(0)$ or $\nu(\infty)$, respectively.

For both Kelvin-and Maxwell-type viscoelastic materials, the upper and lower critical buckling loads are expressible in terms of the elastic critical loads, with the correlations being indicated in Table 8.1. It may be shown that the physical interpretation of the critical temperatures is the same as for the critical buckling loads discussed in [271]. If

$$N_T < \text{ lower critical value}$$

then the deflection decreases with time. If

$$N_T = \text{ lower critical value}$$

then the deflection is constant. For $N_T$ such that

$$\text{lower critical value } < N_T < \text{ upper critical value}$$

the deflection increases with time and, finally, if

$$N_T = \text{ upper critical value}$$

then the deflection is immediately infinite.

In Das [262], the equation governing the deflections of thermally loaded viscoelastic plates is derived as follows: using standard tensor notation, the constitutive relations and kinematic relations are written, following Nowacki [222], as

$$\begin{cases} \sigma_{ij} = \dfrac{E}{1-\nu^2}\{(1-\nu)\epsilon_{ij} + [\nu\epsilon_{kk} - (1+\nu)\alpha_t x_3\tau]\delta_{ij}\} \\ \epsilon_{ij} = -x_3 w_{,ij} \end{cases}$$

where $w_{,ij} = \dfrac{\partial^2 w}{\partial x_i \partial x_j}$ and we sum, as usual, on repeated indices. Thus, with the understanding that $\nu$ is time-dependent,

$$\sigma_{ij} = \frac{-Ex_3}{1-\nu^2}\{(1-\nu)w_{,ij} + [\nu w_{,kk} + (1+\nu)\alpha_t\tau]\delta_{ij}\} \tag{8.128}$$

As the bending moments are given by

$$M_{ij} = \int_{-h/2}^{h/2} x_3\sigma_{ij}dx_3$$

we have, using (8.128),

$$M_{ij} = -K\{(1-\nu)w_{,ij} + [\nu w_{,kk} + (1+\nu)\alpha_t\tau]\delta_{ij}\} \tag{8.129}$$

with $K$ the usual bending stiffness. Assuming that $M_{ij,ij} = 0$, i.e., ignoring the influence of the middle surface forces, we obtain as the differential equation for the deflection of a thermally loaded viscoelastic plate

$$\nabla^4 w + (1+\nu)\alpha_t\,\nabla^2\tau = 0 \tag{8.130}$$

or

$$\nabla^4 w + m(t)\,\nabla^2\tau = 0 \tag{8.131}$$

with

$$m(t) = (1+\nu)\alpha_t \tag{8.132}$$

We note that the temperature distribution in the plate, $T(x_1, x_2, x_3)$, has been taken in [262] to have the form

$$T = \tau_0(x_1, x_2) + x_3\tau(x_1, x_2) \tag{8.133}$$

and that the situation depicted in Fig. 8.12 applies, i.e.,

$$\begin{cases} T_1(x_1, x_2) = T(x_1, x_2, \frac{h}{2}) \\ T_2(x_1, x_2) = T(x_1, x_2, -\frac{h}{2}) \end{cases} \tag{8.134}$$

If $\theta_1$ and $\theta_2$ are, respectively, the temperatures of the medium immediately below and above the plate, then the Newton-type boundary conditions

$$\begin{cases} \lambda \dfrac{\partial T}{\partial x_3}\Big|_{x_3=\frac{h}{2}} = \lambda_1(\theta_1 - T_1) \\ \lambda \dfrac{\partial T}{\partial x_3}\Big|_{x_3=-\frac{h}{2}} = -\lambda_1(\theta_1 - T_1) \end{cases} \tag{8.135}$$

apply where $\lambda$ is the plate thermal conductivity. From the equation of heat conduction and the assumptions listed above, we obtain for the stationary temperature field $\tau$ the following equation (assuming the absence of a heat source):

$$\nabla^2\tau - \frac{12}{h^2}(1+\epsilon)\tau = -\frac{12}{h^3}(\theta_1 - \theta_2) \tag{8.136}$$

where $\epsilon = h\lambda_1/2\lambda$, or

$$\nabla^2\tau - k^2\tau = -\beta \tag{8.137}$$

with

$$k^2 = \frac{12}{h^2}(1+\epsilon) \text{ and } \beta = \frac{12}{h^3}(\theta_1 - \theta_2) = \text{const.} \tag{8.138}$$

We note that, following the analysis in [262], the fourth-order equation (8.131) may be replaced by two equations of the second order: using the fact that

$$-K(1+\nu)\nabla^2 w = M_{11} + M_{22}$$

or

$$\nabla^2 w = -\frac{M}{K}; \ M = \frac{M_{11} + M_{22}}{(1+\nu)} \tag{8.139}$$

From (8.131) we then obtain

$$\nabla^2 M = m(t)K\nabla^2\tau \tag{8.140}$$

so that (8.131) may be replaced by (8.139), (8.140).

Equations (8.139) and (8.140) may be simplified for the special case of a simply supported polygonal plate, i.e., the solution of (8.140) is given by

$$M = m(t)K\tau \tag{8.141}$$

as $M$ and $\tau$ are both zero along the edges of a simply supported plate. Thus, by virtue of (8.139) and (8.141), it follows that it is sufficient to solve

$$\triangledown^2 w + m(t)\tau = 0 \tag{8.142}$$

subject to the condition that $w = 0$ along the edges of the plate. Laplace transforming (8.142) we obtain

$$\triangledown^2 \tilde{W} + \frac{\tilde{m}(s)}{s}\tau = 0 \tag{8.143}$$

with

$$\begin{cases} \tilde{W} = \displaystyle\int_0^\infty we^{-st}dt \\ \tilde{m}(s) = (1 + \tilde{\nu})\alpha_t \end{cases} \tag{8.144}$$

$s$ being the transform parameter. In equations (8.137) and (8.143) we introduce the complex coordinate system given by

$$x_3 = x_1 + ix_2, \quad \bar{x}_3 = x_1 - ix_2 \tag{8.145}$$

thus reducing these equations to the respective forms

$$4\frac{\partial^2 \tau}{\partial x_3 \partial \bar{x}_3} - k^2\tau = -\beta \tag{8.146a}$$

$$4\frac{\partial^2 \tilde{W}}{\partial x_3 \partial \bar{x}_3} + \frac{\tilde{m}(s)}{s}\tau = 0 \tag{8.146b}$$

We now let $x_3 = f(\xi)$ be the function which maps the domain of the plate onto the unit circle so that, in the system of coordinates $(\xi, \bar{\xi})$, (8.146a,b) reduce to

$$4\frac{\partial^2 \tau}{\partial \xi \partial \bar{\xi}} - k^2\tau\frac{dx_3}{d\xi}\frac{d\bar{x}_3}{d\bar{\xi}} = -\beta\frac{dx_3}{d\xi}\frac{d\bar{x}_3}{d\bar{\xi}} \tag{8.147a}$$

$$4\frac{\partial^2 \tilde{W}}{\partial \xi \partial \bar{\xi}} + \frac{\tilde{m}(s)}{s}\tau\frac{dx_3}{d\xi}\frac{d\bar{x}_3}{d\bar{\xi}} = 0 \tag{8.147b}$$

with $\xi = re^{i\theta}$, $\bar{\xi} = re^{-i\theta}$. Assuming that

$$\frac{dx_3}{d\xi} = \frac{d\bar{x}_3}{d\bar{\xi}} = a_1 = \text{const.} \tag{8.148}$$

we obtain from (8.147a)

$$\frac{\partial^2 \tau}{\partial \xi \partial \bar{\xi}} - \mu^2 \tau = -\frac{\beta a_1^2}{4}; \ \mu^2 = \frac{k^2 a_1^2}{4} \tag{8.149}$$

The closed form solution of (8.149) is given in the form

$$\tau = A I_0(2\mu\sqrt{\xi\bar{\xi}}) + \frac{\beta a_1^2}{4\mu^2}$$
$$= A I_0(2\mu r) + \frac{\beta a_1^2}{4\mu^2} \tag{8.150}$$

with $I_0(2\mu r)$ the modified Bessel function of zeroth order and $A$ a constant to be determined by the boundary condition $\tau = 0$ at $r = 1$. A direct computation shows that

$$A = -\frac{\beta a_1^2}{4\mu^2} \cdot \frac{1}{I_0(2\mu)} \tag{8.151}$$

in which case,

$$\tau = \frac{\beta a_1^2}{4\mu^2} \left[ 1 - \frac{I_0(2\mu r)}{I_0(2\mu)} \right] \tag{8.152}$$

To solve (8.147b), we again apply the assumption (8.148), in which case, (8.147b) reduces to

$$\frac{\partial^2 \tilde{W}}{\partial \xi \partial \bar{\xi}} + \gamma^2 \tau = 0 \tag{8.153}$$

with

$$\gamma^2 = \frac{\tilde{m}(s)}{4s} a_1^2 \tag{8.154}$$

Substituting for $\tau$ in (8.153) from (8.152) we obtain

$$\frac{\partial^2 \tilde{W}}{\partial \xi \partial \bar{\xi}} = -\gamma^2 \frac{\beta a_1^2}{4\mu^2} \left[ 1 - \frac{I_0(2\mu r)}{I_0(2\mu)} \right]$$
$$= P + Q I_0(2\mu r) \tag{8.155}$$

where

$$P = -\frac{\gamma^2 \beta a_1^2}{4\mu^2}, \ Q = -\frac{P}{I_0(2\mu)} \tag{8.156}$$

The solution of (8.155) is of the form

$$\tilde{W} = B + Pr^2 + \frac{Q}{\mu^2}I_0(2\mu r) \qquad (8.157)$$

with $B$ determined by the boundary condition $\tilde{W} = 0$ at $r = 1$; applying this boundary condition we have

$$B = -P - \frac{Q}{\mu^2}I_0(2\mu) \qquad (8.158)$$

in which case,

$$\tilde{W} = -P - \frac{Q}{\mu^2}I_0(2\mu) + Pr^2 + \frac{Q}{\mu^2}I_0(2\mu r) \qquad (8.159)$$

The maximum of $\tilde{W}$ occurs at $r = 0$ and has the value

$$(\tilde{W})_{\max} = \frac{\tilde{m}(s)}{s} \cdot \frac{\beta a_1^4}{16\mu^2}\left[1 - \frac{1}{\mu^2} + \frac{1}{\mu^2 I_0(2\mu)}\right] \qquad (8.160)$$

By taking the inverse Laplace transform of (8.160) the maximal plate deflection is obtained.

Finally, we turn our attention to the problem of thermal buckling of elastic-plastic plates; one of the most fundamental and comprehensive treatments of this problem to date appears to be the work of Williams [263] which is based on a variational formulation of the rate problem in elastoplasticity by Neale [264], [265] in which both the strain and stress rates can be varied independently. In the treatment developed in [263], which we will outline below, the field equations that follow from Neale's original analysis [264] are employed to study the thermal buckling of constrained rectangular plates. It is worth noting that, in contrast to the constitutive assumptions of Neale [264], who uses a $J_2$-incremental theory and employs the Ramberg-Osgood relation to describe the hardening behavior, several authors have approached the same basic problem using deformation theories of plasticity; these include the work of Mayers and Nelson [272], who use a modified form of Reissner's variational principle and Turvey [273] who employs a constitutive equation that was proposed by Myszkowski [274].

One advantage of the approach in Neale's work [264] is that when the Ramberg-Osgood hardening exponent is set equal to 3.0, the hardening parameter becomes a constant, and this is the case considered by Williams in [263]; because the interest in [263] is in thermal behavior, a slight modification of the energy potential, as suggested by Neale in [264], is introduced by Williams [263]. The buckling and postbuckling equations in [263] are derived following the approach in Danielson [275].

Following the Lagrangian formulation proposed by Neale [264], the displacement components and relevant Kirchhoff stress components in [263] assume the form (see Fig. 8.13)

$$\begin{cases} u_1 = u(x,y) - zw_{,x}(x,y) \\ u_2 = v(x,y) - zw_{,y}(x,y) \\ u_3 = w(x,y) \end{cases} \tag{8.161}$$

and

$$P_{ij} = \frac{N_{ij}(x,y) + \dfrac{12z}{h^2} M_{ij}(x,y)}{h} \tag{8.162}$$

The equations of equilibrium and the constitutive relations governing $u, v, w, N_{ij}$, etc., are obtained from the requirement that $\delta \mathcal{I}^{(0)} = 0$ where the Lagrangian form of the functional $\mathcal{I}^{(0)}$ is

$$\begin{aligned} \mathcal{I}^{(0)} = \int_{V_0} \left\{ \dot{P}_{ij} \dot{E}_{ij} + \frac{1}{2} P_{ij} v_{k,i} v_{k,j} - W(\dot{\tau}) \right\} dV_0 \\ - \int_{S_F^{(0)}} \dot{F}_i^* v_i dS^{(0)} - \int_{S_V^{(0)}} \dot{F}_i (v_i - v_i^*) dS^{(0)} \end{aligned} \tag{8.163}$$

In (8.163) $S_F^{(0)}, S_V^{(0)}$ are the undeformed areas over which the traction rates $F_i^*$ and velocities $v_i^*$ are prescribed. The potential $W(\dot{\tau})$ is defined such that the time rate of change of Green's strain tensor $E_{ij}$ is

$$\frac{\partial}{\partial t} E_{ij} \equiv \dot{E}_{ij} = \frac{\partial W}{\partial P_{ij}} \tag{8.164}$$

The nominal surface tractions per unit undeformed area $F_i$ are given by

$$F_j = (P_{ij} + P_{ik} u_{j,k}) n_i^{(0)} \tag{8.165}$$

with $n^{(0)}$ the unit normal vector to the undeformed area. Using the simplifications indicated in [264], Williams takes

$$\begin{aligned} \dot{P}_{ij} \dot{E}_{ij} = \dot{P}_{11} \dot{E}_{11} + 2\dot{P}_{12} \dot{E}_{12} + \dot{P}_{22} \dot{E}_{22} \\ P_{ij} v_{k,i} v_{k,j} = P_{11} \dot{w}_1^2 + 2P_{12} \dot{w}_1 \dot{w}_2 + P_{22} \dot{w}_2^2 \end{aligned} \tag{8.166}$$

with

$$E_{ij} = \frac{1}{2}(u_{i,j} + u_{j,i} + u_{3,i} u_{3,j}) \tag{8.167}$$

The effect of a temperature variation is accounted for in Williams [263] by adding the term $\alpha \dot{T}(\dot{P}_{11} + \dot{P}_{22})$ to Neale's [264] form for $W(\dot{\tau})$, i.e.,

$$W(\dot{\tau}) = \frac{1}{2}(c_{11}\dot{P}_{11}^2 + c_{22}\dot{P}_{22}^2) + c_{12}\dot{P}_{11}\dot{P}_{22} + c_{13}\dot{P}_{11}\dot{P}_{12}$$
$$+ c_{23}\dot{P}_{22}\dot{P}_{12} + c_{33}\dot{P}_{12}^2 + \alpha \dot{T}(\dot{P}_{11} + \dot{P}_{22}) \tag{8.168}$$

Following the analysis in Neale [264], the material coefficients $c_{ij}$ in [263] have the form

$$c_{11} = \frac{1}{E} + \frac{G(2P_{11} - P_{22})^2}{9}, \quad c_{22} = \frac{1}{E} + \frac{G(2P_{22} - P_{11})^2}{9}$$

$$c_{12} = -\frac{\nu}{E} + \frac{G(2P_{11} - P_{22})(2P_{22} - P_{11})}{9}, \quad c_{13} = \frac{2GP_{12}(2P_{11} - P_{22})}{3}$$

$$c_{23} = \frac{2GP_{12}(2P_{22} - P_{11})}{3}, \quad c_{33} = \frac{1+\nu}{E} + 2GP_{12}^2$$

$$\tag{8.169}$$

In (8.169), the hardening parameter $G$ is identified using uniaxial stress-strain data and the basic assumption that plastic behavior depends on the stress only through the invariant $J_2$; following Neale [264], $G$ has the form

$$G = \frac{3}{4J_2}\left(\frac{1}{E_t} - \frac{1}{E}\right) \tag{8.170}$$

where $\frac{1}{E_t}$, $E_t = E_t(J_2)$, is the slope of the graph of $E_{11}$ vs. $P_{11}$ at $T =$ const. Introducing the Ramberg-Osgood form

$$E_{11} - \alpha \delta T = \frac{1}{E}P_{11} + \left(\frac{P_{11}}{E_0}\right)^k \tag{8.171}$$

where $\delta T = T - T_0$, $T_0$ the reference temperature, it follows that

$$G = \frac{3k(3J_2)^{(k-1)/2}}{4J_2 E_0^k} \tag{8.172}$$

so that the choice $k = 3$ leads to $G = \frac{27}{4}/E_0^3$; this restriction to $k = 3$ is then followed in the remainder of the development in [263]. Assuming that the effects of bending and stretching are equally important in inducing stress in an elastic-plastic plate, that spatial rates of change are significant over distances of order $a$, and introducing the associated order-of-magnitude estimates

$$\begin{cases} M_{ij} = hN_{ij}\mathcal{O}(1) \\ (u, v, w) = h\left(\dfrac{h}{a}, \dfrac{h}{a}, 1\right)\mathcal{O}(1) \end{cases} \tag{8.173}$$

it may be shown that

$$\int_{-h/2}^{h/2} \dot{F}_j v_j dz \bigg|_{x=\pm a} = \pm\bigg[\dot{N}_{11}\dot{v} + \dot{N}_{12}\dot{v} - \dot{M}_{11}\dot{w}_1 - \dot{M}_{12}\dot{w}_2$$
$$+\dot{w}\frac{\partial}{\partial t}(Q_1 + N_{11}w_1 + N_{12}w_2)\bigg] \tag{8.174a}$$

$$\int_{-h/2}^{h/2} \dot{F}_j v_j dz \bigg|_{y=\pm b} = \pm\bigg[\dot{N}_{12}\dot{u} + N_{22}\dot{v} - \dot{M}_{12}w_1 - \dot{M}_{22}\dot{w}_2$$
$$+\dot{w}\frac{\partial}{\partial t}(Q_2 + N_{12}w_2 + N_{22}w_2)\bigg] \tag{8.174b}$$

where

$$(Q_1, Q_2) = \int_{-h/2}^{h/2}(P_{13}, P_{23})dz \tag{8.174c}$$

In writing down (8.174a) and (8.174b), terms of order $(h/a)^2$ have been neglected. Integrating the bending moment $M_{12}$ by parts we obtain

$$\int_S \dot{F}_i v_i dS = -2[\dot{M}_{12}\dot{w}]\bigg|_{-a}^{a}\bigg|_{-b}^{b}$$
$$+\int_{-b}^{b}[\dot{N}_{11}\dot{u} + \dot{N}_{12}\dot{v} + \dot{V}_1\dot{w} - \dot{M}_{11}\dot{w}_1]\bigg|_{-a}^{a}\, dy \tag{8.175}$$
$$+\int_{-a}^{a}[\dot{N}_{12}\dot{u} + \dot{N}_{22}\dot{v} + \dot{V}_2\dot{w} - \dot{M}_{22}\dot{w}_2]\bigg|_{-b}^{b}\, dx$$

where the net shear terms $V_1, V_2$ are given by

$$V_1 = Q_1 + M_{12,2} + N_{11}w_1 + N_{12}w_2$$
$$V_2 = Q_2 + M_{12,1} + N_{12}w_1 + N_{22}w_2 \tag{8.176}$$

By carrying out the calculation to obtain the Euler equations from the variational equation $\delta\mathcal{I}^{(0)} = 0$, the following results are obtained in [263]:

$$\dot{N}_{\alpha\beta,\beta} = 0 \tag{8.177}$$

$$\frac{\partial}{\partial t}\{M_{\alpha\beta,\alpha\beta} + (N_{\alpha\beta}w_\beta)_{,\alpha}\} = 0 \tag{8.178}$$

and

$$\begin{bmatrix} \dot{\epsilon}_{11}^{(0)} - \dfrac{\dot{N}_T + \dot{N}_{11} - \nu\dot{N}_{22}}{Eh} \\[2mm] \dot{\epsilon}_{22}^{(0)} - \dfrac{\dot{N}_T + \dot{N}_{22} - \nu\dot{N}_{11}}{Eh} \\[2mm] 2\dot{\epsilon}_{12}^{(0)} - \dfrac{2(1+\nu)\dot{N}_{12}}{Eh} \\[2mm] -\dot{w}_{11} - \dfrac{12(\dot{M}_T + \dot{M}_{11} - \nu\dot{M}_{22})}{Eh^3} \\[2mm] -\dot{w}_{22} - \dfrac{12(\dot{M}_T + \dot{M}_{22} - \nu\dot{M}_{11})}{Eh^3} \\[2mm] -2\dot{w}_{12} - \dfrac{24(1+\nu)\dot{M}_{12}}{Eh^3} \end{bmatrix} = \frac{G}{9h^3} \cdot \boldsymbol{G} \begin{bmatrix} \dot{N}_{11} \\[2mm] \dot{N}_{22} \\[2mm] \dot{N}_{12} \\[2mm] \dot{M}_{11} \\[2mm] \dot{M}_{22} \\[2mm] \dot{M}_{12} \end{bmatrix} \tag{8.179}$$

where $N_T$, $M_T$ are given by the usual relations, i.e.,

$$(N_T, M_T) = \alpha E \int_{-h/2}^{h/2} \delta T(1, z) dz$$

and where the hardening matrix $\boldsymbol{G}$ is defined at the end of this section. If we let a superposed asterisk again denote quantities which are prescribed along the boundary of the plate, then the natural boundary conditions assume the following form:

$\dot{N}_{11} = \dot{N}_{11}^*$ or $\dot{u} = \dot{u}^*$

$\dot{N}_{12} = \dot{N}_{12}^*$ or $\dot{v} = \dot{v}^*$

$\dot{M}_{11,1} + 2\dot{M}_{12,2} + \dfrac{\partial}{\partial t}(N_{11}w_1 + N_{12}w_2) = \dot{V}_1^*$ or $\dot{w} = \dot{w}^*$

$\dot{M}_{11} = \dot{M}_{11}^*$ or $\dot{w}_1 = \dot{w}_1^*$

on $x = \pm a$ and $\tag{8.180}$

$\dot{N}_{12} = \dot{N}_{12}^*$ or $\dot{u} = \dot{u}^*$

$\dot{N}_{22} = \dot{N}_{22}^*$ or $\dot{v} = \dot{v}^*$

$\dot{M}_{22,2} + 2\dot{M}_{12,1} + \dfrac{\partial}{\partial t}(N_{12}w_1 + N_{22}w_2) = \dot{V}^*$ or $\dot{w} = \dot{w}^*$

$\dot{M}_{22} = \dot{M}_{22}^*$ or $\dot{w}_2 = \dot{w}_2^*$

at the corners of the plate. Taking

$$\begin{cases} V_1 = M_{11,1} + 2M_{12,2} + N_{11}w_1 + N_{12}w_2 \\ V_2 = M_{22,2} + 2M_{12,1} + N_{12}w_1 + N_{22}w_2 \end{cases} \tag{8.181}$$

we find that

$$\begin{cases} \dot{Q}_1 = \dot{M}_{11,1} + \dot{M}_{12,2} \\ \dot{Q}_2 = \dot{M}_{12,1} + \dot{M}_{22,2} \end{cases} \tag{8.182}$$

and, thus obtain the form which is familiar from classical plate theory, i.e.,

$$\dot{M}_{\alpha\beta,\alpha\beta} = \dot{Q}_{\alpha,\alpha} \tag{8.183}$$

Therefore, although constitutive formulation in [264] differs from earlier formulations of the elastic-plastic deformation problem for plates, one does recover (in rate form) the equilibrium equations which are familiar from classical von Karman plate theory.

We now proceed with the derivation of the equations governing the buckling and postbuckling behavior of heated rectangular plates, assuming that the plate is constrained against displacement along its edges; such a plate, if elastic, will in general buckle if its edges are clamped or, if $M_T = 0$, if its edges are simply supported. In [263], attention is restricted to the case in which $M_T = 0$ and $N_T = $ const. over the surface of the plate (although, in principle, any spatial variation in $N_T$ could be studied); under these restrictions it may be shown that there exists a prebuckled membrane state given by

$$\begin{cases} u = v = w = 0 \\ N_{11} = N_{22} = \mathcal{N}(N_T) \\ N_{12} = 0 \\ M_{11} = M_{22} = M_{12} = 0 \end{cases} \tag{8.184}$$

where

$$\dot{N}_T + \dot{\mathcal{N}}\left(1 - \nu + \frac{2EG\mathcal{N}^2}{9h^2}\right) = 0 \tag{8.185}$$

is a consequence of the constitutive relations. It may be expected that this prebuckled membrane state persists for $N_T < (N_T)_{cr}$ with $(N_T)_{cr}$ corresponding to the critical temperature parameter.

To determine $(N_T)_{cr}$, and the behavior of the plate in a neighborhood of the critical temperature, Williams [263] adopts the procedure given by Danielson [275], i.e., the time variable is replaced by the distance parameter $s$ along the load path measured from the critical point. With $\dfrac{\partial}{\partial t} = \dfrac{\partial}{\partial s}\dfrac{ds}{dt}$, we then assume the expansions

$$
\begin{aligned}
u &= su^{(1)} + s^2 u^{(2)} + \ldots \\
v &= sv^{(1)} + s^2 v^{(2)} + \ldots \\
w &= sw^{(1)} + s^2 w^{(2)} + \ldots
\end{aligned}
\tag{8.186}
$$

and

$$
\begin{aligned}
N_{\alpha\beta} &= N\delta_{\alpha\beta} + sN_{\alpha\beta}^{(1)} + s^2 N_{\alpha\beta}^{(2)} + \ldots \\
M_{\alpha\beta} &= sM_{\alpha\beta}^{(1)} + s^2 M_{\alpha\beta}^{(2)} + \ldots \\
N &= N_{cr} + N^{(1)} + s^2 N^{(2)} + \ldots \\
N_T &= (N_T)_{cr} + sN_T^{(1)} + s^2 N_T^{(2)} + \ldots
\end{aligned}
\tag{8.187}
$$

and obtain from the Euler (equilibrium) equations (8.177)-(8.178)

$$
\left\{
\begin{aligned}
&N_{\alpha\beta,\beta}^{(1)} = 0 \quad N_{\alpha\beta,\beta}^{(2)} = 0 \\
&M_{\alpha\beta,\alpha\beta}^{(1)} + N_{cr}\delta_{\alpha\beta}w_{\alpha\beta}^{(1)} = 0 \\
&M_{\alpha\beta,\alpha\beta}^{(2)} + N_{cr}\delta_{\alpha\beta}w_{\alpha\beta}^{(2)} + (N_{\alpha\beta}^{(1)} + N^{(1)}\delta_{\alpha\beta})w_{\alpha\beta}^{(1)} = 0
\end{aligned}
\right.
\tag{8.188}
$$

Also, from (8.185) it follows that

$$
\left\{
\begin{aligned}
&(N_T)_{cr} = -N_{cr}\left(1 - \nu + \frac{2g^*}{3}\right), \quad g^* = \frac{GEN_{cr}^2}{9h^2} \\
&N_T^{(1)} = -N^{(1)}(1 - \nu + 2g^*) \\
&N_T^{(2)} = -N^{(2)}(1 - \nu + 2g^*) - \frac{2g^* N^{(1)2}}{N_{cr}}
\end{aligned}
\right.
\tag{8.189}
$$

The Euler constitutive relations (8.179) are now rewritten in the form

$$
\begin{aligned}
Eh\dot{\epsilon}_{\alpha\beta}^{(0)} &= \dot{N}_T\delta_{\alpha\beta} + A_{\alpha\beta\gamma\delta}\dot{N}_{\gamma\delta} + B_{\alpha\beta\gamma\delta}\dot{M}_{\gamma\delta} \\
-Eh\dot{w}_{\alpha\beta} &= \frac{12\dot{M}_T}{h^2}\delta_{\alpha\beta} + B_{\alpha\beta\gamma\delta}\dot{N}_{\gamma\delta} + D_{\alpha\beta\gamma\delta}\dot{M}_{\gamma\delta}
\end{aligned}
\tag{8.190}
$$

where

$$
\begin{aligned}
A_{1111} &= A_{11},\ A_{1122} = A_{12},\ A_{1112} = A_{13},\ A_{1131} = A_{14} \\
A_{2211} &= A_{21},\ A_{2222} = A_{22},\ A_{2212} = A_{23},\ A_{2221} = A_{24} \\
A_{1211} &= A_{31},\ A_{1222} = A_{32},\ A_{1212} = A_{33},\ A_{1221} = A_{34} \\
A_{2111} &= A_{41},\ A_{2122} = A_{42},\ A_{2112} = A_{43},\ A_{2121} = A_{44}
\end{aligned}
\tag{8.191}
$$

and the components of the symmetric $\boldsymbol{A}$, $\boldsymbol{B}$, and $\boldsymbol{D}$ matrices are defined at the end of this section.

The strain tensor $\epsilon_{\alpha\beta}^{(0)}$ is decomposed according to

$$\epsilon_{\alpha\beta}^{(0)} = e_{\alpha\beta} + \frac{w_{,\alpha} w_{,\beta}}{2} \tag{8.192}$$

and, furthermore, we write

$$\begin{cases} e_{\alpha\beta} = s e_{\alpha\beta}^{(1)} + s^2 e_{\alpha\beta}^{(2)} + .... \\[2mm] \kappa_{\alpha\beta} \equiv w_{,\alpha\beta} = s \kappa_{\alpha\beta}^{(1)} + s^2 \kappa_{\alpha\beta}^{(2)} + ... \end{cases} \tag{8.193}$$

where

$$\begin{cases} e_{11}^{(k)} = u_{,1}^{(k)}, \quad e_{22}^{(k)} = v_{,2}^{(k)} \\[2mm] e_{12}^{(k)} = e_{21}^{(k)} = \frac{1}{2}\left(u_{,2}^{(k)} + v_{,1}^{(k)}\right) \end{cases} \tag{8.194}$$

The tensor components $A_{\alpha\beta\gamma\delta}$, $B_{\alpha\beta\gamma\delta}$, $D_{\alpha\beta\gamma\delta}$ are also expanded as series in the parameter $s$, i.e.,

$$\begin{aligned} A_{\alpha\beta\gamma\delta} &= A_{\alpha\beta\gamma\delta}^{(0)} + s A_{\alpha\beta\gamma\delta}^{(1)} + s^2 A_{\alpha\beta\gamma\delta}^{(2)} + ... \\ B_{\alpha\beta\gamma\delta} &= s B_{\alpha\beta\gamma\delta}^{(1)} + s^2 B_{\alpha\beta\gamma\delta}^{(2)} + ... \\ D_{\alpha\beta\gamma\delta} &= D_{\alpha\beta\gamma\delta}^{(0)} + s D_{\alpha\beta\gamma\delta}^{(1)} + s^2 D_{\alpha\beta\gamma\delta}^{(2)} + ... \end{aligned} \tag{8.195}$$

Using the expansions (8.193), (8.195) in the constitutive relations (8.190), (8.191) it follows that

$$Ehe_{\alpha\beta}^{(1)} = N_T^{(1)}\delta_{\alpha\beta} + A_{\alpha\beta\gamma\delta}^{(0)}(N_{\gamma\beta}^{(1)} + N^{(1)}\delta_{\gamma\delta})$$

$$Eh\left(e_{\alpha\beta}^{(2)} + \frac{w_\alpha^{(1)} w_\beta^{(1)}}{2}\right) = N_T^{(2)}\delta_{\alpha\beta} + A_{\alpha\beta\gamma\delta}^{(0)}(N_{\gamma\delta}^{(2)} + N^{(2)}\delta_{\gamma\delta})$$

$$+ \frac{1}{2}A_{\alpha\beta\gamma\delta}^{(1)}(N_{\gamma\delta}^{(1)} + N^{(1)}\delta_{\gamma\delta}) + \frac{1}{2}B_{\alpha\beta\gamma\delta}^{(1)}M_{\gamma\delta}^{(1)}$$

$$-Eh\kappa_{\alpha\beta}^{(1)} = D_{\alpha\beta\gamma\delta}^{(0)}M_{\gamma\delta}^{(1)}$$

$$-Eh\kappa_{\alpha\beta}^{(2)} = D_{\alpha\beta\gamma\delta}^{(0)}M_{\gamma\delta}^{(2)} + \frac{1}{2}D_{\alpha\beta\gamma\delta}^{(1)}M_{\gamma\delta}^{(1)} + \frac{1}{2}B_{\alpha\beta\gamma\delta}^{(1)}(N_{\gamma\delta}^{(1)} + N^{(1)}\delta_{\gamma\delta})$$

$$-Eh\kappa_{\alpha\beta}^{(3)} = D_{\alpha\beta\gamma\delta}^{(0)}M_{\gamma\delta}^{(3)} + \frac{2}{3}D_{\alpha\beta\gamma\delta}^{(1)}M_{\gamma\delta}^{(2)} + \frac{1}{3}D_{\alpha\beta\gamma\delta}^{(2)}M_{\gamma\delta}^{(1)}$$

$$+ \frac{2}{3}B_{\alpha\beta\gamma\delta}^{(1)}(N_{\gamma\delta}^{(2)} + N^{(2)}\delta_{\gamma\delta}) + \frac{1}{3}B_{\alpha\beta\gamma\delta}^{(2)}(N_{\gamma\delta}^{(1)} + N^{(1)}\delta_{\gamma\delta})$$

$$\tag{8.196}$$

In (8.196), one has a hierarchy of equations which must be solved to obtain the expansion variables $w^{(k)}$, etc. We note, however, that for displacements and stresses which satisfy the boundary conditions identically, the variational functional $\mathcal{I}^{(0}$ may be simplified so as to read

$$\mathcal{I}^{(0)} = \int_{v_0} [\dot{P}_{ij}\dot{E}_{ij} + \frac{1}{2}P_{ij}v_{k,i}v_{k,j} - W(\dot{\tau})]dV_0 \tag{8.197}$$

Using the order of magnitude estimates (8.173), and restricting attention to stress rates which satisfy the constitutive relations, the variational equation $\delta\mathcal{I}^{(0)} = 0$ simplifies to

$$\int_{-a}^{a}\int_{-b}^{b}\{N_{\alpha\beta}(\delta\dot{e}_{\alpha\beta} + w_{,\alpha}\delta\dot{w}_{,\beta}) + N_{\alpha\beta}\dot{w}_{,\alpha}\delta\dot{w}_{,\beta} - M_{\alpha\beta}\delta\dot{\kappa}_{\alpha\beta}\}\,dxdy = 0 \tag{8.198}$$

The prebuckled state given by $(M_T = 0)$

$$w = 0, \ N_{\alpha\beta} = \mathcal{N}(N_T)\delta_{\alpha\beta}, \ M_{\alpha\beta} = 0 \tag{8.199}$$

must satisfy (8.198), i.e.,

$$\int_{-a}^{a}\int_{-b}^{b}\dot{\mathcal{N}}(N_T)\delta_{\alpha\beta}\delta\dot{e}_{\alpha\beta}dxdy = 0 \tag{8.200}$$

We now substitute the expansions (8.186), (8.187) into both (8.198) and (8.200) and subtract the resulting equations so as to obtain

$$\int_{-a}^{a}\int_{-b}^{b}(N_{\alpha\beta}^{(1)}\delta\dot{e}_{\alpha\beta} + N^*w_{\alpha}^{(1)}\delta_{\alpha\beta}\delta\dot{w}_{\beta} - M_{\alpha\beta}^{(1)}\delta\dot{\kappa}_{\alpha\beta}$$

$$+2s\{N_{\alpha\beta}^{(2)}\delta\dot{e}_{\alpha\beta} + (w_{\alpha}^{(1)}(N_{\alpha\beta}^{(1)} + N^{(1)}\delta_{\alpha\beta}) + N^*w_{\alpha}^{(2)}\delta_{\alpha\beta})\delta\dot{w}_{\beta}$$

$$-M_{\alpha\beta}^{(2)}\delta\dot{\kappa}_{\alpha\beta} + 3s^2N_{\alpha\beta}^{(3)}\delta\dot{e}_{\alpha\beta} + [w_{\alpha}^{(2)}(N_{\alpha\beta}^{(1)} + N^{(1)}\delta_{\alpha\beta}$$

$$+w_{\alpha}^{(1)}(N_{\alpha\beta}^{(2)} + N^{(2)}\delta_{\alpha\beta}) + N^*w_{\alpha}^{(3)}\delta_{\alpha\beta}]\delta\dot{w}_{\beta} - M_{\alpha\beta}^{(3)}\delta\dot{\kappa}_{\alpha\beta}| + ...)dxdy = 0 \tag{8.201}$$

Restricting the strain-rate variations to have the form

$$\delta\dot{e}_{\alpha\beta} = se_{\alpha\beta}^{(k)}, \ \delta\dot{w}_{,\beta} = \dot{s}w_{,\beta}^{(k)}, \ \delta\dot{r}_{\alpha\beta} = \dot{s}\kappa_{\alpha\beta}^{(k)} \tag{8.202}$$

equation (8.201) becomes, as $s \to 0$,

$$\int_{-a}^{a}\int_{-b}^{b}(N_{\alpha\beta}^{(1)}e_{\alpha\beta}^{(k)} + \mathcal{N}w_{,\alpha}^{(1)}\delta_{\alpha\beta}w_{,\beta}^{(k)} - M_{\alpha\beta}^{(k)})dxdy = 0 \tag{8.203}$$

To avoid the possibility that $w^{(k)}$ contains an arbitrary multiple of $w^{(k)}$, Williams [263] imposes the restriction

$$\int_{-a}^{a}\int_{-b}^{b}\delta_{\alpha\beta}w_{,\alpha}^{(k)}w_{,\beta}^{(k)}\,dx\,dy = 0 \tag{8.204}$$

the consequence of which is

$$\int_{-a}^{a}\int_{-b}^{b}(N_{\alpha\beta}^{(1)}e_{\alpha\beta}^{(k)} - M_{\alpha\beta}^{(1)}\kappa_{\alpha\beta}^{(k)})\,dx\,dy = 0 \tag{8.205}$$

On the other hand, if the strain-rate variations are restricted to have the form

$$\delta\dot{e}_{\alpha\beta} = \dot{s}e_{\alpha\beta}^{(1)}, \quad \delta\dot{w}_{,\beta} = \dot{s}w_{,\beta}^{(1)}, \quad \delta\dot{\kappa}_{\alpha\beta} = \dot{s}\kappa_{\alpha\beta}^{(1)} \tag{8.206}$$

then (8.201) becomes, with the aid of (8.203)–(8.205),

$$\left(N^{(1)} + \frac{3sN^{(2)}}{2} + \cdots\right)\int_{-a}^{a}\int_{-b}^{b}(\delta_{\alpha\beta}w_{\alpha}^{(1)}w_{\beta}^{(1)}$$

$$+(N_{\alpha\beta}^{(2)}e_{\alpha\beta}^{(1)} - N_{\alpha\beta}^{(1)}e_{\alpha\beta}^{(2)}) + N_{\alpha\beta}^{(1)}w_{\alpha}^{(1)}w_{\beta}^{(1)} - (M_{\alpha\beta}^{(2)}K_{\alpha\beta}^{(1)} - M_{\alpha\beta}^{(2)}K_{\alpha\beta}^{(2)})$$

$$+\frac{3s}{2}[N_{\alpha\beta}^{(3)}e_{\alpha\beta}^{(1)} - N_{\alpha\beta}^{(1)}e_{\alpha\beta}^{(3)} + N_{\alpha\beta}^{(1)}w_{\alpha}^{(2)}w_{\beta}^{(1)} + N_{\alpha\beta}^{(2)}w_{\alpha}^{(1)}w_{\beta}^{(1)}$$

$$(M_{\alpha\beta}^{(3)}K_{\alpha\beta}^{(1)} - M_{\alpha\beta}^{(1)}K_{\alpha\beta}^{(3)})]dx\,dy = 0$$

$$\tag{8.207}$$

By setting the coefficients of $s^0$ and $s^1$ equal to zero in (8.207), the equations for $N^{(1)}$ and $N^{(2)}$ are obtained; in fact, by anticipating that $e_{\alpha\beta}^{(1)}$ and $N_{\alpha\beta}^{(1)}$ vanish it follows from (8.207) that

$$N^{(1)}\int_{-a}^{a}\int_{-b}^{b}\delta_{\alpha\beta}w_{,\alpha}^{(1)}w_{,\beta}^{(1)}\,dx\,dy$$

$$= \int_{-a}^{a}\int_{-b}^{b}(M_{\alpha\beta}^{(2)}\kappa_{\alpha\beta}^{(2)} - M_{\alpha\beta}^{(1)}\kappa_{\alpha\beta}^{(2)})\,dx\,dy \tag{8.208}$$

It is shown in [263], that, as a consequence of (8.208), $N^{(1)} = 0$, in which case it also follows that $N_T^{(1)} = 0$. From (8.207) one then obtains

$$N^{(2)}\int_{-a}^{a}\int_{-b}^{b}\delta_{\alpha\beta}w_{\alpha}^{(1)}w_{\beta}^{(1)}\,dx\,dy + \int_{-a}^{a}\int_{-b}^{b}\left\{N_{\alpha\beta}^{(2)}w_{\alpha}^{(1)}w_{\beta}^{(1)}\right.$$

$$\frac{M_{\alpha\beta}^{(1)}}{3Eh}[D_{\alpha\beta\gamma\delta}^{(2)}M_{\gamma\delta}^{(1)} + 2B_{\alpha\beta\gamma\delta}^{(1)}(N_{\gamma\delta}^{(2)}\delta_{\gamma\delta})]\Big\} = 0 \tag{8.209}$$

and, thus, the curvature term $N^{(2)}$, and hence $N_T^{(2)}$ can be computed without having to use the third-order equations.

The critical temperature $(N_T)_{cr}$, at which the rectangular plate (which is heated so that $N_T = $ const. and $M_T = 0$) will buckle, may now be determined along with the form of the dependence of the buckling temperature on the hardening parameter $G$; such determinations are limited in [263] to those cases for which an analytical solution exists, i.e., the fully simply supported plate, the clamped simply supported plate, and the long strip. In all cases, we assume that the in-plane displacements vanish on the boundaries.

We begin by noting that the first pair of constitutive relations in (8.196) may be rewritten in the form

$$N_{11}^{(1)} = \frac{Eh/(1+\nu)}{1-\nu+2g^*}\left[(1+g^*)\frac{\partial u^{(1)}}{\partial x} + (\nu - g^*)\frac{\partial v^{(1)}}{\partial y}\right]$$

$$N_{22}^{(1)} = \frac{Eh/(1+\nu)}{1-\nu+2g^*}\left[(\nu - g^*)\frac{\partial u^{(1)}}{\partial x} + (1+g^*)\frac{\partial v^{(1)}}{\partial y}\right] \qquad (8.210)$$

$$N_{12}^{(1)} = \frac{Eh/2}{1+\nu}\left(\frac{\partial u^{(1)}}{\partial y} + \frac{\partial v^{(1)}}{\partial x}\right)$$

and if these relations are then substituted into the in-plane equations of equilibrium it can be shown that both $u^{(1)}$ and $v^{(1)}$ satisfy the biharmonic equation; in conjunction with homogeneous boundary data it may then be concluded that $u^{(1)}, v^{(1)}$, and hence, $e_{\alpha\beta}^{(1)}$ in general vanish. As a consequence of (8.210), it then follows that the $N_{\alpha\beta}^{(1)}$ also vanish; thus, the conditions hold which validate (8.208), in which case, $N^{(1)} = 0$. Using the equation governing transverse equilibrium, and the remaining constitutive relations in (8.196), it follows that $w^{(1)}$ is the nonzero solution of

$$\nabla^2\left(\nabla^2 w^{(1)} - \frac{12}{a^2}\bar{N}_{cr}\left\{\frac{1-\nu+2g^*}{1+g^*}\right\}w^{(1)}\right) = 0 \qquad (8.211)$$

where

$$M_{11}^{(1)} = \frac{-Eh^3/[12(1+\nu)]}{1-\nu+2g^*}[(1+g^*)w_{11}^{(1)} + (1+g^*)w_{22}^{(1)}]$$

$$M_{22}^{(1)} = \frac{-Eh^3/[12(1+\nu)]}{1-\nu+2g^*}[(\nu - g^*)w_{11}^{(1)} + (1+g^*)w_{22}^{(1)}] \qquad (8.212)$$

$$M_{12}^{(1)} = \frac{-Eh^3/12}{1+\nu}w_{12}^{(1)}$$

and we have defined

$$
\begin{cases}
(N_{cr}, (N_T)_{cr}) = \dfrac{Eh^3/a^3}{1+\nu}(\bar{N}_{cr}, (\bar{N}_T)_{cr}) \\[4mm]
g^* = \bar{G}\left(\dfrac{\bar{N}_{cr}}{1+\nu}\right)^2 \\[4mm]
\bar{G} = \dfrac{1}{9}EG\left(\dfrac{Eh^2}{a^2}\right)^2
\end{cases}
\qquad (8.213)
$$

We note that there exists an infinite number of eigenvalues $\lambda_n$ of (8.211) subject to appropriate boundary conditions, such that

$$
\lambda_n^2 = -12\bar{N}_{cr}\left\{\frac{1-\nu+g^*}{1+g^*}\right\} \qquad (8.214)
$$

We also note that as $g^*$ depends on $N_{cr}$ though (8.213), $N_{cr}$ is a solution of the cubic equation

$$
\bar{N}_{cr}^3 + \frac{\lambda_n^2}{24}\bar{N}_{cr}^2 + \frac{(1+\nu)^2}{2\bar{G}}\left\{(1-\nu)\bar{N}_{cr} + \frac{\lambda_n^2}{12}\right\} = 0 \qquad (8.215)
$$

Thus, the minimum buckling stress $\bar{N}_{cr}$ can be determined only after one examines the roots of (8.215), which correspond to the previously determined eigenvalues $\lambda_n$. As Williams [263] notes, the limiting case $\bar{G} = 0$ yields some insight into the dependence $(\bar{N}_T)_{cr}(G)$; if we consider $\lambda_n$ as being known, then (8.215) determines the functional relationship $\bar{N}_{cr}(G)$. Also, as

$$
(\bar{N}_{cr})\Big|_{\bar{G}=0} = -\frac{\lambda_n^2}{12(1-\nu)} \quad \text{and} \quad \frac{dg^*}{d\bar{G}}\Big|_{\bar{G}=0} = \left(\frac{\bar{N}_{cr}(0)}{1+\nu}\right)^2 \qquad (8.216)
$$

it follows, as a consequence of (8.215), that

$$
\frac{d(\bar{N}_T)_{cr}}{d\bar{G}}\Big|_{\bar{G}=0} = \frac{1+3\nu}{(1+\nu)^2} \cdot \frac{(\bar{N}_{cr})^3(0)}{3} \qquad (8.217)
$$

As $(\bar{N}_{cr})(0) < 0$ the effect of the hardening parameter is always to reduce the buckling temperature and the absolute value of the stress from those values which apply for a linearly elastic material in a small neighborhood of $G = 0$.

The solution of (8.211) which is the easiest to secure is the one which corresponds to the case in which the plate is simply supported along all its edges.

Taking

$$w^{(1)} = A_{mn} \sin\left[\frac{n\pi}{2a}(x+a)\right] \sin\left[\frac{m\pi}{2b}(y+b)\right] \qquad (8.218)$$

it follows that the boundary conditions along the edges $x = \pm a$, $y = \pm b$ are identically satisfied and that the eigenvalues are

$$\lambda_{mn}^2 = \frac{\pi^2}{4}(n^2 + m^2\phi^2); \quad \phi = a/b \qquad (8.219)$$

For $\nu = 0.3$, numerical results for the buckling stress and temperatures are reported by Williams in [263] and are summarized in tables 8.2 and 8.3. For all the cases shown, the minimum values of stress and temperature correspond to $m = n = 1$, i.e., a mode shape with one wave in each direction. The effect of the hardening parameter is to reduce the stress and temperature from values which would correspond to $G = 0$; this effect is more pronounced, as noted in [263], for long slender plates. Also, the effect of $G$ is not monotonic: the buckling temperature first decreases with increasing $G$ and then increases.

For the case in which the edges at $x = \pm a$ are simply-supported, while those at $y = \pm b$ are clamped, we may take

$$w^{(1)} = W(\bar{y}) \sin\left[\frac{n\pi}{2a}(x+a)\right]; \quad \bar{y} = \frac{y}{h} \qquad (8.220)$$

so that the boundary conditions along $x = \pm a$ are satisfied identically, while $w(\bar{y})$ must satisfy

$$W'''' + (\beta^2 - \bar{n}^2)W'' - \bar{n}^2\beta^2 W = 0 \qquad (8.221)$$

with

$$\bar{n} = \frac{n\pi b}{2a}; \quad \bar{n}^2 + \beta^2 = \left(\frac{\lambda n}{\phi}\right)^2 \qquad (8.222)$$

The solution of (8.221) has the form

$$W = A_1 \sin h\bar{n}\bar{y} + A_2 \cos h\bar{n}\bar{y} + A_3 \sin \beta\bar{y} + A_4 \cos \beta\bar{y} \qquad (8.223)$$

From the clamped conditions along $\bar{y} = \pm 1$, we infer that either the buckling mode is even ($A_1 = A_3 = 0$) with $\beta$ satisfying

$$\beta \tan \beta = -\bar{n} \tan h\bar{n} \qquad (8.224)$$

or the buckling mode is odd ($A_2 = A_4 = 0$) with

$$\frac{\tan \beta}{\beta} = \frac{\tan h\bar{n}}{\bar{n}} \qquad (8.225)$$

Therefore to determine the eigenvalues $\lambda_n$ the buckling stress and temperature, (8.224) or (8.225) must be solved for a fixed $\bar{n}$, for $\beta(\bar{n})$. Numerical results for this clamped simply supported case are delineated in tables 8.4 and 8.5, for $\nu = 0.3$; in all the cases exhibited, the minimum values of the buckling temperature and stress correspond to the even mode case ($A_1 = A_3 = 0$ in (8.223)) and thus to the smallest root $\beta$ of (8.224). Once again, the general effect of $G$ is to reduce the buckling stress from those values associated with $G = 0$, but the reduction in the buckling temperature due to $G$ appears to occur in higher-aspect ratio plates only and only in a small neighborhood of $G = 0$.

Williams [263] also treats the initial buckling of a long strip undergoing both cylindrical and non-cylindrical deformations; for the case of a strip with simply supported edges the deflection has the form

$$w^{(1)} = A_s \sin \frac{\lambda x}{a} + B_s \cos \frac{\lambda x}{a} \tag{8.226}$$

where $2a$ is the width of the strip, while for a strip with clamped edges at $x = \pm a$

$$w^{(1)} = A_c \sin \frac{\lambda x}{a} + B_c \cos \frac{\lambda x}{a} + \frac{a^2}{\lambda^2}(c_1 x + c_2) \tag{8.227}$$

In the simply supported case we have

$$\text{or} \quad \begin{matrix} B_s \neq 0, \ A_s = 0; \ \lambda = \dfrac{\pi}{2}, \ \dfrac{3\pi}{2}, \ ... \\ A_s \neq 0, \ B_s = 0, \ \lambda = \pi, \ 2\pi, \ ... \end{matrix} \tag{8.228}$$

while in the clamped case we have

$$\text{or} \quad \begin{matrix} B_c \neq 0, \ A_c = 0; \ \lambda = \pi, \ 2\pi, \ ... \\ A_c \neq 0, \ B_c = 0; \ \tan \lambda = \lambda (\lambda = 4.4934, \ 7.7252, \ ...) \end{matrix} \tag{8.229}$$

Numerical results for these two situations are delineated in tables 8.6 and 8.7 with $\nu = 0.3$; for all the cases depicted, the minimum buckling stress and temperature are those which are associated with an even mode. The effect of $G$ is to reduce both the buckling temperature and stress with the effect now monotonic over the range of $G$ considered.

We now want to briefly summarize some of the results obtained in [263] relative to the postbuckling behavior of the plate and the nature of the dependence of that behavior on the hardening parameter $G$. The first parameter which arises in a postbuckling calculation is $w^{(2)}(s)$, which is the initial curvature of the transverse deflection as measured along the loading curve that initiates at the critical point. As $N^{(1)} = 0$, it follows from the equilibrium and constitutive equations that $w^{(2)}$ satisfies the same equation as $w^{(1)}$ and thus, by virtue of the orthogonality relation (8.204), $w^{(1)} = 0$. The next parameter which must be computed is $N^{(2)}$. For the specific case of cylindrical deformation of a long strip, the second-order constitutive equations assume the form

$$Eh(u_1^{(2)} + \frac{1}{2}w_1^{(1)2}) = (1+g^*)N_{11}^{(2)} + (-\nu+g^*)N_{22}^{(2)}$$
$$+ \frac{6g^*(4M_{11}^{(1)2} - M_{11}^{(1)}M_{22}^{(1)} + M_{22}^{(1)2})}{N_{cr}h^2}$$
$$0 = (1+g^*)N_{22}^{(2)} + (-\nu+g^*)N_{11}^{(2)} + \frac{6g^*(4M_{22}^{(1)2} - M_{11}^{(1)}M_{22}^{(1)} + M_{11}^{(1)2})}{N_{cr}h^2}$$
$$(8.230)$$

However,

$$M_{22}^{(1)} = \frac{\nu - g^*}{1+g^*}M_{11}^{(1)}, \quad M_{11}^{(1)} = -\frac{Eh^3}{12(1+\nu)}\frac{1+g^*}{1-\nu+2g^*}w_{11}^{(1)} \quad (8.231)$$

so that

$$Eh(1+g^*)(u_1^{(2)} + \frac{1}{2}w_1^{(1)2}) = (1+\nu)(1-\nu+2g^*)N_{11}^{(2)}$$
$$+ \frac{24(1+\nu)g^*}{N_{cr}h^2(1+g^*)^2}[(1-\nu)(1+3g^*)+\nu^2+3g^*]M_{11}^{(1)2} \quad (8.232)$$

while the second-order equilibrium equation (8.188) requires that $N_{11}^{(2)} =$ const. The form of $N_{11}^{(2)}$ can be obtained by integrating (8.232) over $-a \leq x \leq a$ and using the fact that $u^{(2)}(\pm a) = 0$. In both the clamped and simply supported cases one obtains (with $B_i = B_s$ or $B_c$)

$$N_{11}^{(2)} = -\frac{3N_{cr}B_i^2}{h^2}\left[1 + \frac{4g^*}{(1+g^*)^2}\frac{(1-\nu)(1+3g^*)+\nu^2+3g^{*2}}{1-\nu+2g^*}\right]$$
$$(8.233)$$

The form of $N^{(2)}$ may now be computed by employing (8.209); numerical results for both $N^{(2)}$ and $N_{11}^{(2)}$ (with $\nu = 0.3$) are presented in Tables 8.8 and 8.9; Table 8.8 covers the case of a clamped strip , while Table 8.9 addresses the same situation for a simply supported strip. As

$$\frac{N_{11}}{N_{cr}} = 1 + \frac{s^2(N_{11}^{(2)} + N^{(2)})}{N_{cr}} + \mathcal{O}(s^3) \quad (8.234)$$

and $N_{11}^{(2)} + N^{(2)} = \mathcal{O}(g^*) > 0$, the stress resultant $N_{11}$ decreases in absolute value after buckling as a consequence of hardening; moreover, the effect of hardening is to increase the rate of decrease of $N_{11}$. We indicate in Figs. 8.14 and 8.15, respectively, some numerical results for the second-order stress parameter $N_{22}^{(2)}$ in the simply supported and clamped cases; in contrast to $N_{11}$ we have $N_{22}^{(2)} + N^{(2)} < 0$ so that the stress resultant $N_{22}$ decreases (actually increases in absolute value) after buckling in a manner which is independent of hardening. As noted in [263], at $x = 0$, the centerline of the plate, the effect of hardening is to effect a small decrease in the rate of decrease of $N_{22}^{(2)}$.

We have presented, above, the equations used in [263] to study the buckling of elastoplastic rectangular plates; the model chosen leads to a modified form of von Karman's equations in which the constitutive relations are generalized so as to account for the flow rule. It has been shown that, in for clamped plates and for simply supported plates in which there is no thermal moment, i.e., $M_T = 0$, buckling may occur at temperatures that are substantially below those which are predicted by an elastic analysis; in fact, as the effect of $g^*$ is to reduce the factors which multiply the curvature terms in our equations, the elastoplastic equations essentially describe a classic plate with reduced stiffness. Thus, as Williams [263] explicitly notes, it is not suprising that the equations in question predict a general reduction in the temperature which is necessary to cause buckling. The predictions in [263] are, in general, (qualitative) agreement with the results given by Gerard and Becker [274] for mechanically loaded plates; for example, for long plates which are either clamped or simply supported, respectively, on the unloaded sides, the plasticity reduction factor $\eta$, which is the ratio of the critical buckling loads in the elastic and plastic cases, is given by

$$\eta = \frac{\sigma_{cr}^p}{\sigma_{cr}^e} = f\left[(0.352, 0.500) + (0.324, 0.250)\sqrt{1 + \frac{3E_t}{E_s}}\right] \qquad (8.235)$$

In (8.236),

$$f = \frac{E_s(1 - \nu_e^2)}{E(1 - \nu^2)}, \quad \nu = \frac{1}{2} - \left(\frac{1}{2} - \nu_e\right)\frac{E_s}{E} \qquad (8.236)$$

where $\nu_e$, $E_s$, and $E_t$ are the elastic Poisson's ratio, the secant modulus, and the tangent modulus of the maternal, respectively. For the

Ramberg-Osgood form (8.170)

$$
\begin{cases}
\dfrac{E}{E_t} = 1 + \dfrac{3E}{E_0}\left(\dfrac{P_{11}}{E_0}\right)^2 \equiv 1 + \dfrac{9EJ_2}{E_0^3} \\[3mm]
\dfrac{E}{E_s} = 1 + \dfrac{E}{E_0}\left(\dfrac{P_{11}}{E_0}\right)^2 \equiv 1 + \dfrac{3EJ_2}{E_0^3}
\end{cases}
\tag{8.237}
$$

where $E_t = E_t(J_2)$ and $J_2 = \dfrac{1}{3}P_{11}^2$. However, as $P_{11} = P_{22} = N_{cr}/h$ and $P_{12} = 0$, at the critical point, one obtains from (8.237)

$$
\dfrac{E}{E_t} = 1 + 4g^*, \quad \dfrac{E}{E_s} = 1 + \dfrac{4g^*}{3}
\tag{8.238}
$$

A comparison of the results in [276] and [263] is presented in Table 8.10; in this table, the predictions associated with a long $\left(\phi = \dfrac{a}{b} = 3.5\right)$, (completely) simply supported plate appear in the first column, while those in the second column correspond to the case of a clamped strip. As noted by Williams [263], the general behavior indicated in Table 8.10 is the same for both sets of boundary conditions. Concerning the postbuckling analysis, it should be noted that the condition of loading ($\dot{J}_2 > 0$) is generally violated once buckling has occurred. In fact, to first order, the plane stress form for $J_2$ is given by

$$
J_2 = \dfrac{1}{3}(P_{11}^2 + P_{22}^2 - P_{11}P_{22}) + P_{12}^2
\tag{8.239}
$$

Thus, as

$$
\dot{J}_2 = \left\{\dfrac{1}{3}\left[\dfrac{\partial P_{11}}{\partial s}(2P_{11} - P_{22}) + \dfrac{\partial P_{22}}{\partial s}(2P_{22} - P_{11})\right] + 2P_{12}\dfrac{\partial P_{12}}{\partial s}\right\}\dfrac{ds}{dt},
\tag{8.240}
$$

$$
\begin{cases}
\dfrac{\partial P_{ij}}{\partial s} = \dfrac{12y}{h^3}M_{ij}^{(1)} + \mathcal{O}(s) \\[3mm]
(2P_{11} - P_{22})(2P_{22} - P_{11}) = \dfrac{N_{cr}}{h} + \mathcal{O}(s),
\end{cases}
\tag{8.241}
$$

and

$$
M_{11}^{(1)} + M_{22}^{(1)} = -\dfrac{Eh^3}{12}\left\{\dfrac{w_{11}^{(1)} + w_{22}^{(1)}}{1 - \nu + 2g^*}\right\}
\tag{8.242}
$$

it follows that

$$
\dot{J}_2 = \left\{-\dfrac{EyN_{cr}/3h}{1 - \nu + 2g^*}(w_{11}^{(1)} + w_{22}^{(1)}) + \mathcal{O}(s)\right\}\dot{s}
\tag{8.243}
$$

If the first curvature is positive, i.e., $w_{11}^{(1)} + w_{22}^{(1)} > 0$ then, because $N_{cr} < 0$, there will be unloading on the $y < 0$ (negative) side of the plate; as indicated in [263], this result is consistent with the expectation that the bending moments due to buckling will induce tension in the domain $y < 0$ and thus produce unloading.

We close out this section by delineating the explicit structure of the hardening matrix which appears in (8.179); the $B_{ij}$ are given by

$$G_{11} = (2N_{11} - N_{22})^2 + \frac{12(2M_{11} - M_{22})^2}{h^2}$$

$$G_{12} = (2N_{11} - N_{22})(2N_{22} - N_{11})$$
$$+ \frac{12(2M_{11} - M_{22})(2M_{22} - M_{11})}{h^2}$$

$$G_{13} = 6\left[N_{12}(2N_{11} - N_{22}) + \frac{12M_{12}(2M_{11} - M_{22})}{h^2}\right]$$

$$G_{14} = \frac{24(2N_{11} - N_{22})(2M_{11} - M_{22})}{h^2}$$

$$G_{15} = \frac{12[5(N_{11}M_{22} + N_{22}M_{11}) - 4(N_{11}M_{11} + N_{22}M_{22})]}{h^2} = G_{24}$$

$$G_{16} = \frac{72[N_{12}(2M_{11} - M_{22}) + M_{12}(2N_{11} - N_{22})]}{h^2} = G_{34}$$

$$G_{22} = (2N_{22} - N_{11})^2 + \frac{12(2M_{22} - M_{11})^2}{h^2}$$

$$G_{23} = 6\left[N_{12}(2N_{22} - N_{11}) + \frac{12M_{12}(2M_{22} - M_{11})}{h^2}\right]$$

$$G_{25} = \frac{24(2N_{22} - N_{11})(2M_{22} - M_{11})}{h^2}$$

$$G_{26} = \frac{72[N_{12}(2M_{22} - M_{11}) + M_{12}(2N_{22} - N_{11})]}{h^2} = G_{35}$$

$$G_{33} = 36\left(N_{12}^2 + \frac{12M_{12}^2}{h^2}\right), \quad G_{36} = \frac{864N_{12}M_{12}}{h^2}$$

$$G_{44} = \frac{12[(2N_{11} - N_{22})^2 + 108(2M_{11} - M_{22})^2/5h^2}{h^2}$$

$$(8.244)$$

$$G_{45} = \frac{12}{h^2} bigg[(2N_{11} - N_{22})(2N_{22} - N_{11})$$

$$+ \frac{108}{5h^2}(2M_{11} - M_{22})(2M_{22} - M_{11})\Big]$$

$$G_{46} = \frac{72[N_{12}(2N_{11} - N_{22}) + 108M_{12}(2M_{11} - M_{22})/5h^2]}{h^2}$$

$$G_{55} = \frac{12[(2N_{22} - N_{11})^2 + 108(2M_{22} - M_{11})^2/5h^2}{h^2}$$

$$G_{56} = \frac{72[N_{12}(2N_{22} - N_{11}) + 108M_{12}(2M_{22} - M_{11})/5h^2]}{h^2}$$

$$G_{66} = \frac{432(N_{12}^2 + 108M_{12}^2/5h^2)}{h^2}$$

It should be noted that, in evaluating the integrals leading to the constitutive equations in [263], the author treats $E$ and $G$ as constants. Hence, it is not possible to account for a temperature dependency of $E$ and $G$ without considerable additional complication.

**Remarks:** It may be convenient, as indicated in [263], to express the constitutive equation in terms of the symmetric $A, B$, and $D$ matrices that are defined as follows:

$$A = \begin{bmatrix} 1 + HG_{11} & -\nu + HG_{12} & \dfrac{HG_{13}}{2} & \dfrac{HG_{13}}{2} \\ & 1 + HG_{22} & \dfrac{HG_{23}}{2} & \dfrac{HG_{23}}{2} \\ & & \dfrac{1 + \nu + (HG_{33})/2}{2} & \dfrac{1 + \nu + (HG_{33})/2}{2} \\ (sym) & & & \dfrac{1 + \nu + (HG_{33})/2}{2} \end{bmatrix}$$

(8.245)

$$B = \begin{bmatrix} G_{14} & G_{15} & \dfrac{G_{16}}{2} & \dfrac{G_{16}}{2} \\[2mm] & G_{25} & \dfrac{G_{26}}{2} & \dfrac{G_{26}}{2} \\[2mm] & & \dfrac{G_{36}}{4} & \dfrac{G_{36}}{4} \\[2mm] (sym) & & & \dfrac{G_{36}}{4} \end{bmatrix} \tag{8.246}$$

$$D = \begin{bmatrix} \dfrac{12}{h^2} + HG_{44} & -\dfrac{12\nu}{h^2} + HG_{45} & \dfrac{HG_{46}}{2} & \dfrac{HG_{46}}{2} \\[2mm] & \dfrac{12}{h^2} + HG_{55} & \dfrac{HG_{56}}{2} & \dfrac{HG_{56}}{2} \\[2mm] & & 6\dfrac{1+\nu}{h^2} + \dfrac{HG_{46}}{4} & 6\dfrac{1+\nu}{h^2} + \dfrac{HG_{66}}{4} \\[2mm] (sym) & & & 6\dfrac{1+\nu}{h^2} + \dfrac{HG_{66}}{4} \end{bmatrix} \tag{8.247}$$

where

$$H = \frac{GE}{9h^2} \tag{8.248}$$

Further, in the neighborhood of a critical point, these matrices can be expanded as

$$\begin{aligned} A &= A^{(0)} + sA^{(1)} + s^2 A^{(2)} + \dots \\ B &= sB^{(1)} + s^2 B^{(2)} + \dots \\ D &= D^{(0)} + sD^{(1)} + s^2 D^{(2)} + \dots \end{aligned} \tag{8.249}$$

where

$$A^{(0)} = \begin{bmatrix} 1+g^* & -\nu+g^* & 0 & 0 \\[2mm] & 1+g^* & 0 & 0 \\[2mm] & & \dfrac{1+\nu}{2} & \dfrac{1+\nu}{2} \\[2mm] (sym) & & & \dfrac{1+\nu}{2} \end{bmatrix} \tag{8.250}$$

$$A^{(1)} = H N_{cr} \begin{bmatrix} 2(2N_{11}^{(1)} - N_{22}^{(1)} + N^{(1)}) & (N_{11}^{(1)} + N_{22}^{(1)} + 2N^{(1)}) & 3N_{12}^{(1)} & 3N_{12}^{(1)} \\ & 2(2N_{22}^{(1)} - N_{11}^{(1)} + N^{(1)}) & 3N_{12}^{(1)} & 3N_{12}^{(1)} \\ & & 0 & 0 \\ (sym) & & & 0 \end{bmatrix}$$

(8.251)

$$B^{(1)} = \frac{12 H N_{cr}}{h^2} \begin{bmatrix} 2(2M_{11}^{(1)} - M_{22}^{(1)}) & (M_{11}^{(1)} + M_{22}^{(1)}) & 3M_{12}^{(1)} & 3M_{12}^{(1)} \\ & 2(2M_{22}^{(1)} - M_{11}^{(1)}) & 3M_{12}^{(1)} & 3M_{12}^{(1)} \\ & & 0 & 0 \\ (sym) & & & 0 \end{bmatrix}$$

(8.252)

and

$$(D^{(0)}, D^{(1)}) = \frac{12}{h^2} \cdot (A^{(0)}, A^{(1)})$$

(8.253)

As noted in [263] the general form of the matrix $D^{(2)}$ is unnecessarily cumbersome; thus one may anticipate the result that $N_{ij}^{(1)} = N^{(1)} = 0$ and define $D^{(2)}$ as

$$D^{(2)} = \frac{12H}{h^2} \cdot (D_1^{(2)} + D_2^{(2)} + D_3^{(2)})$$

(8.254)

where

$$D_1^{(2)} = \frac{108}{5h^2} \begin{bmatrix} (2M_{11}^{(1)} - M_{22}^{(1)})^2 & (2M_{11}^{(1)} - M_{22}^{(1)})(2M_{22}^{(1)} - M_{11}^{(1)}) & 3M_{12}^{(1)}(2M_{11}^{(1)} - M_{22}^{(1)}) & 3M_{12}^{(1)}(2M_{11}^{(1)} - M_{22}^{(1)}) \\ & (2M_{22}^{(1)} - M_{11}^{(1)})^2 & 3M_{12}^{(1)}(2M_{22}^{(1)} - M_{11}^{(1)}) & 3M_{12}^{(1)}(2M_{22}^{(1)} - M_{11}^{(1)}) \\ & & 9M_{12}^{(1)2} & 9M_{12}^{(1)2} \\ (sym) & & & 9M_{12}^{(1)2} \end{bmatrix}$$

(8.255)

$$D_2^{(2)} = (N)_{cr} \begin{bmatrix} 2(2N_4^{(1)} - N_{22}^{(2)}) & (N_{11}^{(2)} + N_{22}^{(2)}) & 3N_{12}^{(2)} & 3N_{12}^{(2)} \\ & 2(2N_{22}^{(2)} - N_{11}^{(2)}) & 3N_{12}^{(2)} & 3N_{12}^{(2)} \\ & & 0 & 0 \\ (sym) & & & 0 \end{bmatrix}$$

$$(8.256)$$

and

$$D_3^{(2)} = 2(N)_{cr} N^{(2)} \begin{bmatrix} 1 & 1 & 0 & 0 \\ & 1 & 0 & 0 \\ & & 0 & 0 \\ (sym) & & & 0 \end{bmatrix}$$

$$(8.257)$$

## 8.4    Thermal Bending, Buckling, and Postbuckling Figures, Graphs, and Tables II

Table 8.1   Upper and Lower Critical Temperatures for Viscoelastic Plates [242] (Reprinted from the *Int. J. Solids and Structures*, 16, Jones, R., Mazumdar, J., and Y.K. Cheung, Vibration and buckling of plates at elevated temperatures, 61-70, 1980, with permission from Elsevier Science)

| Material | Lower critical temp. | Upper critical temp. |
|---|---|---|
| Elastic | Standard elastic critical temp. | |
| Maxwell | 0 | Elastic critical temp. |
| Kelvin | Elastic critical temp. | $\infty$ |

Table 8.2 Buckling Temperature for Simply Supported Plates ($\nu = 0.3$) [263] (Copyright 1982. From Neale: Plate equations and applications to buckling of rectangular plates, *J. of Thermal Stresses*, **5**, 207-243, by Williams, H.E. Reproduced by permission of Taylor & Francis, Inc./Routledge, Inc., http://www.routledge-ny.com)

| | $(\bar{N}_T)_{cr}$ | | | |
| | $\bar{G}$ | | | |
| $a/b$ | 0 | 0.1 | 0.2 | 0.3 |
|---|---|---|---|---|
| 0.5 | 0.2570 | 0.2553 | 0.2537 | 0.2523 |
| 1.0 | 0.4112 | 0.4046 | 0.3995 | 0.3954 |
| 1.5 | 0.6683 | 0.6448 | 0.6313 | 0.6225 |
| 2.0 | 1.028 | 0.9657 | 0.9428 | 0.9320 |
| 2.5 | 1.491 | 1.365 | 1.343 | 1.342 |
| 3.0 | 2.056 | 1.854 | 1.857 | 1.890 |
| 3.5 | 2.724 | 2.456 | 2.531 | 2.632 |

Table 8.3 Buckling Stress for Simply Supported Plates
($\nu = 0.3$) [263] (Copyright 1982. From Neale: Plate equations and applications to buckling of rectangular plates, *J. of Thermal Stresses*, 5, 207-243, by Williams, H.E. Reproduced by permission of Taylor & Francis, Inc./Routledge, Inc., http://www.routledge-ny.com)

| $a/b$ | $-\bar{N}_{cr}$ | | | |
| | $\bar{G}$ | | | |
| | 0 | 0.1 | 0.2 | 0.3 |
|---|---|---|---|---|
| 0.5 | 0.3672 | 0.3620 | 0.3573 | 0.3530 |
| 1.0 | 0.5875 | 0.5677 | 0.5518 | 0.5385 |
| 1.5 | 0.9546 | 0.8825 | 0.8360 | 0.8020 |
| 2.0 | 1.469 | 1.265 | 1.168 | 1.104 |
| 2.5 | 2.130 | 1.682 | 1.522 | 1.427 |
| 3.0 | 2.937 | 2.115 | 1.892 | 1.767 |
| 3.5 | 3.892 | 2.562 | 2.280 | 2.129 |

Table 8.4   Buckling Temperature for Mixed Boundary Condition Plates ($\nu = 0.3$) [263] (Copyright 1982. From Neale: Plate equations and applications to buckling of rectangular plates, *J. of Thermal Stresses*, 5, 207-243, by Williams, H.E. Reproduced by permission of Taylor & Francis, Inc./Routledge, Inc., http://www.routledge-ny.com)

| | $(\bar{N}_T)_{cr}$ | | | |
| | $\bar{G}$ | | | |
| $a/b$ | 0 | 0.1 | 0.2 | 0.3 |
|---|---|---|---|---|
| 0.5 | 0.3046(1) | 0.3017(1) | 0.2993(1) | 0.2971(1) |
| 1.0 | 0.7876(1) | 0.7529(1) | 0.7356(1) | 0.7253(1) |
| 1.5 | 1.743(1) | 1.582(1) | 1.569(1) | 1.579(1) |
| 2.0 | 3.146(1) | 2.858(1) | 2.999(1) | 3.166(1) |
| 2.5 | 4.831(2) | 4.696(2) | 5.314(2) | 5.931(2) |
| 3.0 | 6.973(2) | 7.744(2) | 9.583(2) | 11.37(2) |
| 3.5 | 9.491(2) | 12.68(3) | 17.17(3) | 21.54(3) |

Table 8.5 Buckling Stress for Mixed Boundary Condition Plates ($\nu = 0.3$) [263] (Copyright 1982. From Neale: Plate equations and applications to buckling of rectangular plates, *J. of Thermal Stresses*, 5, 207-243, by Williams, H.E. Reproduced by permission of Taylor & Francis, Inc./Routledge, Inc., http://www.routledge-ny.com)

| $a/b$ | $-\bar{N}_{cr}$ | | | |
| | $\bar{G}$ | | | |
| | 0 | 0.1 | 0.2 | 0.3 |
|---|---|---|---|---|
| 0.5 | 0.4351(1) | 0.4266(1) | 0.4192(1) | 0.4126(1) |
| 1.0 | 1.125(1) | 1.016(1) | 0.9532(1) | 0.9091(1) |
| 1.5 | 2.490(1) | 1.883(1) | 1.693(1) | 1.584(1) |
| 2.0 | 4.494(1) | 2.819(1) | 2.507(1) | 2.344(1) |
| 2.5 | 6.902(2) | 3.746(2) | 3.351(2) | 3.156(2) |
| 3.0 | 9.961(2) | 4.806(2) | 4.359(2) | 4.151(2) |
| 3.5 | 13.561(3) | 5.992(3) | 5.524(3) | 5.320(3) |

Table 8.6 Buckling Stress and Temperature for the Simply Supported Strip ($\lambda = \pi/2$) [263] (Copyright 1982. From Neale: Plate equations and applications to buckling of rectangular plates, *J. of Thermal Stresses*, 5, 207-243, by Williams, H.E. Reproduced by permission of Taylor & Francis, Inc./Routledge, Inc., http://www.routledge-ny.com)

| $\bar{G}$ | $-\bar{N}_{cr}$ | $(\bar{N}_T)_{cr}$ |
|---|---|---|
| 0 | 0.2937 | 0.2056 |
| 0.1 | 0.2910 | 0.2047 |
| 0.2 | 0.2885 | 0.2039 |
| 0.3 | 0.2861 | 0.2031 |

Table 8.7 Buckling Stress and Temperature for the Clamped Strip ($\lambda = \pi$) [263] (Copyright 1982. From Neale: Plate equations and applications to buckling of rectangular plates, *J. of Thermal Stresses*, 5, 207-243, by Williams, H.E. Reproduced by permission of Taylor & Francis, Inc./Routledge, Inc., http://www.routledge-ny.com)

| $\bar{G}$ | $-\bar{N}_{cr}$ | $(\bar{N}_T)_{cr}$ |
|---|---|---|
| 0 | 1.175 | 0.8225 |
| 0.1 | 1.054 | 0.7841 |
| 0.2 | 0.9860 | 0.7658 |
| 0.3 | 0.9390 | 0.7553 |

Table 8.8   Curvature Parameters for the Clamped Strip
($\lambda = \pi, \nu = 0.3$) [263] (Copyright 1982. From Neale: Plate
equations and applications to buckling of rectangular plates,
*J. of Thermal Stresses*, 5, 207-243, by Williams, H.E. Reproduced by permission of Taylor & Francis, Inc./Routledge, Inc.,
http://www.routledge-ny.com)

| $\bar{G}$ | $-\bar{N}_{cr}$ | $N^{(2)}/N_{cr}$ | $-N_{11}^{(2)}/N_{cr}$ | $-\dfrac{N_{11}^{(2)} + N^{(2)}}{N_{cr}}$ |
|---|---|---|---|---|
| 0.1 | 1.054 | 3.004 | 3.786 | 0.7822 |
| 0.2 | 0.9860 | 3.069 | 4.279 | 1.2103 |
| 0.3 | 0.9390 | 3.120 | 4.653 | 1.5327 |

Table 8.9   Curvature Parameters for the Simply Supported
Strip $\lambda = \pi/2, \nu = 0.3$) [263] (Copyright 1982.   From
Neale: Plate equations and applications to buckling of rect-
angular plates, *J. of Thermal Stresses*, 5, 207-243, by
Williams, H.E. Reproduced by permission of Taylor & Francis,
Inc./Routledge, Inc., http://www.routledge-ny.com)

| $\bar{G}$ | $-\bar{N}_{cr}$ | $N^{(2)}/N_{cr}$ | $-N_{11}^{(2)}/N_{cr}$ | $-\dfrac{N_{11}^{(2)} + N^{(2)}}{N_{cr}}$ |
|---|---|---|---|---|
| 0.1 | 0.2910 | 2.991 | 3.067 | 0.0762 |
| 0.2 | 0.2885 | 2.984 | 3.131 | 0.1462 |
| 0.3 | 0.2861 | 2.980 | 3.191 | 0.2107 |

Table 8.10  Comparison with Mechanically Loaded Plates
[263] (Copyright 1982.  From Neale: Plate equations and
applications to buckling of rectangular plates, *J. of Thermal Stresses*, 5, 207-243, by Williams, H.E. Reproduced
by permission of Taylor & Francis, Inc./Routledge, Inc.,
http://www.routledge-ny.com)

| $\bar{G}$ | $\eta$ | (a) $\dfrac{N_{cr}(\bar{G})}{N_{cr}(0)}$ | $\eta(c)$ | (b) $\dfrac{N_{cr}(\bar{G})}{N_{cr}(0)}$ |
|---|---|---|---|---|
| 0.1 | 0.6383 | 0.6583 | 0.8982 | 0.8970 |
| 0.2 | 0.5342 | 0.5858 | 0.8366 | 0.8391 |
| 0.3 | 0.4701 | 0.5470 | 0.7928 | 0.7991 |

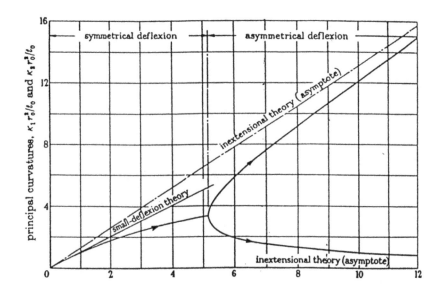

FIGURE 8.1

Variation of principal curvatures with the temperature gradient through the thickness ($k_0 = 0$). [254] . (From Mansfield, E. H., "Bending, Buckling, and Curling of a Heated Thin Plate", Proc. Roy. Soc., Ser. A, 1334, 316, 1962. Reprinted by permission of the Royal Society).

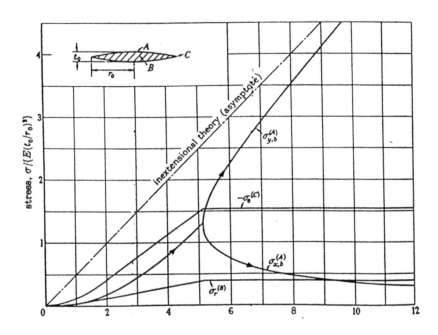

**FIGURE 8.2**
Variation of middle-surface and bending stresses with the temperature gradient through the thickness ($\hat{k}_0 = 0$) [254]. (From Mansfield, E. H., "Bending, Buckling, and Curling of a Heated Thin Plate", Proc. Roy. Soc., Ser. A, 1334, 316, 1962. Reprinted by permission of the Royal Society).

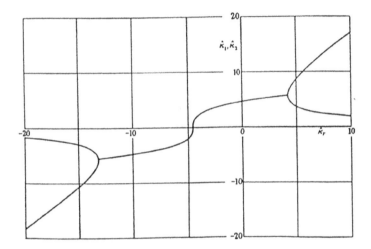

**FIGURE 8.3**
Variation of principal curvatures with the temperature gradient through the thickness ($\hat{k}_0 = \lambda^{1/2}$) [254]. (From Mansfield, E. H., "Bending, Buckling, and Curling of a Heated Thin Plate", Proc. Roy. Soc., Ser. A, 1334, 316, 1962. Reprinted by permission of the Royal Society).

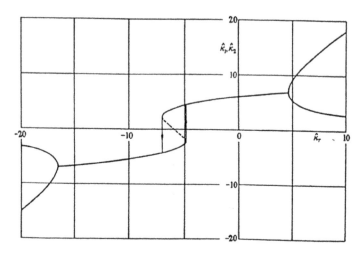

**FIGURE 8.4**
Variation of principal curvatures with the temperature gradient through the thickness ($\hat{k}_0 = 6$) [254]. (From Mansfield, E. H., "Bending, Buckling, and Curling of a Heated Thin Plate", Proc. Roy. Soc., Ser. A, 1334, 316, 1962. Reprinted by permission of the Royal Society).

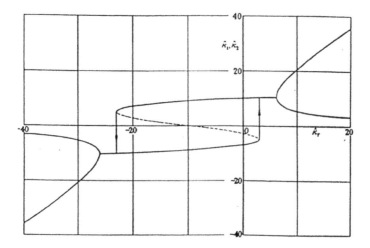

**FIGURE 8.5**
Variation of principal curvatures with the temperature gradient through the thickness ($k_0 = 10$) [254]. (From Mansfield, E. H., "Bending, Buckling, and Curling of a Heated Thin Plate", Proc. Roy. Soc., Ser. A, 1334, 316, 1962. Reprinted by permission of the Royal Society).

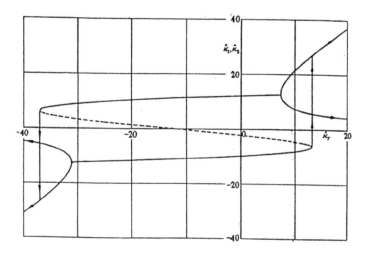

**FIGURE 8.6**
Variation of principal curvatures with the temperature gradient through the thickness ($\hat{k}_0 = 12$) [254]. (From Mansfield, E. H., "Bending, Buckling, and Curling of a Heated Thin Plate", Proc. Roy. Soc., Ser. A, 1334, 316, 1962. Reprinted by permission of the Royal Society).

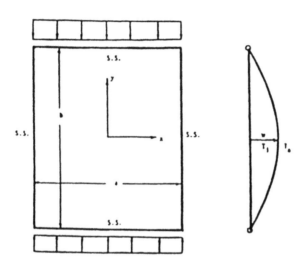

**FIGURE 8.7**
Plate dimensions, applied stresses, and deflections [261].
(Copyright 1981. From Temperature-Dependent Creep Buck-
ling of Plates, J. of Thermal Stresses, 4, 237-247, by Ross, D.A.
and Berke, L. Reproduced by permission of Taylor & Francis,
Inc./Routledge, Inc., http://www.routledge-ny.com)

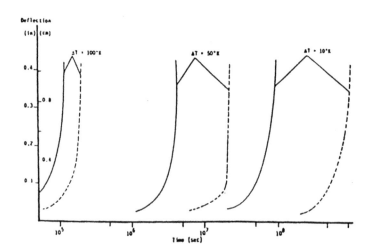

**FIGURE 8.8**
Creep deflections as a function of temperature differential and
creep exponent for an applied stress of **6.9 $M$ $Pa(1ksi)$** [261].
(Copyright 1981. From Temperature-Dependent Creep Buckling of Plates, J. of Thermal Stresses, 4, 237-247, by Ross, D.A.
and Berke, L. Reproduced by permission of Taylor & Francis,
Inc./Routledge, Inc., http://www.routledge-ny.com)

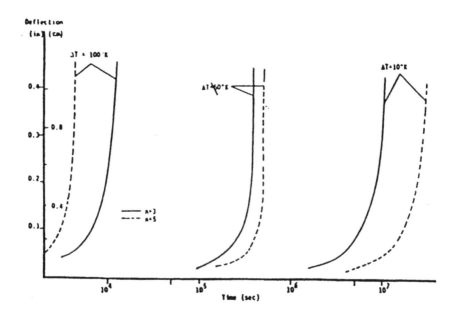

**FIGURE 8.9**
Creep deflections as a function of temperature differential and
creep exponent for an applied stress of $13.8\ M\ Pa(2ksi)$ [261].
(Copyright 1981. From Temperature-Dependent Creep Buckling of Plates, J. of Thermal Stresses, 4, 237-247, by Ross, D.A.
and Berke, L. Reproduced by permission of Taylor & Francis,
Inc./Routledge, Inc., http://www.routledge-ny.com)

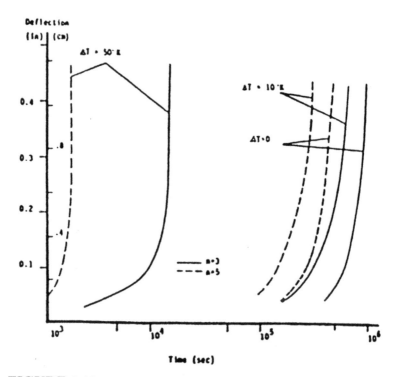

**FIGURE 8.10**
Creep deflections as a function of temperature differential and
creep exponent for an applied stress of 34.5 $M\,Pa(5ksi)$ [261].
(Copyright 1981. From Temperature-Dependent Creep Buckling of Plates, J. of Thermal Stresses, 4, 237-247, by Ross, D.A.
and Berke, L. Reproduced by permission of Taylor & Francis,
Inc./Routledge, Inc., http://www.routledge-ny.com)

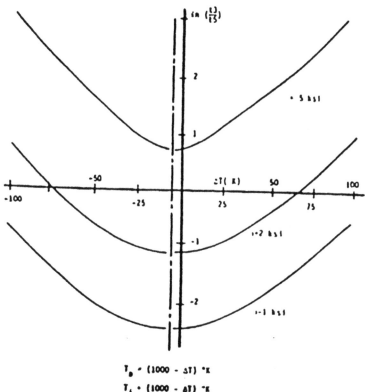

**FIGURE 8.11**
Comparison of predictions for the time until the creep deflection reaches a specified value for creep exponents $n = 3$ and $n = 5$ [261]. (Copyright 1981. From Temperature-Dependent Creep Buckling of Plates, J. of Thermal Stresses, 4, 237-247, by Ross, D.A. and Berke, L. Reproduced by permission of Taylor & Francis, Inc./Routledge, Inc., http://www.routledge-ny.com)

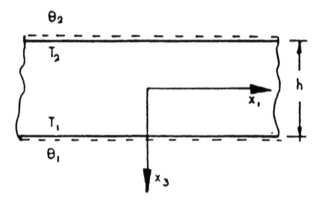

**FIGURE 8.12**
The viscoelastic plate [262]. (Reprinted from Int. J. Mech. Sci., 23, Das, S., Note on Thermal Deflection of Regular Polygonal Viscoelastic Plates, 323-329, 1981, with permission from Elsevier Science)

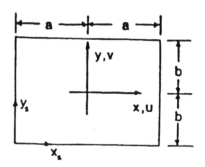

**FIGURE 8.13**
Coordinate system for the elastic-plastic plate [263]. (Copyright 1982. From Neale Plate Equations and Applications to Buckling of Rectangular Plates, J. of Thermal Stresses, 5, 207-243, by Williams, H.E. Reproduced by permission of Taylor & Francis, Inc./Routledge, Inc., http://www.routledge-ny.com)

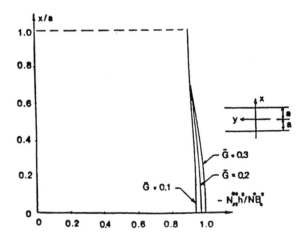

**FIGURE 8.14**
Stress factor for simply supported conditions [263]. (Copyright 1982. From Neale Plate Equations and Applications to Buckling of Rectangular Plates, J. of Thermal Stresses, 5, 207–243, by Williams, H.E. Reproduced by permission of Taylor & Francis, Inc./Routledge, Inc., http://www.routledge-ny.com)

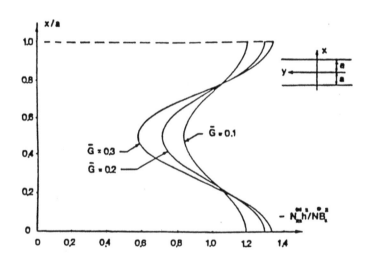

**FIGURE 8.15**
Stress factor for clamped conditions [263]. (Copyright 1982. From Neale Plate Equations and Applications to Buckling of Rectangular Plates, J. of Thermal Stresses, 5, 207-243, by Williams, H.E. Reproduced by permission of Taylor & Francis, Inc./Routledge, Inc., http://www.routledge-ny.com)

# References

[1] Euler, L., "Additamentum I de curvis elasticis, methodus inveniendi lineas curvas maximi minimiui proprietate gaudentes," Bousquent, Lausanne, in *Opera Omnia I*, 24, 1744, 231-297.

[2] von Karman, T., "Festigkeitsprobleme im Maschinenbau," in *Encylopädie der Mathematischen Wissenschaften* IV/4C, 1910, 311-385.

[3] Koiter, W.T., "Over de Stabiliteit van het Elastisch Evenwicht," Thesis, Delft: H.J. Paris, Amsterdam, 1945. English translation issued as Technical Report AFFDL-TR-70-25, Air Force Flight Dynamics Lab, Wright-Patterson Air Force Base, Ohio, 1970.

[4] Koiter, W.T., Elastic Stability, *Z.f.W.*, 9, 1985, 205-210.

[5] Koiter, W.T., "Elastic Stability and Post-Buckling Behavior," in *Nonlinear Problems*, (Ed. R. Langer), Univ. Wisconsin Press, Madison, 1963.

[6] Poincaré, H., "Sur l'equilibre d'une masses fluide animés d'un mouvement de rotation," *Acta Math*, 7, 1885, 259-380.

[7] Liapunov, A., *Probleme Général de la Stabilité du Mouvement,* Princeton University Press, 1947.

[8] Schmidt, E., "Zur Theorie der Linearen und nichtlinearen Integralgleichungen," *Math. Ann.*, 65, 1908, 370-379.

[9] Keller, J.B. and Antman, S.S., *Bifurcation Theory and Nonlinear Eigenvalue Problems*, Benjamin, N.Y., 1969.

[10] Sattinger, D.H., *Topics in Stability and Bifurcation Theory*, Lecture Notes in Math., 309, Springer-Verlag, N.Y., 1973.

[11] Iooss, G. and Joseph, D.D., *Elementary Stability and Bifurcation Theory*, UTM, Springer-Verlag, 1980.

[12] Chow, S-N. and Hale, J.K., *Methods of Bifurcation Theory*, Springer-Verlag, N.Y., 1982.

[13] Golubitsky, M. and Schaeffer, D.G., *Singularities and Groups in Bifurcation Theory, vol. 1*, Springer-Verlag, N.Y., 1984.

[14] Golubitsky, M. and Schaeffer, D.G., *Singularities and Groups in Bifurcation Theory, vol. 2*, Springer-Verlag, N.Y., 1988.

[15] Potier-Ferry, "Foundations of Elastic Postbuckling Theory," of *Lecture Notes in Physics*, vol. 288, Springer-Verlag, N.Y. 1987, 1-82.

[16] Budiansky, B., "Theory of Buckling and Post-Buckling Behavior of Elastic Structures," in *Advances in Applied Mechanics*, vol. 14, Academic Press, N.Y., 1974, 1-65.

[17] Hutchinson, J.W., "Plastic Buckling," in *Advances in Applied Mechanics*, Academic Press, N.Y., 1974, 67-145.

[18] Golubitsky, M. and Schaeffer, "A Theory for Imperfect Bifurcation via Singularity Theory," *Comm. Pure Appl. Math*, 32, 1979, 21-98.

[19] Golubitsky, M. and Schaeffer, "Imperfect Bifurcation in the Presence of Symmetry," *Comm. Math Phys.*, 67, 1979, 205-232.

[20] Thom, R., *Structural Stability and Morphogenesis,* Benjamin, Reading, Mass. 1975.

[21] Antman, S.S., "Bifurcation Problems for Nonlinearly Elastic Structures," in *Symposium on Applications of Bifurcation Theory*, Academic Press, N.Y., 1977, 73-125.

[22] Antman, S.S., *Nonlinear Problems of Elasticity*, Springer-Verlag, N.Y., 1995.

[23] Babuška, I. and Li, L., "The Problem of Plate Modeling: Theoretical and Computational Results," Comp. Meths. in *Appl. Mech. Eng.*, 100, 1992, 249-273.

[24] Batdorf, S.B., "Theories of Plastic Buckling," *J. Aeronaut. Sci.*, 16, 1949, 405-408.

[25] Bauer, L., Keller, H.B., and Reiss, E., "Multiple Eigenvalues Lend to Secondary Buckling," *SIAM Rev.*, 17, 1975, 101-122.

[26] Berger, M.S., *Nonlinearity and Functional Analysis*, Academic Press, N.Y., 1977.

[27] Bijlaard, P.P., "On the Plastic Buckling of Plates," *J. Aeronaut. Sci.*, 17, 1950, 742-743.

[28] Bolotin, V., *Dynamic Stability of Elastic Systems*, Holden-Day Pub., N.Y., 1964.

[29] Brush, D.O. and Almoth, B.O., *Buckling of Bars, Plates, and Shells*, McGraw-Hill, N.Y., 1975.

[30] Budiansky, B., "Dynamic Buckling of Elastic Structures: Criteria and Estimates," *Proc. Int. Conf. Dynamic Stability of Structures*, Northwestern Univ., Evanston, IL, 1965, 83-106.

[31] Budiansky, B. and Hutchinson, J.W., "Dynamic Buckling of Imperfection-Sensitive Structures," in *Proc. Int. Cong. Appl. Mech.*, Munich, XI, 1964, 636-651.

[32] Bulson, P.S., *The Stability of Flat Plates*, American Elsevier, N.Y., 1969.

[33] Chia, C.Y., *Nonlinear Analysis of Plates*, McGraw-Hill, N.Y., 1980.

[34] Chow, S-N., Hale, J.K., and Mallet-Paret, J., "Applications of Generic Bifurcations, I," *Arch. Rat. Mech. Anal.*, 59, 1975, 159-188.

[35] Chow, S-N., Hale, J.K., and Mallet-Paret, J., "Applications of Generic Bifurcations, II," *Arch. Rat. Mech. Anal.*, 62, 1976, 209-236.

[36] Ciarlet, P.G., "A Justification of the von Karman Equations," *Arch. Rat. Mech. Anal.*, 73, 1980, 349-389.

[37] Cohen, G.A., "Effect of a Nonlinear Prebuckling State on the Postbuckling Behavior and Imperfection Sensitivity of Elastic Structures," *AIAA J.*, 6, 1968, 1616-1620.

[38] Dickey, R.W., Bifurcation Problems in Nonlinear Elasticity, *Research Notes in Math.*, 3, Pitman, N.Y., 1976.

[39] Dym, C.L., *Stability Theory and its Applications to Structural Mechanics*, Noordhoff, Amsterdam, 1974.

[40] Fox, D.D., Raoult, A., and Simo, J.C., "A Justification of Nonlinear Properly Invariant Plate Theories," *Arch. Rat. Mech. Anal.*, 124, 1993, 157-199.

[41] Gauss, R.C. and Antman, S.S., "Large Thermal Buckling of Nonuniform Beams and Plates," *Int. J. Solids Structures*, 20, 1984, 979-1000.

[42] Hill, R., "A General Theory of Uniqueness and Stability in Elastic/Plastic Solids," *J. Mech. Phys. Solids*, 6, 1958, 236-249.

[43] Stavsky, Y. and Hoff, N.J., "Mechanics of Composite Structures," in *Composite Engineering Laminates*, A.G.H. Dietz, Ed., MIT Press, Cambridge, MA., 1969, 5-59.

[44] Huseyin, K., *Nonlinear Theory of Elastic Stability*, Sijthoff & Noordhoff, Amsterdam, 1975.

[45] Hutchinson, J.W., "Imperfection Sensitivity in the Plastic Range," *J. Mech. Phys. Solids*, 21, 191-204.

[46] Hutchinson, J.W., "Post-bifurcation Behavior in the Plastic Range," *J. Mech. Phys. Solids*, 21, 1973, 163-190.

[47] Hutchinson, J.W. and Koiter, W.T., "Postbuckling Theory," *Appl. Mech. Rev.*, 23, 1970, 1353-1366.

[48] Knightly, G.H., "An Existence Theorem for the von Kármán u Equations," *Arch. Rat. Mech. Anal.*, 27, 1967, 233-242.

[49] Knightly, G.H., "Some Mathematical Problems from Plate and Shell Theory," in *Nonlinear Functional Analysis and Differential Equations*, Dekker, N.Y., 1976.

[50] Knightly, G.H. and Sather, D., "On Nonuniqueness of Solutions of the von Karman Equations," *Arch. Rat. Mech. Anal.*, 36, 1970, 65-78.

[51] Lekhnitskii, S.G., *Anisotropic Plates*, Gordon and Breach Pub. Co., N.Y., 1968.

[52] Lekhnitskii, S.G., *Theory of Elasticity of an Anisotropic Elastic Body*, Holden-Day, San Francisco, 1963.

[53] List, S., "Generic Bifurcation with Applications to the von Kármán Equations," *J. Diff. Eqns.* 30, 1978, 89-118.

[54] Potier-Ferry, M., "On the Mathematical Foundations of Elastic Stability Theory," *Arch. Rat. Mech. Anal.*, 77, 1982, 55-72.

[55] Potier-Ferry, M., "Perturbed Bifurcation Theory," *J. Diff. Eqns.*, 33, 1979, 112-146.

[56] Rabinowitz, P., "Applications of Bifurcation Theory," in *Proc. Symp. Wisc.*, Oct. 1976, Academic Press, N.Y., 1977.

[57] Rabinowitz, P., "Some Global Results for Nonlinear Eigenvalue Problems," *J. Func. Anal.*, 7, 1971, 487-513.

[58] Sather, D., "Branching of Solutions of Nonlinear Equations," *Rocky Mtn. J. Math*, 3, 1973, 203-250.

[59] Sewell, M.J., "A Survey of Plastic Buckling," in *Stability*, (H. Leipholz, Ed.), Univ. of Waterloo Press, Ontario, 1972, 85-197.

[60] Stakgold, I., "Branching of Solutions of Nonlinear Equations," *SIAM Rev.*, 13, 1971, 289-332.

[61] Stein, M., "Loads and Deformations of Buckled Rectangular Plates," Tech. Rep. R-40, NASA, 1959.

[62] Thompson, J.M.T. and Hunt, G.W., *A General Theory of Elastic Stability*, Wiley, London, 1973.

[63] Thompson, J.M.T. and Hunt, G.W., *Elastic Instability Phenomena*, Wiley, N.Y., 1984.

[64] Timoshenko, S.P. and Gere, J.M., *Theory of Elastic Stability*, 2nd ed., McGraw-Hill, N.Y., 1961.

[65] Timoshenko, S. and Woinowsky-Krieger, S., *Theory of Plates and Shells*, McGraw-Hill, N.Y., 1959.

[66] Troger, H. and Steindl, A., *Nonlinear Stability and Bifurcation Theory, An Introduction for Engineers and Applied Scientists*, Springer-Verlag, N.Y., 1991.

[67] Vanderbauwhede, A., "Generic Bifurcation and Symmetry, with an Application to the von Karman Equations," *Proc. Roy. Soc. Edinburgh*, 81A, 1978, 211-235.

[68] Vanderbauwhede, A., "Generic and Nongeneric Bifurcation for the von Karman Equations," *J. Math Anal. Appl.*, 66, 1978, 550-573.

[69] Föppl, A., *Vorlesungen über Technische Mechanik*, 5, 1907, 132.

[70] Uthgenannt, E. and Brand, R., "Postbuckling of Orthotropic Annular Plates," *J. Appl. Mech.*, 40, 1973, 559-564.

[71] Coffin, D., "Flange Buckling and the Deep-Drawing of Thermoplastic Composite Sheets," Ph.D. thesis, University of Delaware (Mech. Eng.), 1993.

[72] Friedrichs, K.O. and Stoker, J.J., "The Nonlinear Boundary Value Problem of the Buckled Plate," *Amer. J. Math*, 63, 1941, 839-888.

[73] Johnson, M. and Urbanik, T.J., "A Nonlinear Theory for Elastic Plates with Application to Characterizing Paper Properties," *J. Appl. Mech.*, 51, 1984, 146-152.

[74] Johnson, M. and Urbanik, T.J., "Buckling of Axially Loaded, Long Rectangular Paperboard Plates," *Wood and Fiber Science*, 19, 1987, 135-146.

[75] Schleicher, F., "Die Knick Spannungen von eingespannten rechteckigen Platten," *Mitt. Forschungsanstatt.*, Gutchoffnungshütte Konzerno, 1, 1931.

[76] Levy, S., "Buckling of Rectangular Plates with Built-in Edges," *J. Appl. Mech.*, 9, 1942, A 71.

[77] Bryan, G.H., "On the Stability of a Plane Plate under Thrusts in its own Plane with Applications to the Buckling of the Sides of a Ship," *Proc. London Math Soc.*, 22, 1891, 54-67.

[78] Prezemiencki, J.S., "Buckling of Rectangular Plates Under Bi-Axial Compression," *J. Roy. Aero. Soc.*, 59, 1955, 566-568.

[79] Nölke, K. *Der Baningenieur*, 17, 1936, 111.

[80] Stein, M. and Neff, J., "Buckling Stresses of Simply Supported Rectangular Flat Plates in Shear," N.A.C.A. Tech. Note. No. 1222, 1947.

[81] Budiansky, B. and R.W. Connor, " Buckling Stresses of Clamped Rectangular Flat Plates in Shear," N.A.C.A. Tech. Note No. 1559, 1948.

[82] Cook, I.T. and K.C. Rockey, "Shear Buckling of Rectangular Plates with Mixed Boundary Conditions," *Aeronautical Quarterly*, 14, 1963, 349-356.

[83] Batdorff, S.B. and Stein, M., "Critical Combinations of Shear and Direct Stress for Simply Supported Rectangular Flat Plates," N.A.C.A. Tech. Note. No. 1223, 1947.

[84] Seydel, E., "Ausbeulischblast rectackiger Platten," Zeitschrift fur Flugtechnik u. Motorluftsch, 24, 1933, 78-83.

[85] Seydel, E., "Über das Ausbeulen von rechteckigen, isotropen, oder orthogonalanisotropen Platten bei Schubbeanspruchung," *Ingenieur-Archiv*, 4, 1933, 169-191.

[86] Seydel, E., "The Critical Shear Load of Rectangular Plates," NACA Tech. Mem. 705, Washington, D.C. 1933.

[87] Bergman, S.G., *Behavior of Buckled Rectangular Plates under the Action of Shearing Forces*, Victor Pettersons, Stockholm, 1948.

[88] Stein, M., "Behavior of Buckled Rectangular Plates," *J. Eng. Mech. Div., ASCE*, 86, 1960, 59-76.

[89] Stein, M., "Buckling of Long Orthotropic Plates in Combined Shear and Compression," *AIAA Journal*, 23, 1984, 788-794.

[90] Stein, M., "Effects of Transverse Shearing Flexibility on Postbuckling of Plates in Shear," AIAA Journal, 27, 1988, 652-655.

[91] Stein, M., "Loads and Deformations of Buckled Rectangular Plates," NASA Tech. Rep. R-40, 1959.

[92] Stein, M., "Postbuckling of Long Orthotropic Plates Under Combined Loading," *AIAA*, 23, 1985, 1267-1272.

[93] Stein, M., "Analytical Results for Postbuckling Behavior of Plates in Compression and in Shear," *Aspects of the Analysis of Plate Structures* (D.J. Dawe, et al., Eds.), Clarendon Press, Oxford, U.K., 1985, 205-223.

[94] Yamaki, N., "Postbuckling Behavior of Rectangular Plates with Small Initial Curvature Loaded in Edge Compression," *J. Appl. Mech.*, 26, 1959, 407-414.

[95] Walker, A.C., "Flat Rectangular Plates Subjected to a Linearly-Varying Edge Compressive Loading," in *Thin-Walled Structures* (A.H. Chilver, Ed.), J. Wiley and Sons, New York, 1967, 208-247.

[96] Chandra, R. and Raju, B., "Postbuckling Analysis of Rectangular Orthotropic Plates," *Int. J. Mech. Sci.*, 15, 1973, 81-97.

[97] Green, A.E. and J.E. Adkins, *Large Elastic Deformations*, Oxford Press, U.K., 1960.

[98] Panc, V., *Theories of Elastic Plates*, Noordhoff International, The Netherlands, 1975.

[99] Considére, A., "Resistance des pieces comprimes," *Proc. Congr. Int. Construction* 1891, 371.

[100] von Karman, T., "Discussion of Inelastic Column Theory," *J. Aeronaut. Sci*, 14, 1947, 267-268.

[101] Shanley, F.R., "Inelastic Column Theory," *J. Aeronaut. Sci*, 14, 1947, 261-267.

[102] Stowell, E.Z., "A Unified Theory of Plastic Buckling of Columns and Plates," Nat. Adv. Comm. Aeronaut., Rep. 989, 1948.

[103] Tvergaard, V., "Effect of Plasticity on Post-Buckling Behavior," *Lecture Notes in Physics*, vol. 288, Springer-Verlag, N.Y., 1987, 143-183.

[104] Needleman, A. and Tvergaard, V., "An Analysis of the Imperfection Sensitivity of Square Elastic-Plastic Plates under Axial Compression," *Int. J. Solids Structures*, 12, 1976, 185-201.

[105] Sewell, M.J., "A Survey of Plastic Buckling," in *Stability*, Chapter 5, (1972), H. Leipholz, Ed., 85-197, University of Waterloo Press, Ontario.

[106] Brilla, J., "Stability Problems in the Mathematical Theory of Viscoelasticity," in *Equadiff IV, Proceedings*, (Ed. J. Fabera), Springer-Verlag, Berlin 1979, 46-53.

[107] Bozhenov, A., "Stability of a Rectangular Plate of Variable Thickness Compressed in Two Directions," *Prikl. Mekh.*, 10, 1964, 628.

[108] Hamada, M., Inoue, Y., and Hashimoto, H., "Buckling of Simply Supported but Partially Clamped Rectangular Plates Uniformly Compressed in One Direction," *Bull. Jap. Soc. Mech. Eng.*, 10, 1967, 35-40.

[109] Hu, P.C., Lundquist, E.E., and Batdorf, S.B., "Effect of Small Deviations from Flatness on the Effective Width and Buckling of Plates in Compression," N.A.C.A., Tech. Note No. 1124, 1946.

[110] Johns, D.J., "Shear Buckling of Isotropic and Orthotropic Plates - A Review," British Aeronautical Research Council, RAM, 3677, 1970.

[111] Leggett, D.M.A., "The Effect of Two Isolated Forces on the Elastic Stability of a Flat Rectangular Plate," *Proc. Cambridge Phil. Soc.*, 33, 1937, 325-339.

[112] Marguerre, K. and Treffte, E., "Über die Tragfähigkeit eines Längsbelasteten Plattenstreifens nach überschreifen der Beullast," *Zeitschrift für Angewandte Mathematik und Mechanik*, 17, 1937, 85-100.

[113] Ojalvo, M. and Hull, F.W., "Effective Width of Thin Rectangular Plates," *J. Engrg. Mech. Div.*, ASCE, 84, 1958, 1718-1 - 1718-18.

[114] Romeo, G. and Frulla, G., "Nonlinear Analysis of Anisotropic Plates with Initial Imperfections and Various Boundary Conditions Subjected to Combined Biaxial Compression and Shear Loads," *Int. J. Solids Structures*, 31, 1994, 763-783.

[115] Shuleshko, P., "Buckling of Rectangular Plates Uniformly Compressed in Two Perpendicular Directions with One Free Edge and Opposite Edge Elastically Restrained," *J. Appl. Mech.*, 23, 1956, 359-363.

[116] Shuleshko, P., "Buckling of Rectangular Plates with Two Unsupported Edges," *J. Appl. Mech.*, 24, 1957, 537-540.

[117] Thielemann, W., "Contribution to the Problem of Buckling of Orthotropic Plates, with Special Reference to Plywood," N.A.C.A. Tech. Memo. No. 1263, 1950.

[118] Wittrick, W.H., "Correlation Between some Stability Problems for Orthotropic and Isotropic Plates under Bi-Axial and Uni-Axial Direct Stress," *Aeronautical Quarterly*, 4, 1952, 83-93.

[119] Wittrick, W.H., "Some Observations on the Compressive Buckling of Rectangular Plates," *Aeronautical Quarterly*, 14, 1963, 17-30.

[120] Woolley, R.M., Corrick, J.M., and Levy, S., "Clamped Long Rectangular Plate Under Combined Axial Load and Normal Pressure," N.A.C.A. Tech. Note No. 1047, 1946.

[121] Stein, M. and Bains, N., "Postbuckling Behavior of Longitudinally Compressed Orthotropic Plates with Transverse Shearing Flexibility," *AIAA Journal*, 28, 1989, 892-895.

[122] Supple, W.J. and Chilver, A.H., "Elastic Post-Buckling of Compressed Rectangular Plates," in *Thin-Walled Structures* (A.H. Chilver, Ed.), Wiley, N.Y., 1967, 136-152.

[123] Antman, S.S., "Buckled States of Nonlinearly Elastic Plates," *Arch. Rat. Mech. Anal.*, 67, 1978, 111-149.

[124] Berger, M.S., "On von Karman's Equations and the Buckling of a Thin Elastic Plate, I: The Clamped Plate," *Comm. Pure Appl. Math*, 20, 1967, 687-719.

[125] Berger, M.S. and Fife, P.C., "On von Karman's Equations and the Buckling of a Thin Elastic Plate," *Bull. A.M.S.*, 72, 1966, 1006-1011.

[126] Berger, M.S. and Fife, P.C., "Von Karman's Equations and the Buckling of a Thin Elastic Plate, II. Plate with General Edge Conditions," *Comm. Pure Appl. Math*, 21, 1968, 227-241.

[127] Fife, P.C., "Nonlinear Deflection of Thin Elastic Plates Under Tension," *Comm. Pure Appl. Math*, 14, 1961, 81-112.

[128] Knightly, G.H. and Sather, D., "Nonlinear Buckled States of Rectangular Plates," *Arch. Rat. Mech. Anal*, 54, 1974, 356-372.

[129] Negron-Marrero, P.V., "Necked States of Nonlinearly Elastic Plates," *Proc. Roy. Soc. Edin*, 112A, 1989, 277-291.

[130] Bijlaard, P.P., "On the Plastic Buckling of Plates," *J. Aeronaut. Sci.*, 17, 1950, 742-743.

[131] Bijlaard, P.P., "Theory of Plastic Buckling of Plates and Application to Simply Supported Plates Subjected to Bending or Eccentric Compression in their Plane," *J. Appl. Mech.*, 23, 1956, 27-34.

[132] Cicala, P., "On the Plastic Buckling of a Compressed Strip," *J. Aeronaut Sci.*, 17, 1950, 378-379.

[133] Mayers, J. and Budiansky, B., "Analysis of the Behavior of Simply Supported Flat Plates Compressed Beyond Buckling into the Plastic Range," NACA Tech. Note 3368, 1955.

[134] Neale, K.W., "Effect of Imperfections on the Plastic Buckling of Rectangular Plates," *J. Appl. Mech.*, 42, 1975, 115-120.

[135] Kajanto, I., "Finite Element Analysis of Paper Cockling," in *Products of Paper Making, Vol. 1, Trans. 10th. Fundamental Research Symposium*, Oxford, 1989, 237-262.

[136] March, H.W., "Buckling of Flat Plywood Plates in Compression, Shear, or Combined Compression and Shear," U.S. Dept. Agr. Forest Serv., Forest Prod. Lab Rept. 1316, Forest Products Laboratory, Madison, Wis., 1942.

[137] Seo, Y.B., de Oliveira, C., Mark, R.E., "Tension Buckling Behavior of Paper," *J. Pulp Pap. Sci.*, 18, 1992, J55-59.

[138] Smith, R.C.T., "The Buckling of Flat Plywood Plates in Compression," Aust. Council Aeronaut. Rept. ACA-12, Melbourne, 1944.

[139] Antman, S.S., "Buckled States of Nonlinearly Elastic Plates," *Arch. Rat. Mech. Anal.*, 67, 1978, 111-149.

[140] Brilla, J., "Postbuckling Analysis of Circular Viscoelastic Plates," *Mech. Res. Comm.*, 17, 1990, 263-269.

[141] Needleman, A., "Postbifurcation Behavior and Imperfection Sensitivity of Elastic-Plastic Circular Plates," *Int. J. Mech. Sci.*, 17, 1975, 1-13.

[142] Onat, E.T. and D.C. Drucker, "Inelastic Instability and Incremental Theories of Plasticity," *J. Aeronaut. Sci.*, 20, 1953, 181-186.

[143] Friedrichs, K.O. and Stoker, J.J., "Buckling of the Circular Plate Beyond the Critical Thrust," *J. Appl. Mech.*, 9, 1942, A7-A14.

[144] Keller, H.B., Keller, J.B., and Reiss, E.L., "Buckled States of Circular Plates," *Q. Appl. Math*, 20, 1961, 55-65.

[145] Wolkowisky, J.H., "Existence of Buckled States of Circular Plates," *Comm. Pure Appl. Math*, 20, 1967, 549-560.

[146] Reismann, H., "Bending and Buckling of an Elastically Restrained Circular Plate," *J. Appl. Mech.*, 19, 1952, 167-172.

[147] Sherbourne, A.N., "Elastic Postbuckling Behavior of a Simply Supported Circular Plate," *J. Mech. Eng. Sci.*, 3, 1961, 133-141.

[148] Stoker, J.J., *Nonlinear Elasticity*, Gordon and Breach pub., N.Y., 1968.

[149] Bodner, S.R., "The Postbuckling Behavior of a Clamped Circular Plate," *Poly. Inst. Brooklyn*, XII, 1955, 397-401.

[150] Woinowsky-Krieger, S., "Buckling Stability of Circular Plates with Circular Cylindrical Aeolotropy," *Ingenieur-Archiv*, 26, 1958, 129-131.

[151] Mossakowski, J., "Buckling of Circular Plates with Cylindrical Orthotropy" *Arch. Mech. Stos.* (Poland), 12, 1960, 583-596.

[152] Iwinski, T. and Nowinski, J., "The Problem of Large Deflections of Orthotropic Plates," *Arch. Mech. Stos.* (Poland), 7, 1957, 593-602.

[153] Pandalai, K.A.V. and Patel, S.A., "Buckling of Orthotropic Circular Plates," *J. Royal Aerospace Soc.*, 69, 1965, 279-280.

[154] Huang, C.L., "Nonlinear Axisymmetric Flexural Vibration Equations of a Cylindrically Anisotropic Circular Plate," *J. AIAA*, 10, 1972, 1378-1379.

[155] Sherbourne, A.N. and Pandey, M.D., "Postbuckling of Polar Orthotropic Circular Plates—Retrospective," *J. Eng. Mech. Div., ASCE*, 118, 1992, 2087-2103.

[156] Raju, K.K. and Rao, G.V., "Finite-Element Analysis of Postbuckling Behavior of Cylindrically Orthotropic Circular Plates," *Fibre Sci. Tech.*, 19, 1983, 145-154.

[157] Turvey, G.J. and Drinali, H., "Elastic Postbuckling of Circular and Annular Plates with Imperfections," *Proc. 3rd Int. Conf. Composite Struct.*, Appl. Sci. Pub., 1985, 315-335.

[158] Carrier, G.F., "Stress Distribution in Cylindrically Aeolotropic Plates," *J. Appl. Mech. Trans. ASME*, 10, 1943, A117-A122.

[159] Huang, C.L. and Sandman, B.E., "Large Amplitude Vibrations of a Rigidly Clamped Circular Plate," *Int. J. Nonlinear Mech.*, 6, 1971, 451-466.

[160] Swamidas, A.S. and Kunukkasseril, V.X., "Buckling of Orthotropic Circular Plates," *AIAA Journal*, 11, 1963, 1633-1636.

[161] Antman, S.S. and Negrón-Marrero, P.V., "The Remarkable Nature of Radially Symmetric Equilibrium States of Aeolotropic Nonlinearly Elastic Bodies," *J. Elasticity*, 18, 1987, 131-164.

[162] Yamaki, S., "Buckling of a Thin Annular Plate Under Uniform Compression," *J. Appl. Mech.*, 25, 1958, 267-273.

[163] Olsson, R.G., "Über axialsymmetrische Knickung dünner Kriesringplatten," *Ingenieur-Archiv*, 8, 1937, 449-452.

[164] Schubert, "Die Beullast dünner Kreisringplatten, die am Aussen und Innenrand gleichmässigen Druckerfahren," *Zeitschrift Für Angewandte Mathematik und Physik*, 25, 1947, 123-124.

[165] Mansfield, E.H., "On the Buckling of an Annular Plate," *Quart. J. Mech. Appl. Math.*, XIII, 1960, 16-23.

[166] Majumdar, S., "Buckling of a Thin Annular Plate Under Uniform Compression," *AIAA Journal*, 9, 1971, 1701-1707.

[167] Machinek, A. and Troger, H., "Postbuckling of Elastic Annular Plates," *Dynamics and Stability of Systems*, 3, 1988, 79-88.

[168] Machinek, A.K. and Steindl, A., "Behandlung des Beulproblems der Kreisringplatte mit Hilfe eines numerischen Verfahrens," *Zeitschrift für Angewandte Mathematik und Mechanik 66*, 1986, T61-T63.

[169] Uthgenannt, E. and Brand, R., "Buckling of Orthotropic Annular Plates," *AIAA Journal*, 8, 1970, 2102-2104.

[170] Woinowsky-Krieger, S., "ÜberdieBeulsicherheit von Kreisplatten mit Kreiszylindrischer Aeolotropic," *Ingenieur-Archiv*, 26, 1958, 129-133.

[171] Rozsa, M., "Stability Analysis of Thin Annular Plates Compressed Along the Outer or Inner Edge by Uniformly Distributed Radial Forces," *Acta Technica Academiae Scientiarum Hungaricae*, 53, 1966, 359-377.

[172] Meissner, E., "Über das Knicken kreisringförmiger Scheiben," *Schweizerche Bauzeitung*, 101, 1933, 82-89.

[173] Dumir, P.C., Nath, Y., and Gandhi, M., "Non-Linear Axisymmetric Static Analysis of Orthotropic Thin Annular Plates," *Int. J. Non-Linear Mechanics*, 19, 1984, 255-272.

[174] Huang, C.-L., "On Postbuckling of Orthotropic Annular Plates," *Int. J. Non-Linear Mech.*, 10, 1974, 63-74.

[175] Huang, C.-L., "Postbuckling of an Annulus," *AIAA Journal*, 11, 1973, 1608-1612.

[176] Alwar, R.S. and Reddy, B., "Large Deflection Static and Dynamic Analysis of Isotropic and Orthotropic Annular Plates," *Int. J. Non-Linear Mech.*, 14, 1979, 347-359.

[177] Keller, H.B. and Reiss, E.L., "Non-linear Bending and Buckling of Circular Plates," *Proc. 3rd U.S. National Congress of Applied Mech.*, 1958, 375-385.

[178] Suchy, H., Troger, H., and R. Weiss, "A Numerical Study of Mode Jumping of Rectangular Plates," *Zeitschrift für Angewandte Mathematik und Mechanik*, 65, 1985, 71-78.

[179] Yu, T.X. and Johnson, W., "The Buckling of Annular Plates in Relation to the Deep-Drawing Process," *Int. J. Mech. Sci.*, 24, 1982, 175-188.

[180] Geckler, J.W., "Plastiche knicken der wandung von hohlzylindern und einige andern faltungserscheinungen," *Z. Angewandtle Mathematik und Mechanik*, 8, 1928, 341-352.

[181] Senior, B.W., "Flange Wrinkling in Deep-Drawing Operations," *J. Mech. Phys. Solids*, 4, 1956, 235-246.

[182] Dean, W.R., "The Elastic Stability of an Annular Plate," *Proc. Roy. Soc. London, series A*, 106, 1924, 268-284.

[183] Egger, H., "Knickung der Kreisplatte und Kreisringplatte mit veränderlicher Dicke," *Ingenieur-Archiv*, 12, 1941, 190-200.

[184] Federhofer, K., "Knickung der Kreisplatte und Kreisringplatte mit veränderlicher Dicke," *Ingenieur-Archiv*, 11, 1940, 224-238.

[185] Federhofer, K. and Egger, H., "Knickung der auf Scherung beanspruchten kreisringplatte mit veranderlicher Dicke," *Ingenieur-Archiv*, 14, 1943, 155-166.

[186] Nadai, A., "Über das Ausbeulen von kreisförmigen Platten," *Zeitschrif des Vereines Deutscher Ingenieur*, 59, 1915, 169-174.

[187] Ramaiah, G.K. and Vijayakumar, K., "Buckling of Polar Orthotropic Annular Plates Under Uniform Internal Pressure," *J. AIAA*, 12, 1974, 1045-1050.

[188] Reddy, B.S. and Alwar, R.S., "Postbuckling Analysis of Orthotropic Annular Plates," *Acta Mech.*, 39, 1981, 289-296.

[189] Turvey, G.T., "Large Deflection of Tapered Annular Plates by Dynamic Relaxation," *Proc. J. Eng. Mech. Div., ASCE*, 104, 1978, 351-366.

[190] Wempner, G.A. and Schmidt, R., "Large Symmetric Deflections of Annular Plates," *J. Appl. Mech.*, 25, 1958, 449-458.

[191] Keener, J.P. and H.B. Keller, "Perturbed Bifurcation Theory," *Arch. Rat. Mech. Anal.*, 50, 1973, 159-175.

[192] Coan, J.M., "Large Deflection Theory for Plates with Small Initial Curvature Loaded in Edge Compression," *J. Appl. Mech.*, 18, 1951, 143-151.

[193] Levy, S., Goldenburg, D., and Zibritosky, G., "Simply Supported Long Rectangular Plate Under Combined Axial Load and Normal Pressure," N.A.C.A. Tech. Note No. 949, 1944.

[194] Sheinman, I., Frostig, Y., and A. Segal, "Nonlinear Analysis of Stiffened Laminated Panels with Various Boundary Conditions," *J. Composite Materials*, 25, 1991, 634-649.

[195] Minguet, P.J., Dugundji, J., and P. Lagace, "Postbuckling Behavior of Laminated Plates Using a Direct Energy Minimization Technique," *AIAAJ.*, 27 1989, 1785-1792.

[196] Engelstad, S.P., Reddy, J.N., and N.F. Knight, Jr., "Postbuckling Response and Failure Predictions of Graphite-Epoxy Plates Loaded in Compression," *AIAA J.*, 30, 1992, 2106-2113.

[197] Hutchinson, J.W. and B. Budiansky, "Analytical and Numerical Study of the Effects of Initial Imperfections on the Inelastic Buckling of a Cruciform Column," *Proc. IU-TAM Symp. on Buckling of Structures,* Harvard University, June 1974, Springer-Verlag.

[198] Massonnet, Ch., "Le Voilement des Plagues Planes Sollicitèes dan Leurs Plan," Liège, Sept. 1948.

[199] Tani, J. and Yamaki, N., "Elastic Instability of a Uniformly Compressed Annular Plate with Axisymmetric Initial Deflection," *Int. J. Nonlinear Mech.*, 16, 1981, 213-220.

[200] Ramaiah, G.K. and Vijayakumar, K., "Buckling of Polar Orthotropic Annular Plates under Uniform Internal Pressure," *AIAA J.*, 12, 1974, 1045-1050.

[201] Cheo, L.S. and Reiss, E.L., "Secondary Buckling of Circular Plates," *SIAM J. Appl. Math*, 26, 1974, 490-495.

[202] Cheo, L..S. and Reiss, E.L., "Unsymmetric Wrinkling of Circular Plates," *Q. Appl. Math*, 31, 1973, 75-91.

[203] Stroebel, G.J. and Warner, W.H., "Stability and Secondary Bifurcation for von Kármán Plates," *J. Elasticity*, 3, 1973, 185-202.

[204] Bauer, L., Keller, H.B., and E.L. Reiss, "Multiple Eigenvalues Lead to Secondary Bifurcation," *SIAM Review*, 17, 1975, 101-122.

[205] Bauer, L. and Reiss, E.L., "Nonlinear Buckling of Rectangular Plates," *J. Soc. Indust. Appl. Math*, 13, 1965, 603-626.

[206] Matkowsky, B.J. and Putnick, L.J., "Multiple Buckled States of Rectanglar Plates," *Int. J. Non-Linear Mech.*, 9, 1974, 89-103.

[207] Chandrasekhar, S., *Ellipsoidal Figures of Equilibrium*, Yale University Press, New Haven, Conn., 1969.

[208] Morozov, N.F., "Investigation of a Circular Symmetric Compressible Plate with a Large Boundary Load," *Izv. Vyssh, Uchebn, Zaved*, 34, 1963, 95-97.

[209] Yanowitch, M., "Nonlinear Buckling of Circular Elastic Plates," *Comm. Pure Appl. Math*, IX, 1956, 661-672.

[210] Stein, M., "The Phenomenon of Change of Buckling Patterns in Elastic Structures," Tech. Rep. R-39, NASA, 1959.

[211] Rogers, E.H., "Variational Properties of Nonlinear Spectra," *J. Math. Mech.*, *18*, 1968, 479-490.

[212] Levy, S., "Bending of Rectangular Plates with Large Deflections," NACA Rep. 737, 1942.

[213] Schaeffer, D. and Golubitsky, M., "Boundary Conditions and Mode Jumping in the Buckling of a Rectangular Plate," *Comm. Math. Phys*, 69, 1979, 209-236.

[214] Steinrück, H., Troger, H., and Weiss, R., "Mode Jumping of Imperfect Buckled Rectangular Plates," in *vol. 70, Int. Series of Num. Math*, Birkhäuser-Verlag, Basel, 1984, 515-524.

[215] Keller, H.B. and Keener, J.P., "Perturbed bifurcation and Buckling of Circular Plates," in *Lect. Notes Math*, 280, Springer-Verlag, Berlin, 1972, 286-293.

[216] Gosard, M.L., Seide, P., and W.M. Roberts, "Thermal Buckling of Plates," NACA Tech. Note 2771, 1952.

[217] Tauckert, T., "Thermal Stresses in Plates–Statical Problems," in *Thermal Stresses, vol. I.*, R.B. Hetnarski, Ed., (1986), Elsevier Science Publishers.

[218] Biswas, P., "Large Deflections of Heated Elastic Plates," *Applique Mecanique*, 20, (1975), 585-596.

[219] Misra, J.C., "On the Bending of a Heated Rectangular Anisotropic Plate with Two Opposite Edges Simply Supported and Two Others Built In, Due to Different Temperature Distributions on the Plane Surfaces," *Bull. Cal. Math. Soc.*, 63, (1971), 183-190.

[220] Pell, W.H., "Thermal Deflections of Anisotropic Thin Plates," *J. Soc. Indust. Appl. Math.*, IV, (1946), 27-44.

[221] Boley, B.A. and J.H. Weiner, *Theory of Thermal Stresses*, (1960), Wiley Publishing, N.Y.

[222] Nowacki, W., *Thermoelasticity*, (1962), Pergamon Press, Oxford, U.K.

[223] Hetnarski, R.B., "Basic Equations of the Theory of Thermal Stresses," in *Thermal Stresses, vol. I*, R.B. Hetnarski, Ed., (1986), Elsevier Science Publishers.

[224] Kovalenko, A.D., *Thermoelasticity*, (1969), Wolters-Noordhoff Publishing, Gröningen.

[225] Forray, M. and M. Newman, "Bending of Circular Plates Due to Asymmetric Temperature Distribution," *J. of the Aerospace Sciences*, (1961), 773-778.

[226] Misra, J.C., "Note on the Thermal Bending of a Simply Supported Rectangular Plate of Anisotropic Material," *Indian J. Mech. Nath*, 11, (1973), 33-36.

[227] Klosner, J.M. and M.J. Forray, "Buckling of Simply Supported Plates Under Arbitrary Symmetrical Temperature Distributions," *J. of the Aerospace Sciences*, 25, (1958), 181-184.

[228] Forray, M. and M. Newmann, "Axisymmetric Large Deflections of Circular Plates Subjected to Thermal and Mechanical Loads," *J. of the Aerospace Sciences*, 29, (1962), 1060-1066.

[229] Bednarczyk, H. and M. Richter, "Buckling of Plates Due to Self-Equilibrated Thermal Stresses," *J. of Thermal Stresses*, 8, (1985), 139-152.

[230] Bargmann, H.W., "Thermal Buckling of Elastic Plates," *J. of Thermal Stresses*, 8, (1985), 71-98.

[231] Stavsky, Y., "Thermoelastic Stability of Laminated Orthotropic Circular Plates," *Acta Mechanica*, 22, (1975),
31-51.

[232] Boley, S.R., "A Procedure for the Approximate Analysis of Buckled Plates," *J. of the Aerospace Sciences*, 22,
(1955), 336-338.

[233] Heldenfels, R. and W.M. Roberts, "Experimental and
Theoretical Determination of Thermal Stresses in a Flat
Plate," NACA Tech. Note 2769, 1952.

[234] Wilcox, M.W. and L.E. Clemmer, "Large Deflection
Analysis of Heated Plates," *J. Eng. Mech. Div., ASCE*,
6, (1964), 165-189.

[235] Levy, S., "Bending of Rectangular Plates with Large Deflections," NACA Tech. Note 846, 1942.

[236] Biswas, P., "Nonlinear Analysis of Heated Orthotropic
Plates," *Indian J. Pure Appl. Math.*, 12, (1981), 1380-
1389.

[237] Mansfield, E.H., "The Large Deflection Behavior of a
Thin Strip of Lenticular Section," *Quart. J. Mech. Appl.
Math.*, 12, (1959), 421-430.

[238] Sarkar, S.K., "Buckling of a Heated Circular Plate," *Indian. J. Math. Mech.*, V, (1967), 39-42.

[239] Vinokurov, S.G., "Thermal Stresses in Plates and Shells,"
(in Russian), *Izv. Kas. Filyala, AN SSSR* (1953).

[240] Berger, H.M., "A New Approach to the Analysis of Large
Deflection of Plates," *J. Appl. Mech.*, 22, (1955), 465-472.

[241] Nowinski, J.L. and H. Ohnabe, "On Certain Inconsisten-
cies in Berger Equations for Large Deflections of Plastic
Plates," *Int. J. Mech. Sci.*, 14, (1972), 165-170.

[242] Jones, R., Mazumdar, J., and Y.K. Cheung, "Vibration
and Buckling of Plates at Elevated Temperatures," *Int.
J. Solids Structures*, 16, (1980), 61-70.

[243] Banerjee, B. and S. Datta, "A New Approach to an Analysis of Large Deflections of Thin Elastic Plates," *Int. J.
Non-Linear Mech.*, 16, (1981), 47-52.

[244] Basuli, S., "Large Deflection Plate Problems Subjected to Normal Pressure and Heating," *Indian J. Math. Mech.*, VI (1968), 22-30.

[245] Pal, M.C., "Large Deflections of Heated Circular Plates," *Acta Mechanica*, 8, (1969), 82-103.

[246] Biswas, P., "Large Deflections of Heated Orthotropic Plates," *Indian J. Pure Appl. Math.*, 9, (1978), 1027-1032.

[247] Pal, M.C., "Large Amplitude Free Vibration of Circular Plates Subjected to Aerodynamic Heating," *Int. J. Solid Structures*, 6, (1970), 301-314.

[248] Biswas, P., "Thermal Buckling of Orthotropic Plates," *J. Appl. Mech.*, 21, (1996), 361-363.

[249] Pal, M.C., "Static and Dynamic Non-Linear Behavior of Heated Orthotropic Circular Plates," *Int. J. Non-Linear Mech.*, 8, (1973), 489-504.

[250] Williams, M.L., "Large Deflection Analysis for a Plate Strip Subjected to Normal Pressure and Heating," *J. Appl. Mech.*, (1955), 458-465.

[251] Schmidt, R., "On Berger's Method in the Non-Linear Theory of Plates," *J. Appl. Mech.*, 41, (1974), 521-523.

[252] Schmidt, R. and D. DaDeppo, "A New Approach to the Analysis of Shells, Plates, and Membranes with Finite Deflections," *Int. J. Non-Linear Mechanics*, 9, (1974), 409-419.

[253] Mansfield, E.H., "On the Analysis of Elastic Plates of Variable Thickness," Quart. J. Mech. Appl. Math., XV, (1962), 167-183.

[254] Mansfield, E.H., "Bending, Buckling, and Curling of a Heated Thin Plate," *Proc. Roy. Soc. Ser. A.*, *1334*, (1962), 316-329.

[255] Biswas, P., "Buckling of a Heated Thin Annular Circular Plate of Variable Thickness," *Indian J. of Theoretical Physics*, 25, (1977), 107-112.

[256] Mendelson, A. and M. Herschberg, "Analysis of Elastic Thermal Stresses in Thin Plates with Spanwise and

Chordwise Variations of Temperature and Thickness," NACA Tech. Note 3778, 1956.

[257] Z. Thrun, "O Odksztalceniach i naprezeniach temioznyeh w cienkich plytach prostokatnych i kolowych o zmionuej grubosei," *Rozprawy Inzynierskie, tom IV, zeszyt 4*, 523-41 Warsaw 1956.

[258] Olsson, R.G., "Beigung der Rechteckplatte bei linear voranderlicker Biegungssteifigkeit," *Ingenieur-Archiv*, 5 (1934), 363.

[259] Olsson, R.G., "Unsymmetrische Biogung der Kreis-ringplate von quadratisch veranderlicker Steifikeit," *Ingenieur-Archiv*, 10 (1939) 14, 11 (1940), 25, and 13 (1942-3), 147.

[260] Mansfield, E.H., "The Inextensional Theory for Thin Flat Plates," *Quart. J. Mech. Appl. Math.*, 8, (1955), 338-352.

[261] Ross, D.A. and L. Berke, "Temperature-Dependent Creep Buckling of Plates," *J. of Thermal Stresses*, 4, (1981), 237-247.

[262] Das, S., "Note on Thermal Deflection of Regular Polygonal Viscoelastic Plates,," *Int. J. Mech. Sci.*, 23, (1981), 323-329.

[263] Williams, H.E., "Neale Plate Equations and Applications to Buckling of Rectangular Plates," *J. of Thermal Stresses*, 5, (1982), 207-243.

[264] Neale, K.W., "A General Variational Theorem for the Rate Problem in Elasto-Plasticity," *Int. J. Solids Structures*, 8, (1972), 865-876.

[265] Neale, K.W., "Effect of Imperfections on the Plastic Buckling of Rectangular Plates," *J. Appl. Mech.*, 42, (1975), 115-120.

[266] Norton, R.N., *Creep in Steel at High Temperatures*, (1929), 115-120.

[267] Rabotnov, Yu N., *Creep Problems in Structural Members*, (1969), North Holland Publishing, Amsterdam.

[268] Timoshenko, S. and S. Woinowsky-Kreiger, *Theorem of Plates and Shells, 2nd ed.,* (1959), McGraw-Hill, N.Y.

[269] Hoff, N.J. and I.M. Levi, "Short Cuts in Creep Buckling Analysis," *Int. J. Solids Structures,* 8, (1972), 1103-1114.

[270] Ross, D.A., "Creep Buckling of a Flat Plate with Through-Thickness Temperature Variation," NASA Tech. Memo. 81627, 1980.

[271] Deleeuw, S.L. and G.E. Mase, "Behavior of Viscoelastic Plates Under the Action of Inplane Forces," *Proc. 4th U.S. Nat. Cong. Appl. Mech.,* (1962), 999-1005.

[272] Mayers, J. and E. Nelson, "Elastic and Maximum Strength Analysis of Postbuckled Rectangular Plates Based Upon Modified Versions of Reissner's Variational Principle," SUDAAR Report #262, (1966), Stanford University.

[273] Turvey, G.J., "Axisymmetric Elastic-Plastic Flexure of Circular Plates in the Large Deflection Regime," *Proc. Inst. Civil Eng.,* 67, (1979), 81-92.

[274] Myszkowski, J., "Endliche Durchbiegungen Beliebig Eingespannier Dunner Kriesund Kriersring platten im Plastischen Materialbereich," *Ingenieur-Archiv,* 40, (1971), 1-13.

[275] Danielson, D.A., "Theory of Shell Stability," in *Thin-Shell Structures: Theory, Experiment, and Design,* Y.C. Fung and E.E. Sechler, Eds., (1974), Prentice-Hall, N.J.

[276] Gerard, G. and H. Becker, Handbook of Structural Stability, Part I: Buckling of Flat Plates, NACA Tech. Note 3781, 1957.